Biochar as Remediation Tools for Pollutant Removal

Avnish Chauhan · Parul Chaudhary ·
Prabhat K. Chauhan
Editors

Biochar as Remediation Tools for Pollutant Removal

Springer

Editors
Avnish Chauhan
Department of Environmental Science
Graphic Era Hill University
Dehradun, Uttarakhand, India

Parul Chaudhary
School of Agriculture
Graphic Era Hill University
Dehradun, Uttarakhand, India

Prabhat K. Chauhan
Department of Environmental Science
AKS University
Satna, Madhya Pradesh, India

ISBN 978-3-032-16745-3 ISBN 978-3-032-16746-0 (eBook)
https://doi.org/10.1007/978-3-032-16746-0

This Springer imprint is published by the registered company Springer Nature Switzerland AG
The registered company address is: Gewerbestrasse 11, 6330 Cham, Switzerland

Contents

Biochar Production and Characteristics

Manglam Soni, Krupa Kanani, Sonal Singh, Nidhi Shukla, and Kuldip Dwivedi

Abstract Increased population and growing demand for large-scale agricultural production have resulted in unwanted effects of various pollutants. So, it imparts affecting soil fertility and its health. Its ability to potentially enhance soil quality, sequester carbon, and serve as an active electrode material for supercapacitor applications is gaining attention. While mentioning such types of problems, biochar has developed into an environmentally conscious and sustainable option from historical times to the present era, in increasing fertility, health, and productivity of the soil. The chapter highlights the current studies on production and the use of biochar, through diversification of feedstocks like invasive species, agricultural waste, industrial waste, and production approaches ranging from traditional to contemporary pyrolysis practices. Biochar characterizes as a carbon-rich substance renowned for possessing good assets such as water-holding capacity, enhanced nutrient holding, microbial processes, and general promotion of plant growth. Waste produced from post-harvest and agriculture helps reduce greenhouse gas (GHG) emissions, revitalize soil fertility, and increase water holding capacity—goals that biochar can tackle effectively. Through studies, it has been indicated that biochar strengthens cation exchange capacity (CEC), regulates nitrogen dynamics, and promotes a favourable environment for beneficial microbes, providing soil adaptation and improved crop yields for the long term. Properties like temperature, feedstock, and pyrolysis conditions deliberately distress the pyrogenic carbon properties, internal structure, and nutrient dynamics which are both essential to advance soil vigor and maximize the plant efficacy. Approaches like the hydrothermal carbonization can also be used in modifying such properties, specifically for damp biomass. Biochar is not a new

M. Soni
Amity Institute of Biotechnology, Amity University, Gwalior, Madhya Pradesh, India

K. Kanani
AMS Support Officer (Fleming Fund), University of North Carolina Project, Lilongwe, Malawi

M. Soni · S. Singh · N. Shukla · K. Dwivedi (✉)
Department of Environmental Science, Amity School of Life Science, Amity University, Gwalior, Madhya Pradesh, India

N. Shukla
e-mail: nshukla@gwa.amity.edu

A. Chauhan et al. (eds.), *Biochar As Remediation Tools for Pollutant Removal*,
https://doi.org/10.1007/978-3-032-16746-0_1

concept; it has been associated with the archaic times of Amazonian Terra preta soils, where the specific enrichment of soils with biochar has demonstrated a prominent role in sustained enhancements of soil fertility and enhanced stability of carbon for centuries. This chapter inspects numerous plants. Sourced feedstocks for biochar manufacture and summaries leading studies on its physiochemical characterization of biochar and its manufacturing rate.

Keywords Agriculture · Biochar · Microbial · Efficiency · Nutrients · Carbonization

1 Introduction

Biochar has arisen as a key quantifiable in the exciting change near sustainable agriculture habitats, remediation of environmental comprising climate change and energy storage complications. The production of biochar from thermochemical processes, i.e., pyrolysis of biomass under anaerobic conditions,results in biochar containing high carbon stability, porosity in its structure, and adsorption properties; consequently, it is a helpful technology for soil enrichment, carbon sequestration, and pollution remediation. The perception of biochar is not innovative; it has been connected with the early time of Amazonian Terra preta soils, where the occurrence of biochar-enriched soils has indicated its role in long-term fertility improvements and carbon storage (Gyanwali et al. 2024). In the last decade, research has explored massive biochar research regarding application across various sectors, including in soil fertility improvement, industrial disposing management, and advanced energy storage systems (Tao et al. 2020). The problems presented by degradation of soil health, depleting soil productivity, and over reliance on chemical fertilizers have attracted interest in biochar production as a sustainable approach in improving the agriculture production (Zhang et al. 2018a, b: Chauhan et al. 2025).

The properties of biochar can be improved with the incorporation of microorganisms that help in the degradation of organic pollutants and emerging contaminants, even though microbes are more helpful in increasing the interactions between plant root exudates, and along with it, they help in protecting plants against abiotic and biotic stress occurring in the environment (Pathak et al. 2024b). Global agriculture promises to reduce greenhouse gases' (GHGs) emissions, revitalize soil fertility, and increase water-holding capacity goals whereby biochar can tackle effectively. Through studies, it has been indicated that biochar strengthens cation exchange capacity (CEC), regulates nitrogen dynamics, and promotes a favorable environment for beneficial microbes, providing soil adaptation and improved crop yields for a long term (Obia et al. 2017; Pathak et al. 2024b; Zhu et al. 2025). Waste produced from post-harvest and agriculture helps reduce greenhouse gases (GHGs) emissions (Tan et al., 2017). Biochar production proceeds through pyrolysis, which involves heating the biomass between 300 °C and 900 °C under oxygen-limiting environment, resulting in formation of biochar, syngas, and bio-oil (Zhu et al. 2025). Biochar is made through a three-step process as follows: pre-pyrolysis, pyrolysis, and the formation of soil carbon compounds (Ding et al. 2014). During the first cycle or phase, i.e., pre-pyrolysis, biomass is heated up to 200 °C, resulting to the drying of the feedstock and the evaporation of low boiling point volatile compounds. In the second cycle temperature about 200–500 °C, enhanced pyrolytic degradation is more tedious, chiefly by the presence and volatilization of hemicellulose and cellulose structures present in plant-based wastes. The fifth stage finished at as much as 500°C temperature is favourable for the breaking down of more stable complexes, such as lignin, which promotes biochar production, a thermally stable and high-carbon solid product outcome (Askeland et al. 2019). Temperature rise speed and heat temperature beneficial in pyrolysis are key parameters controlling the physical and chemical character of the produced biochar (Ajema and Tafa 2015). These are the conditions driving the most important parameters, including carbon content, surface functional groups, porosity, Cation Exchange Capacity (CEC), and pH. Depending on the variable factors of heat, pyrolysis is mainly classified into slow and fast types:

- Slow pyrolysis, performed at around temperature 250–500 °C with heating rates below 100 °C/minute, with high yields of biochars and the nutrients' and functional groups' holding capacities, is appropriate for soil enrichment and carbon storage.
- Fast pyrolysis method yields supplementary bio-oil and syngas and lower amount of biochar, but it has an evident in potential for bioenergy consumption. It encompasses techniques more than 500 °C with intriguing heating rates (> 1000 °C/ minute).

Temperature plays a pivotal role in pyrolysis methods; it significantly stimulates and enhances the properties of biochar. One important consideration is whether biochar would be suitable for a hypothetical application in pyrolysis temperature. Raising the temperature, for instance, primes the surface area and pH, reduces volatile matter, and modifies the surface functional groups in the biochar. The above-mentioned modifications help in carbon stability, but they may also decline in CEC and nutrient-retention capacity, which could stimulate the material's utility for agronomic resolutions. To design biochar to be helpful in environmental and agricultural results, severe control of the pyrolysis conditions is compulsory, such as temperature, source, and processing time. When making biochar, through the process of pyrolysis, temperature is a significant factor in defining its characteristics. Therefore, high temperatures improve surface area and carbon steadiness, but they also decline the nutrient structure and CEC, which diminish their agronomic worth.

Accordingly, regulatory conditions are important in biochar approach for countless applications (Godlewska et al. 2021). The pyrolysis situations, feedstock, and post-production treatment regulate the physiochemical properties, counting the surface area, porosity, pH, and elemental content of biochar (Ippolito et al. 2020; Leng et al. 2021). The International Biochar Initiative (IBI) defines biochar as a solid material that originates on or after the breakdown of a carbon-based and organic compound in an oxygen-reduced, i.e., anaerobic atmosphere (IBI 2020).

HTC is alternative technique to prepare biochar, specifically when discussing about wet biomass. HTC is an alternative technique to prepare biochar in the presence of water—hydrothermal carbonization—saturated, tight reactor at a lower temperature varies from 180–250 °C and harvests hydrochars that are structurally and chemically different from dry-process biochar (Fang et al. 2018). HTC works in a water-saturated tight reactor at a low-temperature range (180–250 °C) and makes hydrochars, which are dissimilar from dry-processed biochar in both chemically and structurally (Fang et al. 2018). A sequence of reactions, including hydrolysis, dehydration, isomerization, and condensation, that helps in making intermediate compounds like 5-hydroxymethylfurfural.

Hydrothermal carbonization incorporates dealkylation and hydrolysis in lignin-rich biomass, resultant in the making of chemicals such as phenols and catechins. These compounds then continue attaching through crosslinking and repolymerization to form solid hydrochar (Bakraoui et al. 2020). High-moisture waste may be sustainably and economically transformed into stable carbonaceous material for use in environmental applications using this method (Singh et al. 2024). When exposed to high temperatures, biochar often shows carbon stability and reduced volatile matter, which develops carbon sequestration and soil bioremediation more feasible. Conversely, low-temperature biochar recovers soil retention of essential nutrients and labile organic matter, growing microbial activity and plant development (Singh and Dwivedi 2022).

Appropriate selection of biomass feedstock determines properties and functionality of a biomass. Common traditional sources include agriculture residues (e.g., rice husk, corn stover), manure, forest waste, and sewage sludge, and each contributes a unique structural composition and chemical features (Lin et al. 2017; Wang and Wang 2019). The recent breakthroughs have enabled utilization of alternative feedstocks, as aquatic plants, industrial waste as well as invasive species which has shown immense opportunities for waste utilization in a sustainable solution (Rich and Felfel 2015; Manna et al. 2018). Engineered biochar derived though physical or chemical adjustments (Example: metal-doped biochar, functionalized biochar) has explored high performance in catalytic reactions, adsorption, and electrochemical applications (Pathak et al. 2024a; Leng et al. 2025). Due to soil erosion, acidification, organic matter depletion, and salinity which damages the global agricultural sustainability, more reflection on biochar research has strengthened as its role in soil health fixation. Various applications of biochar have shown vitalized soil structure, increased water retention capacity, and enhanced microbial interactions promoting beneficial attention during drought and nutrient-deficient soils (Waheedet al. 2025). Additionally, its porosity offers growth of nitrogen-fixing bacteria and mycorrhizal fungi, which plays an important role in nutrient cycling and organic matter decomposition (Khan 2024). The ability of biochar to heavy metal adsorption and agrochemicals facilitates detoxifying contaminated soils, reducing plant stress, and improving soil biodiversity (Maqbool et al. 2024). Pathak et al. (2024a) determined that plant and soil health can be improved by a carbonaceous chemical known as N-BC (Nano-Biochar), which contributes to the plants' survival in extreme environments and promotes the mitigation of the toxicity of heavy metals and emerging pollutants from different sources. Besides agricultural productivity, biochar has countless supports into climate change efforts and makes sustainable process. Carbon sequestration through the application of biochar is a sustainable long-term means of lowering atmospheric CO_2, since biochar locks up carbon in a stable form that lasts for centuries (Osman et al. 2021). In contrast to untreated biomass, which rapidly degrades and releases CO_2 and CH_4, biochar guarantees stability.

By introducing it in agricultural fields such as paddy fields, manure compost, and the processing of organic waste, it may reduce methane (CH_4) and nitrous oxide (N_2O) emissions, both of which are greenhouse gases and contribute to climate change to a large extent (Jiang et al., 2021). Recent research has prolonged biochar's introduction with renewable energy installations and waste-to-energy efforts in today's time. The concurrent co-production of syngas and bio-oil in the process of pyrolysis offers prospects for the production of biofuels, while the electrochemical properties of biochar have been identified as a promising material in super capacitors, batteries, and hydrogen storage devices (Razzak 2024). Additionally, biochar as the cement, constituents of asphalt, and composite materials enhanced mechanical strength, insulation, and sustainability benefits in making them suitable for usage in the construction (Fawzy et al. 2021). Increasing biochar's application, it is being used more and more quickly in different industries such as applied in wastewater treatment, industrial waste management, and in pollution control. Its advanced adsorption capacity and surface reactivity remove heavy metals like lead (Pb), cadmium (Cd), arsenic (As), organic contaminants, microplastics, and pharmaceutical wastes from contaminated environments. Soni et al. 2025 stated that the microorganisms like fungi, algae and bacteria also used in removal of Heavy metals and other contaminants in wastewater.

The grouping of magnetic biochar and biochar-activated carbon hybrids is much enhanced in the removal of recalcitrant pollutants, considerably emphasizing their role in water treatment systems (Zhang et al. 2020). Despite the delightful advantages, large-scale commercialization of biochar is obstructed due to variability of feedstock issues, lack of economic viability, regulatory issues and lack of adequate long-term field studies. Standardized protocols for characterization of biochar regulatory approval are needed to ensure consistency and quality across various uses. Considering policy programs such as carbon credit and government subsidies, which involves strategic finance, could further promote large-scalebiochar production particularly emphasizing industries as such agriculture, climate action, and waste treatment (Fawzy et al. 2021; Fig. 1). This combination of engineered biochar, interdisciplinary research, and facilitative policies emerges as a reliable and long-term solution. Such an effort is essential to promote soil restoration, to develop carbon-negative solutions for society, and to build momentum towards more effective steps for the environment (Fawzy et al. 2021).

The objectives of this chapter are to explore the evolution of biochar production techniques from traditional practices to modern technologically advanced methods, emphasizing their roles in contaminant removal and environmental remediation. The production of biochar from various other agriculture wastes, invasive plants, etc., is to assess the economic, policy, and regulatory challenges hindering large-scale

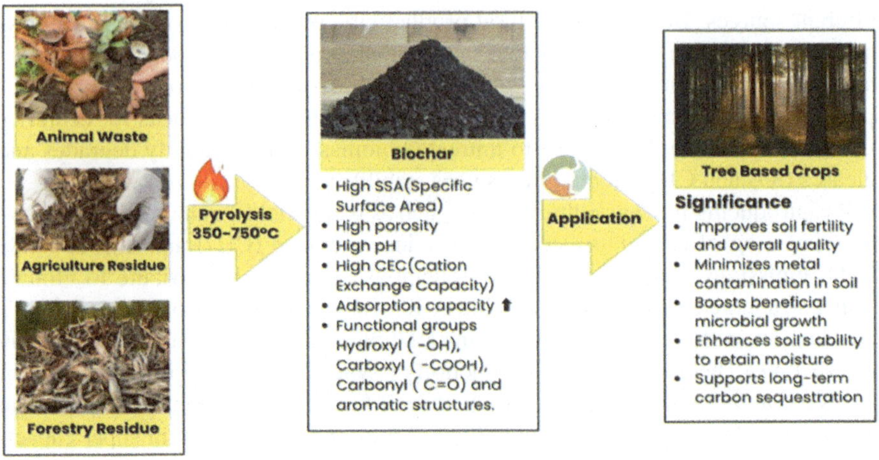

Fig. 1 Production of biochar from various sources and its importance

commercialization of biochar and proposes strategies for standardization and policy support.

The objectives of this chapter are as follows:

1. To explore the evolution of biochar production techniques from traditional practices to modern, technologically advanced methods.
2. Emphasizing biochar's roles in contaminant removal and environmental remediation.
3. The production of biochar from various other agriculture waste, invasive plants, etc., to assess the economic, policies.
4. Regulatory challenges hindering large-scale commercialization of biochar and propose strategies for standardization and policy support.

2 Evolution in Biochar Production from Traditional Method to Modern Technology

The use of biochar dates back approximately 2000 years and has been applied in various regions across the world (Hunt et al. 2010). The introduction of biochar was done by western agriculture in mid of nineteenth century, even though the utilization of biochar was extended in today's time also (Abiven et al. 2014). Tremendous fertile soil name "Terra preta," i.e., the black soil of India was established by ancient cultures, providing the proof of using biochar earlier (O'Neill et al. 2009; Tenenbaum 2009). The amount of various nutrients like potassium (K), phosphorus (P), and nitrogen (N) were known to be highly rich in soil and supports the growth of various plants in various conditions (Chen et al. 2019). Soil enriched with biochar possesses useful characteristics such as a dark color, high mineral content, and nutrient richness that can persist for many years even after heavy tropical rainfall and leaching (Hunt et al. 2010). Biochar can occur naturally in soil worldwide due to the natural process which occurs in forest and grassland fires or by manmade activities in agriculture (Lehmann and Joseph 2012). Traditionally, biochar was produced when stacking it up soil pits, later on supplying with slow burning in aerobic or anaerobic conditions (Gabhane et al. 2020). In the present time also various parts of countries, like Yogyakarta, Indonesia, also depend on the traditional methods, which include predominantly earth mounds with pit kilns covered with portable steel, oil drums, bricks, concrete, and many more to support the structure and with stand high temperatures. Depending on the methods and kiln types, raw materials, processing time, and carbonization, the yield ranges from 22.88% to 35.98%, respectively (Lehmann and Joseph 2012). In modern era, the production of biochar equipment is well control and established where harmful gas emits at slow rates with low level of emission (Bacskai et al. 2019). However, there have been challenges faced during the production of biochar in tropical and rural conditions to obtain same level of efficiencies. Conversely, new technologies are costly compared to conventional techniques and hence are out of reach for local communities who cannot realize their full potential benefits (Peters et al. 2015).

3 Preparation Methods of Biochar

Biochar is typically made through thermochemical approaches such as pyrolysis, hydrothermal carbonization (HTC), gasification, and torrefaction. These procedures thermally transform biomass into carbon-rich biochar, bio-oil, and syngas in the presence of oxygen-limited conditions (Nidheesh et al. 2021). The choice of the methods is heavily supported with factors such as temperature, heating rate, and residence time, which significantly impacts biochar yield and characteristics (Kazemi Shariat Panahi et al. 2020). Pyrolysis is the most broadly adapted method, including carbon-based, i.e., degradation of organic matter at temperatures of normally about 300–900°C, in an environment of the absence of oxygen, i.e., an anaerobic atmosphere (Rangabhashiyam & Balasubramanian 2019). It results in three chief fractions: solid biochar, syngas, and liquid bio-oil. Slow pyrolysis contains moderately minor heating rates (0.1–0.8°C/s) with extreme residence times of up to 25–35 hours, foremost in high production of biochar of roughly 35% (El-Naggar et al. 2019). This happens from increased secondary reactions with continued durations of vapor–solid relations responsible for improved biochar creation (Ponnusamy et al. 2020).

Fast pyrolysis, however, works at temperatures of about 450–600 °C at high reheating rates (10–1000 °C/s) and short residence times of vapors (0.5–2 s) to selectively create bio-oil (~ 75%) and produce lesser biochar (~ 12%) (Tomczyk et al. 2020).

Flash pyrolysis is passed out at ultrashort heating rates of above (> 1000) °C per second at temperatures around 750–1000 °C by very short interaction time of the order of < 0.5 s with the objective of enhancing the making of the bio-oil (Gaurav et al. 2020).

Microwave-assisted pyrolysis (MAP) is indeed a novel technology that facilitates volumetric heating and rapid treatment, producing biochar with a higher surface area (typically 450–800 m^2/g) and better functional groups as compared to conventional methods (Naji and Tye 2022). It depends on electric power, which possibly will verify an interference in its scalability (Ao et al. 2018).

Hydrothermal carbonization (HTC) is an effective technique for processing high-moisture biomass ($\geq$ 70%) without the need of pre-drying (Ha and Lee 2020), using subcritical water at 180–250 °C to convert biomass into hydrochar. During hydrothermal carbonization, the biomass undergoes a series of reactions, as in hydrolysis, dehydration, and condensation, forming 5-hydroxymethylfurfural intermediates that gradually polymerize into hydrochar. Biomass rich in lignin follows a more complex degradation process involving dealkylation and production of phenolic compound (Zhou et al. 2020; Pathak et al. 2024a). Gasification includes a high-temperature process involving partial oxidation of biomass at 800°C, resulting in a gaseous mixture that is composed of CO, CH_4, H_2, and CO_2F (Zaker et al. 2019). Biochar is produced as a by-product (~ 10%) due to the efficient conversion of carbon to gas (Umenweke et al. 2022).

Torre faction is a low-temperature heat treatment that is finished at lower temperatures from the range of 200–300 °C in an inert environment for 10–60 min (Manyà et al. 2020). This step removes the moisture and the light volatiles such that an energy-dense solid residue is obtained that is also water-repelling (Yadav et al. 2022). O:C ratio of torrefied biomass is normally higher than that of true biochar by over 0.4, which may exclude it from being classified as true biochar by the standards of Europe (Daful and Chandraratne 2020).

Choice of appropriate biochar preparation technique is dictated by a variety of factors such as biomass type, moisture content, desired end-product properties, as well as overall energy efficiency. Slow pyrolysis remains the most suitable and most practiced technique of high yield biochar production among existing techniques, while other recent advancements like microwave-assisted pyrolysis and hydrothermal carbonization (HTC) present prospective avenues of development.

4 Biochar Production from Plants

The synthesis of biochar from invasive plant species is characterized by their adaptability, quick reproduction cycles, and aggressive expansion, all of which can have negative effects on the environment, the economy, and human health (Zhang et al. 2019). Not only can biochar be made from invasive plant nuts, but it can also be made from cultivated biomass resources like wheat husk and algae leftovers, which are equally useful feedstocks for biomass production. Therefore, turning these plants

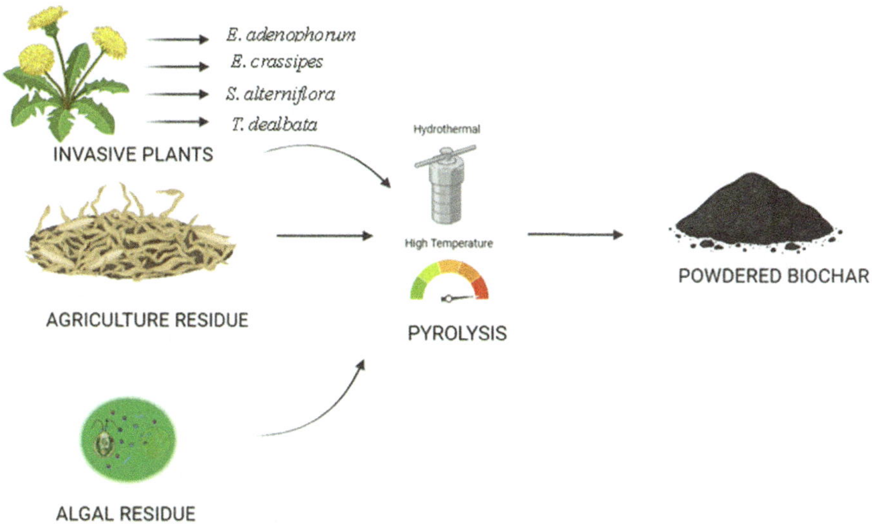

Fig. 2 Biochar production from different sources

(waste/residue) into useful resources, especially by producing biochar, has emerged as a viable and sustainable substitute (Zheng et al. 2014; Cheng et al. 2017). A stable, carbon-rich substantial, biochar is formed by pyrolyzing of biomass with little or no oxygen. Thus, it helps in reducing soil pollutants, increasing carbon capture efficacy, and assisting soil development (Zheng et al. 2019). The manufacturing of biochar from invasive plants, agricultural surplus, and algae leftovers suggests both ecological benefits and economic potential (Fig. 2) (Feng et al. 2018).

4.1 Invasive Plants as Biochar Feedstock for Its Production

Invasive species offer renewable feedstock sources due to their natural distributions, rapid re-colonization, and rejuvenation potential. Manufacture of invasive species as a source of feedstock for biochar manufacture offers economic profits and greener alternatives. Moreover that, the resulting invasive plant-derived biochar (IP-DB) also has characteristic structural and chemical structures that expands its applicability within the environment and agriculture (Fan et al. 2020). At high temperatures, the porosity and carbon content of biochar increase, but simultaneously, it reduces the yield of biochar (Manyà 2012). In biochar from *Eupatorium adenophorum*, for instance, there are improved carbon content and higher steadiness at higher temperatures of the pyrolysis (Fan et al. 2019). *Imperata cylindrica* and *Eupatorium adenophorum biochars* own more complex nutrient quality than biochar of plant (Walter and Rao 2015). At lower temperatures, surface functional groups like hydroxyl (-OH) and carboxyl (-COOH) are not destroyed and show vital functions for adsorption procedures (Hu et al. 2020). Biochar formed at a lower temperature from *Eichhornia crassipes* (water hyacinth) forms functional groups efficiently to chelate Cu (II) ions (Li et al. 2015).

4.2 *Production of Biochar from Agricultural and Algal Biomass*

Cultivated biomass substrates, like wheat husk and algal products, are important feedstocks for biomass construction. Overreliance on chemical fertilizers and high burning of crop residues have immensely contributed to soil pollution, increased greenhouse gases, and ecological conflicts (Mona et al. 2021). In India, cereal crops like wheat, maize, rice, and sugarcane yield 63% of total crop residues, where most of it is combusted (Khare et al. 2021). Biochar from wheat husk pyrolyzed at 350 °C possesses 39.97% conversion efficiency (Sharma et al. 2025). It is alkaline in nature, has 60.6% organic carbon content (Kumar and Bhattacharya 2022), and has a high magnesium (567.66 ppm) and potassium (800 ppm) composition (Tripathi et al. 2014). It raises the soil pH, TOC, potassium, sulfur, and sodium by 8–12% application rates (Alburquerque et al. 2014). It has an improved yield of 76.09% and is composed of 99.6% organic carbon, 99.1% ash, and essential minerals such as magnesium, phosphorus, zinc, and iron. Furthermore, the biochar also improved the soil cation exchange capacity (CEC) and total organic carbon (TOC) and immobilized the cadmium effectively (Michalak et al. 2019).

A comparison between wheat husk and algal biochar reveals marked differences in terms of yield, composition, and performance. Algal biochar exhibited a substantially higher pyrolysis efficiency of 76.09%, meaning that a greater portion of the original biomass is retained as solid product, whereas to 39.97% for wheat husk biochar (Sharma et al. 2025). In terms of organic carbon content, algal biochar contained 99.6%, which is considerably surpassing higher than the 60.6% observed in wheat husk, suggesting greater potential for carbon sequestration and microbial activity (Kumar and Bhattacharya 2022). Differences in Soil improvement were also observed. Wheat husk biochar proved to be more effective in increasing soil pH (from 8.18 to 9.93 at 12% concentration), thereby making it particularly suitable for acidic soil remediation (Sharma et al. 2025). Conversely, algal biochar resulted in only modest pH increase (to 8.3 at 12%). However, algal biochar contributed more to total organic carbon (TOC) in soil, elevating up to 59.1%, compared to 44.1% for wheat husk, and also additionally enhanced cation exchange capacity (CEC) and micronutrient availability (Ullah et al. 2020). Furthermore, in crop growth trials, algal biochar outperformed wheat husk: Moong and Bajra crops treated with algal biochar showed higher shoot lengths, enhanced chlorophyll content, and better overall plant health (Sharma et al. 2025).

4.3 *Comparison of Plant Biomass Feedstocks for Biochar Production*

Agricultural residues like wheat straw and rice husk, along with woody biomass and fruit peels, offer diverse properties in terms of yield, surface area, carbon content, and ash composition. These differences directly influence the functionality of the resulting biochar's determining their suitability for applications as soil conditioners, catalyst supports, or adsorbents. For instance, crop residues such as wheat straw typically yield biochar's well suited for soil enhancement owing to their moderate porosity and nutrient-rich ash. In contrast, materials like peanut shells and hardwoods are preferred where higher adsorption capacity and surface area are required. The pyrolysis temperature plays a critical role in modulating these properties, affecting parameters such as aromaticity, pH, and functional group density. Therefore, selecting appropriate feedstock based on the desired application is crucial in optimizing the environmental and agronomic benefits of biochar. Table1 summarizes several commonly studied plant biomass feedstocks and their specific characteristics.

5 Biochar Characteristics

Biochar has diverse physical and chemical characteristics. Biochar is typically much smaller than soil particles, and its incorporation makes soil lighter, but it helps in increasing soil moisture, maintaining the pH of soil, and increasing nutrient viability with cation exchange capacity. These conditions may vary according to the type of the soil, stage of the soil, and types of biochar, which was applied and the dosage of the biochar. Biochar possesses absorption properties due to various groups like COOH, OH, R-NH2, RCONR, and lactonic groups. The main reason behind the surface functional groups of biochar is biomass and temperature (Qi et al. 2017). In modern technologies, various techniques are available to study the characteristics of biochar through X-rays differentiation (XRD), Fourier transform infrared spectroscopy (FTIR), scanning electron microscopy (SEM), nuclear magnetic resonance spectroscopy (NMR), Raman Spectroscopy, etc. (Sohi et al. 2010; Lehmann & Joseph 2012; Yaashikaa et al. 2020). The acidity of the biochar, which contains mainly O (oxygen) functional groups, e.g., carboxyl, hydroxyl, and carbonyl, helps to alter the hydrophilicity of the soil, which increases the moisture content in the soil. The implication of biochar in soil helps in increasing the cation exchange capacity of soil, thus allowing it to absorb and release nutrients, therefore helping in sustainable agriculture growth. The research currently in publication states that eight indications the bulk density, aggregation, porosity, moisture content, cation exchange capacity, pH, nutrients, and electrical conductivity of the soil are the main markers of how biochar affects the physicochemical characteristics of soil.

Table 1 Plant biomass feedstocks and its specific features

Feedstock	Biochar production methods	Application	Reference
Rice husk	Slow pyrolysis, 2 h at 500 °C, BET surface area: 230.91 m^2/g, pH: ~ 7.99	Removal of cadmium (Cd^{2+}); achieved by 97.1% removal at 20% biochar dosage. Highlighting strong effectiveness for heavy metal immobilization	(Milke et al. 2020)
Wheat straw	Slow pyrolysis, 3 h at 600 °C, BET surface area: 183.3 m^2/g, pH: ~ 9.1	Adsorption of trifluralin (herbicide); achieved 13% removal at 1% biochar dosage. Indicating its potential for moderate pesticide remediation	(Hamzah et al. 2019)
Paper mill sludge	Slow pyrolysis, 2 h at 500–600 °C, BET surface area: 47–50 m^2/g, pH: 8.78–9.17	Adsorption of phenanthrene (PAH); 37.9% removal at 5% dosage. It is applicable for remediation of hydrophobic organic contaminants	(Diacono et al. 2019)
Switchgrass	Fast pyrolysis, 30 s at 600–800 °C, BET surface area: 1.4–17.2 m^2/g, pH: 9.1–11.2	Adsorption of uranium (U); 90% removal at 0.5% dosage. Ideally potential for radioactive pollutant control	(Gwenzi et al. 2017)
Hardwood/ softwood	Typically, slow pyrolysis, ~ 500–700 °C; biochar used as catalyst in biodiesel	Used as biochar catalysts in biodiesel production; converted oil to biodiesel with 81.5–88% yield Indicating strong applicability in biofuel synthesis	(Talebian-Kiakalaieh et al. 2013)

Research has indicated that integration of biochar has resulted in a significant reduction of the density of the soil in bulk while increasing the structure of the aggregates at the same time, more so with sandy clays and Xerisols soil. In addition, correct application of the biochar has been recognized to increase the water-holding capacity of the soil efficiently; with sandy-type soils, the higher the rate of application of the biochar, the higher the porosity (Zhu et al. 2025). The specific surface area (SSA) and pore size of biochar play an significant role in adsorption and regulating the nature of the biochar as hydrophilic or hydrophobic properties (Tan et al. 2017). Normally, biochar possesses more surface area and high porosity; therefore, it has a high sorption property. Biochar obtained the property of porousness through the pyrolysis method, resulting in some water loss during the process of pyrolysis. The biochar of micro size does not absorb pesticides particles despite being charged or possess polarity. The pore size of biochar is checked by scanning electronic microscope (SEM). Surface area helps in absorption of water in biochar, and same temperature plays a crucial role in formation of biochar. Commercial available biochar possesses more surface area with high water sorption capacity as compares to inactivated biochar (Kim et al. 2022). Biochar made from Douglas fir wood (*Pseudotsuga menziesii*) has a porosity of up to 83%; Miscanthus shows 79%, and hazelnut shells are at 62.5% at their lowest pyrolysis temperature between 370 and 400°C (Gray et al. 2014; Liu et al. 2018). The surface area was increased by mixing Fe_2O_3 with premixed banana peel after hydrothermal carbonization. Their output shows sudden increment in the value 407, 504, 451, and 446 m^2/g by combining with 0.05–1.0 gm of the precursor solution, respectively (Rong et al. 2019). Li et al. 2020a, b, performed a study on the Saccharina japonica plant in which the biochar derived from this plant effectively increased with an increase in temperature, resulting in some exciting outcomes (2.9–175 m^2/g), with the maximum surface area attained at 500°C (Li et al. 2020a, b). Rafiq et al. 2016, found that pyrolysis helps in increasing sorption capacity because of blockage of pore blocking substances or they get broken due to thermally cracked and increasing the externally accessible surface area increased (Rafiq et al. 2016). Zhao et al. 2017, reported that pyrolysis helps in increasing surface area and pore volume by a progressive degradation of organic matter like cellulose, lignin, etc., and the formation of vascular bundles or channel structures (Li et al. 2013; Zhao et al. 2017).

During pyrolysis some carbon amorphous substances were also formed through the degradation of cellulose (Zhao et al. 2017). The porosity of soil gets changes due to various factors like pore structure (intrapore), biochar physical characteristics, and its application rate along with pore between soil particles and biochar substances, i.e., inter-pores and soil changes die to absorption and particles size distribution (Zhang and You 2013; Yi et al. 2020). These factors help in increasing or decreasing various factors like porosity, water retention, hydraulic conductivity, and respiration in biochar-enriched soil. Various studies proven that biochar helps in increasing soil porosity and water-holding capacity (Rogovska et al. 2011; Randolph et al. 2017). The pH plays a crucial role in maintaining structure and chemical properties of biochar, typically the pH of biochar is neutral to alkaline, and it depends on the process of making biochar and type of feedstock and functional group present over biochar (Jahirul et al. 2012; Qi et al. 2017). During high temperatures, the acidic group, i.e., bionic acid, breaks down and increases the pH of biochar, and causes a surge in inorganic alkali metal ions, not only does bionic acid increase the pH of biochar, but other functional groups also help in increasing the pH, like carboxylic acid group, acetate ion, hydroxyl, etc. (Yuan et al. 2019), not only bionic acid increased pH of biochar other functional also helps in increasing the pH like carboxylic acid group, acetate ion, hydroxyl, etc. (Li et al. 2017). Zhou et al. 2013, found that biochar's pH values differ with sources; for example, like bamboo has 7.9, sugarcane bagasse 7.5, hickory wood 8.4, and peanut husk 6.9 after some change in chitosan, the values of pH change towards more alkaline 8.2, 8.1, 8.6 & 7.3 correspondingly (Zhou et al. 2013). The pH of corncob biochar at 600 °C shows neutral pH around 7.17, but

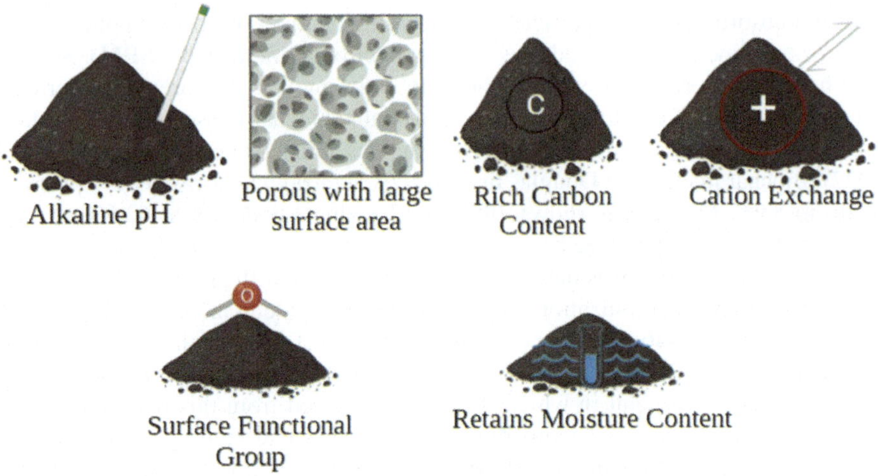

Fig. 3 Various characteristics of biochar

after some modification with magnesium oxide, it reported 10.4 pH (Shi et al. 2021). Various characteristics of biochar are shown in Fig. 3.

6 Environmental Applications of Biochar

6.1 In Remediation

Invasive plant-derived biochars have confirmed a potent capability to eradicate a wide variety of environmental pollutants, including organic and heavy metal toxins. Its performance is varied by the type of vegetation and pyrolysis temperature and preparation process. Spartina alterniflora biochar at 600 °C pyrolysis temperature can remove 83% Cu(II) (Li et al. 2015). Similarly, Eupatorium adenophorum biochar displayed strong physical adsorption of Cu (II) and Cr (VI) (Cornelissen et al. 2016). Apart from that, biochars of another plant such as Solidago and Ambrosia trifida have been abilities to adsorb organic pollutant compounds like trichloroethylene, TNT, and RDX due to their good aromatic structural conformations and huge surface area (Roh et al. 2015). Surface treatment, i.e., magnetic treatment, etc., can suggestively increase adsorption, particularly for contaminants such as arsenic (Zhang et al. 2016). Biochar application is demonstrated in Fig. 4.

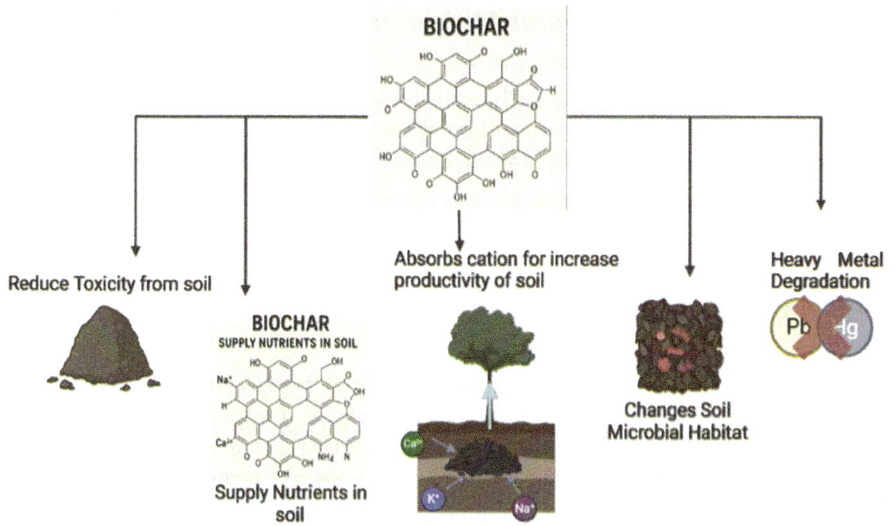

Fig. 4 Application of biochar in various aspects related to soil

6.2 Soil Amendment and Agriculture

Invasive plant-derived biochars (IP-DBs) have also been proven to be extremely operative and proficient in improving physical as well as chemical properties of soil. *Eichhornia crassipes* and *Eupatorium adenophorum* biochars are alleged to be demonstrative of improving soil structure, enhancing water retention capacity, and enhancing nutrient provision. Water hyacinth biochar, for example, enhances enzymatic activity in soil and improves corn development (Mazumder et al. 2020). In the same way, biochar obtained from Eupatorium increases soil pH, potassium content, and the level of phosphorus, thus helping crop yield development, mainly of corn (Pandit et al. 2018). In addition to enhancing fertility, these biochars also block the uptake of poisonous chemicals into plants. For example, Eupatorium has been proved to reduce acetochlor accumulation in corn, a common herbicide. Moreover, it increases water-holding capacity of clay soils (Obia et al. 2017) and was reported to boost pumpkin yields when used in combination with cow urine, pointing toward potential synergies with organic amendments (Schmidt et al. 2015). Biochar also promotes the growth of microbes. Microbes use their cellular metabolisms to convert biodegradable, complicated hazardous chemicals into safe, acceptable end products like CO_2 and H_2O in soil (Singh et al. 2025).

6.3 *Antibiotic Contaminant Mitigation*

Sicyos angulatus biochar, particularly acid-treated with sulfuric or oxalic acid, was reported to effectively reduction in sulfamethazine (SMT) mobility and plant uptake in lettuce. This effect is primarily attributed to its aromaticity and hydrophobic nature (Rajapaksha et al. 2015) (Table 2).

7 Commercialization of Biochar

The supply of biochar for commercialization involves five key steps: feedstock cultivation and availability, feedstock processing, pyrolysis, transportation, and application. Traditionally, the supply chain was much simpler biochar which was produced locally using available biomass through basic pyrolysis techniques, often disregarding gas and liquid byproducts generated during the process. Pyrolysis methods got much fame to produce biofuel and energy directly from biomass but same biochar does not get same importance being biochar are good sources for the improvement of agriculture quality and soil health improvement (Anderson et al. 2017). Biochar required the proper chain and design for commercialization at broad level of market (Zilberman et al. 2019). The commercialization of biochar can be done by integrating with other system like bioenergy systems and connecting various organization throughout the supply chain to effectively deliver a biochar product that meets the objectives of optimizing social value for the entire community and financial gain for farmers and producers (Zilberman et al. 2017). Most studies contend that biochar is more likely to be economical in developing nations, attributing the higher profits in these regions to the low-cost production technologies (Stoves and Kilns) and the possibly higher yield increase in sandy and tropical soils (Pratt and Moran 2010; Dickinson et al. 2015).

8 Conclusion

This chapter helps us to find various aspects of biochar production from ancient time to modern time. It covers the evolution of biochar from ancient practices like the Amazonian terra preta soils to contemporary green uses in agriculture, environmental remediation, and energy storage. Production of biochar from a variety of waste materials, such as farm residues, invasive plants, algae, and many more, shows its variability as a sustainable solution for waste valorization. Biochar's physiochemical characteristics, which are affected by the feedstock, post-production procedures, and pyrolysis conditions, include surface area, porosity, pH value, and elemental composition. These parameters, which are affected by the feedstock, post-production procedures, and pyrolysis conditions, include surface area, porosity, pH value, and

Table 2 Invasive plants used for biochar production and their applications

S. No.	Invasive plant	Biochar application	Remark	Reference
1	*Argemone ochroleuca*	Effective in adsorption of Cr^{6+} from aqueous solutions; good surface area, porosity, and active adsorption sites	Non-functionalized biochar has 26.46% Cr^{6+} removal at pH 2; surface area 8.47 m^2/g	(El-Naggar et al. 2019)
2	*Prosopis juliflora*	Effectively adsorbs bio-dyes like methylene blue and malachite green	Multilingual functional groups assist in removal efficiency	(Bhavsar et al. 2019)
3	*Solidago canadensis*	Adsorbs antibiotics, tetracycline, and enrofloxacin from aqueous medium	Biochar showed significant surface functional groups	(Zhang et al. 2018a, b)
4	*Alternanthera philoxeroides*	Adsorbs dyes and heavy metals including Cr, Cu, Pb; modified biochar improved adsorption	Biochar presence of functional groups such as carboxyl and phenolic improved efficiency	(Yang et al. 2014)
5	*Parthenium hysterophorus*	Decreases deforestation, absorption of plant nutrients	Maximum degradation occurs due to high temperature	(Qurat-ul-Ain et al. 2016)
6	*Thalia dealbata*	Produces higher biochar yield and higher phosphorus and potassium content	The usage of $MgCl_2$ helps in increasing surface area and magnesium is effective in phosphorus retention	(Cui et al. 2016)
7	*Elodea nuttallii*	Enhanced removal of heavy metals and organic toxins including dyes, antibiotics	Produces carbon-rich biochar, pH > 10, surface area ~ 300–340 m^2/g	(Poerschmann et al. 2015)
8	*Lantana camara*	Constructed for natural filters to purify water; suitable for gas purification and wastewater treatment	The surface area and activated biochar can be used in metal ion and pesticide removal	(Kumar et al. 2024a, b)
9	*Eichhornia crassipes*	Adsorbs heavy metals such as Cd, As, and Cr; good cation exchange capacity (CEC)	Biochar is renewable and sustainable, improves soil conditioning, nutrient holding, and metal ion absorption	(Zhang et al. 2015)
10	*Spartina alterniflora*	Decomposes Cr(VI); reduces Cr(VI) and total Cr concentration in contaminated soils	Effective in high-moisture regions and acts as nutrient adsorbent	(Li et al. 2015)

(continued)

Table 2 (continued)

S. No.	Invasive plant	Biochar application	Remark	Reference
11	*Kigelia africana*	Antimicrobial, antioxidant, and antibiofilm activity	Biochar helps microbial and nanomaterial formulation with high antimicrobial adherence	(Kumar et al. 2024a, b)
12	*Ambrosia trifida*	Capable of adsorbing organic explosives such as TNT, RDX, trichloroethylene; also effective in removing Cd(II) from contaminated media	High-temp biochar (700 °C) shows enhanced TCE removal (88.5%) due to greater C-enrichment and aromaticity	(Ahmad et al. 2014)

elemental composition. The physical and chemical properties of biochar are complex. It improves the soil's cation exchange capacity and nutrient readiness by decreasing bulk density, increasing collective constancy, and raising soil moisture content in arid zones. Compared to inactivated biochar, marketable biochar has a greater surface area and a higher capacity for water sorption. In conclusion, this chapter focuses on identifying biochar as a versatile and multifaceted solution that is in line with climate change, sustainable agricultural practices, and environmental fortification. It demands for ongoing research and there is a pressing need of, technological expansion, and policy assistance, to bring its full potential as a fundamental part of the future to life environmentally friendly sustainable plans.

9 Future Prospects

The properties and characteristics of biochar are quite different so it could be an alternative in various applications like water treatment, increase soil fertility, and help to improves plant health and other factors which promotes growth and development of plants and agriculture. Biochar sorption capacity can could also be increased by mixing it with other substances which have higher surface area and are more susceptible to attach with biochar and helps them to enhance more properties of biochar. The incorporation of two invasive plants' biochar could be better options to adopt for increasing the productivity of the plants as well as to increasing the properties of it; at the same time, it helps in adsorption of heavy metals which are being released from new sources or pollutants could be new too. The efficiency of biochar can also be increased by mixing with nanomaterials that can easily enter even in micro places and help the plants to act over it.

Acknowledgements Prof. K. Dwivedi and M. Soni are thankful to Madhya Pradesh Science and Technology (MPCST), Bhopal, for providing fellowship.

Author's Contribution All authors contributed significantly to this work. [Soni] designed the study and coordinated the research activities. [Kanani] performed data collection and analysis. [Singh] provided expertise in methodology and contributed to data interpretation. [Shukla & Dwivedi] assisted with manuscript drafting, revision, and critical intellectual content. All authors reviewed and approved the final manuscript.

References

Abiven S, Schmidt MWI, Lehmann J (2014) Biochar by design. Nat Geosci 7(5):326–327. https://doi.org/10.1038/ngeo2154

Ahmad M et al (2014) Production and use of biochar from buffalo-weed (Ambrosia trifida L.) for trichloroethylene removal from water. J Chem Technol Biotechnol 89(1):150–157. Available at: https://doi.org/10.1002/jctb.4157

Ajema L, Tafa N (2015) (PDF) Effects of biochar application on beneficial soil organism review. ResearchGate. https://www.researchgate.net/publication/324823056_Effects_of_Biochar_Application_on_Beneficial_Soil_Organism_Review

Alburquerque JA, Calero JM, Barrón V, Torrent J, Del Campillo MC, Gallardo A, Villar R (2014) Effects of biochars produced from different feedstocks on soil properties and sunflower growth. J Plant Nutr Soil Sci 177(1), Article 1. https://doi.org/10.1002/jpln.201200652

Anderson NM, Bergman RD, Deborah P-DS (2017) A supply chain approach to biochar systems [Chapter 2] | US Forest Service Research and Development. https://research.fs.usda.gov/treesearch/54270

Ao W, Fu J, Mao X, Kang Q, Ran C, Liu Y, Zhang H, Gao Z, Li J, Liu G (2018) Microwave assisted preparation of activated carbon from biomass: a review. Renew Sustain Energy Rev 92:958–979

Askeland M, Clarke B, Paz-Ferreiro J (2019) Comparative characterization of biochars produced at three selected pyrolysis temperatures from common woody and herbaceous waste streams. PeerJ 7:e6784. https://doi.org/10.7717/peerj.6784

Bacskai I, Madar V, Fogarassy C, Toth L (2019) Modeling of some operating parameters required for the development of fixed bed small scale pyrolysis plant. Resources 8(2):79. https://doi.org/10.3390/resources8020079

Bakraoui M, Karouach F, Ouhammou B, Aggour M, Essamri A, El Bari H (2020) Biogas production from recycled paper mill wastewater by UASB digester: optimal and mesophilic conditions. Biotechnol Rep 25:e00402. https://doi.org/10.1016/j.btre.2019.e00402

Bhavsar N, Ntoumanis N, Quested E, Gucciardi DF, Thøgersen-Ntoumani C, Ryan RM, Reeve J, Sarrazin P, Bartholomew KJ (2019) Conceptualizing and testing a new tripartite measure of coach interpersonal behaviors. Psychol Sport Exerc 44:107–120. https://doi.org/10.1016/j.psychsport.2019.05.006

Chauhan PK, Upadhyay SK, Chauhan A, Bhutiani R, Sikarwar RLS, Tiwari MK (2025) A review on simultaneous use of microbes and biochar for removal of toxic heavy metals: recent updates and future outlooks. Physiol Mol Biol Plants 31(3):1–15. https://doi.org/10.1007/s12298-025-01649-4

Chen W, Meng J, Han X, Lan Y, Zhang W (2019) Past, present, and future of biochar. Biochar 1(1):75–87. https://doi.org/10.1007/s42773-019-00008-3

Cheng S, Zhang S, Zhang L, Xia H, Peng J, Wang S (2017) Microwave-assisted preparation of activated carbon from Eupatorium Adenophorum: effects of preparation parameters. High Temp Mater Process 36(8), Article 8. https://doi.org/10.1515/htmp-2015-0285

Cornelissen G, Pandit NR, Taylor P, Pandit BH, Sparrevik M, Schmidt HP (2016) Emissions and char quality of flame-curtain" Kon Tiki" Kilns for Farmer-Scale charcoal/biochar production. PloS One 11(5), Article 5

Cui X, Dai X, Khan KY, Li T, Yang X, He Z (2016) Removal of phosphate from aqueous solution using magnesium-alginate/chitosan modified biochar microspheres derived from Thalia dealbata. Biores Technol 218:1123–1132. https://doi.org/10.1016/j.biortech.2016.07.072

Daful AG, Chandraratne MR (2020) Biochar production from biomass waste-derived material

Diacono M, Persiani A, Testani E, Montemurro F, Ciaccia C (2019) Recycling agricultural wastes and by-products in organic farming: Biofertilizer production, yield performance and carbon footprint analysis. Sustainability 11(14), Article 14

Dickinson D, Balduccio L, Buysse J, Ronsse F, Van Huylenbroeck G, Prins W (2015) Cost-benefit analysis of using biochar to improve cereals agriculture. GCB Bioenergy 7(4):850–864. https://doi.org/10.1111/gcbb.12180

Ding W, Dong X, Ime IM, Gao B, Ma LQ (2014) Pyrolytic temperatures impact lead sorption mechanisms by bagasse biochars. Chemosphere 105:68–74. https://doi.org/10.1016/j.chemosphere.2013.12.042

El-Naggar A, Lee SS, Rinklebe J, Farooq M, Song H, Sarmah AK, Zimmerman AR, Ahmad M, Shaheen SM, Ok YS (2019) Biochar application to low fertility soils: a review of current status, and future prospects. Geoderma 337:536–554

Fan J, Cai C, Chi H, Reid BJ, Coulon F, Zhang Y, Hou Y (2020) Remediation of cadmium and lead polluted soil using thiol-modified biochar. J Hazard Mater 388:122037

Fan L, Zhou X, Liu Q, Wan Y, Cai J, Chen W, Chen F, Ji L, Cheng L, Luo H (2019) Properties of Eupatorium Adenophora Spreng (Crofton Weed) biochar produced at different pyrolysis temperatures. Environ Eng Sci 36(8), Article 8. https://doi.org/10.1089/ees.2019.0028

Fang J, Zhan L, Ok YS, Gao B (2018) Minireview of potential applications of hydrochar derived from hydrothermal carbonization of biomass. J Ind Eng Chem 57:15–21. https://doi.org/10.1016/j.jiec.2017.08.026

Fawzy S, Osman AI, Yang H, Doran J, Rooney DW (2021) Industrial biochar systems for atmospheric carbon removal: a review. Environ Chem Lett 19(4), Article 4. https://doi.org/10.1007/s10311-021-01210-1

Feng Y, Shu L, Ren Z, Kong Q, Xu F, Wang Q (2018) A mesoporous biochar from bio-invasion alligator weed for adsorption of rhodamine B from aqueous solution. Desalin Water Treat 135:341–350

Gabhane JW, Bhange VP, Patil PD, Bankar ST, Kumar S (2020) Recent trends in biochar production methods and its application as a soil health conditioner: a review. SN Appl Sci 2(7):1307. https://doi.org/10.1007/s42452-020-3121-5

Gaurav GK, Mehmood T, Cheng L, Klemeš JJ, Shrivastava DK (2020) Water hyacinth as a biomass: a review. J Clean Prod 277:122214

Godlewska P, Ok YS, Oleszczuk P (2021) The dark side of black gold: ecotoxicological aspects of biochar and biochar-amended soils. J Hazard Mater 403:123833. https://doi.org/10.1016/j.jhazmat.2020.123833

Gray M, Johnson MG, Dragila MI, Kleber M (2014) Water uptake in biochars: the roles of porosity and hydrophobicity. Biomass Bioenerg 61:196–205. https://doi.org/10.1016/j.biombioe.2013.12.010

Gwenzi W, Chaukura N, Noubactep C, Mukome FN (2017) Biochar-based water treatment systems as a potential low-cost and sustainable technology for clean water provision. J Environ Manage 197:732–749

Gyanwali P et al (2024) Exploring the benefits of biochar: a review of production methods, characteristics, and applications in soil health and environment. Egypt J Soil Sci 64(3). Available at: https://doi.org/10.21608/ejss.2024.270380.1725

Ha JH, Lee I-G (2020) Study of a method to effectively remove char byproduct generated from fast pyrolysis of lignocellulosic biomass in a bubbling fluidized bed reactor. Processes 8(11), Article 11

Hamzah N, Tokimatsu K, Yoshikawa K (2019) Solid fuel from oil palm biomass residues and municipal solid waste by hydrothermal treatment for electrical power generation in Malaysia: a review. Sustainability 11(4), Article 4

Hou C-H, Liu N-L, Hsi H-C (2015) Highly porous activated carbons from resource-recovered Leucaena leucocephala wood as capacitive deionization electrodes. Chemosphere 141:71–79

Hu X, Zhang X, Ngo HH, Guo W, Wen H, Li C, Zhang Y, Ma C (2020) Comparison study on the ammonium adsorption of the biochars derived from different kinds of fruit peel. Sci Total Environ 707:135544

Hunt J, DuPonte M, Sato D, Kawabata A (2010). The basics of biochar: a natural soil amendment. https://doi.org/10.5281/ZENODO.4319331

International Biochar Initiative (IBI). Available online:https://biocharinternational.org/characterizationstandard/ (accessed on 20 October 2020)

Ippolito JA, Cui L, Kammann C, Wrage-Mönnig N, Estavillo JM, Fuertes-Mendizabal T, Cayuela ML, Sigua G, Novak J, Spokas K, Borchard N (2020) Feedstock choice, pyrolysis temperature and type influence biochar characteristics: a comprehensive meta-data analysis review. Biochar 2(4), Article 4. https://doi.org/10.1007/s42773-020-00067-x

Jahirul M, Rasul M, Chowdhury A, Ashwath N (2012) Biofuels production through biomass pyrolysis—a technological review. Energies 5(12):4952–5001. https://doi.org/10.3390/en5124952

Jiang Z, Yang S, Pang Q, Xu Y, Chen X, Sun X, Qi S, Yu W (2021) Biochar improved soil health and mitigated greenhouse gas emission from controlled irrigation paddy field: insights into microbial diversity. J Clean Prod 318:128595

Kazemi Shariat Panahi H, Dehhaghi M, Ok YS, Nizami A-S, Khoshnevisan B, Mussatto SI, Aghbashlo M, Tabatabaei M, Lam SS (2020) A comprehensive review of engineered biochar: production, characteristics, and environmental applications. J Clean Prod 270:122462. https://doi.org/10.1016/j.jclepro.2020.122462

Khan BN (2024) Impact of biochar on soil organisms. In: Biochar—solid carbon for sustainable agriculture. Bentham Science Publishers, pp 29–63. https://www.benthamdirect.com/content/books/9789815238068.chapter-3

Khare P, Deshmukh Y, Yadav V, Pandey V, Singh A, Verma K (2021) Biochar production: a sustainable solution for crop residue burning and related environmental issues. Environ Prog Sustain Energy 40(2), Article 2. https://doi.org/10.1002/ep.13529

Kim H, Zhang G, Chung TCM, Nam C (2022) A role for newly developed sorbents in remediating large-scale oil spills: reviewing recent advances and beyond. Adv Sustain Syst 6(1). https://doi.org/10.1002/adsu.202100211

Kumar A, Bhattacharya T, Shaikh WA, Biswas JK (2024a) Valorization of invasive plant and leaf litter wastes into biochar: production, properties and potential for arsenic removal. Groundw Sustain Dev 24:101066. https://doi.org/10.1016/j.gsd.2023.101066

Kumar V, Arun A, Rajkumar K, Palaniyappan S (2024b) Effect of Kigelia pinnata biochar inclusion on mechanical and thermal properties of curtain climber fiber reinforced epoxide biocomposites. Polym Compos 45(5):4459–4472. https://doi.org/10.1002/pc.28073

Kumar A, Bhattacharya T (2022) Removal of arsenic by wheat straw biochar from soil. Bull Environ Contam Toxicol 108(3), Article 3. https://doi.org/10.1007/s00128-020-03095-2

Lehmann J, Joseph S (Eds) (2012). Biochar for environmental management. Routledge. https://doi.org/10.4324/9781849770552

Leng L, Xiong Q, Yang L, Li H, Zhou Y, Zhang W, Jiang S, Li H, Huang H (2021) An overview on engineering the surface area and porosity of biochar. Sci Total Environ 763:144204. https://doi.org/10.1016/j.scitotenv.2020.144204

Leng L, Zheng H, Shen T, Wu Z, Xiong T, Liu S, Cao J, Peng H, Zhan H, Li H (2025) Engineering biochar from biomass pyrolysis for effective adsorption of heavy metal: an innovative machine learning approach. Sep Purif Technol 361:131592. https://doi.org/10.1016/j.seppur.2025.131592

Li X, Shen Q, Zhang D, Mei X, Ran W, Xu Y, Yu G (2013) Functional groups determine biochar properties (pH and EC) as studied by two-dimensional 13C NMR correlation spectroscopy. PLoS ONE 8(6):e65949. https://doi.org/10.1371/journal.pone.0065949

Li M, Lou Z, Wang Y, Liu Q, Zhang Y, Zhou J, Qian G (2015) Alkali and alkaline earth metallic (AAEM) species leaching and Cu (II) sorption by biochar. Chemosphere 119:778–785

Li J, He F, Shen X, Hu D, Huang Q (2020a) Pyrolyzed fabrication of N/P co-doped biochars from $(NH_4)_3PO_4$-pretreated coffee shells and appraisement for remedying aqueous Cr(VI) contaminants. Biores Technol 315:123840. https://doi.org/10.1016/j.biortech.2020.123840

Li X, Zhang L, Yang Z, Wang P, Yan Y, Ran J (2020b) Adsorption materials for volatile organic compounds (VOCs) and the key factors for VOCs adsorption process: a review. Sep Purif Technol 235:116213. https://doi.org/10.1016/j.seppur.2019.116213

Lin L, Jiang W, Xu P (2017) Comparative study on pharmaceuticals adsorption in reclaimed water desalination concentrate using biochar: impact of salts and organic matter. Sci Total Environ 601:857–864

Liu Z, Dugan B, Masiello CA, Wahab LM, Gonnermann HM, Nittrouer JA (2018) Effect of freeze-thaw cycling on grain size of biochar. PLoS ONE 13(1):e0191246. https://doi.org/10.1371/journal.pone.0191246

Manna MC, Rahman MM, Naidu R, Sahu A, Bhattacharjya S, Wanjari RH, Patra AK, Chaudhari SK, Majumdar K, Khanna SS (2018) Bio-waste management in subtropical soils of India: future challenges and opportunities in agriculture. Adv Agron 152:87–148

Manyà JJ (2012) Pyrolysis for biochar purposes: a review to establish current knowledge gaps and research needs. Environ Sci Technol 46(15), Article 15. https://doi.org/10.1021/es301029g

Manyà JJ, García-Morcate D, González B (2020) Adsorption performance of physically activated biochars for postcombustion CO_2 capture from dry and humid flue gas. Appl Sci 10(1), Article 1

Maqbool Z, Farooq MS, Rafiq A, Uzair M, Yousuf M, Khan MR, Huo S (2024) Unlocking the potential of biochar in the remediation of soils contaminated with heavy metals for sustainable agriculture. Funct Plant Biol 51(2), Article 2. https://doi.org/10.1071/FP23257

Mazumder P, Khwairakpam M, Kalamdhad AS (2020) Bio-inherent attributes of water hyacinth procured from contaminated water body–effect of its compost on seed germination and radicle growth. J Environ Manage 257:109990

Michalak I, Baśladyńska S, Mokrzycki J, Rutkowski P (2019) Biochar from a freshwater macroalga as a potential biosorbent for wastewater treatment. Water 11(7), Article 7

Milke J, Gałczyńska M, Wróbel J (2020) The importance of biological and ecological properties of Phragmites australis (Cav.) Trin. Ex Steud., in phytoremendiation of aquatic ecosystems—The review. Water, 12(6), Article 6

Mona S, Malyan SK, Saini N, Deepak B, Pugazhendhi A, Kumar SS (2021) Towards sustainable agriculture with carbon sequestration, and greenhouse gas mitigation using algal biochar. Chemosphere 275:129856

Naji SZ, Tye CT (2022) A review of the synthesis of activated carbon for biodiesel production: precursor, preparation, and modification. Energy Convers Manag: X 13:100152

Nidheesh PV, Gopinath A, Ranjith N, Akre AP, Sreedharan V, Kumar MS (2021) Potential role of biochar in advanced oxidation processes: a sustainable approach. Chem Eng J 405:126582

O'Neill B, Grossman J, Tsai MT, Gomes JE, Lehmann J, Peterson J, Neves E, Thies JE (2009) Bacterial community composition in brazilian anthrosols and adjacent soils characterized using culturing and molecular identification. Microb Ecol 58(1):23–35. https://doi.org/10.1007/s00248-009-9515-y

Obia A, Børresen T, Martinsen V, Cornelissen G, Mulder J (2017) Effect of biochar on crust formation, penetration resistance and hydraulic properties of two coarse-textured tropical soils. Soil Tillage Res 170:114–121

Osman AI, Hefny M, Abdel Maksoud MIA, Elgarahy AM, Rooney DW (2021) Recent advances in carbon capture storage and utilisation technologies: a review. Environ Chem Lett 19(2), Article 2. https://doi.org/10.1007/s10311-020-01133-3

Pandit NR, Mulder J, Hale SE, Martinsen V, Schmidt HP, Cornelissen G (2018) Biochar improves maize growth by alleviation of nutrient stress in a moderately acidic low-input Nepalese soil. Sci Total Environ 625:1380–1389

Pathak HK, Chauhan PK, Seth CS, Dubey G, Upadhyay SK (2024a) Mechanistic and future prospects in rhizospheric engineering for agricultural contaminants removal, soil health restoration, and management of climate change stress. Sci Total Environ 927:172116. https://doi.org/10.1016/j.scitotenv.2024.172116

Pathak HK, Seth CS, Chauhan PK, Dubey G, Singh G, Jain D, Upadhyay SK, Dwivedi P, Khoo KS (2024b) Recent advancement of nano-biochar for the remediation of heavy metals and emerging contaminants: mechanism, adsorption kinetic model, plant growth and development. Environ Res 255:119136. https://doi.org/10.1016/j.envres.2024.119136

Peters JF, Iribarren D, Dufour J (2015) Biomass pyrolysis for biochar or energy applications? A life cycle assessment. Environ Sci Technol 49(8):5195–5202. https://doi.org/10.1021/es5060786

Poerschmann J, Weiner B, Wedwitschka H, Zehnsdorf A, Koehler R, Kopinke F-D (2015) Characterization of biochars and dissolved organic matter phases obtained upon hydrothermal carbonization of Elodea nuttallii. Biores Technol 189:145–153

Ponnusamy VK, Nagappan S, Bhosale RR, Lay C-H, Nguyen DD, Pugazhendhi A, Chang SW, Kumar G (2020) Review on sustainable production of biochar through hydrothermal liquefaction: physico-chemical properties and applications. Biores Technol 310:123414

Pratt K, Moran D (2010) Evaluating the cost-effectiveness of global biochar mitigation potential. Biomass Bioenerg 34(8):1149–1158. https://doi.org/10.1016/j.biombioe.2010.03.004

Qi F, Dong Z, Lamb D, Naidu R, Bolan NS, Ok YS, Liu C, Khan N, Johir MAH, Semple KT (2017) Effects of acidic and neutral biochars on properties and cadmium retention of soils. Chemosphere 180:564–573. https://doi.org/10.1016/j.chemosphere.2017.04.014

Qurat-ul-Ain, Firdaus-e-Bareen, Shafiq, M (2016) Management of the Parthenium hysterophorus through biochar formation and its application to rice-wheat cultivation in Pakistan. Agric, Ecosyst Environ 235:265–276. https://doi.org/10.1016/j.agee.2016.10.027

Rafiq MK, Bachmann RT, Rafiq MT, Shang Z, Joseph S, Long R (2016) Influence of pyrolysis temperature on physico-chemical properties of corn stover (Zea mays L.) biochar and feasibility for carbon capture and energy balance. PLOS ONE 11(6):e0156894. https://doi.org/10.1371/journal.pone.0156894

Rajapaksha AU, Vithanage M, Ahmad M, Seo D-C, Cho J-S, Lee S-E, Lee SS, Ok YS (2015) Enhanced sulfamethazine removal by steam-activated invasive plant-derived biochar. J Hazard Mater 290:43–50

Randolph P, Bansode RR, Hassan OA, Rehrah Dj, Ravella R, Reddy MR, Watts DW, Novak JM, Ahmedna M (2017) Effect of biochars produced from solid organic municipal waste on soil quality parameters. J Environ Manag 192:271–280. https://doi.org/10.1016/j.jenvman.2017.01.061

Rangabhashiyam S, Balasubramanian P (2019) The potential of lignocellulosic biomass precursors for biochar production: performance, mechanism and wastewater application-a review. https://www.cabidigitallibrary.org/doi/full/https://doi.org/10.5555/20193253377

Razzak SA (2024) Municipal solid and plastic waste derived high-performance biochar production: a comprehensive review. J Anal Appl Pyrol 181:106622. https://doi.org/10.1016/j.jaap.2024.106622

Rich T, Felfel A (2015) An overview of Canadian food loss and waste estimates. Rural Development Institute, Brandon University, Brandon, MB, Canada, Webinars and Speaker Series, p 4

Rogovska N, Laird D, Cruse R, Fleming P, Parkin T, Meek D (2011) Impact of biochar on manure carbon stabilization and greenhouse gas emissions. Soil Sci Soc Am J 75(3):871–879. https://doi.org/10.2136/sssaj2010.0270

Roh H, Yu M-R, Yakkala K, Koduru JR, Yang J-K, Chang Y-Y (2015) Removal studies of Cd (II) and explosive compounds using buffalo weed biochar-alginate beads. J Ind Eng Chem 26:226–233

Rong X, Xie M, Kong L, Natarajan V, Ma L, Zhan J (2019) The magnetic biochar derived from banana peels as a persulfate activator for organic contaminants degradation. Chem Eng J 372:294–303. https://doi.org/10.1016/j.cej.2019.04.135

Schmidt HP, Pandit BH, Martinsen V, Cornelissen G, Conte P, Kammann CI (2015) Fourfold increase in pumpkin yield in response to low-dosage root zone application of urine-enhanced biochar to a fertile tropical soil. Agriculture 5(3), Article 3

Sharma A, Mittal R, Bohat N, Malik K, Dahiya B (2025) Sustainable agriculture and soil enrichment through diverse organic vermicompost synthesized from different organic waste. Int J Environ Sci Technol 22(4), Article 4. https://doi.org/10.1007/s13762-024-05774-7

Shi Q, Zhang H, Shahab A, Zeng H, Zeng H, Bacha A-U-R, Nabi I, Siddique J, Ullah H (2021) Efficient performance of magnesium oxide loaded biochar for the significant removal of Pb^{2+} and Cd^{2+} from aqueous solution. Ecotoxicol Environ Saf 221:112426. https://doi.org/10.1016/j.ecoenv.2021.112426

Singh S, Dwivedi K (2022) Phytoremediation: an alternate way to treat cadmium in soil 3:377–382

Singh S, Bhardwaj AK, Dwivedi K (2024) Advances of plant-assisted synthesized metal nanoparticles and their environmental application. In: Biogenic wastes-enabled nanomaterial synthesis. Springer Nature Switzerland, pp 373–392. https://doi.org/10.1007/978-3-031-59083-2_15

Singh S, Dwivedi K, Chakraborty D, Pathak S, Shukla N (2025) Toxic pollutants in environment: microbial technology in wastewater. In: Sustainable landscape planning and natural resources management. Springer Nature Switzerland, pp 1–7. https://doi.org/10.1007/978-3-031-80845-6_1

Sohi SP, Krull E, Lopez-Capel E, Bol R (2010) A review of biochar and its use and function in soil. In: Advances in agronomy, vol 105. Elsevier, pp 47–82. https://doi.org/10.1016/S0065-2113(10)05002-9

Soni M, Singh S, Gupta S, Dwivedi K, Shukla N (2025) Wastewater management and urban water sustainability. In: Urban sustainability. Springer Nature Singapore, pp 93–105. https://doi.org/10.1007/978-981-96-1140-9_4

Talebian-Kiakalaieh A, Amin NAS, Mazaheri H (2013) A review on novel processes of biodiesel production from waste cooking oil. Appl Energy 104:683–710

Tan Z, Lin CSK, Ji X, Rainey TJ (2017) Returning biochar to fields: a review. Appl Soil Ecol 116:1–11. https://doi.org/10.1016/j.apsoil.2017.03.017

Tao W et al (2020) Formation of persistent free radicals in biochar derived from rice straw based on a detailed analysis of pyrolysis kinetics. Sci Total Environ 715:136575

Tenenbaum DJ (2009) Biochar: carbon mitigation from the ground up. Environ Health Perspect 117(2). https://doi.org/10.1289/ehp.117-a70

Tomczyk A, Sokołowska Z, Boguta P (2020) Biochar physicochemical properties: pyrolysis temperature and feedstock kind effects. Rev Environ Sci Bio/Technol 19(1), Article 1. https://doi.org/10.1007/s11157-020-09523-3

Tripathi DK, Singh VP, Chauhan DK, Prasad SM, Dubey NK (2014) Role of macronutrients in plant growth and acclimation: recent advances and future prospective. In Ahmad P, Wani MR, Azooz MM, Phan Tran L-S (Eds) Improvement of crops in the era of climatic changes. Springer New York, pp 197–216. https://doi.org/10.1007/978-1-4614-8824-8_8

Ullah N, Ditta A, Khalid A, Mehmood S, Rizwan MS, Ashraf M, Mubeen F, Imtiaz M, Iqbal MM (2020) Integrated effect of algal biochar and plant growth promoting rhizobacteria on physiology and growth of maize under deficit irrigations. J Soil Sci Plant Nutr 20(2), Article 2. https://doi.org/10.1007/s42729-019-00112-0

Umenweke GC, Afolabi IC, Epelle EI, Okolie JA (2022) Machine learning methods for modeling conventional and hydrothermal gasification of waste biomass: a review. Bioresour Technol Rep 17:100976

Waheed A, Xu H, Qiao X, Aili A, Yiremaikebayi Y, Haitao D, Muhammad M (2025) Biochar in sustainable agriculture and climate mitigation: mechanisms, challenges, and applications in the circular bioeconomy. Biomass Bioenerg 193:107531. https://doi.org/10.1016/j.biombioe.2024.107531

Walter R, Rao BKR (2015) Biochars influence sweet-potato yield and nutrient uptake in tropical Papua New Guinea. J Plant Nutr Soil Sci 178(3), Article 3. https://doi.org/10.1002/jpln.201400405

Wang J, Wang S (2019) Preparation, modification and environmental application of biochar: a review. J Clean Prod 227:1002–1022
Yaashikaa PR, Kumar PS, Varjani S, Saravanan A (2020) A critical review on the biochar production techniques, characterization, stability and applications for circular bioeconomy. Biotechnol Rep 28:e00570. https://doi.org/10.1016/j.btre.2020.e00570
Yadav M, Joshi C, Paritosh K, Thakur J, Pareek N, Masakapalli SK, Vivekanand V (2022) Reprint of organic waste conversion through anaerobic digestion: a critical insight into the metabolic pathways and microbial interactions. Metab Eng 71:62–76
Yang Y, Wei Z, Zhang X, Chen X, Yue D, Yin Q, Xiao L, Yang L (2014) Biochar from Alternanthera philoxeroides could remove Pb(II) efficiently. Biores Technol 171:227–232. https://doi.org/10.1016/j.biortech.2014.08.015
Yi S, Chang NY, Imhoff PT (2020) Predicting water retention of biochar-amended soil from independent measurements of biochar and soil properties. Adv Water Resour 142:103638. https://doi.org/10.1016/j.advwatres.2020.103638
Yuan P, Wang J, Pan Y, Shen B, Wu C (2019) Review of biochar for the management of contaminated soil: preparation, application and prospect. Sci Total Environ 659:473–490. https://doi.org/10.1016/j.scitotenv.2018.12.400
Zaker A, Chen Z, Wang X, Zhang Q (2019) Microwave-assisted pyrolysis of sewage sludge: a review. Fuel Process Technol 187:84–104
Zhang J, You C (2013) Water holding capacity and absorption properties of wood chars. Energy Fuels 27(5):2643–2648. https://doi.org/10.1021/ef4000769
Zhang F, Wang X, Yin D, Peng B, Tan C, Liu Y, Tan X, Wu S (2015) Efficiency and mechanisms of Cd removal from aqueous solution by biochar derived from water hyacinth (Eichornia crassipes). J Environ Manage 153:68–73
Zhang Z, Chen L, Wang J, Yao J, Li J (2018b) Biochar preparation from Solidago canadensis and its alleviation of the inhibition of tomato seed germination by allelochemicals. RSC Adv 8(40):22370–22375. https://doi.org/10.1039/C8RA03284J
Zhang L, Li Y, Huang J, Liu J, Liu X (2019) Evaluation of the short-term and long-term performance of biological invasion management in the China-Myanmar border region. J Environ Manage 240:1–8
Zhang H, Xiao R, Li R, Ali A, Chen A, Zhang Z (2020) Enhanced aqueous Cr(VI) removal using chitosan-modified magnetic biochars derived from bamboo residues. Chemosphere 261:127694. https://doi.org/10.1016/j.chemosphere.2020.127694
Zhang C, Liu L, Zhao M, Rong H, Xu Y (2018) The environmental characteristics and applications of biochar. Environ Sci Pollut Res 25(22), Article 22. https://doi.org/10.1007/s11356-018-2521-1
Zhang Feng ZF, Wang Xin WX, Ji XiongHui JX, Ma LiJuan ML (2016) Efficient arsenate removal by magnetite-modified water hyacinth biochar. https://www.cabidigitallibrary.org/doi/full/https://doi.org/10.5555/20163366899
Zhao S-X, Ta N, Wang X-D (2017) Effect of temperature on the structural and physicochemical properties of biochar with apple tree branches as feedstock material. Energies 10(9):1293. https://doi.org/10.3390/en10091293
Zheng Z, Xia H, Srinivasakannan C, Peng J, Zhang L (2014) Utilization of Crofton weed for preparation of activated carbon by microwave induced CO_2 activation. Chem Eng Process 82:1–8
Zheng Y, Wang B, Wester AE, Chen J, He F, Chen H, Gao B (2019) Reclaiming phosphorus from secondary treated municipal wastewater with engineered biochar. Chem Eng J 362:460–468
Zhou Y, Gao B, Zimmerman AR, Fang J, Sun Y, Cao X (2013) Sorption of heavy metals on chitosan-modified biochars and its biological effects. Chem Eng J 231:512–518. https://doi.org/10.1016/j.cej.2013.07.036
Zhou J-M, Jiang Z-C, Qin X-Q, Zhang L-K, Huang Q-B, Xu G-L, Dionysiou DD (2020) Efficiency of Pb, Zn, Cd, and Mn removal from karst water by Eichhornia crassipes. Int J Environ Res Public Health 17(15), Article 15

Zhu Z, Zhang Y, Tao W, Zhang X, Xu Z, Xu C (2025) The biological effects of biochar on soil's physical and chemical characteristics: a review. Sustainability 17(5), Article 5. https://doi.org/10.3390/su17052214

Zilberman D, Lu L, Reardon T (2017) Innovation-induced food supply chain design. Food Policy 83:289–297. https://doi.org/10.1016/j.foodpol.2017.03.010

Zilberman D, Lu L, Reardon T (2019) Innovation-induced food supply chain design. Food Policy 83:289–297. https://doi.org/10.1016/j.foodpol.2017.03.010

An Overview of Global Biochar Utilization: Its Role in Soil Health Under Environmental Stress Conditions

Faheem Ahamad, Rakesh Bhutiani, M. Mohana Keerthi, Avnish Chauhan, Prabhat K. Chauhan, and Benjouid Zakaria

Abstract The global population is exceeding 8 billion and will reach up to 9.1 billion by 2050, and maintaining soil health is crucial for sustainable food production. Meeting the intensifying call for food, water, and shelter requires sustainable soil management strategies, where biochar emerges as a promising solution. To feed the continuously growing population, soil health is crucial. Utilization of agrochemicals in the field increases the fertility of soil, but prolonged use makes the soil polluted due to accumulation of chemicals. Similarly, improper management of crops leftover also contributes to air, water, and soil pollution along with greenhouse gas (GHG) emission. Through pyrolysis process, the crops leftover can be converted to a carbon-rich material known as biochar. Biochar is globally utilized for the enhancement of soil structure and fertility and is therefore promoting sustainable agriculture. The current book chapter deals with the preparation of biochar and its utilization. The chapter also focuses on the impacts of biochar on soil health and environmental stress mitigation. To fulfill the objective of the current study, a systematic search was conducted on different research platform, i.e., Scopus, Web of Science, Google Scholar, etc., for literature. The results of the current review unveil the potential of biochar in soil physicochemical and microbial properties' improvement. Biochar also plays an important role in reducing the environmental stress by mitigating the concentration of GHG, absorbing heavy metals, and increasing the salinity tolerance capacity of soil. Varying physicochemical properties, lack of information about its

F. Ahamad (✉) · R. Bhutiani
Limnology and Ecological Modelling Laboratory, Department of Zoology and Environmental Sciences, Gurukul Kangri (Deemed to be University), Haridwar, Uttarakhand, India
e-mail: faheem.ahamad170390@gmail.com

M. Mohana Keerthi
Department of Agronomy, School of Agriculture, SR University Warangal, Telangana, India

A. Chauhan
Department of Environmental Science, Graphic Era Hill University, Dehradun, Uttarakhand, India

P. K. Chauhan
Department of Environmental Science, A.K.S. University, Satna, Madhya Pradesh, India

B. Zakaria
Faculty of Economics and Management, Hassan 1st University, Settat, Morocco

A. Chauhan et al. (eds.), *Biochar As Remediation Tools for Pollutant Removal*,
https://doi.org/10.1007/978-3-032-16746-0_2

long-term utilization, and cost are some of the main challenges faced on commercial application of biochar. The current book chapter will help researchers, agro-planners, and farmers to understand the biochar and its potential in the management of soil fertility and environmental stress globally.

Keywords Population explosion · Soil health · Food scarcity · Crop residues · Biochar economics · Carbon sequestration

1 Introduction

1.1 Global Population

The global population is continuously growing, and it is expected that if current trends continue, the global population will reach up to 9.1 billion by 2050 (Fig. 1). The continuously increasing population needs both shelter and food for their survival. About 28% of the world's ice-free land is covered with pastures and only 12% land is available for agriculture (Ramankutty et al. 2008). The Food and Agriculture Organization (FAO) has projected that by 2050, the global population will reach up to 9.1 billion, requiring a 70% increase in food production to encounter the demand. Nearly all this population growth will occur in developing countries, with urbanization accelerating significantly. Currently 49% of the total population inhabit in urban areas, but by 2050, 70% of the total population will inhabit in urban areas. Currently, the world is facing the problem of less agricultural land, especially fertile, and by 2050, the problem will become horrible. This means that agricultural systems must adapt to ensure food security, requiring investments in sustainable farming, technological advancements, and efficient resource management.

1.2 Need Sustainable Soil Management

Farmers used fertilizers, especially chemical fertilizers like urea, diammonium phosphate (DAP), single superphosphate (SSP), triple superphosphate (TSP), to enhance the soil productivity. Although these fertilizers enhance soil fertility temporarily, prolonged use of these chemical-based fertilizers (agrochemicals) degrades the soil structure, reduces the microbial diversity, and leads to contamination. On the other hand, the whole globe is facing the big challenge of crop residue management as its improper management leads to environmental issues (Wiedner and Glaser 2015) such as air pollution, soil degradation, and increased GHG emissions (Yadav et al. 2023). Similarly, the management of sludge generated from sewage treatment plant (STP) and effluent treatment plant (ETP) is another big challenge, the whole globe is facing currently (Holatko et al. 2025). Therefore, researchers are continuously

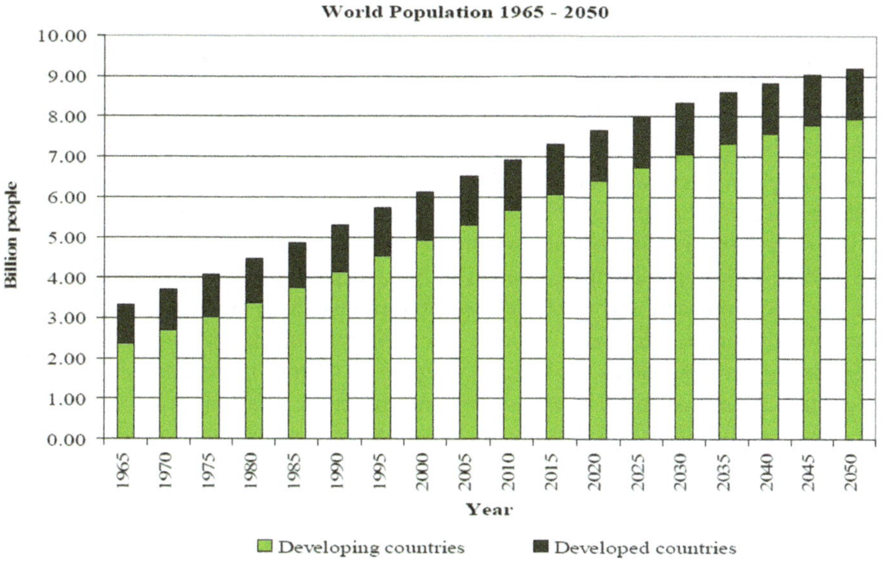

Fig. 1 Increase in the world population from 1965 to 2050 adapted from Food and Agriculture Organization (FAO) report. *Source* Population Division of the Department of Economic and Social Affairs of the United Nations Secretariat (2007)

searching the advanced ways of waste management like use in soil mulching (Salem et al. 2021; Saudy et al. 2021; El-Metwally et al. 2022), use as organic fertilizer (Mubarak et al. 2021), and as biochar (Wang et al. 2022). Among the sustainable soil management practices, biochar emerges as the most promising solution. Through pyrolysis, crop residues can be converted into biochar, a stable carbon-rich material with significant agronomic and environmental benefits (Chagas et al. 2022). Biochar is globally recognized for enhancing soil fertility, improving structure, and promoting sustainable agriculture.

1.3 Historical Perspective on Biochar Utilization

Amazon Basin Terra Preta soils serve as evidence of biochar utilization for soil alteration for millennia. The formation of these highly fertile "dark earths" took place due to unification of carbon-based residues like ceramic wreckage, food waste, and charred biomass into nutrient-deficient tropical soils by the Pre-Columbian indigenous civilizations (Glaser and Birk 2012). In certain parts of the world, slash-and-burn techniques were used in those days for biomass management, which is more destructive and leads to nutrient loss and carbon emission leading to environmental pollution. At the same time in certain parts of the world including parts of Africa, Asia, and Oceania, the traditional slash-and-char techniques have been employed which involve controlled charring of biomass and thereby protect the soil nutrient loss while converting organic material into stable carbon forms. This method not only maintains soil fertility for an extended period but also helps in the reduction of atmospheric carbon by locking the carbon in the form of biochar and therefore considered as a more sustainable alternative for land management.

Although biochar was historically (Fig. 2) utilized as a soil amendment, its widespread adoption declined by the mid of twentieth century (Lehmann and Joseph 2015) with the rise of industrial agriculture and the increased reliance on synthetic fertilizers. Again, the scientific interest in biochar has experienced a resurgence in the late-twentieth and early-twenty-first centuries because of intensified concerns over soil degradation, climate change, and unsustainable waste management. The literature review elucidates the biochar's ability in improving soil physicochemical properties, reduction in GHG emissions, and function as a long-term carbon sink (Lefebvre et al. 2023), reinforcing its potential as a sustainable agricultural and environmental solution (Lehmann and Joseph 2015; Chen et al. 2019; Amonette et al. 2021). Biochar's revival has lifted it from indigenous knowledge practice to a topic of international research and policy discussion. Biochar is now a feasible tool in frameworks for sustainable agriculture, climate mitigation, and the circular economy because of recent developments in pyrolysis technology and feedstock optimization that enable scalable production (Lefebvre et al. 2023; Saletnik and Saletnik 2025).

1.4 Objectives of the Chapter

Keeping in mind the above discussion, the research was conducted to explore the biochar preparation, applications, and its impact on soil health and environmental stress mitigation.

2 Biochar Production and Characteristics

2.1 Raw Materials and Feedstocks

Biochar is a carbon-rich material produced from crop residues, sewage sludge, animal manure, forestry waste, and municipal organic waste with the help of pyrolysis process. Table 1 shows the different materials used for the preparation of biochar by the researchers globally.

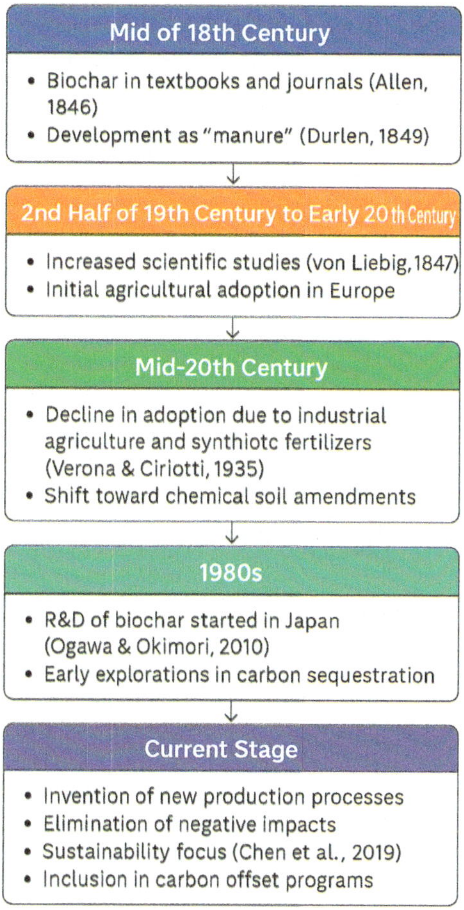

Fig. 2 Flowchart showing the historical perspective of the biochar. *Source* Lehmann and Joseph (2015)

2.2 *Pyrolysis Process and Parameters*

In the pyrolysis process, the crops leftovers are decomposed in anoxic environment at a very high range of temperature. The process is classified in to slow, fast, and flash pyrolysis, gasification, and hydrothermal carbonization (Table 2). In slow pyrolysis process, the crops' leftovers completely decompose at very slow rate yielding primarily biochar and bio-oil and syngas as secondary product (Pahnila et al. 2023). In fast and flash pyrolysis process, the crops leftovers decompose at comparatively high rate yielding bio-oil as primarily product (Wright et al. 2010; Fahmy et al. 2020; Karkach et al. 2023). In fast pyrolysis process, the range of temperature is from 400 to 600 °C, while in flash, the temperature is about 2500 °C. The gasification process converts the crops leftovers primarily into syngas in the limited supply of oxygen. Hydrothermal carbonization (HTC) is a process which takes place at moderate temperature (180–250 °C) and high pressure in watery environments.

Table 1 Raw materials used for biochar preparation and their efficiency

Sr. No.	Feedstock used	Significant impacts on soil health	References
1	Wood chips	Improved microbial activity and soil structure, increases WHC	Latini et al. (2019), Rechberger et al. (2019)
2	Wheat straw	Enhances nutrient availability and plant growth, reduces heavy metal uptake	Latini et al. (2019)
3	Saw dust	Increases organic matter, improved microbial activity, and soil respiration	Holatko et al. (2025)
4	Sewage sludge	Contains high nitrogen and phosphorus content, enhances soil enzyme activity and nutrient cycling	Holatko et al. (2025)
5	Zeolite-enriched biochar	Improves soil pH, nutrient retention, and restrains microbial enzyme activity	Holatko et al. (2025)
6	Corn stover	Enhances organic matter, soil aeration, and WHC	Zafeer et al. (2024)
7	Rice husk	Improves soil pH, microbial activity, and reduces heavy metal toxicity	Latini et al. (2019), Wijitkosum and Jiwnok (2020)
8	Coconut shell	Increases soil carbon content, enhances nutrient retention and soil structure	Khan et al. (2024)
9	Poultry litter	Boosts nitrogen availability, microbial diversity, and crop yield	Khan et al. (2024), Freitas et al. (2020)
10	Sugarcane bagasse	Enhances soil fertility and WHC, supports beneficial microbes	
11	Acacia wood and coconut shell	Carrier for microbial inoculants	Saranya et al. (2011)
12	Bermuda grass	Improves effluent characteristics	Muhammad et al. (2022)
13	Paper mill sludge	For the absorption of pentachlorophenol (PCP) from artificially prepared solution	Devi and Saroha (2015)
14	Dairy manure	Environmental remediation	Cao and Harris (2010), Xie et al. (2022)

2.3 Physicochemical Properties of Biochar

The appropriate biochar among the different types of available biochar for particular bioremediation process is selected based on the physicochemical properties of the biochar. Almost all the physicochemical properties depend on the type of feedstocks, type of pyrolysis along with temperature and other process-based conditions. Color, porosity, surface area, electrical conductivity (EC), stability, etc., are some physical properties of biochar (Kamarudin et al. 2022), while cation exchange capacity (CEC), carbon content, volatile content, and functional group are some of chemical

Table 2 Different type of pyrolysis processes

Sr. No.	Type of pyrolysis process	Temperature (°C)	Residence time	Primary product	Secondary product	References
1	Slow	300–600	Longer	Biochar	Bio-oil and syngas	Pahnila et al. (2023)
2	Fast	400–600	Shorter	Bio-oil	Biochar	Wright et al. (2010), Karkach et al. (2023)
3	Flash	2500	Shorter	Bio-oil	Biochar	Fahmy et al. (2020)
4	Gasification	> 700		Syngas	Biochar	Yao et al. (2016)
5	Hydrothermal carbonization	180–250	Longer	Hydrochar	Gases (CO_2, CH_4, H_2)	González-Arias et al. (2022)

parameters. Biochar prepared at higher temperature reflect higher surface area and porosity and higher porosity which helps in increased time of water content (Scott et al. 2014). A higher number of pores in the biochar also increase the aeration in the soil which results in the increase growth of plants (Liu et al. 2017). The pores present on the surface of biochar helps in the stabilization of microbial colonies and therefore helps in restoration of soil (Pietikainen et al. 2000; Liu et al. 2017; Wong and Ogbonnaya 2021). Moreover, the biochar prepared at higher temperature shows the increased stability and decreased reactivity due to less volatile content (Tomczyk et al. 2020).

3 Global Utilization of Biochar

3.1 Application in Different Agricultural Systems

Due to multiple benefits like carbon sequestration, GHG reduction, managed soil physicochemical properties, and as a biosorbent in soil, air, and water pollution remediation, the biochar caught the global attention (Wu et al. 2024a). In different types of agricultural system (Fig. 3), i.e., cereal cropping, tree cropping, horticulture, pasture cropping, agroforestry, organic farming, biochar plays different types of role (Luo et al. 2025).

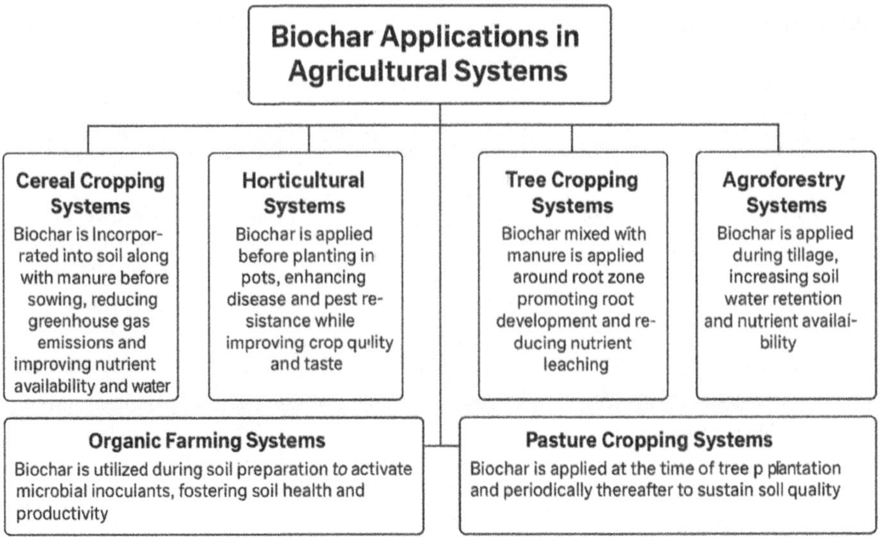

Fig. 3 Applications of biochar in different agricultural systems. *Source* (Hossain et al. 2020; Schmidt et al. 2021; Wu et al. 2024b; Luo et al. 2025)

3.2 Regional Case Studies on Biochar Adoption

3.2.1 India: Integrating Biochar into Sustainable Agriculture

In India, the utilization of biochar is gaining momentum through region-specific initiatives aimed at sustainable agriculture and biomass waste management (Sharma et al. 2024). Although there is no comprehensive national policy dedicated exclusively to biochar, programs such as the National Bio-Energy Mission and the National Mission on Sustainable Agriculture (NMSA) support practices that align with its application (https://mnre.gov.in/en/bio-energy/). In agriculturally intensive states like Punjab and Haryana, where the open burning of crop residues contributes to severe air quality issues, pyrolysis-based biochar production is being explored as a cleaner alternative. Preliminary studies have indicated that combining biochar with compost and organic fertilizers can significantly improve soil organic carbon levels, enhance water retention capacity, and boost crop productivity.

3.2.2 China: Scaling Biochar Through National Programs

China has emerged as a global leader in biochar research and implementation, integrating its use into national strategies for rural development and climate change mitigation. Government-supported programs have facilitated the establishment of industrial-scale facilities that convert agricultural residues especially maize stalks and rice husks into biochar (Wijitkosum and Jiwnok 2020). Deng et al. (2024) reported that application of biochar to soil enhances the structure of the soil, microbial diversity, and reduction in nitrogen leaching from soil to groundwater and therefore helps in both soil and groundwater management.

3.2.3 Sub-Saharan Africa: Low-Cost Biochar for Soil Restoration

African countries focused on the low-cost and effective soil management solution and started converting the crops leftover into biochar to restore the soil quality and to enhance the crop production (Koné and Galiegue 2023). The people mostly utilize the locally available waste, carbon material, crops residue, invasive species of plants for biochar preparation. In these countries, the use of biochar along with animal manure to enhance and maintain the soil fertility is promoted by the non-governmental organizations (NGO's) and agencies working in the field of development and landscaping (Ndambi et al. 2019). Due to all these efforts, the notable changes were reported in soil fertility (increased), crop yields (increased), and drought resilient (increased) (Mashamaite et al. 2024). However, the use of biochar at commercial level is still a limitation due to lack of awareness among common man and technical expertise.

3.2.4 Europe: Research-Driven Adoption and Policy Integration

European countries like Austria, Germany, and Switzerland have implemented governing guidelines and quality standards for the safe and effective utilization of biochar (Schmidt et al. 2016). These countries are continuously using and exploring the biochar in agriculture, carbon sequestration, and construction industry (Abenavoli et al. 2016). The establishment of the European Biochar Certificate (EBC) ensures the environmental safety during the production, utilization, and disposal of biochar. Moreover, the commitment of European Union emphasized on the use of biochar for the sustainable management of land resources to achieve circular economy and carbon neutral status. On the other hand, the acceptance among the general public is fueled due to advance innovations in biochar preparation and cost reduction as well as supportive policy measures.

3.3 Policy and Market Trends

During the recent years, the biochar caught the global attention as it simultaneously manages the many environmental problems like climate change, agricultural and other solid waste management, and soil degradation. Intergovernmental Panel on Climate Change (IPCC) recognized biochar as negative emissions technology due to its carbon sequestration potential.

Biochar is a component of waste reclamation, soil health, or climate mitigation policies in certain European Members (EU) member states like China, and Australia. In India, currently there is no specific policy regarding the utilization of biochar, but the National Mission for Sustainable Agriculture (NMSA) and the National Biofuel Policy (2018) are providing indirect support, as both policies focus on the conversion of waste material to reduce carbon footprint. The European Biochar Certificate (EBC) of the EU ensures the transparency in safety and quality of biochar.

The global market of biochar is continuously growing due to its remarkable benefits in forestry, construction, agriculture, and environmental remediation. In Agricultural sector, biochar helps in the restoration and maintaining the soil quality. Similarly in the livestock husbandry, biochar is utilized as feed additive and as an absorbent for the treatment of waste (Gracia et al. 2022). Owing to its potential of carbon sequestration, the market of carbon credit trading using biochar is expanding and so is the market of biochar. With the increasing global attention towards climate change and carbon neutrality, biochar is projected as a practical approach by the investors and businessmen. Moreover, the transparent quality control procedures, incentives on the utilization of biochar, and awareness and incentives to farmers are required to further boost the biochar market.

4 Biochar and Soil Health

4.1 Impact of Biochar on Soil Physical Properties

The physical properties of soil are greatly impacted by the utilization of biochar in agricultural system. The decrease in the bulk density of soil amended with biochar is reported due to increase in the pore spaces (Githinji 2014). Similarly, the biochar-amended soil shows enhanced aeration and increased water retention capacity and therefore helps in the increase root development and seed germination (Aslam et al. 2014). Application of biochar enhances the pore spaces between soil particles and therefore increases the porosity of the soil from 13 to 52% (Głąb et al. 2016; Alghamdi 2018). Highly porous soil increases the infiltration rate and retains the water for a longer duration as compared to less porous soil (Aslam et al. 2014). Addition of biochar to the soil increases the WHC and infiltration rate of the soil especially in sandy soil. Buchkina et al. (2017) reported the 34.9% increase in the WHC of loam soil and more than 55% increase in case of clayey soil. Zheng et al. (2019) reported the 26.9% increase in the total porosity and 70.2% increase in the air permeability coefficient of the soil.

4.2 Impact on Soil Chemical Properties (pH, Nutrients, CEC)

Biochar addition poses the impacts on soil chemical properties, i.e., pH, nutrient availability, organic carbon, cation exchange capacity (CEC), and immobilization of heavy metals and toxicants. Slight increase in the pH of the soil was reported after the application of wheat straw (Wu et al. 2019), bamboo, hardwood, rice straw (Tarin et al. 2019), and chicken manure (Clark et al. 2019)-based biochar. Werner et al. (2018) reported that the increase or decrease in the pH value of the soil depends on the initial properties of the soil. Ghorbani et al. (2019) observed 9–20% increase in the CEC value of the soil after the application of rice husk-based biochar. Similarly, Clark et al. (2019) and Yadav et al. (2018) observed the increase in the CEC value on the use of chicken manure and winter grass, respectively. Change in the CEC value is directly related to the change in nutrient availability and water retention capacity of the soil. Biochar-amended soil shows more than 25% increase in the organic matter (OM) content (Tarin et al. 2019; Ghorbani et al. 2019). Biochar fixes the carbon in the soil, therefore helps in the reduction of carbon content in the atmosphere, and thereby helps in reducing the global warming (Schofield et al. 2019). On the other hand, the application of biochar enhances the leaching of nitrogen and phosphorous from the soil to the groundwater causing groundwater pollution and at the same time surface water pollution (Cheng et al. 2020). Zheng et al. (2019) reported the 15.07% increase in organic carbon soil and 0.23 unit increase in the pH coefficient of the soil. Rizwan et al. (2018) reported an increase in the pH and at the same time reduction in the availability of cadmium in the soil. Shwe et al. (2024) also reported an increase in pH and saturated electrical conductivity (EC) of the soil while reduction in soil adsorption ratio (SAR) and exchangeable sodium percentage (ESP).

4.3 Improvement in Soil Biological Activity

Application of biochar in the soil affects the soil biological activity like microbial biomass and diversity, enzyme activity, microbial relationships, soil flora and fauna, and pathogenic diversity. Porous nature of biochar provides the suitable habitat for microorganism development, growth, and colonization. Positive and negative opinions were reported by several authors with reference to microbial population. Some authors reported a significant increase in the microbial population leading to the increase in available organic matter of the soil (Van Zwieten et al. 2010; Jien and Wang 2013; Buchkina et al. 2017). Irmak Yilmaz and Ergun (2019) reported 12.7 times increase in the dehydrogenase enzymatic activity present in soil amended with biochar in comparison to control. However, Paz-Ferreiro et al. (2012) reported only 2.8 times increase after the application of sewage sludge-based biochar. Zheng et al. (2019) reported more than 11.0% increase in the soil microbial biomass carbon. Increase in the microbial biomass in soil favors the solubility of the nutrients and makes them available for the plants.

5 Biochar Under Environmental Stress Conditions

According to Lazarus' environmental stress theory (1966), "stress" is the term used to describe an individual's response to any external event. Accordingly, the term "Environmental Stress" describes the external environmental elements that interfere with an organism's capacity to maintain homeostasis, which has an impact on its ability to survive and reproduce (Gatersleben and Griffin 2016). Here in reference to current study, the term "Environmental Stress" refers to the external factors or stressors which affect the health of the soil which results in reduced crop production. These stresses can be biological (illness, competition), chemical (pollution, poisons), or physical (droughts, extreme temperatures). Storms and wildfires are examples of natural stressors, whereas habitat destruction and climate change are examples of human-caused stressors. Biochar is globally reported as "healing agent" for soil recovery under different physical, chemical, and biological stressors (Ali et al. 2017; Rizwan et al. 2018; Khan et al. 2021; Chi et al. 2024).

5.1 *Role in Drought Mitigation and Water Retention*

Biochar is helpful in enhancing the water-holding capacity of the soil and therefore plays an important role in reducing the drought stress. Li et al. (2021) reported that the addition of biochar reduced the space present between soil particles converting the water drainable pores into retention pores. Liu et al. (2017) studied the size, shape, and porosity of the biochar and their influence on WHC of the soil and concluded that WHC depends on the physicochemical and biological properties of the soil. Other characteristics of biochar like hydrophobicity help in water retention in the soil for a long duration. Dengxiao et al. (2024) reported that simultaneous application of biochar and water retaining agent increased the WHC of the soil and drought resistance in Fluvisols. Rakib et al. (2025) conducted an experiment at different biochar concentrations and watering levels and concluded that in biochar-amended soil the crop shows better growth and better WHC and nutrient retention. Kim et al. (2021) reported that the addition of biochar to the soil decreases the pore space between soil particles and converts macropores into mesopores which enhances the WHC of the soil and makes more water available for plants for a long period. It also improves drought tolerance by regulating stress-responsive proteins, nutritional homeostasis, and plant water uptake (Wu et al. 2023).

5.2 *Biochar as a Buffer Against Soil Salinity*

Salinity is another abiotic stress and major threat to total crop produce (Wu et al. 2023). Biochar is a useful soil additive for reducing soil salinity. It increases water retention, strengthens soil structure, and lessens the harm that salt stress causes to plants. To reduce salt buildup in the root zone, biochar increases soil organic matter, enhances CEC, and stimulates microbial activity (De Vasconcelos 2020; Wu et al. 2023; Wu et al. 2024a, b). Moreover, excess sodium ions can be absorbed by biochar, which stops them from obstructing plant growth and nutrient intake. In saline soils, it has been demonstrated to improve seed germination, nutrient availability, and general plant resistance. Biochar improves soil aeration and balances soil pH, which makes it easier for plants to flourish even in high-salinity environments (Huang et al. 2023). Wang et al. (2024) reported the positive impact on salinity in biochar-amended soil and 13.20% reduction in EC of the soil. However, Fernandes et al. (2019) reported an increase in the salinity after the application poultry litter-based biochar. Lee et al. (2022) also reported the increase in salinity as well as salt migration and leaching. Due to leaching, the salt was found higher in the lower layers of the soil hence making the soil suitable for plants' growth. Dahlawi et al. (2018) studied the use of biochar for salt-affected soil treatment and reported that biochar can reclaim the salt-affected soil; however, its efficiency depends on the physicochemical properties of the biochar which in turn depends on the feedstock used for biochar preparation and type of pyrolysis process used. Shwe et al. (2024) also reported the reduction SAR value, soluble chloride value, and ESP indicating toward reduction in salinity of the soil.

5.3 Alleviation of Heavy Metal Toxicity

Accumulation of heavy metals in the soil puts stress on the growth of the crops and in the long term on the health of human beings and animals. Biochar application is one of the effective methods for reducing the toxicity of heavy metals in soils. Due to porous nature, high surface area, and cation exchange capacity, the biochar binds the pollutants and heavy metals and therefore makes them unavailable to plants and also helps them in the remediation from the ecosystem (Dhiman et al. 2020). Similarly, the biochar-amended soil shows in the increase of pH which reduces the solubility of hazardous metals in the soil and thus helps in reducing the damage to flora (Haiying et al. 2022). Biochar also helps in the development of microbial population in the soil, which enhances the breakdown and stabilization of heavy metals. The remediation of heavy metals from the medium with the help of biochar varies with biochar and pyrolysis process used for the preparation of biochar (Cheng et al. 2020; Khan et al. 2022).

Plants grown in biochar-amended soil show enhanced activity of antioxidants thereby making the plants more sensitive to heavy metal stress (Khan et al. 2022; Haiying et al. 2022). Moulick et al. (2025) reported the reduction of heavy metal stress in the crops grown in biochar-amended soil, and the author recommends the use of nano-biochar for sustainable agriculture. Alami-Milani et al. (2024) reported that chromium stress significantly reduces the growth of Rapeseed plants under chromium stress. The biochar application enhances the uptake of potassium, phosphorus, calcium, nitrogen, and magnesium from the soil due to better root growth catalyzed by glycine betaine and indole acetic acid.

5.4 Biochar for Soil Carbon Sequestration

The use of biochar to improve soil sustainability and carbon sequestration has gained international recognition. High carbon content of biochar (50–90%) makes it significant carbon reservoir or sink, and it also could lock up carbon in soil for centuries, possibly up to 2000 years, which makes it a valuable tool for reducing GHG concentrations and halting climate change (Li and Chan 2022). Certain physicochemical properties of biochar like particle size, surface area, porosity, pH, and CEC have impact on its carbon sequestration property (Li and Tasnady 2023). Fine particle biochar exhibits large inter and intra contact surface which helps in the enhanced adsorption of GHG and carbon capturing (Liao and Thomas 2019; Jung et al. 2019) Smaller particle size exhibits large surface area which provides a greater number of active sites for binding of carbon particle (Budai et al. 2013) and microbial colonization (Singh and Kookana 2009) and thereby helps in increased carbon sequestration. Similarly high porosity of biochar provides high number of sites for carbon binding and microbial colonization. Different pore sizes help with the adsorption of different materials like micropores helps in the adsorption of various GHG and materials (Table 2). Greater microbial activity, enhanced WHC, and greater nutrient retention are some of its advantages for soil health (Shyam et al. 2025). Furthermore, biochar creates a physical barrier that influences the microbiological habitat of the soil, promotes the development of stable soil aggregates, and helps stop carbon loss (Table 3). However, factors including the kind of raw material used, the pyrolyzing conditions, and the application method all affect the overall impact of biochar (Li and Tasnady 2023). Biochar is reported to have alkaline pH (Zhao et al. 2014), and soil amended with biochar shows an increase in the pH (Domingues et al. 2020) making the soil fit for the growth of plants and thereby enhancing the biomass production which leads to carbon storage. Majumder et al. (2019) reported that conversion of crop residues into biochar helps in sustainable and cost-effective management of carbon present in the atmosphere (absorption of GHG), lithosphere (retention of soil carbon for a long period in the soil), and biosphere (carbon present the crop residues convert into biochar). Wang et al. (2023) studied the release of GHG from the biochar-amended soil and reported an offset of about 2.1%, 13.1%, 1.6%, and 5.3% in the release of CO_2, CH_4, N_2O, and CO_{2e}, respectively. Lefebvre et al. (2023) studied the quantification of biochar impacts on the environment and concluded that biochar possesses the potential of about 7% reduction of carbon dioxide of the total GHG emissions (Figs. 4 and 5).

Table 3 Description of particle classification, size, and respective adsorbed materials

Sr. No.	Class	Pore size	Absorbed materials	References
1	Micropores	< 2 mm	Carbon dioxide (CO_2) Methane (CH_4) Nitrous oxide (N_2O)	Memetova et al. (2022)
2	Mesopores	2–50 mm	Water Organic substances Nutrients	Atkinson et al. (2010)
3	Macropores	> 50 mm	Air Water Nutrients	Yazdani et al. (2019)

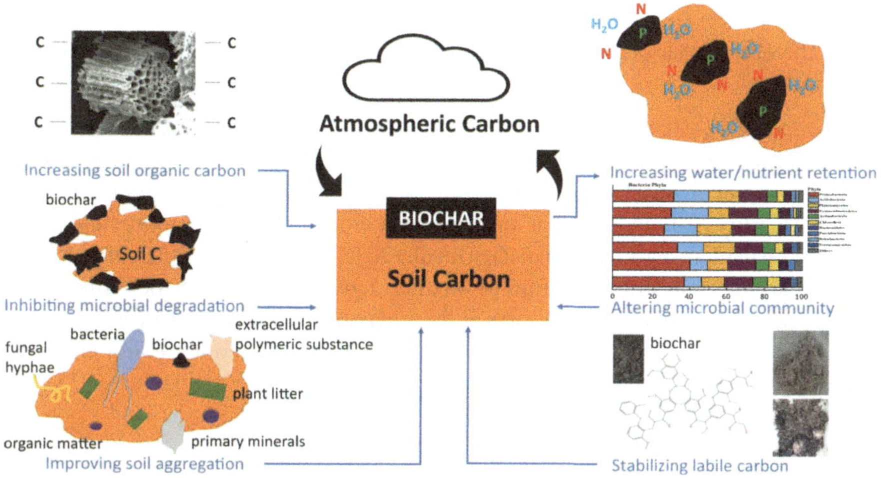

Fig. 4 Major effects of biochar on enhancing soil carbon sequestration. *Source* Taken from Li and Tasnady (2023)

6 Challenges and Limitations

Biochar has emerged as a promising amendment for enhancing soil health; however, its effectiveness under conditions of environmental stress is often constrained by various challenges and limitations.

6.1 Variability in Raw Materials

The composition and physicochemical properties depend on the feedstocks used for biochar preparation. Variability in the biochar properties can lead to inconsistent effects on soil health. Besides this, biochar properties vary with the type of

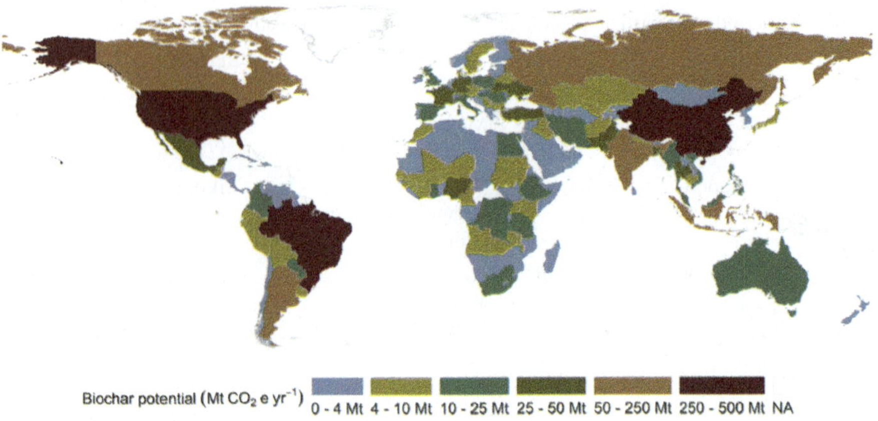

Fig. 5 Global carbon dioxide removal (CDR) potential ($MtCO_2$/year) of biochar. *Source* Lefebvre et al. (2023)

pyrolysis process (slow, fast, or flash pyrolysis, gasification) used (Li and Tasnady 2023). Freitas et al. (2020) studied the composition of biochar with reference to feedstock variability and reported that animal-based biochar possessed more nutrients in comparison to plant-based biochar. Zhao et al. (2013) reported that nutrient properties and the carbon sequestration nature of biochar depend on the feedstock used, while the pH, recalcitrant nature, and surface area depend on the treatment temperature. Availability of suitable feedstock for biochar preparation is another major challenge impacting its commercialization and global application (Al Masud et al. 2023).

6.2 Unintended Environmental Risks and Potential Toxicity

Biochar can interact with environmental media, potentially releasing pollutants that may threaten ecosystems. Gwenzi et al. (2015) studied the challenges in the production and application of biochar in Sub-Saharan Africa and reported that public health and environmental risk are the challenges which must be explored properly before the use of biochar. Some biochar formulations may contain harmful substances that could negatively impact plants, microorganisms, and human health. Ahmed et al. (2024) reported the presence of hazardous substances like volatile organic carbon (VOC) in the biochar that may cause harm to the environment and people. Feedstocks like municipal solid waste and sewage sludge contain different hazardous materials like polyaromatic hydrocarbons (PAHs), polystyrene, and heavy metals (lead, cadmium, chromium, mercury) that may be released either during biochar preparation stage or during application stage into the environment (Ndirangu et al. 2019; Hu et al. 2022). Biochar absorbed the pollutants from the environment and reduced their mobility, and therefore, the nature of the pollutants remains the same, i.e., toxic and therefore some quantity remains available for the plants (Cheng et al. 2020; Wang et al. 2020; Ke et al. 2023). Brtnicky et al. (2021) reported that addition of biochar in the soil increases the pH value of the soil that results into the precipitation of the nutrient from the soil, impacting the fertility.

6.3 Application Challenges

The effectiveness of biochar varies based on soil type, climate, and stress conditions, making it difficult to standardize its use. Freitas et al. (2020) reported that nutrient value of biochar depends on the type of soil and climate. Dong et al. (2025) reported that release of nutrients and toxic substances from the biochar depends on the medium in which biochar is applied. Application of biochar to the soil, add or increase the concentration of some toxic substances that in the long term become harmful to the health of soil flora and fauna (Brtnicky et al. 2021; Juriga et al. 2021). High production cost is another big challenge of biochar application for environmental management (Mohammadi et al. 2020).

6.4 Limited Long-Term Studies

While biochar shows excellent results, more research is needed to fully understand its long-term effects on soil health and environmental sustainability. The biochar applied interacts with environmental components, i.e., soil, sir, and water. Most of the studies available reported a positive side of biochar application; however, few studied reported negative impacts. Although a lot of studies are available regarding the production, use, and impacts of the environment but none of them explored the long-term impacts of biochar utilization on air water and soil (Dong et al. 2025). Vijayaraghavan (2019) studied the current practices of feedstock modification for biochar preparation and recommended further research to assess the suitability of biochar for different uses. Tisserant and Cherubini (2019) declared that biochar improves the soil properties and simultaneously puts low risk on the environment and recommended the fate analysis of biochar in aqueous medium and emission as black carbon.

7 Conclusion

The current research was carried out to explore preparation, application, and its impact on soil health and environmental stress mitigation. A systematic search was performed using keywords like biochar, pyrolysis process, soil health, challenges in biochar application, biochar potential on the databases such as Scopus, Web of Science, and Google Scholar. It was highlighted in most literature that soil amended with biochar displays the improvement in physicochemical properties. Besides this, biochar is mitigating environmental stress by reducing the concentration of heavy metals, salinity, and GHG emissions of soil. One of the biggest challenges in biochar applications is the varying properties of the feedstock and thereby biochar which increases the uncertainties in outcomes of the application and therefore restricts its global use. Another major challenge is the presence of toxic elements in biochar which slowly releases into the environment and poses a threat to the health of ecosystems. The unavailability of long-term and field study about biochar impact on the environment is also a constraint. There is an urgent need for large-scale and cost-effective biochar production for the commercialization of biochar application. At the same time, development of biochar policies and involvement of NGOs and research institutions can further boost the research, development, and utilization of biochar.

8 Future Perspectives and Research Directions

8.1 Use of Innovative Techniques for Biochar Production and Application

As discussed in the current book chapter that besides positive effect of biochar, it also has a negative impact on the environment. Therefore, new innovations are required for biochar productions and applications focusing on large-scale production and minimized negative impacts on application. Gasification and low-temperature pyrolysis are some of the techniques focusing on increased efficiency and scalability. Uniformity in biochar characteristics is another major challenge which can be mitigated using real-time monitoring systems during the production of biochar. Reduction in the cost of biochar production using the pilot scale production plant and waste agro-residues as feedstock material can make it a real sustainable solution for agriculture and environmental management.

8.2 Integrating Biochar in Sustainable Agriculture Policies

The potential of fertilizers based on biochar to improve nutrient availability, minimize GHG emissions, and improve soil health is becoming more widely acknowledged. In India, currently the policies referring to sustainable utilization and management of resource and conversion of waste especially the crop residues into biochar are in popular Sharma et al. (2024). Extensive utilization of biochar in the conventional farming may happens due to circular economy model approach of governments and authorities (Keerthi 2024). Excessive production cost, variable properties of biochar, and variable output depending on the physicochemical properties of the soil are some of obstacles in the way of widespread utilization of biochar (Luo et al. 2025). There is an urgent need of research focusing on the improvement of biochar formulations and production efficacy. Also, the need of enduring studies exploring the agricultural and environmental benefits is on peak. As currently there is no policy in India regarding the production, utilization, and disposal of biochar, therefore collaboration between agronomic experts and government representatives is necessary to encourage the biochar utilization.

8.3 Use of Biochar for Climate Change Mitigation

Biochar helps in carbon sequestration by absorbing atmospheric carbon and fixing the carbon available in the soil (Chaudhry et al., 2025), and therefore plays a crucial role in the climate change mitigation policies. Biochar can absorb about 3 billion metric tons of projected carbon dioxide globally. Biochar-amended soil enhances the growth of plants which absorb the atmospheric carbon dioxide and release the oxygen in the atmosphere. The release of non-nitrogen and carbon-based nitrous oxide and methane also reduces up to 50% in biochar-amended soil (Lehmann et al. 2021). Besides the agricultural sector, biochar is an effective additive in building materials and helps in reducing the carbon footprint and ecological impacts of the construction industry. Excessive production cost, variable biochar properties, and variable efficacy based on soil physicochemical properties are some of challenges in the way of widespread utilization of biochar (Luo et al. 2025). Besides this, the lack of national and international level policy concerning the production, quality of raw materials and prepared biochar, utilization, final disposal, and environmental impacts is a major hurdle.

Acknowledgements The authors would like to express heartfelt thanks to the Department of Zoology and Environmental Science, Gurukula Kangri (Deemed to be University), Haridwar, India, and SAM (India) Builtwell Pvt. Ltd., for providing support and resources.

Author's Contribution FA and AC conceptualized this chapter, FA and MMK designed table and figures and wrote the initial stage of chapter, and RB, AC, and PKC critically edited the chapter and contribution to constructive recommendations before the submission. All the authors have no conflict of interest, and the work is not related to animal ethical clearance.

References

Abenavoli LM, Longo L, Proto AR, Gallucci F, Ghignoli A, Zimbalatti G, Russo D, Colantoni A (2016) Characterization of biochar obtained from olive and hazelnut prunings and comparison with the standards of European Biochar Certificate (EBC). Procedia Soc Behav Sci 223:698–705. https://doi.org/10.1016/j.sbspro.2016.05.244

Ahmed SF, Mehejabin F, Chowdhury AA, Almomani F, Khan NA, Badruddin IA, Kamangar S (2024) Biochar produced from waste-based feedstocks: mechanisms, affecting factors, economy, utilization, challenges, and prospects. GCB Bioenergy 16(8):e13175. https://doi.org/10.1111/gcbb.13175

Al Masud MA, Shin WS, Sarker A, Septian A, Das K, Deepo DM, Iqbal MA, Islam ARMT, Malafaia G (2023) A critical review of sustainable application of biochar for green remediation: research uncertainty and future directions. Sci Total Environ 904:166813. https://doi.org/10.1016/j.scitotenv.2023.166813

Alami-Milani M, Aghaei-Gharachorlou P, Davar R, Rashidpour A, Torabian S, Farhangi-Abriz S (2024) Biochar solutions: Slow and fast pyrolysis effects on chromium stress in rapeseed roots. Plant Physiol Biochem 216:109197. https://doi.org/10.1016/j.plaphy.2024.109197

Alghamdi AG (2018) Biochar as a potential soil additive for improving soil physical properties—a review. Arab J Geosci 11(24):766. https://doi.org/10.1007/s12517-018-4056-7

Ali S, Rizwan M, Qayyum MF, Ok YS, Ibrahim M, Riaz M, Arif MS, Hafeez F, Al-Wabel MI, Shahzad AN (2017) Biochar soil amendment on alleviation of drought and salt stress in plants: a critical review. Environ Sci Pollut Res 24:12700–12712. https://doi.org/10.1007/s11356-017-8904-x

Amonette JE, Blanco-Canqui H, Hassebrook C, Laird DA, Lal R, Lehmann J, Page-Dumroese D (2021) Integrated biochar research: a roadmap. J Soil Water Conserv 76(1):24A–29A. https://doi.org/10.2489/jswc.2021.1115a

Aslam Z, Khalid M, Aon M (2014) Impact of biochar on soil physical properties. Sch J Agric Sci 4(5):280–284

Atkinson CJ, Fitzgerald JD, Hipps NA (2010) Potential mechanisms for achieving agricultural benefits from biochar application to temperate soils: a review. Plant and Soil 337:1–18. https://doi.org/10.1007/s11104-010-0464-5

Brtnicky M, Datta R, Holatko J, Bielska L, Gusiatin ZM, Kucerik J, Hammerschmiedt T, Danish S, Radziemska M, Mravcova L, Fahad S (2021) A critical review of the possible adverse effects of biochar in the soil environment. Sci Total Environ 796:148756. https://doi.org/10.1016/j.scitotenv.2021.148756

Buchkina NP, Balashov EV, Simansky V, Igaz D, Horak J (2017) Changes in biological and physical parameters of soils with different texture after biochar application. SelskokhozyaistvennayaBiologiya (Agricultural Biology) 52(3):471–477. https://doi.org/10.15389/agrobiology.2017.3.471eng

Budai A, Zimmerman AR, Cowie AL, Webber JBW, Singh BP, Glaser B, Masiello CA, Andersson D, Shields F, Lehmann J, Camps Arbestain M (2013) Biochar carbon stability test method: an assessment of methods to determine biochar carbon stability. Int Biochar Initiat 1:1–20. https://doi.org/10.1071/9781486305100

Cao X, Harris W (2010) Properties of dairy-manure-derived biochar pertinent to its potential use in remediation. Bioresour Technol 101(14):5222–5228. https://doi.org/10.1016/j.biortech.2010.02.052

Chagas JKM, de Figueiredo CC, Ramos M.L.G. (2022) Biochar increases soil carbon pools: evidence from a global meta-analysis. J Environ Manag 305:114403. https://doi.org/10.1016/j.jenvman.2021.114403

Chaudhry UK, Asim A, Iqbal J, Ashraf A, Yaqoob S (2025) Biochar for climate change adaptation and mitigation. In: Biochar revolution. Springer, Cham, pp. 83–97. https://doi.org/10.1007/978-3-031-73154-9_5

Chen W, Meng J, Han X, Lan Y, Zhang W (2019) Past, present, and future of biochar. Biochar 1:75–87. https://doi.org/10.1007/s42773-019-00008-3

Cheng S, Chen T, Xu W, Huang J, Jiang S, Yan B (2020) Application research of biochar for the remediation of soil heavy metals contamination: a review. Molecules 25(14):3167. https://doi.org/10.3390/molecules25143167

Chi W, Nan Q, Liu Y, Dong D, Qin Y, Li S, Wu W (2024) Stress resistance enhancing with biochar application and promotion on crop growth. Biochar 6(1):43. https://doi.org/10.1007/s42773-024-00336-z

Clark M, Hastings MG, Ryals R (2019) Soil carbon and nitrogen dynamics in two agricultural soils amended with manure-derived biochar. J Environ Qual 48(3):727–734. https://doi.org/10.2134/jeq2018.10.0384

Dahlawi S, Naeem A, Rengel Z, Naidu R (2018) Biochar application for the remediation of salt-affected soils: Challenges and opportunities. Sci Total Environ 625:320–335. https://doi.org/10.1016/j.scitotenv.2017.12.257

Deng X, Teng F, Chen M, Du Z, Wang B, Li R, Wang P (2024) Exploring negative emission potential of biochar to achieve carbon neutrality goal in China. Nat Commun 15(1):1085. https://doi.org/10.1038/s41467-024-45314-y

Dengxiao Z, Hongbin J, Wenjing Z, Qingsong Y, Zhihang M, Haizhong W, Wei R, Shiliang L, Daichang W (2024) Combined biochar and water-retaining agent application increased soil water retention capacity and maize seedling drought resistance in Fluvisols. Sci Total Environ 907:167885. https://doi.org/10.1016/j.scitotenv.2023.167885

Devi P, Saroha AK (2015) Effect of pyrolysis temperature on polycyclic aromatic hydrocarbons toxicity and sorption behaviour of biochars prepared by pyrolysis of paper mill effluent treatment plant sludge. Bioresour Technol 192:312–320. https://doi.org/10.1016/j.biortech.2015.05.084

Dhiman S, Handa N, Sharma N, Kaur R, Ohri P, Bhardwaj R (2020) Role of biochar in heavy metal toxicity in plants. Cell Mol Phytotoxicity Heavy Metals 349–371. https://doi.org/10.1007/978-3-030-45975-8_17

Domingues RR, Sánchez-Monedero MA, Spokas KA, Melo LC, Trugilho PF, Valenciano MN, Silva CA (2020) Enhancing cation exchange capacity of weathered soils using biochar: feedstock, pyrolysis conditions and addition rate. Agronomy 10(6):824. https://doi.org/10.3390/agronomy10060824

Dong M, Jiang M, He L, Zhang Z, Gustave W, Vithanage M, Niazi NK, Chen B, Zhang X, Wang H, He F (2025) Challenges in safe environmental applications of biochar: Identifying risks and unintended consequence. Biochar 7(1):1–20. https://doi.org/10.1007/s42773-024-00412-4

El-Metwally IM, Saudy HS, Elewa TA (2022) Natural plant by-products and mulching materials to suppress weeds and improve sugar beet (*Beta vulgaris* L.) yield and quality. J Soil Sci Plant Nutr 22(4):5217–5230. https://doi.org/10.1007/s42729-022-00997-4

Fahmy TY, Fahmy Y, Mobarak F, El-Sakhawy M, Abou-Zeid RE (2020) Biomass pyrolysis: past, present, and future. Environ, Dev Sustain 22:17–32. https://doi.org/10.1007/s10668-018-0200-5

Fernandes JD, Chaves LH, Mendes JS, Chaves IB, Tito GA (2019) Alterations in soil salinity with the use of different biochar doses. Revista De CiênciasAgrárias 42(1):89–98

Freitas AM, Nair VD, Harris WG (2020) Biochar as influenced by feedstock variability: implications and opportunities for phosphorus management. Front Sustain Food Syst 4:510982. https://doi.org/10.3389/fsufs.2020.510982

Garcia B, Alves O, Rijo B, Lourinho G, Nobre C (2022) Biochar: production, applications, and market prospects in Portugal. Environments 9(8):95. https://doi.org/10.3390/environments9080095

Gatersleben B, Griffin I (2016) Environmental stress. In: Handbook of environmental psychology and quality of life research. Springer International Publishing, Cham, pp 469–485. https://doi.org/10.1007/978-3-319-31416-7_25

Ghorbani M, Asadi H, Abrishamkesh S (2019) Effects of rice husk biochar on selected soil properties and nitrate leaching in loamy sand and clay soil. Int Soil Water Conserv Res 7(3):258–265. https://doi.org/10.1016/j.iswcr.2019.05.005

Githinji L (2014) Effect of biochar application rate on soil physical and hydraulic properties of a sandy loam. Arch Agron Soil Sci 60(4):457–470. https://doi.org/10.1080/03650340.2013.821698

Głąb T, Palmowska J, Zaleski T, Gondek K (2016) Effect of biochar application on soil hydrological properties and physical quality of sandy soil. Geoderma 281:11–20. https://doi.org/10.1016/j.geoderma.2016.06.028

Glaser B, Birk JJ (2012) State of the scientific knowledge on properties and genesis of Anthropogenic Dark Earths in Central Amazonia (terra preta de Índio). Geochimica et Cosmochimica acta 82:39–51. https://doi.org/10.1016/j.gca.2010.11.029

González-Arias J, Sánchez ME, Cara-Jiménez J, Baena-Moreno FM, Zhang Z (2022) Hydrothermal carbonization of biomass and waste: a review. Environ Chem Lett 1–11. https://doi.org/10.1007/s10311-021-01311-x

Gwenzi W, Chaukura N, Mukome FN, Machado S, Nyamasoka B (2015) Biochar production and applications in sub-Saharan Africa: opportunities, constraints, risks and uncertainties. J Environ Manag 150:250–261. https://doi.org/10.1016/j.jenvman.2014.11.027

Haiying TANG, Shubin WANG, Ying LIU, Hassan MU, Ying SONG, Huang G, Hashem M, Alamri S, Mostafa YS (2022) Biochar: a promising soil amendment to mitigate heavy metals toxicity in plants. NotulaeBotanicae Horti Agrobotanici Cluj-Napoca 50(3):12778–12778. https://doi.org/10.15835/nbha50312778

Holatko J, Kucerik J, Mustafa A, Lonova K, Siddiqui MH, Naveed M, Hammerschmiedt T, Kintl A, Malicek O, Chorazy T, Baltazar T (2025) Influence of biochar feedstock blends on soil enzyme activity, nutrient cycling, lettuce biomass accumulation and photosynthesis. BMC Plant Biology 25(1):323. https://doi.org/10.1186/s12870-025-06352-w

Hossain MZ, Bahar MM, Sarkar B, Donne SW, Ok YS, Palansooriya KN, Kirkham MB, Chowdhury S, Bolan N (2020) Biochar and its importance on nutrient dynamics in soil and plant. Biochar 2:379–420. https://doi.org/10.1007/s42773-020-00065-z

https://www.fao.org/fileadmin/templates/wsfs/docs/expert_paper/How_to_Feed_the_World_in_2050.pdf

Hu Y, Tong K, Guo Q, Zhang B, Jiao L, Li M, Zhang Z (2022) Effect of controlled temperature and biomass addition on the formed environmental persistent free radicals (EPFRs) in sewage sludge-based biochar from pyrolysis treatment. J Anal Appl Pyrolysis 162:105460. https://doi.org/10.1016/j.jaap.2022.105460

Huang K, Li M, Li R, Rasul F, Shahzad S, Wu C, Shao J, Huang G, Li R, Almari S, Hashem M (2023) Soil acidification and salinity: the importance of biochar application to agricultural soils. Front Plant Sci 14:1206820. https://doi.org/10.3389/fpls.2023.1206820

Irmak Yilmaz F, Ergun YA (2019) Impact of biochar and animal manure on some biological and chemical properties of soil. Appl Ecol Environ Res 17:8865–8876. https://doi.org/10.15666/aeer/1704_88658876

Jien SH,Wang CS (2013) Effects of biochar on soil properties and erosion potential in a highly weathered soil. Catena 110:225–233. https://doi.org/10.1016/j.catena.2013.06.021

Jung S, Park YK, Kwon EE (2019) Strategic use of biochar for CO_2 capture and sequestration. J CO_2 Util 32:128–139. https://doi.org/10.1016/j.jcou.2019.04.012

Juriga M, Aydın E, Horák J, Chlpík J, Rizhiya EY, Buchkina NP, Balashov EV, Šimanský V (2021) The importance of initial application and reapplication of biochar in the context of soil structure improvement. J Hydrol Hydromech 69(1):87–97. https://doi.org/10.2478/johh-2020-0044

Kamarudin NS, Dahalan FA, Hasan M, An OS, Parmin NA, Ibrahim N, Hamdzah M, Zain NAM, Muda K, Wikurendra EA (2022) Biochar: a review of its history, characteristics, factors that influence its yield, methods of production, application in wastewater treatment and recent development. Biointerface Res Appl Chem 12(6):7914–7926. https://doi.org/10.33263/BRIAC126.79147926

Karkach B, Tahiri M, Haibi A, Bouya M, Kifani-Sahban F (2023) Review on fast pyrolysis of biomass for biofuel production from date palm. Appl Sci 13(18):10463. https://doi.org/10.3390/app131810463

Ke Y, Zhang F, Zhang Z, Hough R, Fu Q, Li YF, Cui S (2023) Effect of combined aging treatment on biochar adsorption and speciation distribution for Cd (II). Sci Total Environ 867:161593. https://doi.org/10.1016/j.scitotenv.2023.161593

Keerthi MM (2024) Harnessing the power of biochar: a sustainable solution for agriculture and climate change mitigation. Environ Rep Int J. https://doi.org/10.51470/er.2024.6.2.05

Khan Z, Khan MN, Zhang K, Luo T, Zhu K, Hu L (2021) The application of biochar alleviated the adverse effects of drought on the growth, physiology, yield and quality of rapeseed through regulation of soil status and nutrients availability. Ind Crop Prod 171:113878. https://doi.org/10.1016/j.indcrop.2021.113878

Khan Z, Fan X, Khan MN, Khan MA, Zhang K, Fu Y, Shen H (2022) The toxicity of heavy metals and plant signaling facilitated by biochar application: implications for stress mitigation and crop production. Chemosphere 308:136466. https://doi.org/10.1016/j.chemosphere.2022.136466

Khan S, Irshad S, Mehmood K, Hasnain Z, Nawaz M, Rais A, Gul S, Wahid MA, Hashem A, Abd_Allah EF, Ibrar D (2024) Biochar production and characteristics, its impacts on soil health, crop production, and yield enhancement: a review. Plants 13(2):166. https://doi.org/10.3390/plants13020166

Kim YJ, Hyun J, Yoo SY, Yoo G (2021) The role of biochar in alleviating soil drought stress in urban roadside greenery. Geoderma 404:115223. https://doi.org/10.1016/j.geoderma.2021.115223

Koné S, Galiegue X (2023) Potential development of biochar in Africa as an adaptation strategy to climate change impact on agriculture. Environ Manag 72(6):1189–1203. https://doi.org/10.1007/s00267-023-01821-0

Latini A, Bacci G, Teodoro M, Gattia DM, Bevivino A, Trakal L (2019) The impact of soil-applied biochars from different vegetal feedstocks on durum wheat plant performance and rhizospheric bacterial microbiota in low metal-contaminated soil. Front Microbiol 10:2694. https://doi.org/10.3389/fmicb.2019.02694

Lazarus R (1966) Psychological stress and the coping process. McGraw Hill, New York

Lee X, Yang F, Xing Y, Huang Y, Xu L, Liu Z, Holtzman R, Kan I, Li Y, Zhang L, Zhou H (2022) Use of biochar to manage soil salts and water: effects and mechanisms. Catena 211:106018. https://doi.org/10.1016/j.catena.2022.106018

Lefebvre D, Fawzy S, Aquije CA, Osman AI, Draper KT, Trabold TA (2023) Biomass residue to carbon dioxide removal: quantifying the global impact of biochar. Biochar 5(1):65. https://doi.org/10.1007/s42773-023-00258-2

Lehmann J, Joseph S (2015) Biochar for environmental management: an introduction. In: Biochar for environmental management. Routledge, pp 1–13

Lehmann J, Cowie A, Masiello CA, Kammann C, Woolf D, Amonette JE, Cayuela ML, Camps-Arbestain M, Whitman T (2021) Biochar in climate change mitigation. Nat Geosci 14(12):883–892. https://doi.org/10.1038/s41561-021-00852-8

Li S, Chan CY (2022) Will biochar suppress or stimulate greenhouse gas emissions in agricultural fields? Unveiling the dice game through data syntheses. Soil Syst 6(4): 73. https://doi.org/10.3390/soilsystems6040073

Li S, Tasnady D (2023) Biochar for soil carbon sequestration: Current knowledge, mechanisms, and future perspectives. C 9(3):67. https://doi.org/10.3390/c9030067

Li L, Zhang YJ, Novak A, Yang Y, Wang J (2021) Role of biochar in improving sandy soil water retention and resilience to drought. Water 13(4):407. https://doi.org/10.3390/w13040407

Liao W, Thomas SC (2019) Biochar particle size and post-pyrolysis mechanical processing affect soil pH, water retention capacity, and plant performance. Soil Syst 3(1):14. https://doi.org/10.3390/soilsystems3010014

Liu Z, Dugan B, Masiello CA, Gonnermann HM (2017) Biochar particle size, shape, and porosity act together to influence soil water properties. Plos one 12(6):e0179079. https://doi.org/10.1371/journal.pone.0179079

Luo P, Zhang W, Xiao D, Hu J, Li N, Yang J (2025) Biochar-based fertilizers: advancements, applications, and future directions in sustainable agriculture—a review. Agronomy 15(5):1104. https://doi.org/10.3390/agronomy15051104

Majumder S, Neogi S, Dutta T, Powel MA, Banik P (2019) The impact of biochar on soil carbon sequestration: Meta-analytical approach to evaluating environmental and economic advantages. J Environ Manag 250:109466. https://doi.org/10.1016/j.jenvman.2019.109466

Mashamaite CV, Motsi H, Manyevere A, Poswa SB (2024) Assessing the potential of biochar as a viable alternative to synthetic fertilizers in sub-Saharan Africa smallholder farming: a review. Agronomy 14(6):1215. https://doi.org/10.3390/agronomy14061215

Memetova A, Tyagi I, Karri RR, Kumar V, Tyagi K, Memetov N, Zelenin A, Pasko T, Gerasimova A, Tarov D, Dehghani MH (2022) Porous carbon-based material as a sustainable alternative for the storage of natural gas (methane) and biogas (biomethane): a review. Chem Eng J 446:137373. https://doi.org/10.1016/j.cej.2022.137373

Mohammadi A, Khoshnevisan B, Venkatesh G, Eskandari S (2020) A critical review on advancement and challenges of biochar application in paddy fields: environmental and life cycle cost analysis. Processes 8(10):1275. https://doi.org/10.3390/pr8101275

Moulick D, Mukherjee A, Mahanta S, Chowdhara B, Pattnaik BK, Mukhopadhyay AK, Gujre N, Shree J, Santra SC, Hazra S (2025) Nano-composites and nano-biochar: a novel solution to alleviate heavy metal stress in crops. In: Nanomaterials and nano-biochar in reducing soil stress. Apple Academic Press, pp 183–224. https://doi.org/10.1201/9781003503453-9

Mubarak M, Salem EM, Kenawey MK, Saudy HS (2021) Changes in calcareous soil activity, nutrient availability, and corn productivity due to the integrated effect of straw mulch and irrigation regimes. J Soil Sci Plant Nutr 21(3):2020–2031. https://doi.org/10.1007/s42729-021-00498-w

Muhammad N, Ge L, Chan WP, Khan A, Nafees M, Lisak G (2022) Impacts of pyrolysis temperatures on physicochemical and structural properties of green waste derived biochars for adsorption of potentially toxic elements. J Environ Manag 317:115385. https://doi.org/10.1016/j.jenvman.2022.115385

Ndambi OA, Pelster DE, Owino JO, De Buisonjé F, Vellinga T (2019) Manure management practices and policies in sub-Saharan Africa: implications on manure quality as a fertilizer. Front Sustain Food Syst 3:29. https://doi.org/10.3389/fsufs.2019.00029

Ndirangu SM, Liu Y, Xu K, Song S (2019) Risk evaluation of pyrolyzed biochar from multiple wastes. J Chem 2019(1):4506314. https://doi.org/10.1155/2019/4506314

Pahnila M, Koskela A, Sulasalmi P, Fabritius T (2023) A review of pyrolysis technologies and the effect of process parameters on biocarbon properties. Energies 16(19):6936. https://doi.org/10.3390/en16196936

Paz-Ferreiro J, Gasco G, Gutiérrez B, Mendez A (2012) Soil biochemical activities and the geometric mean of enzyme activities after application of sewage sludge and sewage sludge biochar to soil. Biol Fertil Soils 48:511–517. https://doi.org/10.1007/s00374-011-0644-3

Pietikäinen J, Kiikkilä O, Fritze H (2000) Charcoal as a habitat for microbes and its effect on the microbial community of the underlying humus. Oikos 89(2):231–242. https://doi.org/10.1034/j.1600-0706.2000.890203.x

Rakib MH, Chowdhury MIH, Hossain T, Sadnan MWM, Hossain SMS (2025) Biochar as a drought mitigation tool to enhance growth and water retention in *Pinus caribaea* M. seedlings. Ceylon J Sci 54(2). https://doi.org/10.4038/cjs.v54i2.8600

Ramankutty N, Evan AT, Monfreda C, Foley JA (2008) Farming the planet: 1. Geographic distribution of global agricultural lands in the year 2000. Glob Biogeochem Cycles 22(1). https://doi.org/10.1029/2007gb002952

Rechberger MV, Kloss S, Wang SL, Lehmann J, Rennhofer H, Ottner F, Wriessnig K, Daudin G, Lichtenegger H, Soja G, Zehetner F (2019) Enhanced Cu and Cd sorption after soil aging of woodchip-derived biochar: what were the driving factors? Chemosphere 216:463–471. https://doi.org/10.1016/j.chemosphere.2018.10.094

Rizwan M, Ali S, Abbas T, Adrees M, Zia-ur-Rehman M, Ibrahim M, Abbas F, Qayyum MF, Nawaz R (2018) Residual effects of biochar on growth, photosynthesis and cadmium uptake in rice (*Oryza sativa* L.) under Cd stress with different water conditions. J Environ Manag 206:676–683. https://doi.org/10.1016/j.jenvman.2017.10.035

Salem EM, Kenawey KM, Saudy HS, Mubarak M (2021) Soil mulching and deficit irrigation effect on sustainability of nutrients availability and uptake, and productivity of maize grown in calcareous soils. Commun Soil Sci Plant Anal 52(15):1745–1761. https://doi.org/10.1080/00103624.2021.1892733

Saletnik A, Saletnik B (2025) Technology–economy–policy: biochar in the low-carbon energy transition—a review. Appl Sci. https://doi.org/10.3390/app15115882

Saranya K, Krishnan PS, Kumutha K, French J (2011) Potential for biochar as an alternate carrier to lignite for the preparation of biofertilizers in India. Int J Agric Environ Biotechnol 4(2):167–172

Saudy HS, El-Bially M, Ramadan KA, Abo El-Nasr EK, Abd El-Samad GA (2021) Potentiality of soil mulch and sorghum extract to reduce the biotic stress of weeds with enhancing yield and nutrient uptake of maize crop. GesundePflanzen 73(4):555–564. https://doi.org/10.1007/s10343-021-00577-z

Schmidt HP, Bucheli T, Kammann C, Glaser B, Abiven S, Leifeld J (2016) European biochar certificate-guidelines for a sustainable production of biochar

Schmidt HP, Kammann C, Hagemann N, Leifeld J, Bucheli TD, Sánchez Monedero MA, Cayuela ML (2021) Biochar in agriculture–a systematic review of 26 global meta-analyses. GCB Bioenergy 13(11):1708–1730. https://doi.org/10.1111/gcbb.12889

Schofield HK, Pettitt TR, Tappin AD, Rollinson GK, Fitzsimons MF (2019) Biochar incorporation increased nitrogen and carbon retention in a waste-derived soil. Sci Total Environ 690:1228–1236. https://doi.org/10.1016/j.scitotenv.2019.07.116

Scott HL, Ponsonby D, Atkinson CJ (2014) Biochar: an improver of nutrient and soil water availability-what is the evidence? CABI Reviews (2014):1–19. https://doi.org/10.1079/PAVSNNR20149019

Sharma P, Abrol V, Sharma N, Sharma R, Chadha D, Anand S, Khenrab S, Shabir H, Singh P, Kumari S, Verma D (2024) Policies and strategies for sustainable use of biochar in Indian agriculture. BioResources 19(3). https://doi.org/10.15376/biores.19.3.sharma

Shwe KK, Chaopayao N, Yampracha S (2024) Mitigation of soil salinity by addition of different rice straw biochar doses in salt-affected acid soil. Environ Res Commun 6(9):095028. https://doi.org/10.1088/2515-7620/ad79bf

Shyam S, Ahmed S, Joshi SJ, Sarma H (2025) Biochar as a soil amendment: implications for soil health, carbon sequestration, and climate resilience. Discov Soil 2(1):18. https://doi.org/10.1007/s44378-025-00041-8

Singh N, Kookana RS (2009) Organo-mineral interactions mask the true sorption potential of biochars in soils. J Environ Sci Health Part B 44(3):214–219. https://doi.org/10.1080/03601230902728112

Tarin MWK, Fan LL, Shen L, Lai JL, Tayyab M, Sarfraz R, Chen LY, Ye J, He TY, Rong JD, Chen LG (2019) Effects of different biocharsammendments on physiochemical properties of soil and root morphological attributes of Fokeniahodginsii (Fujian Cypress). Appl Ecol Environ Res 17(5). https://doi.org/10.15666/aeer/1705_1110711120

Tisserant A, Cherubini F (2019) Potentials, limitations, co-benefits, and trade-offs of biochar applications to soils for climate change mitigation. Land 8(12):179. https://doi.org/10.3390/land8120179

Tomczyk A, Sokołowska Z, Boguta P (2020) Biochar physicochemical properties: pyrolysis temperature and feedstock kind effects. Rev Environ Sci Bio/Technol 19(1):191–215. https://doi.org/10.1007/s11157-020-09523-3

De Vasconcelos ACF (2020) Biochar effects on amelioration of adverse salinity effects in soils. In: Applications of biochar for environmental safety. IntechOpen. https://doi.org/10.5772/intechopen.92464

Vijayaraghavan K (2019) Recent advancements in biochar preparation, feedstocks, modification, characterization and future applications. Environ Technol Rev 8(1):47–64. https://doi.org/10.1080/21622515.2019.1631393

Wang B, Zhang Y, Zhu D, Li H (2020) Assessment of bioavailability of biochar-sorbed tetracycline to *Escherichia coli* for activation of antibiotic resistance genes. Environ Sci Technol 54(20):12920–12928. https://doi.org/10.1021/acs.est.9b07963

Wang L, Ok YS, Tsang DC, Alessi DS, Rinklebe J, Mašek O, Bolan NS, Hou D (2022) Biochar composites: emerging trends, field successes and sustainability implications. Soil Use Manag 38(1):14–38. https://doi.org/10.1111/sum.12731

Wang L, Chen D, Zhu L (2023) Biochar carbon sequestration potential rectification in soils: synthesis effects of biochar on soil CO_2, CH_4 and N_2O emissions. Sci Total Environ 904:167047. https://doi.org/10.1016/j.scitotenv.2023.167047

Wang X, Ding J, Han L, Tan J, Ge X, Nan Q (2024) Biochar addition reduces salinity in salt-affected soils with no impact on soil pH: a meta-analysis. Geoderma 443:116845. https://doi.org/10.1016/j.geoderma.2024.116845

Werner S, Kätzl K, Wichern M, Buerkert A, Steiner C, Marschner B (2018) Agronomic benefits of biochar as a soil amendment after its use as wastewater filtration medium. Environ Pollut 233:561–568. https://doi.org/10.1016/j.envpol.2017.10.048

Wiedner K, Glaser B (2015) Traditional use of biochar. In: Biochar for environmental management. Routledge, pp 15–37

Wijitkosum S, Jiwnok P (2020) Applying rice husk biochar for carbon sequestration in building-integrated agriculture for sustainable urban agriculture. Mater Int 2(4):0477–0490. https://doi.org/10.33263/materials24.477490

Wong JW, Ogbonnaya UO (2021) Biochar porosity: a nature-based dependent parameter to deliver microorganisms to soils for land restoration. Environ Sci Pollut Res 28:46894–46909. https://doi.org/10.1007/s11356-021-14803-8

Wright MM, Daugaard DE, Satrio JA, Brown RC (2010) Techno-economic analysis of biomass fast pyrolysis to transportation fuels. Fuel 89:S2-S10. https://doi.org/10.1016/j.fuel.2010.07.029

Wu L, Wei C, Zhang S, Wang Y, Kuzyakov Y, Ding X (2019) MgO-modified biochar increases phosphate retention and rice yields in saline-alkaline soil. J Clean Prod 235:901–909. https://doi.org/10.1016/j.jclepro.2019.07.043

Wu Y, Wang X, Zhang L, Zheng Y, Liu X, Zhang Y (2023) The critical role of biochar mitigates the adverse impacts of drought and salinity stress in plants. Front Plant Sci 14:1163451. https://doi.org/10.3389/fpls.2023.1163451

Wu B, Yang H, Li S, Tao J (2024) The effect of biochar on crop productivity and soil salinity and its dependence on experimental conditions in salt-affected soils: a meta-analysis. Carbon Res 3(1):56. https://doi.org/10.1007/s44246-024-00138-9

Wu P, Fu Y, Vancov T, Wang H, Wang Y, Chen W (2024) Analyzing the trends and hotspots of biochar's applications in agriculture, environment, and energy: a bibliometrics study for 2022 and 2023. Biochar 6(1):78. https://doi.org/10.1007/s42773-024-00370-x

Xie Y, Wang L, Li H, Westholm LJ, Carvalho L, Thorin E, Yu Z, Yu X, Skreiberg Ø (2022) A critical review on production, modification and utilization of biochar. J Anal Appl Pyrolysis 161:105405. https://doi.org/10.1016/j.jaap.2021.105405

Yadav SPS, Bhandari S, Bhatta D, Poudel A, Bhattarai S, Yadav P, Ghimire N, Paudel P, Paudel P, Shrestha J, Oli B (2023) Biochar application: a sustainable approach to improve soil health. J Agric Food Res 11:100498. https://doi.org/10.1016/j.jafr.2023.100498

Yadav V, Khare P, Deshmukh Y, Shanker K, Nigam N, Karak T (2018) Performance of biochar derived from Cymbopogon winterianus waste at two temperatures on soil properties and growth of *Bacopa monneri*. Commun Soil Sci Plant Anal 49(22):2741–2764. https://doi.org/10.1080/00103624.2018.1538371

Yao D, Hu Q, Wang D, Yang H, Wu C, Wang X, Chen H (2016) Hydrogen production from biomass gasification using biochar as a catalyst/support. Bioresour Technol 216:159–164. https://doi.org/10.1016/j.biortech.2016.05.011

Yazdani MR, Duimovich N, Tiraferri A, Laurell P, Borghei M, Zimmerman JB, Vahala R (2019) Tailored mesoporous biochar sorbents from pinecone biomass for the adsorption of natural organic matter from lake water. J Mol Liq 291:111248. https://doi.org/10.1016/j.molliq.2019.111248

Zafeer MK, Menezes RA, Venkatachalam H, Bhat KS (2024) Sugarcane bagasse-based biochar and its potential applications: a review. Emergent Mater 7(1):133–161. https://doi.org/10.1007/s42247-023-00603-y

Zhao L, Cao X, Mašek O, Zimmerman A (2013) Heterogeneity of biochar properties as a function of feedstock sources and production temperatures. J Hazard Mater 256:1–9. https://doi.org/10.1016/j.jhazmat.2013.04.015

Zhao L, Zheng W, Cao X (2014) Distribution and evolution of organic matter phases during biochar formation and their importance in carbon loss and pore structure. Chem Eng J 250:240–247. https://doi.org/10.1016/j.cej.2014.04.053

Zheng Y, Han X, Li Y, Yang J, Li N, An N (2019) Effects of biochar and straw application on the physicochemical and biological properties of paddy soils in Northeast China. Sci Rep 9(1):16531. https://doi.org/10.1038/s41598-019-52978-w

Van Zwieten L, Kimber S, Morris S, Chan KY, Downie A, Rust J, Joseph S, Cowie A (2010) Effects of biochar from slow pyrolysis of papermill waste on agronomic performance and soil fertility. Plant Soil 327:235–246. https://doi.org/10.1007/s11104-009-0050-x

Biochar and Soil Microbial Activity

Gaurav Pandey, Avnish Chauhan, Radhika Khanna, Man Vir Singh, Rachan Karmakar, Khalid Stanikzai, and Kaiss Redouane

Abstract It has been historically challenging for researchers and others to rehabilitate damaged soils, such as those that have been chemically treated, used for infrastructure or mining, which can be costly, time-intensive, and invasive. Especially, in recent years, biochar has not only emerged as an attractive alternative, but also a cost effective and sustainable one. The carbonaceous material produced by the pyrolysis of biomass in the absence of excess oxygen is carbon rich and confers several benefits. By retaining carbon in the soil, soil carbon sequestration through biochar has potential climate change mitigation benefits and may also contribute to enhanced soil fertility and structure. This paper focuses on numerous applications of biochar in soil improvement including nutrient immobilization, pH control, and healthy microbial population stimulation. Via wear sorption, precipitation, and electrostatic interactions, biochar can immobilize contaminants, but also be a carrier of specific species of microorganisms. Importantly, these factors (feedstock type, pyrolysis temperature and heating time) are interwoven with the properties of biochar and their use.

G. Pandey (✉) · R. Khanna
Chemistry and Bioprospecting Division, Forest Research Institute, Dehradun, Uttarakhand, India
e-mail: gauravpandeyfri@gmail.com; jc_em@fridu.edu.in

A. Chauhan
Department of Environmental Sciences, Graphic Era Hill University, Dehradun, Uttarakhand, India

M. V. Singh
School of Applied and Life Sciences, UIT, Uttaranchal University, Dehradun, Uttarakhand, India

R. Karmakar
Department of Environmental Science, Graphic Era (Deemed to be University), Dehradun, Uttarakhand, India

K. Stanikzai
Department of Forestry and Natural Resources, Kabul University, Kabul, Afghanistan

K. Redouane
Faculty of Economics and Management, Hassan 1st University, Settat, Morocco

A. Chauhan et al. (eds.), *Biochar As Remediation Tools for Pollutant Removal*,
https://doi.org/10.1007/978-3-032-16746-0_3

Keywords Biochar · Heavy metal · Remediation · Microbial diversity · Soil carbon

1 Introduction

As problems related to erosion, nutrient runoff, and loss of organic matter looms over agricultural land productivity, soil fertility has become a major global concern. This is a concern that is magnified by the fast approaching 8 billion global population (Ortiz-Bobea et al. 2021). Soil nutrient loss is a risk to food security globally, because healthy soils provide important ecosystem services (McBratney et al. 2017) and nearly 99% of our food supply depends on them (Kopittke et al. 2019; Ortiz-Bobea et al. 2021; Chauhan et al. 2025). A number of amendments have been tried in different crops for conservation and reclamation of degraded lands over time. Biochar is now a candidate and sustainable alternative among them. Pyrolysis involves heating organic materials, like animal manure or solid agricultural residues, under high temperatures and less oxygen, so that biochar is formed (Lehmann 2007; Ge et al. 2020). Slow pyrolysis carried out between temperatures of 350 °C and 700 °C is reported to be the best way to produce biochar for environmental and soil health (Lehmann 2007); the feedstock used, moisture content, pyrolysis temperature, and time all have a large impact on the properties of the biochar produced (Fig. 1).

The feedstock material to make biochar can come from various sources including animal manures, agricultural by-products (e.g., corn stalk, rice husk), agro-forestry residues (e.g., wood, bark, sawdust, nut shells), up to industrial materials creeping into the use, such as carbon black and bagasse (Das et al. 2021). These materials are often high in carbon but relatively low in hydrogen and oxygen as a consequence of their high cellulose and lignin content. Because of its porous structure, numerous functional groups, and large specific surface area, biochar is a good soil conditioner. It enhanced the capacity of soil to store nutrients, reduced soil compaction and helped in the retention of water (Liu et al. 2018; Chauhan et al. 2025). Biochar with a larger pH value and fixed carbon content can be obtained with higher temperature (600–700 °C) and lower fixed carbon content and acidic functional group can be produced with low temperature pyrolysis (100–300 °C), which has more porous structure and higher adsorption ability (Sun et al. 2017). This porous carbon material laden with organic matter was found particularly applicable as adsorbent in agricultural based and remediation contexts, as it was demonstrated to have potential to abstract environmental pollutants as plastics host pesticides, herbicides, and heavy metals (Ahmed et al. 2016; Amonette et al. 2021).

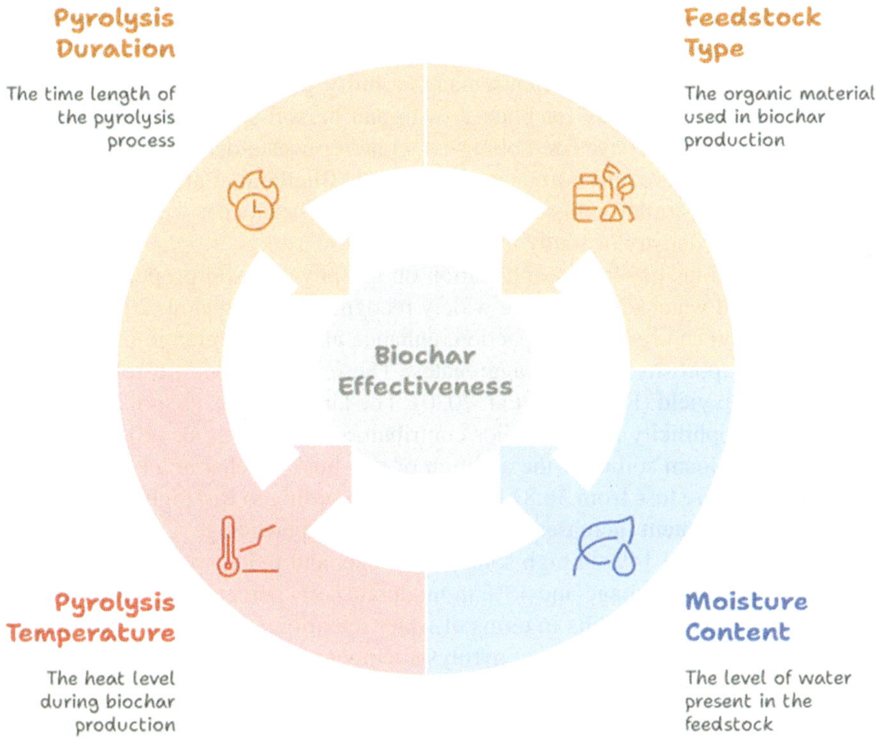

Fig. 1 Factors influencing biochar effectiveness

2 Contribution of Biochar to Soil Physical Improvement

The effect of biochar on soil physical properties is particularly pronounced in sandy soil where the addition of biochar can significantly improve water holding capacity and decrease bulk density (Das et al. 2021; Joseph et al. 2021). But the type of biochar feedstock and the inherent qualities of the soil can influence how much the soil is improved. Notably, previous studies have proven that soil aggregate stability, a key indicator of healthy soil structure, is further enhanced BY biochar (Islam et al. 2021; Yan et al. 2022). More recently, when 3% (w/w, and pH 8.3–9.7) wood chip, vineyard prunings, and wheat straw derived biochar were applied to soils, coarse-textured soils showed greater increasing improvement in structure compared to fine-textured soils (Burrell et al. 2016).

Application of wheat straw biochar (pH 10.35 at the rate between 2 and 40 t/ha) has shown to enhance aggregation and raise crop yields for upland red soil that is cultivated with crops like potato and rapeseed. These benefits are usually accentuated in sandy soils and with higher levels of biochar (commonly above 10 t/ha). It should be noted that the positive effects on aggregation and structure generally also disappear for 2 years (Verheijen et al. 2019). A common impact of the application of biochar is at reduced bulk density which contributes to enhance the habitat for root growth and reduce the compaction of soil (Bhattarai et al. 2015). The ability of biochar to decrease BD will contribute to nutrient retention and mobility, given the fact that the average optimum BD of individuals for plant growth can be soil-specific (Palladino et al. 2022). Similar benefits have been observed with crop waste-derived biochar, which enhances soil structure and mitigates compaction (Bhattarai et al. 2015; Amonette et al. 2021). In a similar study by Igalavithana (2017), biochar from maize straw reduced the bulk density of sandy loam soils.

The implications of biochar application on bio-physical soil properties, namely compaction and water retention, are widely recognised (Wang et al. 2019). Indeed, biochar addition can reduce compaction, enhance also water retention, and overall increase soil's porosity and soil aggregates. These two factors are related to the increase in crop yield (Razzaghi et al. 2020). The large specific area of the biochar, and their hydrophilicity are the major contributors. Oladele et al. (2019) reported that sandy clay loam soil after the addition of rice husk biochar at a rate of 12 t/ha decreased moisture loss from 36.87 to 32.94%. According to Razzaghi et al. (2020), available water content increased by 45% in low soils percentage, 21% in moderate soils percentage, and 14% in high soils percentage, and field capacity increased by 51% in low soils percentage and 13% in moderate soils percentage. Hussaina et al. (2020) studied that the results in terms of water retention are influenced by various parameters such as feedstock type, pyrolysis temperature, particle size, type of soil and degree of compaction (Fig. 2).

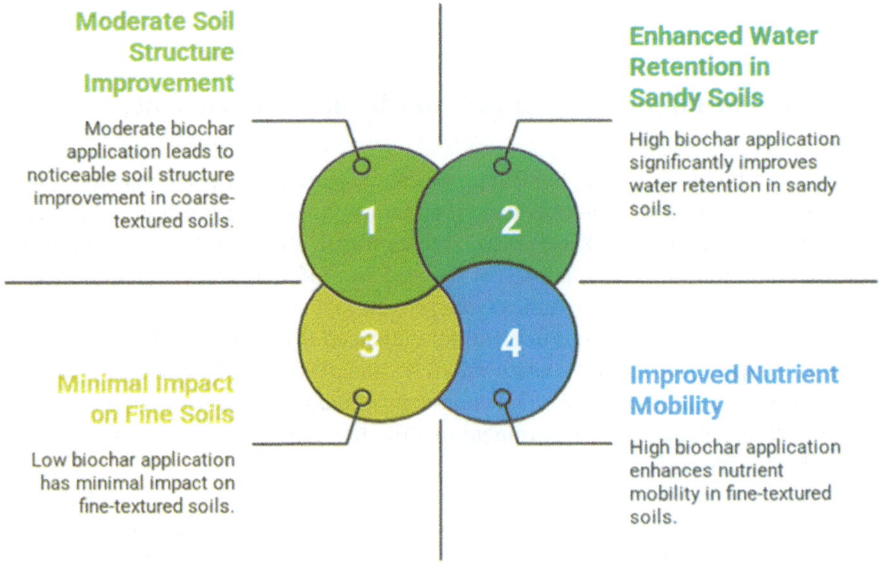

Fig. 2 Biochar effects on soil properties

3 Contribution of Biochar to Soil Chemical Composition

Acid soils are a worldwide issue, impacted by both anthropogenic and natural factors (Dai et al. 2017). Lack of sufficient acidity in the soil environment can also cut down on productivity, as plant growth is exquisitely pH sensitive, in part because they lock up important nutrients such as phosphorus in a form less available to plants (Rouphael et al. 2016). A similar increase in the pH of acidic soils (alfisols, ultisols, and latosols) was observed following the application of biochar (Fig. 3) and studies suggested by Jeffery et al. (2021) (Kamali et al. 2022).

The increase in pH may result from various mechanisms:

Functional groups present on the biochar surface, like carboxyl and hydroxyl groups, interact with basic cations of the soil, acting as pH elevators (Xu et al. 2012). Alkaline compounds in ash content of biochar, which depends upon the source, also help in neutralizing the acidity (Onwuka and Osodeke 2012). Application of biochar to acidic soils increases dissolved alkaline cations such as sodium, potassium,

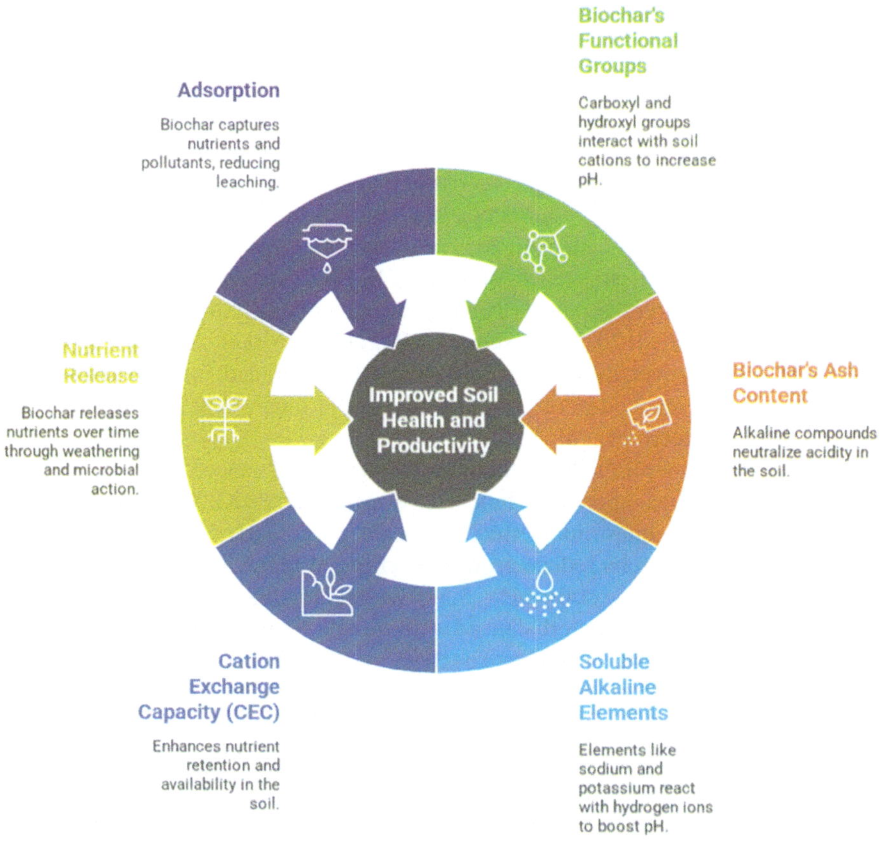

Fig. 3 Biochar impact on soil improvement

calcium and magnesium, which can neutralize soil aluminum and H ions and increase the pH of the soil (Brennan et al. 2014; Enders et al. 2012).

Biochar not only acts on pH regulation; it also enhances soil CEC (Cation exchanging capacity), thereby allowing for enhancing nutrient availability associated with decreasing nutrient losses caused by leaching (Xu et al. 2016). According to Liang et al. (2006), a greater CEC allows the soil to retain more important nutrients such as calcium, potassium, ammonium, and magnesium. According to Khan et al., application of maize straw biochar increased CEC from 12.9 to 15.6, 17.4 and 19.2 mol/kg at 4, 12 and 36 t/ha application rates, respectively Hossain et al. (2020). According to Hossain et al. (2010), biochar increased charge density and CEC by 20–40%. According to Chintala et al. (2014) the presence of surface anions which help to retain nutrients makes biochar suitable for both acidic and alkaline soils.

Biochar initially experiences a series of reactions in soil that comprises leaching of the biochar in the initial one month, activation of the surface in one to six months and stabilization for at least in six months (Joseph et al. 2021). These interactions are beneficial to soil health and the environment as they increase nutrient availability and retain carbon and nitrogen in the soil for a longer duration (Verheijen et al. 2019). As a result of its porous structure and functionality groups, biochar serves as an excellent adsorbent to adsorb nutrients and pollutants, which can prevent nutrients and pollutants leaching out, reduce runoff and maintain soil fertility (Carey et al. 2015; Yuan et al. 2019; Pathak et al. 2024b).

GenBiochar FarmShort Description Biochar slowly provides nutrients due to microbe activity and weathering. Over thousands of years, while its carbon can remain stable, it can permit continuous soil property amelioration and prolonged carbon sequestration (Oliveira et al. 2017; Pathak et al. 2024a). By reducing nutrient losses and enhancing fertilizer use efficiency, biochar can also assist in achieving goals for both agricultural productivity and environmental sustainability (Joseph et al. 2010; Zimmerman et al. 2011; Kocsis et al. 2022) (Joseph et al. 2021; Lehmann and Joseph 2015). Moreover, the application of biochar to degraded, infertile and nutrient deficient soils has resulted in observing a substantial increase in crop health, yield and soil fertility (Zeeshan et al. 2022).

This benefit is not all the time, though. Results are influenced by crop, soil type and biochar properties (specific to biochar) (Kocsis et al. 2022; Weber and Quicker 2018). The nutrient sequestration effect of biochar, particularly N, could inhibit plant establishment, and thereby reduce growth experienced in the first month after application especially on acidic soils (pH ~ 5.5; Joseph et al. 2021). Yet, such temporary downsides are generally counterbalanced by the longer term benefits including higher short-term nutrient retention and uptake (Biederman and Harpole 2013). Altogether, the biochar could enhance crop yields, especially in degraded soils. It can raise the base saturation (Glaser et al. 2002; Major et al. 2010), improve the efficiency of nutrients in the rooting zone, and raise the fertiliser utilization efficiency in plants (Steiner et al. 2007; Laird et al. 2010).

4 Contribution of Biochar to Soil Biological Functions

Biochar's contribution to the biological quality of soils in its positive influence on soil biology is as large as its effects on the physical and chemical soil properties (Fig. 4). The alkaline nature as well as the porous structure of biochar enriched soil porosity, water holding capacity and nutrient available conditions in the soil are known to be potentially conducive to microbial growth and biological activity (Beheshti et al. 2018; Gorovtsov et al. 2020).

When biochar is field applied, it will interact with the mixture of various microbes, and the roots in the soil. These positive interactions are not only observed for unpolluted soils, but also from polluted or decomposed ones, as it was found in the heavy metal or microplastic contaminated soil treatments too, and justify biochar as a soil improver for the long term (Joseph et al. 2010; Gorovtsov et al. 2020; Palansooriya et al. 2022). Biochar enhances soil respiration as well as the stimulation of microbial growth by the carbon and diversity of microbial biomass and other similar studies (Rombola et al. 2019; Diatta et al. 2020). Biochar-treated soils can alter the biological quality of the surrounding soil, thereby attenuating soil-borne pathogens and plant systemic resistance (Tan et al. 2022).

According to Warnock et al. (2007), biochar acts to sequester carbon, enhance soil fertility and support desirable organisms, including mycorrhizal fungi. The organisms serve to stabilize the ecosystem and have developed strong microbial communities. It has also been demonstrated that application of biochar to the soil enhances the activity of plant-beneficial microorganisms. For example, a combination of rhizobia

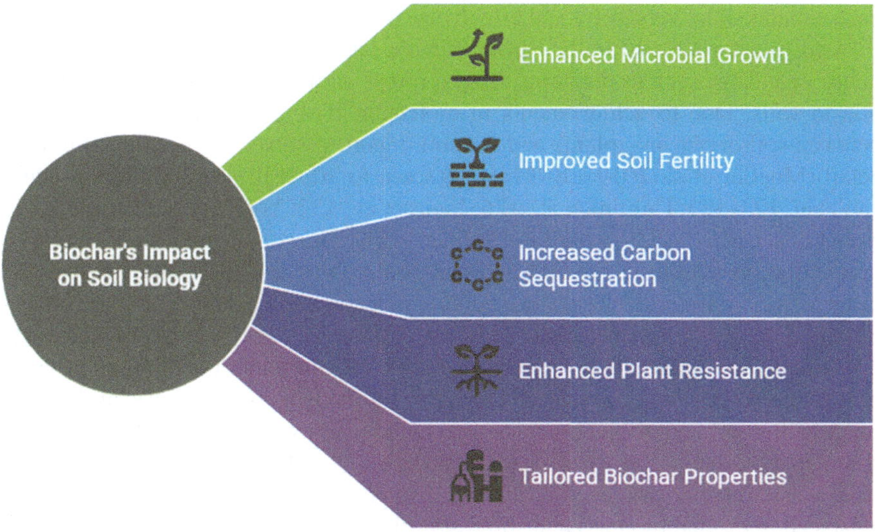

Fig. 4 Biochar's multifaceted impact on soil biology

inoculation and biochar co-application was found to improve soil organic matter, soil micronutrient (Fe, Zn, Mn, and Cu) and soil health status of lentils (Ali et al. 2022).

In contrast, effects of biochar on soil biology are variable depending on various factorssuch as temperature of pyrolysis occurs and feedstock (Sun et al. 2017; Wang et al. 2021). However, its ability to induce soil aggregation and its alkalinity are the main ways by which it can contribute to a biologically active environment (Joseph et al. 2021). Biochar has the potential to contribute to improve soil fertility through enhancement of nutrient availability and beneficial microbe activation (Das et al. 2021; Doran and Zeiss 2000). It also influences processes by which clay minerals and SOM as nutrient exchange media and reservoirs interact (Amonette et al. 2021). Microbially diverse soils are more stable under environmental perturbations and agricultural management pressure (Steinbeiss et al. 2009; Diatta et al. 2020; Palladino et al. 2022; Palansooriya et al. 2022).

Many key soil processes, including decomposition of organic matter, nutrient cycling, agglomeration, disease control, hormone regulation and even detoxification of toxic compounds are also highly dependent on microbes. Microorganisms directly enhance nutrient availability by increasing populations of beneficial microbes in the soil. Higher biological activities of biochar applied soils reverse the fixated nutrient and the nutrients are also made available because mineralization by microorganisms increases nutrient content (Beheshti et al. 2018; Jones et al. 2012). Levels, structure and biochemical properties of soil organic carbon affect the extent of these benefits.

With the recent advances in biochar production technology (Zhou et al. 2021), the properties of the biochar can be adjusted for specific environmental applications, (e.g., wastewater treatment, heavy metal and contaminant removal, solid waste management and remediation of contaminated soil). These technological innovations have enhanced the versatility and effectiveness of using biochar in soil amendment.

Also, plant roots naturally harbor a wide variety of extension plant roots in microorganisms, such as, bacteria, fungi, protists, archaea, and microfauna, in addition to soils. Due to achievements in molecular biology, nowadays the relations between soil organisms, plants and the soil organics matrix can be studied in more detail (Mueller et al. 2019). Biochar is known to affect these communities (Intizzari et al. 122–123; Ugolini et al., 126; Janus et al., 132; Yoon et al., 139; Kahn 144; Kahnt & Scharko 147; Borchard and Siemens 150). The beneficial impacts of biochar production on alteration in soil microbial community and its related processes in different soils have been shown by a number of studies, e.g., chicken manure biochar produced at 500 °C enhanced microbial activity in sedimentary alfisols (Meier et al. 2017) and corn straw biochar produced at 500 °C promoted populations of gram-positive bacteria, fungi, and microbial biomass in sandy loam soils (Lu et al. 2014). In calcaric fluvisols, microbial biomass, enzymatic activity and microbial activity were stimulated by the cotton straw biochar produced at 450 °C (Liao et al. 2016). Date palm waste biochar produced at 300 °C also enhanced soil organic matter, albeit, enhanced microbial biomass and soil respiration, as well as heavy metal adsorption (Al-Wabel et al. 2019).Biochar derived from wood of *Gliricidia sepium* and heated at 300 °C and 900 °C was reported to increase the total soil enzymatic activity and sorb the heavy metal when applied in serpentine soils (Bandara et al. 2017). At

900 °C, another biochar increased the number of beneficial plant growth bacteria, microbial biomass, numbers of bacteria and fungi, and heavy metal removal (Herath et al. 2017).

Addition of glucose enhanced gram-positive bacteria populations in forest and arable soil (Steinbeiss et al. 2009), and the addition of maize straw biochar (400 °C) also increased the ratio of AM fungi to saprotrophic fungi in silt loam (Luo et al. 2017). Application of pre-washed oak pellet biochar (550 °C) to clay and sandy loam soils increased phospholipid fatty acid biomass by 40–64% (Awad et al. 2018), while phosphate-solubilizing and carbon-degrading bacteria were favored in silt-loam soils by the Pinus radiata biochar (Anderson et al. 2011). Application of biochar from Pinus sp. together with dairy and bull manure (500 °C) resulted in an increase in both microbial biomass and activity in fertile mollisol soils (Kolb et al. 2009). However, soil organic carbon mineralization was increased and microbial biomass was slightly increased in clay loam soils following application of rice straw biochar (550 °C) (Pan et al. 2016), and soil organic matter and phospholipid fatty acid biomass were stimulated (Gomez et al. 2014) with sawdust (550 °C) biochar addition to sandy loams.

For example, it is found that sugarcane bagasse biochar (450 °C) decreased heavy metal bioavailability while boosting actinomycetes by 280%. This trend was again observed for biochar made from pine cone and vegetable waste at 200 °C which was found to absorb heavy metals while increasing microbial abundance (Igalavithana et al. 2017), meaning that the biochar material was still biologically active while retaining its chemical adsorption capacity. Bacterial (+28%), actinobacterial (+62%) and gram-negative bacterial (+27%) population were significantly increased in flinty clay loam by Willow leaf biochar (470 °C) application and its effects on the N cycling process (Prayogo et al. 2014). Willow wood/swine manure biochar produced at 350 °C enhanced microbial biomass and both gram-positive and gram-negative bacteria in temperate sandy loam (Ameloot et al. 2013).

5 Biochar's Contribution to the Remediation of Metals, Metalloids, and Pollutants

Heavy metals and metalloids pollution persist as one of the most serious global threats of soil health and food safety and environmental sustainability (Zhang et al. 2013; Megharaj and Naidu 2017). Biochar causes greater long-term changes in the remediation of soil, in relation to the traditional method (such as leaching, soil washing or excavation) more ecologically sound, also often more expensive, lengthy and invasive (Fig. 5). Biochar can decrease the solubility and bioavailability of heavy metals through fixing them in soil matrix, which reduces the threat to the environment and human health (Liu et al. 2018; Hue 2020). Considering its special properties-porosity and high surface area, alkaline pH, particle size, high water retention capacity and

Fig. 5 Biochar's contribution to the remediation of metals, metalloids, and pollutants

structural stability (Joseph et al. 2010, 2021), biochar is most suited for improving the quality of poor or degraded soils (Diatta et al. 2020).

Characteristics of biochar, such as the physical (e.g., density, pore structure), chemical (e.g., electrical conductivity) and biological (e.g., cation exchange capacity [CEC]) properties, all influence the efficacy of biochar as a remedial agent (Jones et al. 2012; Amonette et al. 2021). High-temperature pyrolysis (550–750 °C) facilitates the conversion of biochar metals (e.g., Cd^2+, Zn^2+, Pb^2+) to a less toxic metal oxide form, leading to advanced biochar application in supporting phytoremediation strategies (Huang et al. 2018). The surface functional groups, pH, ion-exchange capacity and the inner pore structure of biochar play a much more important role in immobilization of metals in general (Qiu et al. 2022).

Biochar's assist with remediation is mainly through heavy metal bioavailability reduction. This can be achieved by ion-exchange, pH adjustment and adsorption, which limit the movement of metal through soil (Zhang et al. 2013; Qiu et al. 2022). In addition to its enhanced sorption for contaminants, biochar's overall porosity and heterogeneous charge enhance its ability to retain nutrients, water, and organic matter (Zhang et al. 2013; Brown et al. 2006). Surface oxygenation which occurs during biochar aging in soil generates negatively charged functional groups (such as hydroxyl, carboxyl, carbonyl, and phenol) that will increase the pH and CEC of the soil (Cheng et al. 2006). Moreover, biochar releases beneficial cations (potassium, calcium, and magnesium), which ameliorate the soil acidity and enhance availability of cationic and anionic nutrients (potassium, phosphorus, and nitrogen) from soil to plants (Chintala et al. 2014). Biochar is capable of effectively immobilizing metals by adsorption, precipitation and cation-exchange, and extractability of the metals can be reduced by biochars containing high phosphate and carbonate content (Cao et al. 2009).

Nanobiochar and biochar-metal oxide composites are emerging technologies with even greater potential for adsorption of pollutants, however, further research is needed to understand their long-term impacts on the environment (Zhang et al. 2020). Another important parameter that influences the remediation capacity of the biochar-soil is the pH because with the increase in pH from 3 to 8, the surface of biochar is more negative which facilitates the adsorption of the metal cations (Tan et al. 2015; Xu et al. 2011). Biochars produced at higher temperatures usually have a higher point of zero charge influencing the ability to retain anionic pollutants and their CEC (Banik et al. 2018; Fidel et al. 2018). The aromatic structure in biochar can be used to attract metal cations through ion exchange and electrical forces (Vithanage et al. 2017).

Nevertheless, environmental conditions affect the efficacy of ion exchange. Ion exchange is limited in low pH soils and in high ionic strength conditions, competition among ions can reduce the exchange efficiency (Fidel et al. 2018). Commonly, the adsorption of metals, including Ca, Mg, K and Na, by biochar increases with CEC because of the ability of the biochar to adsorb these metals (Hassan and Carr 2018). Despite the fact that the biochar has shown some sorption and desorption ability towards chromate, arsenate and other contaminants, their environmental risk profile has been often mitigated thanks to the use of biochar in combination with phytoremediation, although it may increase their mobility.

Another significant treatment mechanism that biochar can support is metal precipitation. Most biochars are alkaline and elevate the hydroxyl concentration of soil solutions, thereby promoting the precipitation of metals such as Mn and Al (Case et al. 2012). The mineral components, naturally occurring or man-made, e.g., phosphates, sulfates, carbonates also precipitate them out of the solution in a stable, firmly bound, non-corrosive form. Indeed, biochar offers potential to form stable complexes with metals (e.g., mercury, zinc, lead, copper and cadmium) depending on feedstock and pyrolysis conditions (Vithanage et al. 2017; Ngatia et al. 2019; Cheng et al. 2020).

Owing to its functional groups, biochar is also involved in redox reaction besides these reaction processes. Its redox potential, increasing with aromatic and electron donor character according to the rise of the pyrolysis temperature, is mainly due to the phenolic groups (Chacon et al. 2020; Xu et al. 2019). This pattern of redox activity is particularly relevant to Hg, as nonoptimal conditions can lead to more toxic accumulation of methyl Hg (Xing et al. 2022).

The results of long-term experiments also show that biochar can reduce the bioavailability of heavy metals to plants. For instance, the accumulation of cadmium in rice fields dropped significantly following 3 years of 40 t/ha biochar application (Chen et al. 2016). Even if used only once, the use a couple of times lead to small increases. Concentration of cadmium and lead in edible parts of vegetables were decreased to safe level in pakchoi field with 3 t/ha biochar application (Nie et al. 2018). Rice-straw biochar (RB) and its alkali-modified biochar (MRB) decreased soil available Zn concentration by 28.96% and 36.86%, respectively Wang et al. (2021). According to Liang et al. (2006) the leaching of Pb, Cu, and Cd could be decreased with maxima of 97, 8%, 100% and 77, 2% with biochar from Sphagnum. The cadmium content in brown rice also decreased by 25.1% and 34.6% when sulfur- and iron-treated biochars were applied (Park et al. 2016). These results suggest that the biochar is an effective approach to manipulating the bioavailability of metals and their environmental risk by affecting soil parameters including Ox, pH, organic C and metal sorption capacity.

5.1 Alleviating the Toxicity of Contaminants Toward Microbial Populations

To remediate the harmful effect on soil microorganisms of soil heavy metals is important. For instance, willow biochar pyrolyzed at 700 °C has also been found to alleviate the microbial death rates in contaminated soils (Farooq et al. 2022; Nawaz et al. 2022). Biochar can immobilize metals including Ni, Co, Cr, Mn, Cd, and Al in the soil due to its porous structure and make the metals unavailable (Zainab et al. 2021; Saleem et al. 2022; Chauhan et al. 2024b). The fact that biochar decreases the toxicity of leachates for sensitive organisms, such as *Vibrio fischeri*, and enhances the reproduction of soil fauna, such as *Folsomia candida*, means not only that it protects the soil microbiome, but also that it stimulates plant growth (Zielinska and Oleszczuk 2016).

Biochar also preserves helpful microbial colonies. For instance, it has been reported that application of rice straw biochar can decrease the toxicity of copper, zinc, lead, and cadmium up to 68% and promote the recovery of the nitrogen fixing bacteria such as *Bradyrhizobium japonicum* (Seneviratne et al. 2016). Biochar minimizes the interaction between soil and microorganisms, in the presence of metals, which will enhance subsequently soil fertility and health.

5.2 Changing the Living Conditions for Microbial Communities

Biochar is able to indirectly modify microbial habitats (besides reducing metal toxicity) by improving soil structure, decreasing compaction, and increasing aeration and water holding capacity (Abel et al. 2013). Apart from pH modification of soil, the more stable environment can shield microbes from detrimental dry–wet cycles, which are common phenomena in natural conditions (Liang et al. 2006). Even small increases in pH (by 0.2–0.3 units) can greatly modify the structure of the microbial community, promoting the bacterial populations and decreasing the toxicity of metals such as aluminum (Qian et al. 2013).

Biochar has been demonstrated to enhance the enzymes, and the soil activities of enzymes provide valuable information regarding the microbial health and organic material decompositions (Sinsabaugh et al. 2008; Chauhan et al. 2024a). It enhances the activity of phosphatases (e.g., enzymes that release phosphorus from organic compounds) (Sinsabaugh et al. 2008). The acid phosphatase activity increased by 27.85% after a seven-week application of biochar compared with untreated soils (Mao et al. 2022). It has also been shown that biochar enhances urease and dehydrogenase activities by 27% and 7.4–39%, respectively (Pandey et al. 2016; Jia et al. 2017). Protease (8.4%), urease (13.9%), alkaline phosphatase (81.3%), catalase (21.7%) and sucrase (150.5%) were all increased as well Song et al. (2022).

What is interesting is that factors such as pyrolysis temperature and biochar type determine the influence of biochar on enzymatic activity. According to Liao et al. (2016), biochar pyrolyzed at a lower temperature (e.g., 300 °C) sufficiently affected the activation of soil enzymes more than the biochar pyrolyzed at a higher temperature (500 °C). These results indicated that the biochar selection is reliant on both soil and application purposes for soils (Liao et al. 2016). By and large, the biochar not only enhances nutrient recycling, but also leads to ecological agriculture, as it improves soil enzyme activities and microbial processes.

5.3 *Emerging Opportunities for Biochar in Agro-Environmental Practices*

There is much to be said for biochar as an inexpensive green technology to improve soil health, capture carbon, clear pollutants from water and air and reduce greenhouse gases. However, as process outcomes are directly controlled by feedstock and soil properties and input parameters in biochar technology, it is important that the specificities of biochar properties are taken into account to fully exploit this potential.

Nano-biochar is also the result of the most recent technology development, in particular as regard to nanotechnology that causes the strong enhancement of the surface area and of its reactivity. They may work because bamboo has these properties and can absorb impurities from soil and water. Although nano-biochar shows great potential, further investigation is required to obtain a better insight into the long-term impact of employing nano-biochar for climate mitigation and carbon sequestration.

Customized for specific applications, biochar's natural metal, microplastic, and emerging contaminant-retaining properties could become an important part of the treatment toolbox for wastewater, landfill leachate, or farm runoff. Dedicated filters for different pollutants may be needed to fully achieve such benefits. The importance of biochar might become crucial in reducing environmental risks and safeguarding human health as hazardous chemical use becomes more prevalent worldwide (Amonette et al. 2021; Joseph et al. 2021).

Fertilizers combined with biochar may increase nutrient use efficiency and decrease emissions. Furthermore, we can promote biochar as a terrestrial carbon sink in agriculture and environmental management by quantifying the carbon sequestration potential of the biochar at the ecosystem scale and incorporating it as part of carbon market efforts.

6 Conclusion

Biochar (BC), the carbonaceous material derived from pyrolysis of lingo-cellulosic feedstocks, such as animal manures, forestry residues, agriculture biomass, can be considered for soil remediation. It has strong potential to be applied in the restoration of degraded land because it improves bio-chemical functions, macrostructure and fertility incrementally. Biochar also increases nutrient use, microbial activity, water supply and pH buffering, especially in sandy soil.

Even though biochar can be mixed with other conventional and organic amendments, and can be applied in various rates and methods, the use of biochar to realize positive impacts on different agro-environments can be considerably improved. Due to their high cation exchange capacity and capacity to immobilize contaminants, biochar is also considered the most promising adsorbent to remove (in particular heavy metals and metalloids and) new pollutants (like microplastics) in soils.

Recent activity has led to the metal or nutrient biochars which can be used to address particular environmental problems at sites, e., municipal or industry wastes. Further researches will be needed to understand completely the environmental resilience, long-term impacts, and role of biochar in a carbon market and in a cost–benefit perspective.

With proper feedstock and matching pyrolysis and site-specific application, biochar may be a sustainable way to restore soil fertility, support sustainable agriculture, and curtail environmental degradation.

Acknowledgements The authors would like to express heartfelt thanks to the Chemistry and Bioprospecting Division, Forest Research Institute, Dehradun, Uttarakhand, India, and Department of Environmental Sciences, Graphic Era Hill University, Dehradun, Uttarakhand, India, for their invaluable support and resources.

Authors' Contributions GP and AC conceptualized this chapter, RK and GP designed table, figures and wrote the initial stage of chapter, GP, and AC, critically edited the initial draft and contribution to constructive recommendations before the submission. All the authors have no any conflict of interest and the work is not related to animal ethical clearance.

References

Abel S, Peters A, Trinks S (2013) Impact of biochar and hydrochar addition on water retention and water repellency of sandy soil. Geoderma 202–203:183–191. https://doi.org/10.1016/j.geoderma.03.003

Ahmed MB, Zhou JL, Ngo HH, Guo W (2016) Insight into biochar properties and its cost analysis. Biomass Bioenergy 76–86. https://doi.org/10.1016/j.biombioe.2015.11.002

Ali A, Ahmad W, Munsif F, Khan A, Nepal J, Wójcik Gront E (2022) Residual effect of finely-ground biochar inoculated with bio-fertilization impact on productivity in a lentil–maize cropping system. Agronomy 12:2036. https://doi.org/10.3390/agronomy12092036

Al-Wabel MI, Usman ARA, Al-Farraj AS (2019) Date palm waste biochars alter a soil respiration, microbial biomass carbon, and heavy metal mobility in contaminated mined soil. Environ Geochem Health 41:1705–1722

Ameloot N, De Neve S, Jegajeevagan K (2013) Short-term CO2 and N2O emissions and microbial properties of biochar amended sandy loam soils. Soil Biol Biochem 57:401–410

Amonette JE, Blanco-Canqui H, Hassebrook C, Laird DA, Lal R, Lehmann J (2021) Integrated biochar research: a roadmap. J Soil Water Conserv 76:24A-29A. https://doi.org/10.2489/jswc.2021.1115A

Anderson CR, Condron LM, Clough TJ (2011) Biochar induced soil microbial community change: implications for biogeochemical cycling of carbon, nitrogen and phosphorus. Pedobiologia 54:309–320. https://doi.org/10.1016/j.pedobi.2011.07.005

Awad YM, Ok YS, Abrigata J (2018) Pine sawdust biomass and biochars at different pyrolysis temperatures change soil redox processes. Sci Total Environ 625:147–154

Bandara T, Herath I, Kumarathilaka P (2017) Role of woody biochar and fungal-bacterial co-inoculation on enzyme activity and metal immobilization in serpentine soil. J Soils Sediments 17:665–673

Banik C, Lawrinenko M, Bakshi S, Laird DA (2018) Impact of pyrolysis temperature and feedstock on surface charge and functional group chemistry of biochars. J Environ Qual 43(3):452–461

Beheshti M, Etesami H, Alikhani HA (2018) Effect of different biochars amendment on soil biological indicators in a calcareous soil. Environ Sci Pollut Res 25:14752–14761. https://doi.org/10.1007/s11356-018-1682-2

Bhattarai B, Neupane J, Dhakal SP, Nepal J, Gnyawali B, Timalsina R (2015) Effect of biochar from different origin on physio-chemical properties of soil and yield of garden pea (Pisumsativum L.) at paklihawa, rupandehi, Nepal. World. J Agric Res. https://doi.org/10.12691/wjar-3-4-3

Biederman LA, Harpole WS (2013) Biochar and its effects on plant productivity and nutrient cycling: a meta-analysis. GCB Bioenergy 2:202–214

Brennan A, Jimenez EM, Puschenreiter M, Alburquerque JA, Switzer C (2014) Effects of biochar amendment on root traits and contaminant availability of maize plants in a copper and arsenic impacted soil. Plant Soil 379:351–360

Brown RA, Kercher AK, Nguyen TH, Nagle DC, Ball WP (2006) Production and characterization of synthetic wood chars for use as surrogates for natural sorbents. Org Geochem 37:321–333. https://doi.org/10.1016/j.orggeochem.2005.10.008

Burrell LD, Zehetner F, Rampazzo N, Wimmer B, Soja G (2016) Long-term effects of biochar on soil physical properties. Geoderma 282:96–102. https://doi.org/10.1016/j.geoderma.2016.07.019

Cao X, Ma L, Gao B, Harris W (2009) Dairy-manure derived biochar effectively sorbs lead and atrazine. Environ Sci Technol 43:3285–3291. https://doi.org/10.1021/es803092k

Carey DE, McNamara PJ, Zitomer DH (2015) Biochar from pyrolysis of biosolids for nutrient adsorption and turfgrass cultivation. Water Environ Res 87:2098–2106. https://doi.org/10.2175/106143015X14362865227391

Case SDC, Mc Namara NP, Reay DS, Whitaker J (2012) The effect of biochar addition on N_2O and CO_2 emissions from a sandy loam soil—the role of soil aeration. Soil Biol Biochem 51:125–134

Chacon FJ, Sanchez Monedero MA, Lezama L, Cayuela ML (2020) Enhancing biochar redox properties through feedstock selection, metal preloading and post-pyrolysis treatments. Chem Eng J 395:125100

Chauhan PK, Upadhyay SK, Rajput VD (2024) Fostering plant growth performance under drought stress using rhizospheric microbes, their gene editing, and biochar. Environ Geochem Health 46:41. https://doi.org/10.1007/s10653-023-01823-1

Chauhan PK, Upadhyay SK, Chauhan A, Bhutiani R, Sikarwar RLS, Tiwari MK (2025) A review on simultaneous use of microbes and biochar for removal of toxic heavy metals: recent updates and future outlooks. Physiol Mol Biol Plants 31(3):1–15. https://doi.org/10.1007/s12298-025-01649-4

Chen D, Guo H, Li R, Li L, Pan G, Chang A (2016) Low uptake affinity cultivars with biochar to tackle Cd-tainted rice—a field study over four rice seasons in Hunan, China. Sci Total Environ 541:1489–1498

Cheng CH, Lehmann J, Thies JE, Burton SD, Engelhard MH (2006) Oxidation of black carbon by biotic and abiotic processes. Org Geochem 37:1477–1488. https://doi.org/10.1016/j.orggeochem.2006.06.022

Cheng S, Chen T, Xu W, Huang J, Jiang S, Yan B (2020) Application research of biochar for the remediation of soil heavy metals contamination: a review. Molecules 25:3167

Chintala R, Mollinedo J, Schumacher TE, Malo DD, Julson JL (2014) Effect of biochar on chemical properties of acidic soil. Arch Agron Soil Sci 60:393–404

Chuahan PK, Pathak HK, Dubey G, Sharma H, Upadhyay SK (2024b) Impact of bacillus cereus SPB-10 on growth promotion of wheat (*Triticum aestivum* L.) under arsenic contaminated soil. Curr Microbiol. https://doi.org/10.1007/s00284-024-03673-1

Dai Z, Zhang X, Tang C, Muhammad N, Wu J, Brookes PC (2017) Potential role of biochars in decreasing soil acidification—a critical review. Sci Total Environ 581–582:601–611

Das S, Mohanty S, Sahu G, Rana M, Pilli K (2021) Biochar: a sustainable approach for improving soil health and environment. In: Soil erosion—current challenges and future perspectives in a changing world. Intech Open, London, UK. https://doi.org/10.5772/intechopen.97136

Diatta AA, Fike JH, Battaglia ML, Galbraith JM, Baig MB (2020) Effects of biochar on soil fertility and crop productivity in arid regions: a review. Arab J Geosci 13:595. https://doi.org/10.1007/s12517-020-05586-2

Doran JW, Zeiss MR (2000) Soil health and sustainability: managing the biotic component of soil quality. Appl Soil Ecol 15:3–11. https://doi.org/10.1016/S0929-1393(00)00067-6

Enders A, Hanley K, Whitman T, Joseph S, Lehmann J (2012) Characterization of biochars to evaluate recalcitrance and agronomic performance. Biores Technol 114:644–653

Farooq TH, Rafy M, Basit H (2022) Morpho-physiological growth performance and phytoremediation capabilities of selected xerophyte grass species towards Cr and Pb stress. Front Plant Sci 13:997120. https://doi.org/10.3389/fpls.2022.997120

Fidel RB, Laird DA, Spokas KA (2018) Sorption of ammonium and nitrate to biochars is electrostatic and pH-dependent. Sci Rep 8:17627

Ge S, Foong SY, Ma NL, Liew RK, Wan Mahari WA, Xia C (2020) Vacuum pyrolysis incorporating microwave heating and base mixture modification: an integrated approach to transform biowaste into eco-friendly bioenergy products. Renew Sustain Energy Rev 127:109871. https://doi.org/10.1016/j.rser.2020.109871

Glaser B, Lehmann J, Zech W (2002) Ameliorating physical and chemical properties of highly weathered soils in the tropics with charcoal—a review. Biol Fertil Soils 35:219–230. https://doi.org/10.1007/s00374-002-0466-4

Gomez JD, Denef K, Stewart CE (2014) Biochar addition rate influences soil microbial abundance and activity in temperate soils. Eur J Soil Sci 65:28–39

Gorovtsov AV, Minkina TM, Mandzhieva SS, Perelomov LV, Soja G, Zamulina IV (2020) The mechanisms of biochar interactions with microorganisms in soil. Environ Geochem Health 42:2495–2518. https://doi.org/10.1007/s10653-019-00412-5

Hassan MM, Carr CM (2018) A critical review on recent advancements of the removal of reactive dyes from dyehouse effluent by ion-exchange adsorbents. Chemosphere 209:201–219

Herath I, Iqbal MCM, Al-Wabel MI (2017) Bioenergy-derived waste biochar for reducing mobility, bioavailability, and phytotoxicity of chromium in anthropized tannery soil. J Soils Sediments 17:731–740

Hossain MK, Strezov V, Chan KY, Nelson PF (2010) Agronomic properties of wastewater sludge biochar and bioavailability of metals in production of cherry tomato (Lycopersiconesculentum). Chemosphere 78:1167–1171

Hossain MZ, Bahar MM, Sarkar B, Donne SW, Ok YS, Palansooriya KN (2020) Biochar and its importance on nutrient dynamics in soil and plant. Biochar 2:379–420

Huang H, Yao W, Li R, Ali A, Du J, Guo D (2018) Effect of pyrolysis temperature on chemical form, behavior and environmental risk of Zn, Pb and Cd in biochar produced from phytoremediation residue. Biores Technol 249:487–493. https://doi.org/10.1016/j.biortech.2017.10.020

Hue N (2020) Biochar for maintaining soil health. In: Giri B, Varma A (eds) Soil health. Springer, Cham, Switzerland, pp 21–46

Hussaina RK, Ravia K, Ankit G (2020) Influence of biochar on the soil water retention characteristics (SWRC): potential application in geotechnical engineering structures. Soil Tillage Res 204:104713

Igalavithana AD, Lee SE, Lee YH (2017) Heavy metal immobilization and microbial community abundance by vegetable waste and pine cone biochar of agricultural soils. Chemosphere 174:593–603

Islam MU, Jiang F, Guo Z, Peng X (2021) Does biochar application improve soil aggregation? A meta-analysis. Soil Tillage Res 209:104926. https://doi.org/10.1016/j.still.2020.104926

Jeffery S, Verheijen FGA, vander Velde M, Bastos AC (2021) A quantitative review of the effects of biochar application to soils on crop productivity using meta-analysis. Agric, Ecosyst Environ 144:175–187

Jia W, Wang B, Wang C, Sun H (2017) Tourmaline and biochar for the remediation of acid soil polluted with heavy metals. J Environ Chem Eng 5:2107–2114

Jones DL, Rousk J, Edwards Jones G, DeLuca TH, Murphy DV (2012) Biochar-mediated changes in soil quality and plant growth in a three-year field trial. Soil Biol Biochem 45:113–124. https://doi.org/10.1016/j.soilbio.2011.10.012

Joseph SD, Camps Arbestain M, Lin Y, Munroe P, Chia CH, Hook J (2010) An investigation into the reactions of biochar in soil. Soil Res 48:501. 10.1071SR10009

Joseph S, Cowie AL, Van Zwieten L, Bolan N, Budai A, Buss W (2021) How biochar works, and when it doesn't: a review of mechanisms controlling soil and plant responses to biochar. GCB Bioenergy 1731–1764. https://doi.org/10.1111/gcbb.12885

Kamali M, Sweygers N, Al-Salem S, Appels L, Aminabhavi TM, Dewil R (2022) Biochar for soil applications-sustainability aspects, challenges and future prospects. Chem Eng J 428:131189

Kocsis T, Ringer M, Biro B (2022) Characteristics and applications of biochar in soil–plant systems: a short review of benefits and potential drawbacks. Appl Sci 8:4051

Kolb SE, Fermanich KJ, Dornbush ME (2009) Effect of charcoal quantity on microbial biomass and activity in temperate soils. Soil Sci Soc Am J 73:1173–1181

Kopittke PM, Menzies NW, Wang P, McKenna BA, Lombi E (2019) Soil and the intensification of agriculture for global food security. Environ Int 132:105078. https://doi.org/10.1016/j.envint.2019.105078

Laird D, Fleming P, Wang B, Horton R, Karlen D (2010) Biochar impact on nutrient leaching from a Midwestern agricultural soil. Geoderma 158:436–442. https://doi.org/10.1016/j.geoderma.2010.05.012

Lehmann J, Joseph S (2015) Biochar for environmental management: science, technology and implementation. Routledge, Abingdon, Oxfordshire, United Kingdom

Lehmann J (2007) Bio-energy in the black. Front Ecol Environ 381–387. https://doi.org/10.1890/1540-9295(2007)

Liang B, Lehmann J, Solomon D, Kinyangi J, Grossman JO, Neill B (2006) Black carbon increases cation exchange capacity in soils. Soil Sci Soc Am J 70:1719–1730

Liao N, Li Q, Zhang W (2016) Effects of biochar on soil microbial community composition and activity in drip-irrigated desert soil. Eur J Soil Biol 72:27–34. https://doi.org/10.1016/j.ejsobi.2015.12.008

Liu G, Zheng H, Jiang Z, Zhao J, Wang Z, Pan B (2018) Formation and physicochemical characteristics of nanobiochar: insight into chemical and colloidal stability. Environ Sci Technol 10369–10379. https://doi.org/10.1021/acs.est.8b01481

Lu W, Ding W, Zhang J (2014) Biochar suppressed the decomposition of organic carbon in a cultivated sandy loam soil: a negative priming effect. Soil Biol Biochem 76:12–21

Luo S, Wang S, Tian L (2017) Long-term biochar application influences soil microbial community and its potential roles in semiarid farmland. Appl Soil Ecol 117:10–15

Major J, Rondon M, Molina D, Riha SJ, Lehmann J (2010) Maize yield and nutrition during 4 years after biochar application to a Colombian savannaoxisol. Plant Soil 333:117–128. https://doi.org/10.1007/s11104-010-0327-0

Mao Q, Hu B, Agathokleous E, Wang L, Koike T, Ma M (2022) Biochar application improves karstic lime soil physicochemical properties and enzymes activity and enhances sweet tea seedlings physiological performance. Sci Total Environ 830:154815

McBratney AB, Morgan CL, Jarrett LE (2017) The value of soil's contributions to ecosystem services. In: Global soil security. Springer, Cham, pp 227–235. https://doi.org/10.1007/978-3-319-43394-3-20

Megharaj M, Naidu R (2017) Soil and brownfield bioremediation. Microb Biotechnol 5:1244–1249

Meier S, Curaqueo G, Khan N (2017) Chicken-manure-derived biochar reduced bioavailability of copper in a contaminated soil. J Soils Sediments 17:741–750

Mueller CW, Carminati A, Kaiser C (2019) Rhizosphere functioning and structural development as complex interplay between plants, microorganisms and soil minerals. Front Environ Sci 130

Nawaz H, Ali A, Saleem MH (2022) Comparative effectiveness of EDTA and citric acid assisted phytoremediation of Ni contaminated soil by using canola (Brassica napus). Braz J Biol 82:261785. https://doi.org/10.1590/1519-6984.261785

Ngatia LW, Grace JM, Moriasi D, Bolques A, Osei GK Taylor RW (2019) Biochar phosphorus sorption-desorption: potential phosphorus eutrophication mitigation strategy. In: Abrol V, Sharma P, eds. Biochar—an imperative amendment for soil and the environment. IntechOpen, London, UK

Nie C, Yang X, Niazi NK, Xu X, Wen Y, Rinklebe J (2018) Impact of sugarcane bagasse-derived biochar on heavy metal availability and microbial activity: a field study. Chemosphere 200:274–282

Oladele S, Adeyemo A, Awodun M, Ajayi A, Fasina A (2019) Effects of biochar and nitrogen fertilizer on soil physicochemical properties, nitrogen use efficiency and upland rice (*Oryzasativa*) yield grown on an Alfisol in southwestern Nigeria. Int J Recycl Org Waste Agric (IJROWA) 295–308

Oliveira FR, Patel AK, Jaisi DP, Adhikari S, Lu H, Khanal SK (2017) Environmental application of biochar: current status and perspectives. Biores Technol 246:110–122. https://doi.org/10.1016/j.biortech.08.122

Onwuka MI, Osodeke VE (2012) Effects of agricultural wastes ash on nutrients concentration and yield of maize (*Zea mays*) in haplic acrisol of Southeastern Nigeria. Agro-Sci 3:1–11

Ortiz-Bobea A, Ault TR, Carrillo CM, Chambers RG, Lobell DB (2021) Anthropogenic climate change has slowed global agricultural productivity growth. Nat Clim Chang 306–312. https://doi.org/10.1038/s41558-021-01000-1

Palansooriya KN, Sang MK, Igalavithana AD, Zhang M, Hou D, Oleszczuk P (2022) Biochar alters chemical and microbial properties of microplastic-contaminated soil. Environ Res 209:112807. https://doi.org/10.1016/j.envres.2022.112807

Palladino M, Romano N, Pasolli E, Nasta P (2022) Developing pedotransfer functions for predicting soil bulk density in Campania. Geoderma 412:115726

Pan F, Li Y, Chapman SJ (2016) Microbial utilization of rice straw and its derived biochar in a paddy soil. Sci Total Environ 559:15–23

Pandey V, Patel A, Patra DD (2016) Biochar ameliorates crop productivity, soil fertility, essential oil yield and aroma profiling in basil (*Ocimum basilicum* L.). Ecol Eng 90:361–366

Park JH, Lee SJ, Lee ME, Chung JW (2016) Comparison of heavy metal immobilization in contaminated soils amended with peat moss and peatmoss-derived biochar. Environ Sci Process Impacts 18:514–520

Pathak HK, Chauhan PK, Seth CS, Dubey G, Upadhyay SK (2024a) Mechanistic and future prospects in rhizospheric engineering for agricultural contaminants removal, soil health restoration, and management of climate change stress. Sci Total Environ 927:172116. https://doi.org/10.1016/j.scitotenv.2024.172116

Pathak HK, Seth CS, Chauhan PK, Dubey G, Singh G, Jain D, Upadhyay SK, Dwivedi P, Khoo KS (2024b) Recent advancement of nano-biochar for the remediation of heavy metals and emerging contaminants: mechanism, adsorption kinetic model, plant growth and development. Environ Res 255:119136. https://doi.org/10.1016/j.envres.2024.119136

Prayogo C, Jones JE, Baeyens J, Bending GD (2014) Impact of biochar on mineralisation of C and N from soil and willow litter and its relationship with microbial community biomass and structure. Biol Fertil Soils 50:695–702

Qian L, Chen B, Hu D (2013) Effective alleviation of aluminumphytotoxicity by manure-derived biochar. Environ Sci Technol 47:2737–2745. https://doi.org/10.1021/es3047872

Qiu M, Liu L, Ling Q, Cai Y, Yu S, Wang S (2022) Biochar for the removal of contaminants from soil and water: a review. Biochar 4:19. https://doi.org/10.1007/s42773-022-00146-1

Razzaghi F, Obour PB, Arthur E (2020) Does biochar improve soil water retention? A systematic review and meta-analysis. Geoderma 361:114055. https://doi.org/10.1016/j.geoderma.114055

Rombola AG, Fabbri D, Baronti S, Vaccari FP, Genesio L, Miglietta F (2019) Changes in the pattern of polycyclic aromatic hydrocarbons in soil treated with biochar from a multiyear field experiment. Chemos 219:662–670

Rouphael Y, Rea E, Cardarelli M, Bitterlich M, Schwarz D, Colla G (2016) Can adverse effects of acidity and aluminum toxicity be alleviated by appropriate rootstock selection in cucumber? Front Plant Sci 7:1283

Saleem K, Asghar MA, Saleem MH (2022) Chrysotile-asbestos-induced damage in panicum virgatum and phleumpretense species and its alleviation by organic-soil amendment. Sustainability 4:10824. https://doi.org/10.3390/su141710824

Seneviratne M, Weerasundara L, Ok YS (2016) Phytotoxicity attenuation in Vignaradiata under heavy metal stress at the presence of biochar and N fixing bacteria. J Environ Manage 186:293–300. https://doi.org/10.1016/j.jenvman.07.024

Sinsabaugh RL, Lauber CL, Weintraub MN, Ahmed B, Allison SD, Crenshaw C (2008) Stoichiometry of soil enzyme activity at global scale. Ecol Lett 11:1252–1264

Song X, Li H, Song J, Chen W, Shi L (2022) Biochar/vermicompost promotes hybrid Pennisetum plant growth and soil enzyme activity in saline soils. Plant Physiol Biochem 183:96–110

Steinbeiss S, Gleixner G, Antonietti M (2009) Effect of biochar amendment on soil carbon balance and soil microbial activity. Soil Biol Biochem 41:1301–1310

Steiner C, Teixeira WG, Lehmann J, Nehls T, de Macedo JLV, Blum WEH (2007) Long term effects of manure, charcoal and mineral fertilization on crop production and fertility on a highly weathered central Amazonian upland soil. Plant Soil 291:275–290. https://doi.org/10.1007/s11104-007-9193-9

Sun J, He F, Pan Y, Zhang Z (2017) Effects of pyrolysis temperature and residence time on physicochemical properties of different biochar types. Acta Agric Scand, Soil Plant Sci 12–22. https://doi.org/10.1080/09064710.2016.1214745

Tan X, Liu Y, Zeng G, Wang X, Hu X, Gu Y (2015) Application of biochar for the removal of pollutants from aqueous solutions. Chemosphere 125:70–85

Tan S, Narayanan M, Thu Huong DT, Ito N, Unpaprom Y, Pugazhendhi A (2022) A perspective on the interaction between biochar and soil microbes: a way to regain soil eminence. Environ Res 214:113832. https://doi.org/10.1016/j.envres.113832

Verheijen FGA, Zhuravel A, Silva FC, Amaro A, Ben Hur M, Keizer JJ (2019) The influence of biochar particle size and concentration on bulk density and maximum water holding capacity

of sandy versus sandy loam soil in a column experiment. Geoderma 347:194–202. https://doi.org/10.1016/j.geoderma.2019.03.044
Vithanage M, Herath I, Almaroai YA, Rajapaksha AU, Huang L, Sung JK (2017) Effects of carbon nanotube and biochar on bioavailability of Pb, Cu and Sb in multi-metal contaminated soil. Environ Geochem Health 39:1409–1420
Wang J, Shaheen SM, Jing M, Anderson CWN, Swertz AC, Wang SL (2021) Mobilization, methylation, and demethylation of mercury in a Paddy soil under systematic redox changes. Environ Sci Technol 55:10133–10141
Wang D, Li C, Parikh SJ, Scow KM (2019) Impact of biochar on water retention of two agricultural soils-a multi-scale analysis. Geoderma 185–191. https://doi.org/10.1016/j.geoderma.01.012
Warnock DD, Lehmann J, Kuyper TW, Rillig MC (2007) Mycorrhizal responses to biochar in soil—concepts and mechanisms. Plant Soil 300:9–20. https://doi.org/10.1007/s11104-007-9391-5
Weber K, Quicker P (2018) Properties of biochar. Fuel 217:240–261. https://doi.org/10.1016/j.fuel.2017.12.054
Xing Y, Wang J, Kinder CES, Yang X, Slaný M, Wang B (2022) Rice hull biochar enhances the mobilization and methylation of mercury in a soil under changing redox conditions: Implication for Hg risks management in paddy fields. Environ Int 168:107484
Xu R, Xiao S, Yuan J, Zhao A (2011) Adsorption of methyl violet from aqueous solutions by the biochars derived from crop residues. Biores Technol 102:10293–10298
Xu RK, Zhao AZ, Yuan JH, Jiang J (2012) PH buffering capacity of acid soils from tropical and subtropical regions of China as influenced by incorporation of crop straw biochars. J Soils Sediments 12:494–502
Xu G, Zhang Y, Sun J, Shao H (2016) Negative interactive effects between biochar and phosphorus fertilization on phosphorus availability and plant yield in saline sodic soil. Sci Total Environ 568:910–915
Xu X, Huang H, Zhang Y, Xu Z, Cao X (2019) Biochar as both electron donor and electron shuttle for the reduction transformation of Cr(VI) during its sorption. Environ Pollut 244:423–430
Yan M, Li X, Liu Y, Li Y, He L, Zhang J (2022) Biochar enhanced soil aggregation and C-related enzyme activity in post-mining land on the Loess Plateau, China. Land Degrad Dev 1054–1061. https://doi.org/10.1002/ldr.4230
Yuan P, Wang J, Pan Y, Shen B, Wu C (2019) Review of biochar for the management of contaminated soil: preparation, application and prospect. Sci Total Environ 659:473–490. https://doi.org/10.1016/j.scitotenv.12.400
Zainab N, Khan AA, Azeem MA (2021) PGPR-mediated plant growth attributes and metal extraction ability of *Sesbaniasesban* L. in industrially contaminated soils. Agronomy 21:1820. https://doi.org/10.3390/agronomy11091820
Zeeshan M, Ahmad W, Hussain F, Ahamd W, Numan M, Shah M (2022) Phytostabalization of the heavy metals in the soil with biochar applications, the impact on chlorophyll, carotene, soil fertility and tomato crop yield. J Clean Prod 255:120318
Zhang X, Wang H, He L, Lu K, Sarmah A, Li J (2013) Using biochar for remediation of soils contaminated with heavy metals and organic pollutants. Environ Sci Pollut Res 20:8472–8483. https://doi.org/10.1007/s11356-013-1659-0
Zhang A, Li X, Xing J, Xu G (2020) Adsorption of potentially toxic elements in water by modified biochar: a review. J Environ Chem Eng 8:104196. https://doi.org/10.1016/j.jece.2020.104196
Zhou Y, Qin S, Verma S, Sar T, Sarsaiya S, Ravindran B (2021) Production and beneficial impact of biochar for environmental application: a comprehensive review. Biores Technol 337:125451. https://doi.org/10.1016/j.biortech.125451
Zielinska A, Oleszczuk P (2016) Bioavailability and bioaccessibility of polycyclic aromatic hydrocarbons (PAHs) in historically contaminated soils after lab incubation with sewage sludge-derived biochars. Chemosphere 163:480–489. https://doi.org/10.1016/j.chemosphere.08.072

Zimmerman AR, Gao B, Ahn MY (2011) Positive and negative carbon mineralization priming effects among a variety of biochar-amended soils. Soil Biol Biochem 43:1169–1179. https://doi.org/10.1016/j.soilbio.02.005

Sustainable Approach in Agriculture: Application of Biochar for Plant Growth Under Environmental Stress Condition

Aanchal Saklani, Shivam Attri, Kamal Kant Joshi, Avnish Chauhan, Muneesh Sethi, and Satyam Chaudhri

Abstract Agriculture is the backbone of any economy. Although agriculture is essential to the development of human civilization, increasing productivity frequently results in biodiversity loss and environmental deterioration. Particularly in developing countries like India, eco-friendly farming that prioritizes ecosystem services is becoming a practical approach to attaining sustainable livelihoods and global food security. Biochar is one such promising remedy as the stresses of soil erosion, climate change, and food insecurity increase. It is a carbon-rich substance made by pyrolyzing biomass. In addition to reducing pollution and improving soil health, it also promotes plant growth in stressful environments. The high porosity, alkalinity, and nutrient retention capacity of biochar improve microbial activity, soil fertility, and water-holding capacity. Additionally, it lowers nitrous oxide emissions by up to 50% and contributes significantly to carbon sequestration. Research indicates that soils treated with biochar can boost crop yields by as much as 30%. Notwithstanding these benefits, issues with cost-effectiveness, context-specific performance, and large-scale production continue to prevent widespread adoption. In order to create sustainable and climate-resilient food systems, this chapter examines the many uses of biochar, including improving soil health, reducing abiotic stress, and cleaning up pollutants. It also argues for its incorporation into national agricultural policies and circular bioeconomy frameworks.

Keywords Biochar · Sustainable agriculture · Soil fertility · Pollution remediation · Circular bioeconomy

A. Saklani · S. Attri (✉) · K. K. Joshi · A. Chauhan
Department of Environmental Sciences, Graphic Era Hill University, Dehradun, Uttarakhand, India
e-mail: ttrshvm@gmail.com

M. Sethi
Research and Development, COER University, Roorkee, Uttarakhand, India

S. Chaudhri
State Key Laboratory of Tibetan Plateau Earth System, Resources and Environment (TPESRE), Institute of Tibetan Plateau Research, Chinese Academy of Sciences, Beijing, China

A. Chauhan et al. (eds.), *Biochar As Remediation Tools for Pollutant Removal*,
https://doi.org/10.1007/978-3-032-16746-0_4

1 Introduction

Because environmental stress disrupts normal plant functions, it directly affects crop production, making it one of the largest challenges facing global agriculture (Šola et al. 2025). In addition to living (biotic) factors like pests and plant diseases, stress can also result from non-living (abiotic) factors like drought, flooding, high temperatures, radiation, and low soil nutrients. These conditions disrupt the growth processes of crops, including pollination, flower bud formation, protein production, and nutrient absorption. Consequently, yields are decreased. Abiotic stress alone is thought to be the cause of 51–82% of annual agricultural losses worldwide. Farmers can implement more effective strategies to sustain productivity in the face of climate change by developing a deeper understanding of how stress impacts crops (Oshunsanya et al. 2019).

Rising global temperatures, altered rainfall patterns, and an increase in the frequency of extreme weather events are some of the ways that climate change exacerbates these issues, which have an adverse effect on crop productivity and livestock health (Raza et al. 2024). Farming systems and the surrounding ecosystems are also under stress from problems like pest resistance, biodiversity loss, soil degradation, and water shortages. By emitting greenhouse gases that exacerbate climate change, agriculture itself exacerbates the issue. Small-scale farmers and vulnerable communities are further disadvantaged by these environmental pressures, which frequently result in erratic supply chains and volatile market prices. Global collaboration, cutting-edge technologies, and inclusive policies that give sustainability and resilience in agriculture top priority are all necessary to address these interrelated issues.

Historically, agriculture has been the foundation of many economies, particularly in developing nations. However, contemporary farming methods have drastically changed ecological systems, impacting soil health, carbon storage, and the nitrogen cycle (Adams et al. 2025). Organic farming, which stays away from synthetic pesticides, growth hormones, chemical fertilizers, and additives, is one promising substitute. This approach lessens environmental damage while simultaneously improving food quality. Traditional farming's drawbacks can be further addressed by combining organic methods with contemporary technology. By strengthening agricultural systems' resistance to climate change, such strategies enhance farmers' well-being while advancing the global objective of sustainable agriculture. For example, organic farming encourages conservation and responsible land management, strengthens ecosystems against drought and temperature changes, and lessens soil erosion (Gamage et al. 2023; Attri et al. 2025). Additionally, it is typically less expensive, which helps farmers adjust to climate uncertainty.

In addition to organic approaches, biochar has recently attracted great interest as a sustainable solution. Biochar is a carbon-rich product made from materials such as crop residues, animal waste, or municipal organic waste. Its unique properties such as high carbon content, large surface area, stability, and high cation exchange capacity make it highly useful for improving soil fertility and environmental management (Sharma et al. 2025). Its characteristics largely depend on the type of feedstock used and the production process, which may involve pyrolysis, gasification, or hydrothermal carbonization. Biochar can also be modified using acids, alkalis, oxidizing agents, or gases to suit specific applications (Dong et al. 2025). These technological developments highlight how innovative materials can contribute to sustainable farming practices and mitigate the environmental impacts of agriculture. Applications for biochar include removing contaminants from contaminated soil and water habitats and as a soil additive for crop cultivation. However, extensive laboratory-scale research has been done on how the addition of biochar affects the bulk density, porosity, water-holding capacity, acidity, and mineral content of the soils. Numerous research studies have looked into the characteristics of biochar made from different feedstocks under different pyrolysis circumstances in order to assess the impact of residence time, heating rate, and peak operating temperature. To encourage the commercialization of biochar for large-scale applications, it is essential to thoroughly review the technical, environmental, economic, and social benefits and drawbacks of biochar for various soil applications (such as soil amendment and/ or decontamination) (Kamali et al. 2022).

2 Biochar: Properties and Mechanisms of Action

Although the productivity of agriculture is still restrained by soil erosion and nutrient leaching that causes hunger in the world (Jones and Healey 2010), the problem is compounded by the growing manifestations of these events and a high population density as well as urban sprawl (Lal 2009). In this context, there is an urgent need to re-evaluate structures and management of agroecosystems in order to advance in solving trade-off issues like nutrient imbalances, steep water inefficiency, and the supply–demand emerging problems of agricultural products (Lal 2013). A soil-friendly way of tackling this problem is through the restoration of organic nutrients in soil, which helps in maintaining organic matter and shows direct and positive outcomes on both physical as well as chemical properties of soil (Girmay et al. 2008). Organic matter sources can vary from near (e.g., plant residues) to far (animal manure, household waste, and industrial byproduct) (Ali et al. 2011). Zha et al., who hypothesized that biochar is an important soil amendment that can promote soil fertility, and aid in the stabilization of heavy metals and pollutants in soil as well (Zhuet al. 2017). Biochar generally refers to "biomass-derived black carbon" or "charcoal" with potential to act as a carbon sink over time (Lehmann et al. 2006). First reported in the pre-Columbian Amazonia (500–9000 years BP) as the use of soil amendments that gave rise to "terra preta" with more fertile high pH soils from the acidic and less fertile soils of the region (Lehman et al. 2007), biochar has also been referred to as "agrichar" due to its "combined use to improve agricultural production and address climate issues" (Abewa et al. 2014; Laine 2012; Yao et al. 2012). Some studies that have highlighted the improvement of soil properties associated with the use of biochar have gone beyond the soil carbon sequestration aspects of biochar application, but the extent of these effects seems to be dependent on the type of feedstock and temperature used for pyrolysis (Mukherjee and Zimmerman 2013). Overall, several studies have pointed out that several chemical properties of soils, including soil pH, cation exchange capacity (CEC), organic carbon (OC), electrical conductivity (EC), and extractable nutrients were enhanced by a range of biochars (Wang et al. 2013). Similarly, different surface characteristics including bulk densities and many aspects of soil structure were changed by soil amendments (i.e., biochar) (Blanco-Canqui 2017). Furthermore, it was stated that biochar amendments could improve soil water retention capacity (Razzaghi et al. 2020) and change soil composition, quantity, and microbial community activity in soils (Antonangelo et al. 2025) that provide rich aeration with water giving the surrounding similar characteristics of both (Ameloot et al. 2013). The application of biochar to soil is important for changing microbial communities (Ullah et al. 2020), which may fundamentally vary soil functions like organic matter development or aging and nutrient cycling. These modifications in microbial operation can either directly or indirectly be conducive to soil fitness for plant growth, either by possessing properties in physicochemical soil or by biological factors (Blanco-Canqui 2017). The role of soils hydraulic and structural qualities is significant in functions such as root penetration and nutrient availability, soil aeration, and compaction; water absorption, control of erosion, movement of water, heat, and gas from large scale (flyover) to micro scale of soil properties (directly). Furthermore, biochar application can improve plant growth in acidic soils by altering the soil pH (Syuhada et al. 2016).

2.1 Biochar Physical Properties

Biochar, a carbon-rich material formed through the pyrolysis of organic biomass, has proven to be highly beneficial in helping plants withstand environmental stresses like salinity, drought, and heavy metal contamination. This is largely due to its distinctive physical and chemical characteristics. For instance, biochar is highly porous and has a large surface area, which means it can hold water and air more effectively in the soil, crucial when water is limited or when salt levels are high. Its ability to retain and slowly release essential nutrients such as potassium, calcium, and magnesium (thanks to its high cation exchange capacity) helps maintain plant nutrition even under tough conditions. Moreover, biochar tends to be alkaline, which can help balance out acidic soils and neutralize some of the negative effects of salt and toxic metals. Its surface is also rich in reactive groups like carboxyl and hydroxyl, which can bind and trap heavy metals, keeping them from being absorbed by plant roots. In addition to these chemical interactions, biochar also improves soil structure, making it looser and crumblier, which helps roots grow more easily and allows water to move more freely. These physical improvements create a better environment for beneficial microbes to thrive, especially those that support plant health by improving nutrient uptake, regulating stress hormones, and activating genes that help the plant cope with harsh conditions.

But what really makes biochar unique is how its properties can be fine-tuned depending on how it's made. For example, when looking at how biochar interacts with soil, its physical structures like surface area, pore size, and total pore volume play a major role. Research has shown that higher pyrolysis temperatures generally lead to biochar with a larger surface area and more pores, especially when woody materials are used as the starting biomass. A great example comes from studies on Douglas fir (*Pseudotsuga menziesii*), where wood-derived biochar reached an impressive surface area of 1103.6 square meters per gram. That's significantly higher than what's found in many other types of biochar. Its bark and root also showed high surface areas, 1025.4 and 967.1 m^2/g, respectively. Interestingly, this increase in surface area becomes more noticeable as the pyrolysis temperature rises, showing just how sensitive biochar's performance is to its production conditions. When it comes to porosity, Douglas fir biochar again stood out, with an interior porosity of 83% at a relatively low temperature of 370 °C over 2 h, outperforming even miscanthus-based biochar, which reached 79% under more intense conditions (450 °C for 1 h). These physical characteristics make biochar a highly effective amendment for improving soil health, enhancing water and nutrient dynamics, locking away harmful contaminants, and ultimately supporting stronger, healthier plants that are better able to survive in difficult environments (Table 1).

2.2 *Biochar Chemical Properties*

The quality characteristics of biochar are controlled by various technological factors such as the temperature of pyrolysis, the nature of the type of material used, the time of heat treatment, and the rate of heating. These factors determine the values of pH, the ash content, and the amount of carbon and macronutrients in the final product (Fu et al. 2011). In all cases, an increment in the temperature of processing leads to a decrease in the amount of biochar produced when the material is changed. Among the materials under study, it is dry sewage sludge that gives the highest biochar yield when the temperature changes in a certain way (Zhang et al. 2015). At temperatures ranging from 300–450 °C, the highest yield of all materials, such as date palm, wheat straw, rice straw, and eucalypts, was shown (Fernandes et al. 2020; Naeem et al. 2014; Usman et al. 2015; Table 2) and (Fig. 1).

3 Environmental Stress Conditions in Agriculture

The IPCC's statement that food prices will increase by 2050 because of climate change's dramatic impact on crop yields is undeniable. As a result of high CO_2 levels, expected increased temperatures and crop yields. Changes in the amount

Table 1 Biochar physical properties under different stress conditions

Biochar feedstock	Pyrolysis temperature °C	Bulk density g cm^{-3}	Particle density g cm^{-3}	Porosity	Type of porosity%	SSA BET m^2 g^{-1}	Heating rate °C min^{-1}	Residence time h	Location
Douglas Fir	370	0.262	1.55	83	Intra porosity	153	NA	1	Western Oregon, USA
	500	0.236	1.57	84.9	[1-bulk density/ particle density) × 100]	229	NA	1	
	620	0.254	1.71	85.1		280	NA	1	
Hazelnut shells	370	0.554	1.48	62.5	Intra porosity	58.7	NA	1	Western Oregon, USA
	500	0.557	1.57	64.5	[1-bulk density/ particle density) × 100]	161	NA	1	
	620	0.505	1.64	69.2		211	NA	1	
Maize stalk	400	0.12	0.35	60.2	Intra porosity	12.9	10	4	Sikkim, India
	500	0.12	0.37	62.7	[1-bulk density/ particle density) × 100]	23.2			
	600	0.13	0.46	65.9		43.9			

(continued)

Table 1 (continued)

Biochar feedstock	Pyrolysis temperature °C	Bulk density g cm^{-3}	Particle density g cm^{-3}	Porosity	Type of porosity%	SSA BET m^2 g^{-1}	Heating rate °C min^{-1}	Residence time h	Location
Lantana camara	400	0.15	0.4	53.18	Intra porosity	12.7	10	4	Sikkim, India
	500	0.15	0.42	55.68	[1-bulk density/ particle density) × 100]	22.1			
	600	0.16	0.51	58.88		41.4			
Pine needles	400	0.14	0.43	46.96	Intra porosity	12.5	10	4	Sikkim, India
	500	0.14	0.45	49.46	[1-bulk density/ particle density) × 100]	21.8			
	600	0.15	0.54	56.6		40.2			
Black gram crop	400	0.13	0.37	48.87	Intra porosity	12.1	10	4	Sikkim, India
	500	0.13	0.39	51.37	[1-bulk density/ particle density) × 100]	20.4			
	600	0.14	0.48	54.57		38.9			

(continued)

Table 1 (continued)

Biochar feedstock	Pyrolysis temperature °C	Bulk density g cm^{-3}	Particle density g cm^{-3}	Porosity	Type of porosity%	SSA BET m^2 g^{-1}	Heating rate °C min^{-1}	Residence time h	Location
Sewage waste	400	1.22	1.76	31	Intra porosity	N/A	5	4	
Pine	400	0.41	1.45	72	[1-bulk density/ particle density) × 100]	N/A	5	4	Texas, USA
Mesquite	400	0.52	1.45	64		N/A	5	4	
Miscanthus	400	0.32	1.51	79		N/A	5	4	

SSA: specific surface area (SSA) m^2 g^{-1}; N/A: not available (Gray et al., 2014; Liu et al., 2018)

Table 2 Biochar chemical properties under different pyrolysis conditions

Feedstock Type	PT °C	C%	Yield%	AC%	O/C	H/C	pH	EC (dS m^{-1})	TN%	TP%	TK%	RT min	HR °C min^{-1}	Location
Rice straw	300	45.2	45.0	31.0	N/A	N/A	8.0	2.35	1.15	0.11	3.6	20.0	– 3.0	Faisalabad Pakistan
Rice straw	400	55.5	36.0	40.0	N/A	N/A	9.7	2.85	0.98	0.13	4.1	20.0	– 3.0	Faisalabad Pakistan
Rice straw	500	63.0	32.0	52.0	N/A	N/A	10.4	3.35	0.85	0.14	4.8	20.0	– 3.0	Faisalabad Pakistan
Wheat straw	300	51.7	48.0	25.0	N/A	N/A	7.7	1.78	1.38	0.26	3.0	20.0	– 3.0	Faisalabad Pakistan
Wheat straw	400	62.0	38.0	37.0	N/A	N/A	8.8	2.15	0.94	0.3	3.4	20.0	– 3.0	Faisalabad Pakistan
Wheat straw	500	66.2	33.0	46.0	N/A	N/A	9.4	2.06	0.85	0.34	3.6	20.0	– 3.0	Faisalabad Pakistan
Date palm	300	57.99	49.97	14.42	0.27	0.84	8.32	3.26	0.54	N/A	N/A	240.0	5.0	Riyadh, Saudi Arabia
Date palm	400	66.87	36.54	16.34	0.13	0.63	9.25	3.28	0.45	N/A	N/A	240.0	5.0	Riyadh, Saudi Arabia
Date palm	500	72.3	32.38	19.68	0.05	0.36	9.59	3.5	0.41	N/A	N/A	240.0	5.0	Riyadh, Saudi Arabia
Date palm	600	72.89	30.88	20.71	0.03	0.28	9.57	3.59	0.39	N/A	N/A	240.0	5.0	Riyadh, Saudi Arabia
Dry sewage sludge	300	37.15	64.28	37.42	0.27	1.41	6.2	3.3	6.17	1.043	0.225	0.225	nan	Shanghai, China
Dry sewage sludge	400	35.02	56.55	49.17	0.15	1.02	7.5	0.4	4.96	1.156	0.248	0.248	nan	Shanghai, China

(continued)

Table 2 (continued)

Feedstock Type	PT °C	C%	Yield%	AC%	O/C	H/C	pH	EC (dS m^{-1})	TN%	TP%	TK%	RT min	HR °C min^{-1}	Location
Dry sewage sludge	500	30.08	55.26	57.42	0.15	0.67	8.1	0.8	3.54	1.663	0.275	0.275	nan	Shanghai, China
Dry sewage sludge	600	26.48	53.41	64.23	0.14	0.45	10.2	1.54	1.83	2.083	0.283	0.283	nan	Shanghai, China
Dry sewage sludge	700	27.23	46.66	66.66	0.05	0.39	11.9	2.001	2.001	2.001	0.291	0.291	nan	Shanghai, China
Eucalyptus wood	450	74.96	42.76	0.6	0.25	0.05	N/A	0.12	1.25	0.016	0.182	60.0	1.7	São Paulo, Brazil
Eucalyptus wood	550	79.25	38.0	0.74	0.22	0.06	N/A	0.09	1.49	0.016	0.146	60.0	1.7	São Paulo, Brazil
Eucalyptus wood	650	87.06	34.69	0.63	0.12	0.018	N/A	0.04	1.0	0.023	0.164	60.0	1.7	São Paulo, Brazil
Eucalyptus wood	750	87.74	34.27	0.84	0.11	0.019	N/A	0.16	1.06	0.017	0.21	60.0	1.7	São Paulo, Brazil

PT: pyrolysis temperature; AC: ash content; RT: residence time (minutes); C%: carbon content; N/A: not available; HR: heating rate (°C minute). (Fernandes et al., 2020; Naeem et al., 2014; Usman et al., 2015)

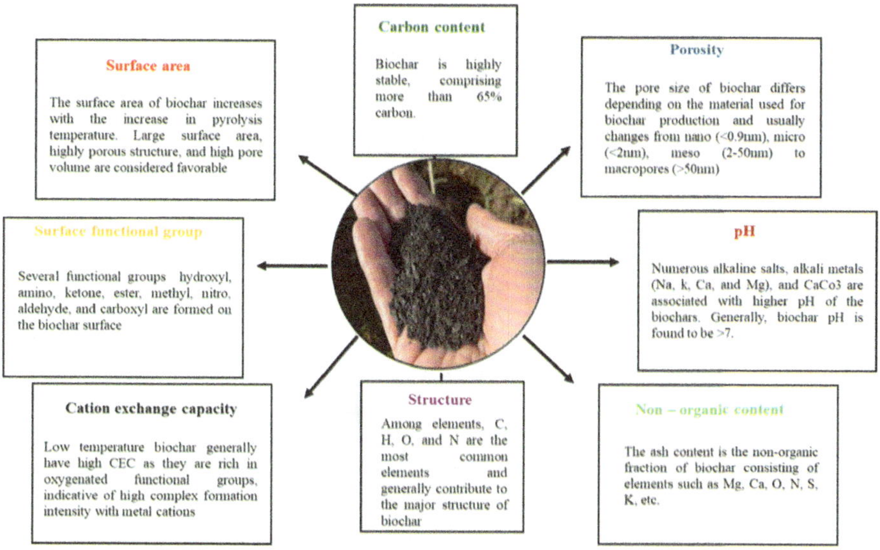

Fig. 1 Physicochemical properties of biochar

and distribution of rainfall are expected, leading to water scarcity and elevation. Excessive precipitation can lead to crop damage and erosion, flooding, while too little rainfall can result in drought and impact crop yields negatively. Drought is considered a major abiotic stressor that directly affects the yield, development, and quality of crops, and as the global population grows, it, like other problems, is only expected to become more severe (Chowdhary and Sharma 2017)., Waterlogging, on the other hand, creates an anaerobic condition in which the oxygen in the soil is reduced due to the water filling these spaces and displacing the gases in a plant's body, which wreaks havoc on its chemical and physiological activities (Hartman and Tringe 2019). It has been demonstrated in various experimental studies that flooding plants can have a same effect on plant ethylene manufacture, where it is significantly increased (Tamang and Fukao 2015). Drought and waterlogging are two of the most severe abiotic stresses that limit crop production and threaten global food security (Bin Rahman and Zhang 2016).

4 Factors Affecting Soil Microbiome

The any microorganism population participates in the determination of the fertility status and health of the soil. The important chemical and physical characteristics of the soil can be easily altered by the external factors like the drought, the floods, and chemicals, respectively, which in turn influence the biological activity of the soil. Even though the characteristics of the soil are determinative of the microbial community, numerous biotic and abiotic variables, including parent material, farming techniques, climate change, and current land use, have an impact on these qualities. The investigation of the microbial community of the soil is challenging because they are vulnerable to abiotic stresses like changes in pH of soil, salinity, UV radiation, temperature, as well as rainfall, which in turn results in changes in the microbial community.

4.1 Drought

The intensity of a drought is identified as a significant constraint to agriculture and is noted to be the most damaging climatic event affecting global food security. Soil moisture regime microbial communities and activity as well as diversity (Denardin et al. 2020; Flórián et al. 2019) were major determinants. Indeed, high temperature is accompanied with droughts causing detrimental impact on soil microbial biomass (Flórián et al. 2019) and populations (Sheik et al. 2011). This can result in the soil moisture under drought conditions to be depleted below the level sufficient for microbial activity (Liu et al. 2009). Less labile carbon and nitrogen in the root surface will result in the receding of *Verrucomicrobia, Proteobacteria, and Acidobacteria* during drought (Castro et al. 2019; Keiblinger et al. 2010). These phyla are known to be sensitive to nitrogen imbalances and are heterotrophic. In addition, a drought results in a reduction of almost 30% in microbial respiration levels due to low soil moisture, and differences between the estimates of C immobilization and N mineralization suggest disturbances in microbial cellular functions (Schimel 2018). The warming associated with droughts can cause a decrease in the abundance of 16S rRNA genes in soil microbes (Meisner et al. 2018).

Biochar enhances soil moisture retention and plant drought tolerance. Uzoma et al. (2011) found that cow manure biochar applied at 10 t/ha improved maize yield by 140% under drought conditions, primarily due to improved water-holding capacity. Mukherjee and Lal (2013) further confirmed that biochar increased soil water retention by 18–37%, depending on the feedstock and pyrolysis conditions, delaying drought stress symptoms in plants (Table 3).

4.2 Heavy Metals

The use of the land, such as over-reliance on agrochemicals, greatly affects the soil microbial community and their functional richness, leading to soil quality deterioration and raising concern on how intensive use of land and pesticide can affect both human and the environment. Soil characteristics like temperature, pH are important drivers when it comes to the effect of the microbial community, and it is known that the unique metals exert a great impact on the microbial community. These metals are released to the physical environment through natural weathering of the mineral

Table 3 Role of biochar in lessening salinity, drought, heavy metal, and temperature stress

Stress type	Biochar type	Crop/system	Key observations	Reference
Salinity	Wheat straw (5% w/w)	Maize	↑ Growth (24%), ↓ Na^+ (34%)	Akhtar et al. 2015
	Peanut hull	Barley	↓ Soil salinity, ↑ germination	Yuan et al. 2011
Drought	Cow manure (10 t/ha)	Maize	↑ Yield by 140%, ↑ water retention	Uzoma et al. 2011
	Wood biochar	General soils	↑ Water retention (18–37%)	Mukherjee and Lal 2013
Heavy metals	Green waste (5% w/w)	Contaminated soil	↓ Pb (90%), ↓ Cd (70%)	Beesley et al. 2011
	Rice husk	Rice	↓ Pb uptake (65%), ↑ biomass (30%)	Ahmad et al. 2014
Temperature extremes	Pine wood biochar	Lettuce	↑ Photosynthesis, ↓ leaf temp	Thomas and Gale (2015)
	General biochar	Cold-prone soils	Thermal buffering, ↑ root performance	Joseph et al. 2010

rocks, which release metal in large quantities into the environment, but human activities like mining, industrial processes, fertilizer production, and wastewater discharge contribute greatly to the excessive deposition of metals into the soil. Too much of these fertilizers result in an environmental issue such as greenhouse gas release, too much of phosphorous loss, which increases the risk of lakes and rivers being affected by heavy emissions of industrial materials (Yang et al. 2021; Wang et al. 2016; Guo et al. 2010). While trace elements such as zinc (Zn), iron (Fe), and manganese (Mn) are indispensable for growth, development, and metabolic activities in trace amount, some heavy metals like cadmium (Cd), lead (Pb), chromium (Cr), and mercury (Hg) possess the ability to disturb biological functions. These metals are able to injure cell membranes, enzyme activities and human tissues such as DNA (Abdu et al. 2017). For example, supplementation of the growth medium with Zn has induced the deformity of *Gluconacetobacterdiazotrophicus* and generated pleomorphic cells associated with aggregates that could interfere with the chelation of Zn by the cell (Reis et al. 2015). The heavy metal cadmium (Cd), poison of soil, disturbs enzyme activities damages membrane structure and interference with enzyme synthesis in cells (Zhang et al. 2018; Martín-Folgar and Martínez-Guitarte 2017).

Heavy metals in soil are harmful to both crops and human health. Biochar, however, can act as a filter by locking these metals away and preventing plants from absorbing them. In one study, adding green-waste biochar reduced the bioavailability of lead by up to 90% and cadmium by 70%. Another experiment using rice husk biochar showed that rice plants absorbed 65% less lead and produced 30%

more biomass. These results suggest that biochar not only reduces toxicity but also improves overall plant growth in contaminated soils (Table 3).

4.3 Salinity

Today one of the greatest problems of soil degradation is salinization across the globe. Salinity from agricultural input use is a source for some forms of degradation in agro-systems accordingly. The application of sewage sludge, manure and city garden waste for several times may promote the salts build-ups in soil (Ding et al. 2020). Salinity (high osmotic pressure, toxic ions and nutrient unbalances Iqbal et al. 2020) adversely affects the growth of plants and soil microbial activity. That is because as it impairs the water relationships within plant tissues leading to nutrient and ion imbalance, inducing toxicity with accumulation of chloride (Cl^-) or sodium (Na^+) in both plants, soil, Otlewska et al. 2020; Evelin et al. 2019; Kamran et al. 2019; Hussain et al. 2019). There is soil salinity (content of salt in soil), which affects the soil processes, osmotic pressure of soil and the sodium fraction in exchange complex (sodicity) that defines structural stability of the soil system. Sodicity would accretion from salinity over time. Banerjee and coworkers showed that the presence of soluble salts could reduce the ground water solute potential (making it more negative), thereby draining hydration out of cells and potentially kill microorganisms or roots (Yan et al. 2015). An increase in the salinity triggers the theory of "immediate osmotic phase" or in other words osmotic stress initiates water withdrawal in soil in just minutes which is followed by "slow ion toxicity phase" or hyperosmotic stress (Ransom & Bowen Jr). The one of this is defined as high level of contaminated ions that decreases cell separation and growth rate, very difficult for roots as well as root associated microbial community (Gamalero et al. 2020).

Biochar significantly mitigates the effects of salinity by improving soil structure and reducing sodium uptake. For example, Akhtar et al. (2015) reported that applying 5% wheat straw biochar in saline soils increased maize plant height by 24% and reduced Na^+ uptake by 34%. Similarly, Yuan et al. (2011) observed improved barley germination and growth under saline irrigation using peanut hull biochar, attributing the benefits to Na^+ adsorption and improved K^+/Na^+ ratios (Table 3).

4.4 Temperature Extremes

Global temperature fluctuations are because of changes in climate; CO_2 level as well as precipitation patterns. Soil temperature has the effect of affecting colonization by Arbuscular Mycorrhizal Fungi under a range having different C:N ratio. Research conducted by Frater et al. (2018) reveal that Arbuscular Mycorrhizal Fungi colonization increased concomitantly with increase temperature and pH where it was done in Central United State; however, Goicoechea (2020) discussed that going by to other Mediterranean decreased whenever the temperature grew. Increase in temperature seems to assist Arbuscular mycorrhizal fungi via putting plant roots elongation allowing better root colonization, though at lower temperature nutrients acquisition by Arbuscular mycorrhizal fungi decreased showing in reduced mycorrhizal colonization (Frater et al. 2018; Jerbi et al. 2020). Thus, AMF of different strains have different optima for growth and development.

Biochar buffers soil against extreme temperatures by improving moisture and root health. Thomas and Gale (2015) observed increased photosynthesis and reduced leaf temperature in lettuce grown in heat-stressed conditions using pine wood biochar. Joseph et al. (2010) found that biochar-insulated soils had lower temperature fluctuations, supporting better root growth in cooler environments (Table 3) and (Fig. 2).

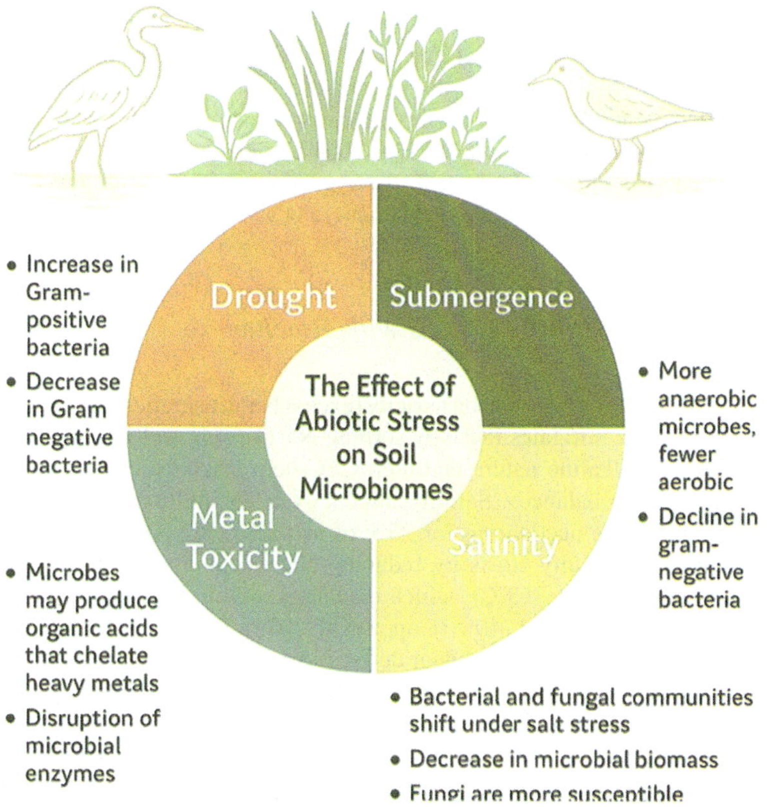

Fig. 2 Impact of alterations in environmental factors on the composition, health, and overall well-being of soil microorganisms

5 Application of Biochar for Plant Growth Under Stress Conditions

5.1 Biochar and Drought Tolerance

Biochar amendments are widely reported to improve plant drought resilience by improving soil water retention and plant water-use effectiveness. Due to its porous structure and high CEC, biochar increases soil moisture-holding capacity and aggregate stability (Chi et al. 2024). For example, maize under deficit irrigation with 5 $t{\cdot}ha^{-1}$ biochar showed 26% higher grain yield and 33% higher water productivity compared to no-biochar controls (Naeem et al. 2024). For instance, in northwestern China, biochar application reduced irrigation needs by 15–20% while boosting maize yields by 30% under water-scarce conditions. Meta-analyses and reviews confirm that biochar generally enhances photosynthesis, leaf water potential, and growth of droughty crops (Rathinapriya et al. 2025). In South Asia, field trials in Nepal found that repeated biochar application significantly raised maize yields under rainfed (water-limited) conditions by improving soil moisture and nutrients (Vista and Pandit 2023). Regional studies in South and East Asia report that biochar-amended crops show stronger drought tolerance (e.g., through higher stomatal conductance and chlorophyll) than untreated plants (Rathinapriya et al. 2025).

5.2 Mitigating Salinity Effects with Biochar

Salinity stress impairs plants via ion toxicity (excess Na^+), osmotic stress, and nutrient imbalance. Biochar mitigates these by sorbing Na^+, raising soil Ph, and improving soil structure. Its alkaline nature and high CEC help leach excess salts and maintain favorable ionic balances. Salinity affects over 900 million hectares globally, with China and India accounting for 25% of salt-affected soils (Lee et al. 2022). Biochar alleviates salinity stress by reducing Na^+ accumulation through its high cation exchange capacity (CEC), which displaces sodium ions and decreases soil electrical conductivity by 4.1–40% (Coppa et al. 2024). Biochar also often increases soil microbial activity and antioxidant defenses under salinity. Studies from China and India report that in saline soils, biochar-treated plants have higher chlorophyll and osmolyte levels than untreated plants. One pot study found biochar-amended tomato seedlings maintained 24% greater leaf water content under combined salt and drought stress than controls (Zhang et al. 2023). In the Yellow River Delta of China, applying biochar (6% w/w) reduced soil salinity by 35% and increased rice yields by 22% (Lee et al. 2022). Similarly, in Gujarat, India, biochar combined with gypsum lowered soil pH from 9.2 to 7.8, restoring fertility in sodic soils. Biochar's alkaline nature and porous structure enhance salt leaching and root aeration, particularly in coastal regions of Bangladesh and Nepal. However, excessive biochar application can exacerbate surface salinity due to capillary action, necessitating optimized dosing.

These regional studies highlight biochar's potential to reclaim saline soils while balancing nutrient uptake, such as increasing the K^+/Na^+ ratio by 21.1%, critical for maintaining metabolic functions in crops like rice (Coppa et al. 2024).

5.3 *Biochar in Heavy Metal Immobilization*

Biochar effectively immobilizes toxic heavy metals in contaminated soils, reducing plant uptake. Its large surface area and functional groups adsorb metals (e.g., Cd, Pb, Zn, Cu) and convert them to less soluble fractions (Awad et al. 2021). Long-term Chinese field trial revealed that wheat straw biochar (40 $t{\cdot}ha^{-1}$) compacts the rice Cd concentration by ~ 32% over three years (Sun et al. 2023). Pot experiments similarly show that adding garden-waste biochar (400 °C pyrolysis) at 6% rate can lower soil extractable Pb, Cd, Zn, and Cu by tens of percent. In one study, 6% biochar cut TCLP-extractable Pb by ~ 68% and Cd by ~ 50% compared to unamended soil (Awad et al., 2021). High pyrolysis temperatures (500–700 °C) enhance surface functional groups (−COOH, −OH), enabling chemisorption of metals. In Nepal's Kathmandu Valley, rice husk biochar reduced Cd uptake in rice grains by 60% (Ghandali et al. 2024). Modified biochars, such as MnO_2- and ZnO-nanoparticle-enriched variants, further enhance immobilization efficiency by 40–50% through stronger electrostatic interactions and precipitation of metal hydroxides (Ghandali et al. 2024).

5.4 *Role in pH Regulation and Nutrient Uptake*

By virtue of its alkalinity and cation exchange capacity, biochar often raises soil pH, especially in acidic soils, thereby improving nutrient availability. Numerous trials in Nepal's acidic hill soils have demonstrated that biochar can shift pH into a more neutral range. For example, Vista and Pandit (2023) showed that applying 15–50 $t{\cdot}ha^{-1}$ biochar to pH 4.6–4.9 soils raised the pH to 5.0–5.6. In China's acidic red soils, biochar increased wheat yields by 25% by elevating pH and available phosphorus (Hou et al. 2022). Biochar's slow nutrient release, particularly potassium, calcium, and magnesium from its ash content, reduces fertilizer dependency in nutrient-poor soils of Nepal 5. However, in neutral soils, biochar may reduce micronutrient uptake (e.g., Cu, Fe, Zn), necessitating feedstock-specific applications (Rodríguez et al., 2022). Overall, biochar's ability to stabilize pH and reduce nutrient leaching enhances plant-available N, P, and K. It improved nutrient homeostasis under stress: drought or salt-stressed crops with biochar show higher tissue N, P, K concentrations than those without.

5.5 *Biochar and Plant Hormonal Modulation*

Emerging evidence indicates that biochar can influence plant hormone levels and signaling under stress. In drought and salinity conditions, plants naturally accumulate abscisic acid (ABA) to regulate stomata. Biochar amendments have been observed to modulate this hormonal response: for instance, in maize under salinity/drought, biochar-treated plants showed significantly lower ABA in leaves and xylem than controls (Zhang et al. 2023). By lowering stress-induced ABA, biochar may help maintain stomatal opening and photosynthesis under moderate stress. Transcriptomic studies also show that biochar exposure upregulates growth-promoting hormones: auxin, cytokinins, and gibberellins are systemically increased in biochar-amended tomato roots and shoots. This is consistent with reports of larger root systems and higher cytokinin activity in biochar-grown plants. Additionally, biochar's enhancement of beneficial microbes (PGPR) can indirectly boost plant hormones: many PGP bacteria produce IAA or cytokinins, further stimulating plant growth. In practical terms, combined treatments of biochar with gibberellic acid (GA_3) have shown synergistic effects: e.g., maize under cadmium stress had 86% higher germination with BC + GA_3 than control (Anwar et al. 2024). Biochar appears to tilt the hormonal balance toward growth even under stress, by tempering ABA-mediated closure and enhancing auxin/GA-driven growth pathways (Zhang et al. 2023).

6 Case Studies

6.1 *Punjab: Mitigating Crop Residue Burning Through Biochar Production*

In 2018, the Punjab Agricultural University, in collaboration with the Department of Agriculture, launched a biochar initiative aimed at curbing the widespread practice of rice straw burning. Medium-scale pyrolysis units (2–5 tons/day capacity) were established in key agricultural districts, utilizing primarily rice straw and wheat residue. Farmer cooperatives managed biomass collection and received compensation. Biochar was applied to wheat fields at a rate of 5–10 tons per hectare. The interference led to a 30% lessening in crop residue burning, a 15–20% growth in wheat yields, and sequestered approximately 3.5 tons CO_2 equivalent per hectare. Over 200 local jobs were created in biomass logistics and processing.

6.2 *Tamil Nadu: Valorization of Coconut Waste Through Cooperative Biochar Systems*

In 2019, the Tamil Nadu Agricultural University partnered with coconut producer companies to convert coconut shells and fronds into biochar using continuous slow pyrolysis with heat recovery systems. Operated through a producer-owned cooperative model, the project integrated biochar into organic farming and marketed soil amendments under the brand "CocoChar." Annually, 12,000 tons of coconut waste were processed, leading to a 22% yield increase in vegetables and annual revenues of ₹3.5 crore (approximately US$470,000). The initiative also contributed to a reduction of 5600 tons CO_2 equivalent in greenhouse gas emissions.

6.3 *Gujarat: Industrial Symbiosis Between Sugar Mills and Agriculture*

The Gujarat Cleaner Production Centre initiated a project in 2020 that repurposed sugar mill boilers to co-produce biochar from sugarcane bagasse and trash. Pyrolysis gases were captured for internal power generation. Biochar was subsidized and distributed to contracted sugarcane farmers. The project transformed 50,000 tons of waste annually, improved soil organic carbon from 0.4 to 0.9%, and increased sugarcane yields by 18%. Soil water retention also improved, decreasing irrigation needs by 15%, while participating sugar mills experienced energy savings of approximately 20%.

6.4 *Uttarakhand: Community-Based Pine Needle Biochar for Forest Fire Mitigation*

In 2017, the Uttarakhand Renewable Energy Development Agency (UREDA) initiated a decentralized biochar production model to address pine needle accumulation and forest fire risks. Village cooperatives operated small-scale kilns using pine needles, supported by capacity-building efforts and market development for biochar-based soil amendments and filtration products. The initiative reduced forest fires by 40%, processed 8000 tons of biomass annually, and supported income generation for 1200 rural households. Estimated carbon sequestration was 2900 tons CO_2 equivalent per year.

6.5 Jharkhand: Biochar Integration in Tribal Agriculture

Supported by the Indian Council of Agricultural Research (ICAR), the Jharkhand State Livelihood Promotion Society initiated a tribal-focused biochar project in 2019. Using improved kilns adapted from traditional charcoal methods, tribal communities processed forest biomass and crop residues. Biochar application, combined with indigenous knowledge, restored acidic soils (raising pH from 4.5–5.0 to 5.5–6.0) and boosted millet and pulse production by 25%. The project benefited 3500 tribal families, enhanced traditional ecological practices, and reduced chemical fertilizer dependency by 30%.

6.6 Karnataka: Urban Green Waste to Biochar Model

In 2021, the Bangalore Municipal Corporation, in partnership with the Indian Institute of Science, implemented a decentralized system to convert urban garden waste into biochar using containerized pyrolysis units. Located at waste transfer stations, these units processed 18 tons of garden waste daily. The resulting biochar was utilized in city parks and gardens, integrated with rainwater harvesting systems. The intervention improved soil water retention, reduced municipal landscaping water use by 25%, and increased urban tree survival rates from 60 to 85%.

7 Biochar Integration with Other Sustainable Techniques

7.1 Biochar with Organic Amendments and Compost

Integrating biochar with organic soil amendments (compost, farmyard manure, vermicompost, etc.) can synergistically improve soil fertility and crop performance under stress. Organic amendments supply a broad spectrum of nutrients (N, P, K, micronutrients) and labile carbon, while biochar's high surface area and porosity adsorb and gradually release these nutrients and retain moisture. This co-application often leads to higher yields and better economic returns than either amendment alone (Sharma et al. 2021). Applying biochar together with organic amendments (e.g., manure, compost) often yields synergistic benefits. Biochar's highly porous carbon matrix adsorbs nutrients and water from the compost, reducing leaching and prolonging nutrient availability (Nobile et al. 2022). For example, in the Indian subtropics, adding 2 t ha^{-1} biochar to a 50:50 mix of farmyard manure and vermicompost significantly boosted soil water retention and knolkhol (Brassica gongyloides) yield. Biochar alone raised yield ~ 7.8% above the no-biochar control, and the biochar + compost treatment produced 6–12% higher yield than compost alone. This dual amendment also improved infiltration (by 4–38%), aggregate stability, and hydraulic conductivity relative to sole compost (Sharma et al. 2021).

Likewise, European and other field trials show compost–biochar mixtures can rival mineral fertilizers. In maize–wheat rotations in France, application of municipal green-waste compost (8 $t{\cdot}ha^{-1}$) plus biochar (4 $t{\cdot}ha^{-1}$) achieved wheat and maize yields comparable to full P/K fertilizer, while substantially increasing soil organic C (up to + 53%) (Nobile et al. 2022). South Asia's smallholders already rely heavily on manure and compost, so co-applying biochar can magnify traditional practices. In India's mid-hills, farmers often apply farmyard manure and compost; adding biochar could further improve the sandy-loam soils' water and nutrient holding capacity under monsoon rainfall. In China, datasets demonstrate modest average yield gains with biochar, suggesting local adaptation. Although published case studies in Nepal are limited, the country's degraded hill soils (often low in P and organic matter) are likely to benefit from combined biochar–compost amendments to restore fertility and moisture. Overall, integration of biochar with organic fertilizers consistently improves soil C and plant productivity across diverse environments (Pump et al. 2019; Sharma et al. 2021).

7.2 *Synergistic Use with Mycorrhizae and Biofertilizers*

Biochar can greatly enhance the efficacy of beneficial soil microorganisms—arbuscular mycorrhizal fungi (AMF), rhizobia, and other biofertilizer bacteria—by providing habitat, retaining nutrients, and mitigating stresses. The porous matrix of biochar serves as a refuge for microbes and adsorbed nutrients, enabling stronger symbioses. For example, applying biochar with an AMF inoculant often increases root colonization rates. In maize, combined biochar + AMF treatments produced the highest colonization (~ 58% of root length)—about 12–16% points more than AMF alone (Meng et al. 2024). Generally, co-applying biochar with inoculated AMF or with native AMF populations promotes drought and nutrient stress tolerance. For instance, under water deficit, biochar + AMF-treated plants often outperform plants with only one amendment or none, showing taller growth and higher photosynthesis rates. Beyond fungi, biochar also boosts performance of nitrogen-fixing and phosphate-solubilizing bacteria (biofertilizers). In barley under drought, co-applying a drought-tolerant PGPR (*Serratia odorifera*) with biochar increased shoot length by 37%, fresh weight by 52%, and dry weight by 62% compared to untreated controls (Gul et al. 2023).

In China's Yellow River Basin, biochar-AMF combinations increased wheat biomass by 58% and microbial biomass carbon by 21%, reducing irrigation needs by 20% in saline-affected soils (Bo et al. 2023). Similarly, rice paddies inoculated with AMF and biochar in Jiangsu Province sequestered 1.2 tons of carbon per hectare annually while lowering methane emissions by 18%. Under heavy metal stress, biochar-AMF systems immobilize contaminants like arsenic and cadmium through adsorption and fungal hyphae encapsulation. A trial in Hunan Province reduced cadmium bioavailability in rice grains by 65%, addressing food safety challenges in polluted regions (Bo et al. 2023). Soil N, P, and K levels were also significantly higher with the combined treatment. It suggests biochar helps sustain beneficial microbes and their plant growth effects during stress. In legume systems, biochar's liming and CEC effects create a better environment for rhizobia; studies report increased nodule formation and N-fixation when biochar is added to seed inoculation. Overall, integrating biochar with mycorrhizal or bacterial inoculants forms a "biochar-based biofertilizer" approach: nutrients released by microbes are efficiently retained and cycled, improving nutrient-use efficiency under stress.

7.3 Role in Climate-Smart Agriculture

Biochar is a potent component of climate-smart agriculture (CSA) because it simultaneously sequesters carbon, mitigates greenhouse gases (GHGs), and builds resilience to environmental stress. By converting biomass waste (rice straw, wood chips, manure) into recalcitrant carbon, biochar locks CO_2 in soil for decades to centuries (Nazim et al. 2025). In fact, a single ton of biochar can sequester roughly 2.0–2.6 tons of CO_2 equivalent, offsetting emissions from field burning or decomposition. Globally, it has been estimated that swapping slash-and-burn through slash-and-char could annually offset ~ 0.2 Pg C (~ 12% of land-use change emissions) (Vijay et al. 2021). Global syntheses report that biochar typically increases crop yields (on average) and soil organic carbon. For example, a China-wide meta-analysis (866 trials) found that biochar applications raised staple-crop yields by a mean of ~ 8.8% over controls (Yu et al. 2023), with larger gains (up to + 26%) when using neutral-pH, low-C/N biochars. Globally, meta-analyses find yield boosts in the range 5–51% and soil carbon increases of 12–102% (Li et al. 2024), indicating strong carbon sequestration potential. These increments help meet CSA goals: higher yields (productivity) are achieved without extra land, and more soil C and nutrient retention means soils buffer against droughts and floods. Moreover, biochar-amended soils often emit less N_2O and retain more moisture, providing further climate adaptation (Li et al. 2024).

In Asia specifically, climate vulnerability is acute, so biochar's CSA benefits are especially relevant. India and Nepal both emphasize climate-resilient farming in national policies; biochar can contribute by improving monsoon-water retention and reducing fertilizer dependence. In Nepal's fragile slopes, for instance, biochar + compost could build soil organic matter, reducing erosion and enhancing drought tolerance in terrace farms. In China, major research programs are already exploring biochar for carbon farming. Chinese trials show that moderate biochar rates (10–20 $t \cdot ha^{-1}$) maximize wheat and corn yield benefits on acidic soils, while 15–25 $t \cdot ha^{-1}$ is optimal for rice (Yu et al. 2023). A recent review on biochar in climate-smart agriculture emphasizes its multifaceted role: Biochar improves nutrient-use efficiency and beneficial soil–microbe interactions, and helps meet CSA criteria of productivity, adaptation, and mitigation. It restores degraded soils, curbs pollutant run-off, and is compatible with zero-tillage or cover cropping practices. In India, China, Nepal, and beyond, integrating biochar into farming can thus enhance yields and income while also storing carbon and lowering emissions (Li et al. 2024).

8 Limitations and Challenges

While the advantages of biochar application to agroecosystems have been extensively studied, concerns about its possible adverse impacts have also been brought forward (Lehmann et al. 2011). The possibility of biochar being toxic can also be a result of processing (production process and choice of raw material to produce biochar). Though it is effective in reducing N_2O emissions with biochar, the production requires a possible increase in NOx emissions since NOx is a precursor (Liu et al. 2019). Ammonia (NH_3), which has potential emission during production, can lead to air pollution and particle formation. Biochar has fine carbon particles, and handling such substances may cause health hazards, for example, respiratory problems. Additionally, it is a medium for noxious organic pollutants and heavy metals that may hinder plant growth (Lehmann et al. 2011). Therefore, the selection of phenolic feedstock is crucial for biochar manufacturing as waste-derived materials can also contain toxic elements such as arsenic or Lead (El-Naggar et al. 2019). This should be of particular concern when large-scale adoption is to be facilitated among farmers who likely have little control on the source of biomass. For example, biochar produced from Miscanthus was shown to contain elevated levels of trace elements and polycyclic aromatic hydrocarbons (PAHs) (Oleszczuk et al. 2013). Polycyclic aromatic hydrocarbons (PAHs) are expected to enhance in plants above the control levels along with their possible inducing of oxidative stress by the biochar addition. The development of PAHs appears due to the formation of aromatic rings during biomass pyrolysis, followed by its accumulation (Rombolà et al. 2019). Research of Wang et al. (2019) performed the lifetime cancer risk assessment on PAH in biochar-amended soils-grown plants and concluded that there is virtually no risk to human health.

There has been one case of toxicity induced by biochar, as biochar impacted nitrate reduction in soil leachate over the course of a 25-day study. This indicated that nitrogen may be adsorbed onto the biochar surface, thus reducing the inorganic nitrogen supply and plant growth for a possible competition (Novak et al. 2010). Biochar has been reported to impact the leaching of phosphate and organic carbon from soil, further reducing their availability for plants (Lipczyńska-Kochany 2018).

9 Challenges

In recent years, biochar is a promising soil amendment that has been shown to have an impact in reducing the greenhouse gas emissions and stabilizing carbon within soil structure (Wang et al. 2018; Zhao et al. 2020). Despite this, there are still hurdles to widespread availability of large-scale applications that follow sustainability guidelines. Until now, numerous laboratory and pilot-scale studies have been led around the world to investigate biochar' soil amendments to crops production and soil remediation potentials (Ashiq et al. 2020; Yang et al. 2020).

Example, Fidel et al. (2019), performed field trial of small-scale with biochar produced from mixed wood gasification on greenhouse gas emissions, principally CO_2 and N_2O from loess soils (Frechen 2011), a specific type of soil derived from fine silt that has been blown in by the wind. These results demonstrated a significant decrease in N_2O emissions (38–56%), especially in the biochar-treated soils particularly for continuous corn systems. Still, they did not find a lasting < 60 days > reduction of CO_2 emissions. It is essential for sustainability to control the pyrolysis process. Pyrolysis generally has to do with energy demand and these often results in greenhouse gas (GHG) emissions especially when using fossil or non-renewable feedstock. Hence, it is a critical reason to investigate and set up properly green energy sources for the production of biochar from biomass. New concepts as in situ production of concentrated solar energy, i.e., through the use of solar tower technology have been proposed to increase sustainability aspect of biochar production (Telsnig et al. 2017; Zheng et al. 2015; Xu et al. 2015 for greenhouse transpiration).

It is equally important to fully evaluate the potential human health hazards and toxicological risk of biochar for calibration of production processes that could reduce toxicity, perhaps more important for the welfare of workers producing it (and farmers applying the material) in a safe manner. Furthermore, definitive studies on the environmental fate of biochar as a function of time and especially long-term impacts on soil properties are required. The movement of it in the soil and the possibility of leaching into water bodies such as groundwater (Kamali et al. 2022) should also be investigated.

10 Conclusion

Biochar offers a sustainable, scalable, and eco-friendly solution to address the growing challenges of environmental stress and food insecurity in agriculture. As climate change intensifies and resources like soil and water become increasingly strained, the need for resilient and regenerative farming practices is more critical than ever. This chapter has highlighted the multifunctional benefits of biochar, including its ability to enhance soil structure, retain moisture, buffer pH, improve nutrient availability, and support beneficial microbial communities, all of which contribute to stronger plant growth under stress conditions.

Biochar also aligns with circular economy principles when produced from agricultural or organic waste, reducing environmental burdens while improving soil health. Its potential to sequester carbon and lower greenhouse gas emissions further adds to its value in climate-smart agriculture.

To maximize its impact, strategic policy frameworks are essential. For instance, India's National Mission for Sustainable Agriculture (NMSA) supports biochar integration in rainfed farming systems, while China has successfully used biochar in large-scale soil remediation. These examples demonstrate that with supportive policies and field-level adoption, biochar can transition from research to real-world impact.

Recommendations include promoting biochar through national soil health initiatives, offering subsidies or incentives for adoption, and ensuring quality control standards. Farmer training and extension services should be strengthened to encourage knowledge transfer, especially in regions facing the greatest environmental stress. With coordinated efforts across policy, research, and practice, biochar can play a transformative role in building resilient agroecosystems and ensuring food security in a changing climate.

11 Future Prospects

Despite most relevance to short-term biochar emissions, extensive long-term field studies are diligently required for detailed understanding on the benefits and physical application of biochar for farmers. The next steps for research will be to identify the ideal rates and application methods of biochar for a variety of crops and soil types and evaluate its lasting effect on soil health as well as plant health. Biochar was proven to increase antioxidant enzyme activity; however, its role on the molecular expression level has not been studied extensively. The effect of biochar on gene expression is also not clear at this time. In addition, biochar properties vary with the feedstock type and pyrolysis temperature, so future studies should evaluate the biochar in a pot experiment, comparing the effects of husk-, wood- and manure-derived biochar (with different temperatures) on one single crop. Additionally, biochar is shown to be a low-cost measure in improving soil fertility and sustainable crop at large, even for the farmers worldwide. For agriculture to effectively use biochar as an organic soil amendment, it must be designed and used appropriately using process optimization. In trying to foster for better future generations toward climate-smart agro-technologies, it is necessary to improve the global food security scheme through biochar's implementation against the inorganic chemicals and pesticides. Future research strategies should examine related problems to fill knowledge gaps, particularly when completing long-term experiments with biochar for organic and sustainable options for improved sustainability.

Acknowledgements The authors would like to express heartfelt thanks to the Department of Environmental Sciences, Graphic Era Hill University, Dehradun, Uttarakhand, India, and Dean (Research and Development), COER University, Roorkee, UK, India, for their invaluable support and resources.

Author's Contributions AS and SA conceptualized this chapter, KKJ and AC designed table, figures, and wrote the initial stage of chapter, MS and AC critically edited it and contributed constructive recommendations before the submission. All the authors have no any conflict of interest, and the work is not related to animal ethical clearance.

References

Abdu N, Abdullahi AA, Abdulkadir A (2017) Heavy metals and soil microbes. Environ Chem Lett 15(1):65–84

Abewa A, Yitaferu B, Selassie YG, Amare T (2014) The role of biochar on acid soil reclamation and yield of Teff (Eragrostis tef [Zucc] Trotter) in Northwestern Ethiopia. J Agric Sci 6(1):1

Adams Z, Modi AT, Kuria SK (2025) Multidimensional perspective of sustainable agroecosystems and the impact on crop production: a review. Agriculture 15(6):581

Ahmad M, Rajapaksha AU, Lim JE, Zhang M, Bolan N, Mohan D, Ok YS (2014) Biochar as a sorbent for contaminant management in soil and water: a review. Chemosphere 99:19–33

Akhtar SS, Andersen MN, Liu F (2015) Biochar enhances yield and quality of tomato under reduced irrigation. Agric Water Manag 138:37–44

Ali K, Munsif F, Zubair M, Hussain Z, Shahid M, Din IU, Khan N (2011) Management of organic and inorganic nitrogen for different maize varieties. Sarhad J Agric 27(4):525–529

Ameloot N, De Neve S, Jegajeevagan K, Yildiz G, Buchan D, Funkuin YN, Prins W, Bouckaert L, Sleutel S (2013) Short-term CO_2 and N_2O emissions and microbial properties of biochar amended sandy loam soils. Soil Biol Biochem 57:401–410

Antonangelo JA, Sun X, Eufrade-Junior HDJ (2025) Biochar impact on soil health and tree-based crops: a review. Biochar 7(1):51

Anwar T, Qureshi H, Akhtar MS, Siddiqi EH, Fatimah H, Zaman W, Seleiman MF (2024) Enhancing maize growth and resilience to environmental stress with biochar, gibberellic acid and rhizobacteria. Front Plant Sci 15:1396594

Ashiq W, Nadeem M, Ali W, Zaeem M, Wu J, Galagedara L, Thomas R, Kavanagh V, Cheema M (2020) Biochar amendment mitigates greenhouse gases emission and global warming potential in dairy manure based silage corn in boreal. Environ Pollut 265:114869. https://doi.org/10.1016/j.envpol.2020.114869

Attri S, Saklani A, Joshi KK, Sharma P, Chauhan A, Goswami R, Flores-García M (2025) Seeds of sustainability: promoting food security through eco-friendly agriculture. In Sustainable synergy: harnessing ecosystems for climate resilience. Springer Nature Switzerland, Cham, pp 83–97

Awad M, El-Sayed MM, Li X, Liu Z, Mustafa SK, Ditta A, Hessini K (2021) Diminishing heavy metal hazards of contaminated soil via biochar supplementation. Sustainability 13(22):12742

Beesley L, Moreno-Jiménez E, Gomez-Eyles JL, Harris E, Robinson B, Sizmur T (2011) A review of biochars' potential role in the remediation, revegetation and restoration of contaminated soils. Environ Pollut 159(12):3269–3282

Bin Rahman ANMR, Zhang J (2016) Flood and drought tolerance in rice: opposite but may coexist. Food Energy Secur 5:76–88

Blanco-Canqui H (2017) Biochar and soil physical properties. Soil Sci Soc Am J 81(4):687–711

Bo X, Zhang Z, Wang J, Guo S, Li Z, Lin H, Zou J (2023) Benefits and limitations of biochar for climate-smart agriculture: a review and case study from China. Biochar 5(1):77

Castro SP, Cleland EE, Wagner R, Sawad RA, Lipson DA (2019) Soil microbial responses to drought and exotic plants shift carbon metabolism. ISME J 13:1776–1787

Chaves Fernandes BC, Ferreira Mendes K, Dias Júnior AF, Silva Caldeira VP, Silva Teófilo TM, Severo Silva T, Valadão Silva D (2020) Impact of pyrolysis temperature on the properties of eucalyptus wood-derived biochar. Materials 13(24):5841

Chi W, Nan Q, Liu Y, Dong D, Qin Y, Li S, Wu W (2024) Stress resistance enhancing with biochar application and promotion on crop growth. Biochar 6(1):43

Chowdhary K, Sharma S (2017) Potential of fungal endophytes in plant growth and disease management. In: Plant-microbe interactions in agro-ecological perspectives, vol. 1. Springer, Singapore, pp 291–310

Coppa E, Quagliata G, Venanzi R, Bruschini A, Bianchini L, Picchio R, Astolfi S (2024) Potential use of biochar as a mitigation strategy for salinity-related issues in tomato plants (Solanum lycopersicum L.). Environments 11(1):17

Das SK, Ghosh GK, Avasthe RK, Sinha K (2021) Compositional heterogeneity of different biochar: effect of pyrolysis temperature and feedstocks. J Environ Manage 278:111501

Denardin LGdO, Alves LA, Ortigara C, Winck B, Coblinski JA, Schmidt MR, Carlos FS, de Toni CAG, Camargo FAdO, Anghinoni I et al (2020) How different soil moisture levels affect the microbial activity. Ciência Rural 50

Ding Z, Kheir AMS, Ali MGM, Ali OAM, Abdelaal AIN, Lin X, Zhou Z, Wang B, Liu B, He Z (2020) The integrated effect of salinity, organic amendments, phosphorus fertilizers, and deficit irrigation on soil properties, phosphorus fractionation and wheat productivity. Sci Rep 10:2736

Dong M, Jiang M, He L, Zhang Z, Gustave W, Vithanage M, He F (2025) Challenges in safe environmental applications of biochar: identifying risks and unintended consequence. Biochar 7(1):1–20

El-Naggar A, Shaheen SM, Hseu Z-Y, Wang S-L, Ok YS, Rinklebe J (2019) Release dynamics of As Co, and Mo in a biochar treated soil under pre-definite redox conditions. Sci Total Environ 657:686–695. https://doi.org/10.1016/j.scitotenv.2018.12.061

Evelin H, Devi TS, Gupta S, Kapoor R (2019) Mitigation of salinity stress in plants by arbuscular mycorrhizal symbiosis: current understanding and new challenges. Front Plant Sci 10:470

Fidel R, Laird D, Parkin T (2019) Effect of biochar on soil greenhouse gas emissions at the laboratory and field scales. Soil Systems 3(1):8

Flórián N, Ladányi M, Ittzés A, Kröel-Dulay G, Ónodi G, Mucsi M, Szili-Kovács T, Gergócs V, Dányi L, Dombos M (2019) Effects of single and repeated drought on soil microarthropods in a semi-arid ecosystem depend more on timing and duration than drought severity. PLoS ONE 14(12):e0219975

Frater PN, Borer ET, Fay PA, Jin V, Knaeble B, Seabloom E, Sullivan L, Wedin DA, Harpole WS (2018) Nutrients and environment influence arbuscular mycorrhizal colonization both independently and interactively in Schizachyriumscoparium. Plant Soil 425(1):493–506

Frechen M (2011) Loess in Europe: guest editorial. E&G Quat Sci J 60(1):3–5

Fu P, Yi W, Bai X, Li Z, Hu S, Xiang J (2011) Effect of temperature on gas composition and char structural features of pyrolyzed agricultural residues. Biores Technol 102(17):8211–8219

Gamage A, Gangahagedara R, Gamage J, Jayasinghe N, Kodikara N, Suraweera P, Merah O (2023) Role of organic farming for achieving sustainability in agriculture. Farming Syst 1(1):100005

Gamalero E, Bona E, Todeschini V, Lingua G (2020) Saline and arid soils: impact on bacteria, plants, and their interaction. Biology 9(11):334

Ghandali MV, Safarzadeh S, Ghasemi-Fasaei R, Zeinali S (2024) Heavy metals immobilization and bioavailability in multi-metal contaminated soil under ryegrass cultivation as affected by ZnO and MnO_2 nanoparticle-modified biochar. Sci Rep 14(1):10684

Girmay G, Singh BR, Mitiku H, Borresen T, Lal R (2008) Carbon stocks in Ethiopian soils in relation to land use and soil management. Land Degrad Dev 19(4):351–367. https://doi.org/10.1002/ldr.844

Goicoechea N (2020) Mycorrhizal fungi as bioprotectors of crops against verticillium wilt a hypothetical scenario under changing environmental conditions. Plants 9(9):1–15

Gray M, Johnson MG, Dragila MI, Kleber M (2014) Water uptake in biochars: the roles of porosity and hydrophobicity. Biomass Bioenerg 61:196–205. https://doi.org/10.1016/j.biombioe.2013.12.010

Gul F, Khan IU, Rutherford S, Dai ZC, Li G, Du DL (2023) Plant growth promoting rhizobacteria and biochar production from Parthenium hysterophorus enhance seed germination and productivity in barley under drought stress. Front Plant Sci 14:1175097

Guo JH, Liu XJ, Zhang Y, Shen JL, Han WX, Zhang WF, Christie P, Goulding KWT, Vitousek PM, Zhang FS (2010) Significant acidification in major Chinese croplands. Science 327(5968):1008–1010

Hartman K, Tringe SG (2019) Interactions between plants and soil shaping the root microbiome under abiotic stress. Biochem J 476:2705–2724

Hou J, Pugazhendhi A, Sindhu R, Vinayak V, Thanh NC, Brindhadevi K, Yuan D (2022) An assessment of biochar as a potential amendment to enhance plant nutrient uptake. Environ Res 214:113909

Hussain S, Shaukat M, Ashraf M, Zhu C, Jin Q, Zhang J (2019) Salinity stress in arid and semi-arid climates: effects and management in field crops. In: Climate change and agriculture. IntechOpen

Iqbal S, Hussain S, Abdul Qayyaum M, Ashraf M (2020) The response of maize physiology under salinity stress and its coping strategies. In: Plant stress physiology. IntechOpen

Jerbi M, Labidi S, Lounès-Hadj Sahraoui A, Chaar H, Jeddi FB (2020) Higher temperatures and lower annual rainfall do not restrict, directly or indirectly, the mycorrhizal colonization of barley (Hordeum vulgare L.) under rainfed conditions. PLOS ONE 15(11):e0241794

Jones DL, Healey JR (2010) Organic amendments for remediation: putting waste to good use. Elements 6(6):369–374

Joseph S, Camps-Arbestain M, Lin Y, Munroe P, Chia CH, Hook J, Lehmann J (2010) An investigation into the reactions of biochar in soil. Aust J Soil Res 48(6–7):501–515

Kamali M, Sweygers N, Al-Salem S, Appels L, Aminabhavi TM, Dewil R (2022) Biochar for soil applications-sustainability aspects, challenges and future prospects. Chem Eng J 428:131189

Kamran M, Parveen A, Ahmar S, Malik Z, Hussain S, Chattha MS, Saleem MH, Adil M, Heidari P, Chen J (2019) An overview of hazardous impacts of soil salinity in crops, tolerance mechanisms, and amelioration through selenium supplementation. Int J Mol Sci 21(1):148

Keiblinger KM, Hall EK, Wanek W, Szukics U, Hämmerle I, Ellersdorfer G, Böck S, Strauss J, Sterflinger K, Richter A et al (2010) The effect of resource quantity and resource stoichiometry on microbial carbon-use efficiency. FEMS Microbiol Ecol 73(3):430–440

Laine J (2012) Perspective of the preparation of agrichars using fossil hydrocarbon coke. Renew Sustain Energy Rev 16(8):5597–5602

Lal R (2009) Soils and world food security. Soil till Res 102:1–4. https://doi.org/10.1016/j.still.2008.08.001

Lal R (2013) Food security in a changing climate. Ecohydrol Hydrobiol 13(1):8–21. https://doi.org/10.1016/j.ecohyd.2013.03.006

Lee X, Yang F, Xing Y, Huang Y, Xu L, Liu Z, Zhou H (2022) Use of biochar to manage soil salts and water: effects and mechanisms. Catena 211:106018

Lehmann J, Gaunt J, Rondon M (2006) Bio-char sequestration in terrestrial ecosystems—a review. Mitig Adapt Strateg Glob Chang 11(2):403–427. https://doi.org/10.1007/s11027-005-9006-5

Lehmann J, Kaempf N, Woods WI, Sombroek W, Kern DC, Cunha TJF (2007) Classification of Amazonian dark earths and other ancient anthropic soils. In: Lehmann J, Kern DC, Glaser B, Woods WI (eds) Amazonian dark earths: origin, properties, management. Springer, New York, pp 77–102

Lehmann J, Rillig MC, Thies J, Masiello CA, Hockaday WC, Crowley D (2011a) Biochar effects on soil biota review. Soil Biol Biochem 43(9):1812–1836

Lehmann J, Rillig MC, Thies J, Masiello CA, Hockaday WC, Crowley D (2011b) Biochar effects on soil biota—a review. Soil Biol Biochem 43(9):1812–1836. https://doi.org/10.1016/j.soilbio.2011.04.022

Li X, Wu D, Liu X, Huang Y, Cai A, Xu H, Zhang W (2024) A global dataset of biochar application effects on crop yield, soil properties, and greenhouse gas emissions. Sci Data 11(1):57

Lipczyńska-Kochany E (2018) Effect of climate change on humic substances and associated impacts on the quality of surface water and groundwater: a review. Sci Total Environ 640–641:1548–1565

Liu Q, Liu B, Zhang Y, Hu T, Lin Z, Liu G, Wang X, Ma J, Wang H, Jin H et al (2019) Biochar application as a tool to decrease soil nitrogen losses (NH_3 volatilization, N_2O emissions, and N leaching) from croplands: options and mitigation strength in a global perspective. Glob Change Biol 25:2077–2093

Liu W, Zhang Z, Wan S (2009) Predominant role of water in regulating soil and microbial respiration and their responses to climate change in a semiarid grassland. Glob Change Biol 15(1):184–195

Liu Z, Dugan B, Masiello CA, Wahab LM, Gonnermann HM, Nittrouer JA (2018) Effect of freeze-thaw cycling on grain size of biochar. PLoS ONE 13(1):e0191246

Martín-Folgar R, Martínez-Guitarte JL (2017) Cadmium alters the expression of small heat shock protein genes in the aquatic midge Chironomus riparius. Chemosphere 169:485–492

Maziarka P, Wurzer C, Arauzo PJ, Dieguez-Alonso A, Mašek O (2021) Do you BET on routine? The reliability of N_2 physisorption for the quantitative assessment of biochar's surface area. Chem Eng J 418:129234

Mbow CC, Rosenzweig LG, Barioni TG, Benton M, Herrero M, Krishnapillai E, Liwenga P, Pradhan MG, Rivera-Ferre T, Sapkota FN et al (2019) Special report on climate change, desertification, land degradation, sustainable land management, food security, and greenhouse gas fluxes in terrestrial ecosystems; IPCC: Geneva, Switzerland, 2019 437–550

Meisner A, Jacquiod S, Snoek BL, Ten Hooven FC, van der Putten WH (2018) Drought legacy effects on the composition of soil fungal and prokaryote communities. Front Microbiol 9:294

Meng L, Cheng Z, Wang Y, Li S, Clarke N (2024) Arbuscular mycorrhizal fungal interacted with biochar and enhanced phosphate-solubilizing microorganism abundance and phosphorus uptake in maize. Agronomy 14(8):1678

Mukherjee A, Zimmerman AR (2013) Organic carbon and nutrient release from a range of laboratory-produced biochars and biochar–soil mixtures. Geoderma 193:122–130

Mukherjee A, Lal R (2013) Biochar impacts on soil physical properties and greenhouse gas emissions. Agronomy 3(2):313–339

Naeem MA, Khalid M, Arshad M, Ahmad R (2014) Yield and nutrient composition of biochar produced from different feedstocks at varying pyrolytic temperatures. Pak J Agric Sci 51(1):75–82

Naeem MB, Jahan S, Rashid A, Shah AA, Raja V, El-Sheikh MA (2024) Improving maize yield and drought tolerance in field conditions through activated biochar application. Sci Rep 14(1):25000

Nazim M, Ghafoor A, Hussain A, Tabassum M, Nawaz A, Ahmad M, Ali M (2025) Biochar as a climate-smart agricultural practice: reducing greenhouse gas emissions and promoting sustainable farming. Phyton-Int J Exp Bot 94(1):65–99

Nobile C, Lebrun M, Védère C, Honvault N, Aubertin ML, Faucon MP, Houben D (2022) Biochar and compost addition increases soil organic carbon content and substitutes P and K fertilizer in three French cropping systems. Agron Sustain Dev 42(6):119

Novak JM, Busscher WJ, Watts DW, Laird DA, Ahmedna MA, Niandou MA (2010) Short-term CO_2 mineralization after additions of biochar and switchgrass to a Typic Kandiudult. Geoderma 154(3–4):281–288

Oleszczuk P, Jośko I, Kuśmierz M (2013) Biochar properties regarding to contaminants content and ecotoxicological assessment. J Hazard Mater 260:375–382

Oshunsanya SO, Nwosu NJ, Li Y (2019) Abiotic stress in agricultural crops under climatic conditions. Sustain Agric, For Environ Manag 71–100

Otlewska A, Migliore M, Dybka-Stępień K, Manfredini A, Struszczyk-Świta K, Napoli R, Białkowska A, Canfora L, Pinzari F (2020) When salt meddles between plant, soil, and microorganisms. Front Plant Sci 11:1429

Pump C, Keiblinger KM, Scheiblauer E, Johnen S, Lehto NJ, Soja G, Zehetner F (2019) Temporal changes in the efficiency of biochar-and compost-based amendments on copper immobilization in vineyard soils. Soil Systems 3(4):78

Rathinapriya P, Maharajan T, Jothi R, Prabakaran M, Lee IB, Yi PH, Jeong ST (2025) Unlocking biochar impacts on abiotic stress dynamics: a systematic review of soil quality and crop improvement. Front Plant Sci 15:1479925

Raza A, Safdar M, Adnan Shahid M, Shabir G, Khil A, Hussain S, Akram HMB (2024) Climate change impacts on crop productivity and food security: an overview. Transform Agric Manag Sustain Futur: Clim Chang Mach Learn Perspect 163–186
Razzaghi F, Obour PB, Arthur E (2020) Does biochar improve soil water retention? A systematic review and meta-analysis. Geoderma 361:114055
Reis VM, Teixeira KR, dos S (2015) Nitrogen fixing bacteria in the family Acetobacteraceae and their role in agriculture. J Basic Microbiol 55(8):931–949
Rodríguez-Vila A, Atuah L, Abubakari AH, Atorqui DW, Abdul-Karim A, Coole S, Sizmur T (2022) Effect of biochar on micronutrient availability and uptake into leafy greens in two urban tropical soils with contrasting soil pH. Front Sustain Food Syst 6:821397
Rombolà AG, Fabbri D, Baronti S, Vaccari FP, Genesio L, Miglietta F (2019) Changes in the pattern of polycyclic aromatic hydrocarbons in soil treated with biochar from a multiyear field experiment. Chemosphere 219:662–670
Schimel JP (2018) Life in dry soils: effects of drought on soil microbial communities and processes. Annu Rev Ecol Evol Syst 49:409–432
Sharma P, Abrol V, Sharma V, Chaddha S, Rao CS, Ganie AQ, Mansoor S (2021) Effectiveness of biochar and compost on improving soil hydro-physical properties, crop yield and monetary returns in inceptisol subtropics. Saudi J Biol Sci 28(12):7539–7549
Sharma S, Mukherjee S, Bolan S, de Figueiredo C, Fachini J, Chang SX, Bolan N (2025) Biochar as a potential nutrient carrier for agricultural applications. Curr Pollut Rep 11(1):1–36
Sheik CS, Beasley WH, Elshahed MS, Zhou X, Luo Y, Krumholz LR (2011) Effect of warming and drought on grassland microbial communities. ISME J 5(11):1692–1700
Šola I, Poljuha D, Pavičić I, Jurinjak Tušek A, Šamec D (2025) Climate change and plant foods: the influence of environmental stressors on plant metabolites and future food sources. Foods 14(3):416
Solomon D, Lehmann J, Thies J, Schäfer T, Liang B, Kinyangi J, Neves E, Petersen J, Luizao F, Skjemstad J (2007) Molecular signature and sources of biochemical recalcitrance of organic C in Amazonian Dark Earths. GeochimCosmochim Acta 71(9):2285–2298
Sun X, Wang J, Zhang M, Liu Z, Lan Y, Meng J (2023) Effects of biochar on the Cd uptake by rice and the Cd fractions in paddy soil: a 3-year field experiment. Agronomy 13(5):1335
Syuhada AB, Shamshuddin J, Fauziah CI, Rosenani AB, Arifin A (2016) Biochar as soil amendment: impact on chemical properties and corn nutrient uptake in a Podzol. Can J Soil Sci 96(4):400–412
Tamang BG, Fukao T (2015) Plant adaptation to multiple stresses during submergence and following desubmergence. Int J Mol Sci 16:30164–30180
Telsnig T, Weinrebe G, Finkbeiner J, Eltrop L (2017) Life cycle assessment of a future central receiver solar power plant and autonomous operated heliostat concepts. Sol Energy 157:187–200. https://doi.org/10.1016/j.solener.2017.08.018
Thomas SC, Gale N (2015) Biochar and forest restoration: a review and meta-analysis of tree growth responses. New Forest 46:931–946
Ullah S, Liang H, Ali I, Zhao Q, Iqbal A, Wei S, Shah T, Yan B, Jiang L (2020) Biochar coupled with contrasting nitrogen sources mediated changes in carbon and nitrogen pools, microbial and enzymatic activity in paddy soil. J Saudi Chem Soc 24(11):835–849
Usman ARA, Abduljabbar A, Vithanage M, Ok YS, Ahmad M, Ahmad M, Al-Wabel MI (2015) Biochar production from date palm waste: charring temperature-induced changes in composition and surface chemistry. J Anal Appl Pyrol 115:392–400
Uzoma KC, Inoue M, Andry H, Fujimaki H, Zahoor A, Nishihara E (2011) Effect of cow manure biochar on maize productivity under sandy soil condition. Soil Use Manag 27(2):205–212
Vijay V, Shreedhar S, Adlak K, Payyanad S, Sreedharan V, Gopi G, Aravind PV (2021) Review of large-scale biochar field-trials for soil amendment and the observed influences on crop yield variations. Front Energy Res 9:710766
Vista SP, Pandit NR (2023) Biochar for improving crop production in Nepal. SAARC J Agric 21(1):227–238

Voesenek LACJ, Colmer TD, Pierik R, Millenaar FF, Peeters AJM (2006) How plants cope with complete submergence. New Phytol 170:213–226
Wang Y, Hu Y, Zhao X, Wang S, Xing G (2013) Comparisons of biochar properties from wood material and crop residues at different temperatures and residence times. Energy Fuels 27(10):5890–5899
Wang J, Odinga ES, Zhang W, Zhou X, Yang B, Waigi MG, Gao Y (2019) Polyaromatic hydrocarbons in biochars and human health risks of food crops grown in biochar-amended soils: a synthesis study. Environ Int 130:104899
Wang S, Zhou Y, Han S, Wang N, Yin W, Yin X, Gao B, Wang X, Wang J (2018) Carboxymethyl cellulose stabilized ZnO/biochar nanocomposites: enhanced adsorption and inhibited photocatalytic degradation of methylene blue. Chemosphere 197:20–25. https://doi.org/10.1016/j.chemosphere.2018.01.022
Wang W, Wang H, Feng Y, Wang L, Xiao X, Xi Y, Luo X, Sun R, Ye X, Huang Y et al (2016) Consistent responses of the microbial community structure to organic farming along the middle and lower reaches of the Yangtze River. Sci Rep 6, Article 35046
Xu X, Rao Z, Wu J, Zhou Y, He D, Liu Y (2015) Preparation and characterization of mullite-silicon carbide heat absorbing ceramics for solar thermal tower plant. J Wuhan Univ Technology—Materials Sci Ed 30(1):27–32. https://doi.org/10.1007/s11595-015-1094-9
Yan N, Marschner P, Cao W, Zuo C, Qin W (2015) Influence of salinity and water content on soil microorganisms. Int Soil Water Conserv Res 3(4):316–323
Yang T, Lupwayi N, Marc SA, Siddique KHM, Bainard LD (2021) Anthropogenic drivers of soil microbial communities and impacts on soil biological functions in agroecosystems. Glob Ecol Conserv 27:e01521
Yang W, Feng G, Miles D, Gao L, Jia Y, Li C, Qu Z (2020) Impact of biochar on greenhouse gas emissions and soil carbon sequestration in corn grown under drip irrigation with mulching. Sci Total Environ 729:138752
Yao Y, Gao B, Zhang M, Inyang M, Zimmerman AR (2012) Effect of biochar amendment on sorption and leaching of nitrate, ammonium, and phosphate in a sandy soil. Chemosphere 89(11):1467–1471
Yu BH, Ji JM, Wang LX, Liu J, Gao H, Liu D (2023) Meta analysis on yield effect of biochar for staple crops in China. Huan jingkexue = Huanjingkexue 44(1):520–530
Yuan JH, Xu RK, Zhang H (2011) The forms of alkalis in the biochar produced from crop residues at different temperatures. Biores Technol 102(3):3488–3497
Zhang J, Lü F, Zhang H, Shao L, Chen D, He P (2015) Multiscale visualization of the structural and characteristic changes of sewage sludge biochar oriented towards potential agronomic and environmental implication. Sci Rep 5:9406
Zhang W, Wei J, Guo L, Fang H, Liu X, Liang K, Siddique KH (2023) Effects of two biochar types on mitigating drought and salt stress in tomato seedlings. Agronomy 13(4):1039
Zhang X, Li M, Yang H, Li X, Cui Z (2018) Physiological responses of Suaeda glauca and Arabidopsis thaliana in phytoremediation of heavy metals. J Environ Manage 223:132–139
Zhao Y, Song M, Cao Q, Sun P, Chen Y, Meng F (2020) The superoxide radicals' production via persulfate activated with $CuFe_2O_4$@Biochar composites to promote the redox pairs cycling for efficient degradation of o-nitrochlorobenzene in soil. J Hazard Mater 400:122887
Zheng H, Yu X, Su Y, Riffat S, Xiong J (2015) Thermodynamic analysis of an idealised solar tower thermal power plant. Appl Therm Eng 81:271–278
Zhu X, Chen B, Zhu L, Xing B (2017) Effects and mechanisms of biochar-microbe interactions in soil improvement and pollution remediation: a review. Environ Pollut 227:98–115

Adsorption and Removal Strategies of Biochar in Organic and Inorganic Pollutants Under Stress Conditions: Salinity, Drought, and Heavy Metals

Govinda Pal and Haider F. Mahmood

Abstract Biochar has been demonstrated to have a high porosity, special surface properties, and chemical stability, which can significantly benefit the adsorption of organic and inorganic pollutants. However, some environmental stimuli such as salinity, drought, and heavy metal stress might influence biochar's remediation efficiency largely. This chapter reviews the biochar adsorption software, modified methods, and application fields for soil and water pollutants under stressed conditions. We study the effects of soil, salinity, drought, and heavy metals on characteristics as well as ion exchange and adsorption capacity of biochar, revealing that the exchange-controlling technologies had been promoted as well as suppressed in the presence of these materials. The chapter then focuses on the potential of biochar to reduce heavy metal toxicity through ion exchange, complexation, and precipitation mechanisms. We report some of the recent technical contributions in biochar production processes (mainly based on chemical/physical modification principles) that may support to produce efficient biochar output even under hostile environments. We estimate and evaluate the capacity of biochar to improve fertility in soil systems through unwashable nutrient retention, but also consumed microbial population support even in unfavorable environments, focusing particularly on bio-film formation and resistance to adverse living conditions in per-genetic environment which cannot be suitably addressed by the traditional strategy based on humus. In addition, the issues about large-scale application of biochar and future research direction for enhancing the potential of biochar as sustainable remediation technology for degraded ecosystems are also discussed in this chapter.

Keywords Sorption mechanisms · Soil remediation · Water purification · Biochar modification · Microbial interaction

---

G. Pal
Department of Agricultural Engineering, Haldia Institute of Technology, Haldia, West Bengal, India

H. F. Mahmood (✉)
Technical College/Al-Mussaib, Al-Furat Al-Awsat Technical University, Babylon, Iraq
e-mail: haider.fawzi@atu.edu.iq

A. Chauhan et al. (eds.), *Biochar As Remediation Tools for Pollutant Removal*,
https://doi.org/10.1007/978-3-032-16746-0_5

1 Introduction

Soil and water pollution are important environmental challenges that can severely affect ecosystem health, agricultural sustainability, and human well-being worldwide (Chauhan et al. 2025). The pollution of these essential resources is rapidly increasing as a result of uncontrolled industrialization, agricultural activities, waste disposal techniques, and climatic changes. Soil degradation has been estimated to cover about 33% of the terrestrial land on earth, with heavy metals, polychlorinated biphenyls, and pesticide pollution being dominant (Anetor et al. 2022). Simultaneously, more than 80% of the global wastewater is discharged into the environment without being properly treated. It contains various industrial effluents, agricultural runoff, and ingested/urban waste that negatively affect water quality and aquatic ecosystems (Larrain et al. 2023). Though air pollution is one of the crucial environmental health problems, here in this chapter, it is considered and emphasizes for developing the suitable approaches to solve the environmental issues for soil and water pollution control under different environmental stresses.

Industrial effluent, agricultural runoff, and improper waste disposal lead to water pollution; this damages not only aquatic populations' lives but also human life. Others include heavy metals, nitrates, microplastics, and pharmaceutical residues that accumulate in water bodies and end up bioaccumulating in marine organisms, thus having potential toxic effects on human health (Anetor et al. 2022). In areas where people cannot access clean water, waterborne diseases such as cholera and dysentery are common. Moreover, the depletion of freshwater resources affects agricultural irrigation, leading to lower crop productivity and an increase in food insecurity. Yet, soil contamination represents yet another significant environmental challenge as persistent organic pollutants (POPs), pesticides, and heavy metals impair soil fertility, dysregulate the microbial diversity, and bioaccumulate in the food chain, consequently endangering human health and agricultural productivity (Anetoret al. 2022).

Climate change compounds environmental stress caused by pollution and worsens the intensity of extreme weather events, including hurricanes, wildfires, heatwaves, and floods. Climate change is already increasing global temperatures and modifying precipitation patterns, resulting in an increase in drought and freshwater deficits, and exacerbating pollution (Larrain et al. 2023; Chauhan et al. 2025). Melting glaciers and rising seas endanger coastal communities, uproot populations, and cause economic instability. Moreover, climate change also causes habitat destruction, which leads to species extinction and biodiversity loss (Singh and Singh 2017). Disruptions of these ecosystems weaken nature's resilience, impairing the ability of forests, wetlands, and oceans to absorb carbon dioxide and control climate patterns, which further drives global warming.

Nor are the health implications of pollution trivial, and this is especially true in low- and middle-income countries with healthcare systems poorly prepared to handle pollution-measurement illnesses. Long-term exposure to air and water contaminants is associated with respiratory infections, lung cancer, cardiovascular disease, and neurological disease (Meo 2024). The disproportionate public health burden falls on vulnerable populations, such as children, the elderly, and people with pre-existing conditions. Moreover, pollution comes with economic burdens, including rising health care costs, reduced labor productivity, and long-term challenges to development. As a result, by 2060, air pollution alone is on schedule to become 1% of GDP globally, with the largest economic costs in the developing world (Larrain et al. 2023).

However, pollution has become a global problem in many countries and areas around the world. Pollution is not always moving plants, the development of science and technology will bring scouts but also enemy. Falling emissions—and health hazards: A mix of clean energy technologies, tough environmental laws, and more sustainable urbanization could ultimately mean fewer emissions—and less risk to our health. For example: Maybe we could fuel ourselves with solar or wind (or perhaps hydroelectric power, rather than fossil fuels), so at least one fewer greenhouse gas and cleaner air. Smart monitoring of pollution and enforcement of clean air will be addressed through balloon tracking using IoT enabled with AI. Drawing from the concept of sustainability in human resource management within circular economies, where employers serve as managers and waste is reduced, reused, and recycled, how can industrial pollution resulting from waste be minimized? This is just one example: Ecosystem restoration and reforestation projects can provide additional measures of biodiversity protection, not to mention carbon sequestration.

To solve global warming or energy crisis—and any reasonable degree of long-term sustainability, we want at one extreme end: science exploration, and also intervention by policy + tech leap all the way to other extreme end. The only realistic way of tackling the pollution damage is through more international cooperation and stricter controls on polluting practices and investment in "green" technology. But if you're more intelligent about how societies grow sustainably and achieve responsible resource management, all that pollution doesn't need to be a headache.

Biochar may constitute an alternative solution to remediate polluted soil, and from the application side, its use is still confined on how to mitigate stressed conditions (salinity, drought or metal contamination). Owing to its intrinsic characters of physicochemical indicators, which could manage soil and mitigate environmental stress factors, the components were investigated innumerably as bioorganic fertilizers in sustainable agriculture and ecological restoration. In order to indicate the operational quality of that biochar, this performance will be labeled as the efficiency for removing or immobilizing the pollutant under certain environmental situations. Biochar performance is level of adsorption and removal efficiency, and biochar mechanism is just ponder the physical–chemical "thing happening between contaminant(s) and soil factors [or ecosystem]"! HTT/LT biochars (which had higher CEC, functional groups and porosity) are able to adsorb pollutants in a contaminated soil/water system and hold water content in the soil matrix supports which also enhances microbial activity and are potential for crop growth under stress conditions. Taken together, these properties govern the performance of biochar for environmental remediation and provide a new perspective for understanding this puzzle game of polluted pollution.

Thus, one of the main functions of biochar-rich (BC-rich) in environmental management and remediation is related to its high strong adsorbing capacity to soil heavy metals. As one of the highly toxic persistent soil pollutants, cadmium (Cd), lead (Pb), and arsenic (As), new kind of ICTP have imperiled the quality of soil and agriculture production deeply. Pyrolysis of the BC may highly reduce the bioavailability and toxicity of the HM by converting them from weakly acid extractable to stable residue form and lower in plant uptake (Qu et al. 2025).

The presence of large surface area and cation exchange capacity (CEC) in biochar also makes it an excellent adsorbent so it can be utilized for the sequestration of heavy metals which is a potential amendment for enhanced agricultural productivity and reduction in soil contamination (Xu et al. 2024).

In addition to the binding of heavy metals, biochar is crucial for mitigating drought through its water retention properties in soil. Drought stress results in a poor crop yield and soil fertility as well due to impaired water and nutrient uptake. Biochar is porous compared to natural organic matter and decrease in bulk density may increase water content holding capacity and water supply to plants (Bhattacharya et al. 2024). In reality, biochar addition is believed in soil to have beneficial effects both on the fertility of soil and the status of nutrients in the latter as well as the respiration activity of (soil) microorganisms and for plants under water stress. Due to the fact that biochar application can also be useful to improve soil properties, it is a good solution for drought management and sustainable agricultural productivity (Lohar et al. 2024).

Biochar comes in handy for cleaning up contamination, as well—it can reduce soil salinity. High soluble salts are responsible for reduced crop yield and fertility, so soil salinity is one of the most troubled characteristics in the fields. Biochar, in particular when inoculated with beneficial microorganisms, may improve saline soil properties such as reducing EC, increasing organic matter and bettering soil structure for plant growth (Table 2). It is subsequently modulated by biochar, with improved soil pH and CEC (cation exchange capacity) (on the one hand), and the harbor for halotolerant bacteria against accumulation of salts under salt stress. This resistance is also enhanced by the ability of biochar to interact with other components of the soil biome and microbial consortia are one of these.

Opportunities and constraints of the use of biochar on a large scale as a mitigation solution for pollution and soil resilience issues are presented here. The influence of biochar in the improvement of soil and in the pollution is controversial and depends on (inter)al. 2014): a range of parameters like the feedstock (the material used to produce biochar), the pyrolysis conditions, which play a role during its production, as well as soil characteristics. This complicates the site-specific treatment, which has to be adapted to local environmental factors and soil properties. In this way, to improve the recovery of OPB from soil and maximize its use, it is necessary research that studies application rates and treatment methods suitable for the agroecosystem in question and for the scenario of pollutants in which are applied this biomass. Figure 1 shows the manufacture and application of biochar.

The factors promoting the adsorption of organic and inorganic pollutants onto biochar under various environmental stress conditions require further elucidation, as

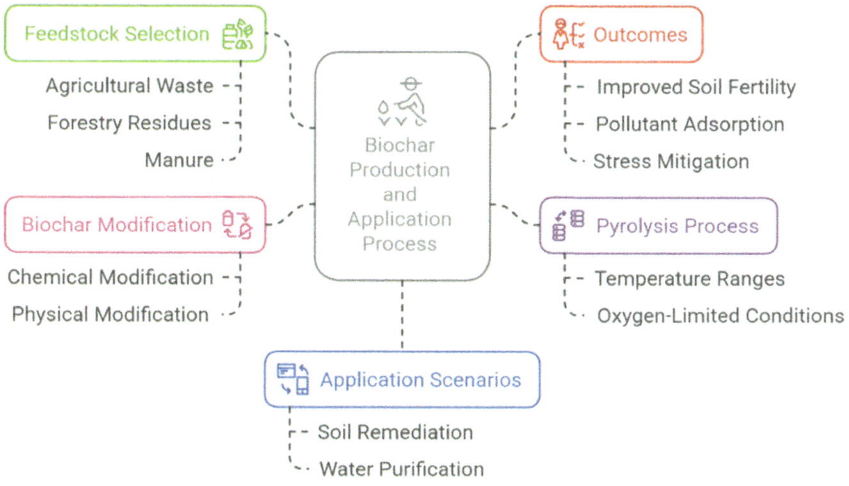

Fig. 1 Biochar production and application process

the interactions between the environment, biochar, and pollutants are highly complex. The ion exchange and electrostatic interactions between pollutants and biochar are significantly influenced by stress conditions such as salinity, drought, and heavy metal contamination. It is essential to highlight these adsorption mechanisms under such extreme conditions to evaluate biochar efficiency in field interventions, particularly where multiple environmental stressors act concurrently.

The physicochemical properties of biochar can be affected by environmental factors, leading to alterations in the pollutants adsorption capacity and structural stability. However, the effectiveness of biochar in pollution immobilization on a large time scale depends on the explicit insight into its stability and reactivity in various environmental and edaphic compartments. Thus, it is important to pay attention to the conditions where biochar effectiveness upon application to soil may depend on stress factors (e.g., aridity and salinity).

Yet there remains a major challenge in achieving the full capacity of biochar as a "one-stop-shop" total-ecosystem service package for environmentally troubled soils under conditions ranging across polluted-acid/smelter soil through to resilient soils. Yet, such differences were mainly attributed to biomass feedstock and pyrolytic conditions as well as the varying physicochemical properties of the target soils. Consequently, site-specific investigations are additional and also the micro-environmental local variables should be considered. The application rates and schedules of treatment could be applied to other agroecosystem conditions through additional research. This has rationalized the necessity for adapting its use in practice, understanding how biochar interacts with organic or/and inorganic contaminants under varying stress (salinity, drought metal pollution, etc.) conditions.

The physicochemical properties of biochar, which consequently affect the pollutant removal capacity of biochar, are markedly influenced by environmental stressors. Changes in surface characteristics, porosity, functional groups, and biochar stability under stress conditions may affect biochar interaction with contaminants. These impacts must be evaluated under real-world conditions to determine optimal rates and establish performance benchmarks that facilitate biochar use in stressed environments. Knowledge on how these stressors affect biochar's efficacy will guide better application strategies in areas affected by pollution or climate stresses.

In response to these issues, modification methods have been applied to improve the functional properties of biochar. These approaches aim to enhance their adsorption efficiency, structural stability, and degradation resistance under harsh environmental conditions. Biochars can be modified chemically, physically, and biologically to increase their surface area, add active functional groups, and enhance their overall reactivity. These pathways could guide the design of tailored biochar compositions that can function effectively in unfavorable and variable environments, broadening the scope of their application in environmental cleanup.

While the use of biochar for large-scale soil and water remediation is promising, its practical application requires further investigation to address practical limitations, including long-term toxicity, ecological trade-offs, and economic viability. Fulfilling this role for sustainable and climate-resilient agriculture can only be realized as the result of numerous field trials, constant monitoring, and the formulation of context-based best practices. This will require future research to address the current knowledge gaps, optimize modification methods, and develop standardization protocols that will facilitate the integration of biochar into different agroecological systems.

2 Biochar: Properties and Adsorption Mechanisms

Due to its high surface area, porosity, and tunability of surface chemistry, biochar is an efficient and inexpensive adsorbent for a wide variety of pollutants, such as heavy metals and pesticides. Adsorption takes place through electrostatic, $\pi-\pi$, and hydrophobic interactions, which are affected by pyrolysis conditions. It's environmentally gentler, benefiting the treatment of wastewater, but large-scale use must be optimized further.

2.1 Physical and Chemical Properties Influencing Pollutant Adsorption

Biochar, pyrolytically produced carbon-rich material from biomass, has unique physical and chemical properties, which contribute especially to the adsorption of pollutants. The characteristic properties of carbon, like high surface area, porosity, and functional groups, render it an effective adsorbent for the removal of organic pollutants/contaminants and heavy metals. It can also be improved/augmented using various modification techniques (e.g., chemical activation, addition of metals or heteroatoms), improving the physicochemical properties, and enhancing the adsorption potential of biochar (Beljin et al. 2024a; Han et al. 2024).

Carbonaceous physicochemical character is regarded as one of the fundamental approaches to adsorption capacity of raw biochar. A key unique feature is the high specific surface of material, which can reach to 1500 m^2/g, resulting in a great number of active sites for adsorption and retention of dioxic pollutants. The properties of biochar with different pore size categorized as micro, meso, and macropores enable it to trap and immobilize the contaminants within its structure, thus serving useful in improving the overall adsorption capacity (Beljin et al. 2024b; Han et al. 2024). Furthermore, the small size of fine particles has a larger specific surface area, and thus, there is a higher adsorption space for adsorbate on the adsorbing agent. Upside of biochar from sludge at particle size range 100–400: Simple biochars promoted with high loadings of heavy metals and organic contaminants-process, property fitting points out the significance (Agoe et al. 2024).

Another useful chemical feature of biochar is that it intrinsically has a capacity to take up pollutants. The functional groups on biochar surface including hydroxyl (–OH), carboxyl (–COOH) and carbonyl (–C=O) groups lead to the hydrogen bonding, electrostatic interaction, and complexation binding with contaminants. These functional groups would improve the physical attachment of pollutants in biochar, due to their compatibility (Han et al. 2024; Htet et al. 2024). Activating or doping metals and heteroatoms are also modifications to improve biochar's adsorption ability. These changes afford biochar an increased surface area and number of active sites, which coupled to specific interactions with the pollutants in the environment make it a very good adsorber (Han et al. 2024; Htet et al. 2024).

Because of the complexity of the biofiltration systems and different mechanisms of pollutant removal by biochar, the process of performance monitoring incorporates both in the selection of variables. One main process is pore filling, where contaminants are physically trapped in biochar's micro- and mesopores. Biochar has been used to study the mechanisms behind pharmaceutical adsorption, including oxytetracycline; as a highly porous, structured material, it retains contaminants in much the same way (Nguyen et al. 2024). In addition, Van der Waals force retention of organic molecules on biochar surfaces is the basis for the elimination of hydrophobic pollutants in the aqueous and soil environment. In addition, ion exchange and surface complexation may also occur concurrently with the overall process (Nguyen et al. 2024), while such intermolecular interactions may result in synergistic enhancement during adsorption.

Even though the sorption of pollutants using biochar has great potential, many aspects must be studied before large-scale application in water purification and soil remediation is optimized. The short-term changes of environmental factors such as pH, temperature, contemporaneous ions have a dominant impact on the adsorption performance of biochar. However, maximizing this type of potential in pollution remediation in practice will require more work to assess within production processes (e.g., efficiency), activation technologies (e.g., scaling), and post-application performance (e.g., long-term efficacy) (Beljin et al. 2024b; Han et al. 2024).

2.2 *Surface Area, Porosity, and Functional Groups*

Biochar has recently received great attention as an eco-friendly adsorbent because of its distinctive physicochemical properties, including high specific surface area, extensive porosity, and abundant surface functional groups, enabling it to adsorb various pollutants. These characteristics depend heavily on the synthetic conditions, mainly on the pyrolysis temperature, manner of feedstock, and technique of activation. Therefore, adjusting these parameters is important to improve the biochar performance in environmental remediation processes as an adsorbent. Deng et al. (2024) utilize the correlations of biochar's structural properties and chemical functionalities to predict its adsorption capacity, thus offering the universal application of biochar for pollution mitigation.

Based on biochar adsorption capacity, one of the most important factors affecting the adsorption capability of biochar is its specific surface area and porosity. The adsorptive behavior was mainly controlled by the micro and mesopore volume of biochar. The results from the experimental study confirmed that temperature was able to adjust pore volume/surface roughness that is helpful in adsorption capacity and so on. It is a particularly significant when pore sizes are optimized for CO_2 capture, facilitating gas adsorption and storage as a biochar-based strategy to mitigate atmospheric greenhouse gas emissions (Deng et al. 2024; He et al. 2024).

Furthermore, the chemical structure of biochar, in particular functional groups, may also play a role in affecting the adsorption mechanism. The special functional groups (carboxyl [$-COOH$] and amine group [$-NH_2$]) on the surface of FA molecule can effectively promote the adsorption of pollutants through the hydrogen bonding, electrostatic interaction, and surface complexation. The resultant biochars containing abundant o/n groups were more effective toward pharmaceuticals and emerging pollutants in this study. The chemical composition of biochars, such as aromatic to aliphatic carbon ratio, also affords the stability and reactivity with contaminants. The enriched aromatic moieties induce more semi-chemosensing and stronger hydrophobic cerebral interaction and can allow biochar to coat the APs with higher efficiency in the removal of them from aquatic microcosms (Xue et al. 2024; Hu et al. 2024).

Despite the favorable properties of biochar, soluble limitation was found to limit adsorption. However, its effectiveness may be influenced by surface function, morphological distribution, or competitive adsorbates in the complicated environmental matrix. Although biochar contains numerous advantages, the heterogeneity of biochar with different physicochemical properties must correlate with the choice of feedstock and pyrolysis conditions; hence, future studies are recommended to modify its structural and chemical properties for various applications. The novel biochar modification technology (activation of metal oxides, ligand functionalization) was also identified as a viable option in increasing removal capacity and applicability in environmental remediation (Deng et al. 2024; He et al. 2024).

2.3 Adsorption Kinetics and Thermodynamics

Biochar is a very carbonized material produced from the pyrolysis of biomass and has emerged as one of the latest inputs for sustainable agricultural practices, especially with respect to interaction between surface area and different functional groups. This makes it to be a good adsorbent of xenobiotics (heavy metals, dyes). Functional groups found on biochar give hydrophobic and hydrophilic sites which are considered for adsorption property [13–15]. Biochar has diverse physicochemical properties and they greatly influence its adsorptive characteristics; they vary due to the pyrolysis condition and source of biomass. Hence, investigating both adsorption mechanisms and broad-spectrum thermodynamic descriptions is very necessary to maximize the application efficiency of biochar in pollution controlling as well as ensuring its environmental persistence (Gotore et al. 2024).

The activated carbon possesses a high surface area and porosity; surface area and porosity are one of the most important factors affecting the adsorbent capacity of biochars (17). What u r saying is quite common; in general, biochar is supposed to have high surface area and porous volume @ higher pyrolysis temperature so that this biochar can adsorb such a pollutant. As a result of these structural properties, the capture adsorption sites were well formed, making them especially important in trapping pollutants in the aqueous environment such as heavy metals and organic contaminants (Gotore et al. 2024).

Additionally, the surface chemistry of biochar heavily influences the adsorption characteristics, such as oxygen-rich functional groups (AFGs) like carboxyl (–COOH) and hydroxyl (–OH). Functional groups of biochar play an important role in the binding of pollutants through the hydrogen bonding, electrostatic attraction, and complexation mechanisms (Li et al. 2024).

Multiple adsorption mechanisms can be derived from biochar according to the nature of the pollutant and the environmental conditions. Having said that, it should be noticed that the attachment of charged pollutants to biochar is highly sensitive to surface charge through electrostatic interactions, which directly depend upon the pH of the solution. At lower pH values, positively charged biochar surface enhances anionic pollutant adsorption, and negatively charged surface at higher pH values increases the retention of cationic contaminants in the biochar (Khanzada et al. 2024; Li et al. 2024). Adsorption mechanisms including ion exchange and surface complexation also enhance heavy metal uptake since metal ions in the solution are exchanged with ions absorbed to biochar's surface, reducing metal availability as well as toxicity (Khanzada et al. 2024). For organic pollutants, such as dyes, the adsorption of biochar to contaminants is driven by π–π interactions (interaction between the aromatic rings in biochar and those on pollutants) (Li et al. 2024).

In addition, studies on the kinetics and thermodynamics of biochar adsorption provide insights into biochar's efficiency, practicality, and feasibility for use in real-life scenarios. The pseudo-second-order kinetics is often observed in adsorption kinetics studies with biochar, suggesting that the chemisorption process predominates in the mechanism of pollutant uptake. Most importantly, this indicates that the slowest process of rate-limiting is chemical interactions between adsorbate and biochar surface, while physical diffusion is not (Li et al. 2024; Masuku et al. 2024; Jiang et al. 2024). Overall, adsorption processes were spontaneous in terms of thermodynamics, with negative Gibbs free energy (ΔG) coefficients. The adsorption mechanisms can be endothermic or exothermic depending on the application of the pollutant as well as on biochar properties, suggesting that it is necessary to optimize operational conditions according to the application (Li et al. 2024; Masuku et al. 2024).

Although biochar has significant potential as an adsorbent for pollutants, many challenges must be addressed, particularly in aspects of the regeneration and reusability of biochar. The biochar performance may diminish over multiple adsorption–desorption cycles, and low-cost regeneration methods need to be explored. This concept can be taken a step further by optimizing the pyrolysis conditions as well as making new modifications to the biochar (impregnation with metal oxides, functionalization with specific ligands) to increase the adsorption capacity and selectivity of the biochar. Addressing these themes will be crucial in improving the pragmatic applicability of applications, and by extension, the sustainability of biochar as an effective adsorbent for environmental remediation (Khanzada et al. 2024; Masuku et al. 2024).

3 Organic Pollutant Removal Using Biochar Under Stress Conditions

As energy demands continue to rise, and global climate change becomes a critical concern, interest in the use of biochar as a potential sorbent for removing organic pollutants under varying environmental stress conditions has been growing. This application is due to its high surface area, porous structure, and diversified functional groups, which could improve adsorption and degradation of contaminants as well. These characteristics make biochar particularly efficient in the treatment of recalcitrant organic pollutants, like PAHs and chlorinated compounds. However, multiple factors such as physicochemical properties of biochar, the modified contents, and environmental considerations have influence on the efficiency of pollutant (Beljin et al. 2024b).

Biochar removes organic pollutants via adsorption primarily. Biochar is well-known for its abundant porosity and a majority of active sites which are available to bind with pollutants, this results in high adsorption rates. PAHs and chlorinated organic compounds (e.g., Beljin et al. 2024a, b; Song et al. 2024). These results reveal that biochar would have a strong adsorption affinity to a broad range of hydrophobic pollutants with the adsorption efficiencies > 90%. In addition to physical adsorption, the surface functional groups existing at the biochar's surface are also significantly contributed to chemical interactions with organic pollutants. These interactions include π–π e. d. a., H-bonding and van der Waals forces that will increase the retention of pollutants at biochar surface (Song et al. 2024).

In addition to adsorption, biochar can serve as a catalyst, particularly when chemically modified. Due to this; functionalized biochar, particularly derived from water hyacinth, could facilitate AOPs for the degradation of organic pollutants. These modified biochars could have catalytic activity to produce ROS that can degrade macromolecular organic pollutants into less toxic small molecules (Hung et al. 2023). Impregnation of biochar with transition metals or metal oxides improves the catalytic activity of electron transfer reactions which consequently plays a vital role in decomposition of pollutants (Hung et al. 2023).

Apart from adsorption, catalytic properties of biochar may be found, particularly if it is chemically modified. Because of this, functional biochar, in particular water hyacinth-derived biochar, triggers advanced oxidation process to promote organic pollutants removal. These modified biochars can produce singlet oxygen, and this molecule is a reactant for breaking macromolecular organic pollutants into nontoxic molecules. Source: Hung et al. Biochar doping with transition metals or metal oxides increases the efficiency of catalyzing the electron transfer reactions that drive pollutant decomposition (Hung et al. 2023).

Although it is a promising biochar, without facing the difficulties of applying to changed environment differently. In specific environmental applications, factors such as pH, temperature, and the presence of competing ions can significantly influence adsorption capacity. Evidence indicates that biochar performance fluctuates across different environments; therefore, further research is required to establish optimal application rates and identify the specific conditions under which biochar is most effective in field settings (Beljin et al. 2024a). Moreover, different modifications have been studied to improve the adsorption capacity and stability of biochar. Besides, adding phosphoric acid to biochar could increase the surface area of modified biochar and then it tends to possess more functional groups which can efficiently remove various pollutants (Guo et al. 2023).

While biochar has promising potential in organic pollutant removal, the practical application of biochar should be enlarged and acceptable implementation under various environmental scenarios are yet to be achieved. Further work should be focused on solving these limitations, including the optimization of the modification method and durability in real applications. Faster work in these directions will facilitate practical applications of biochar for environmental remediation and enhance the contributions of its sustainable pollution control (Guo et al. 2023; Beljin et al. 2024b).

Under different environmental stress conditions, the various adsorption mechanisms are shown in Fig. 2. Notably, salinity mainly reduced biochar adsorption by

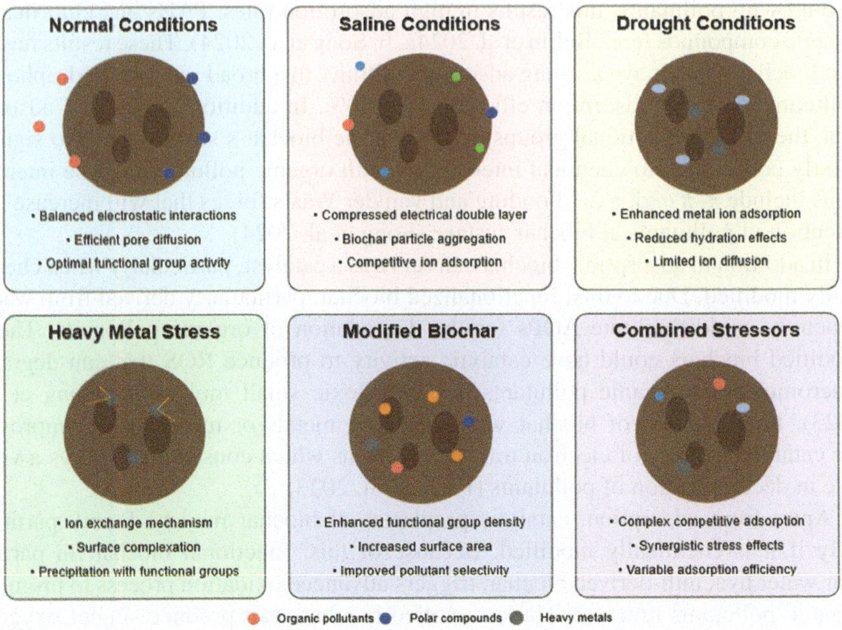

Fig. 2 Biochar adsorption mechanisms under different environmental stressors

the electrical double layer compression and the competitive adsorption of the ions, while the drought conditions enhanced the adsorption of metal ions via the decreased hydration effects and limited ion diffusion.

3.1 Mechanisms of Adsorption (Hydrophobic Interactions, π–π Stacking, Electrostatic Interactions)

Benefiting from its physicochemical properties and mass/complex adsorption roles, biochar is a promising candidate to immobilize organic contaminants in aqueous actual environments. Due to its high specific surface area, porous structure, and the existence of adjustable functional groups, it can efficiently absorb various pollutants and may have potential value in water purification applications. Meanwhile, biochar has many kinds of adsorption mechanism, such as hydrophobic effect, π–π stacking and electrostatic interaction (Sochacki et al. 2024), which make the different datatypes and structural characteristics of adsorption material produce different efficiencies for this type of adsorbents.

The hydrophobic effect is one of the most important adsorption-driving force between organic pollutants and biochar. The naturally hydrophobic character along with the fully developed pore structure of biochar may also facilitate it to absorb nonpolar organic pollutants significantly. It has also been established that organic pollutants are sorbed to spruce-beach biochar via ISR which is induced predominantly hydrophobically. They are much more effective in adsorption of hydrophobic organic compounds, and they remove these compounds from the aqueous solution (Sochacki et al. 2024).

Meanwhile, π–π stacking is not only hydrophobic adsorption but also one of the significant adsorption mechanisms for most aromatic compounds. In addition, the aromatic surface of biochar can be π–π stacked with some pollutants using benzenes rings or other aromatic structures for adsorption. This phenomenon is frequently reported in the literature as a dominant mechanism which contributes to the very high binding affinity of organic pollutants occurring on the biochar surface (showcasing once again that biochars are perfect adsorbents for aromatic pollutants) (Sochacki et al. 2024).

However, the presence of the required charged groups in biochar increases the adsorption potential through electrostatic attractions/repulsions. A high density of oxygen functional groups, particularly –COOH and –OH, yields negative surface charges for the adsorption of positively charged pollutants. On the other hand, protonated amine ($-NH3^{+}$) groups in biochar will yield electrostatic attraction to negative contaminants. It suggests that the mechanism of electrostatic attraction is beneficial to remove polar and ionizable organic matter, e.g., the chlorinated organics (Song et al. 2024).

Although biochar is a promising option in this regard, its large-scale adoption in wastewater treatment is hindered by various challenges, among which are the variability of environmental conditions. More research is needed to determine the impact of factors such as solutions pH, ionic strength, and competition from other ions in the solution to determine the conditions under which biochar adsorption capacity can be improved in order to increase effectiveness in actual applications. Additionally, biochar's regeneration and reusability pose major challenges that need to be overcome to improve its sustainable application for wastewater treatment in systems (Beljin et al. 2024b). It will be necessary to overcome these barriers to make biochar an integrated sustainable water purification technology first and use it to gain the maximum positive effect on the environment.

3.2 Influence of Salinity on Organic Pollutant Sorption Capacity

Salinity is an intricate factor linked to the biochar sorption capacity for organic pollutants removal, which is associated with the physicochemical properties of biochar, the targeted pollutants, and the ionic composition of the salt environment. However, biochar has been observed to have huge performance differences when subjected to changes in salinity, and this has a high effect on both the adsorption capacity and the overall removal efficiency of biochar-based adsorbents. The dominance of these effects is related to the competition of biochar surface properties and solution ionic strengths that can enhance or suppress sorption scenarios (Brunk et al. 1997).

Biochar's sorption capability is significantly impacted by salinity due to modifications in surface interactions and dissolved organic matter (DOM) layer deposition on sorbent surfaces. The interaction of biochar with organic contaminants, especially hydrophobic molecules, such as polycyclic aromatic hydrocarbons (PAHs), can differ in relation to ionic strength as a function of the medium. In our study, the salinity produced the compression effect of the electrical double layer, which probably strengthened the hydrophobic interaction between adsorbate and the biochar surface, thus promoting the adsorption of some organic pollutants. In particular, dissolved salt can facilitate the aggregation of biochar particles at high concentration, thus decreasing the availability of active adsorption sites and limiting the pollutant uptake capacity of biochar (Brunk et al. 1997).

The impact of salinity on biochar adsorption also varies significantly depending on the conditions under which the biochar was produced, most importantly, the temperature of pyrolysis. The high-temperature process pyrolysis biochar has the potential for high adsorption capacity through diversified pore space development, high surface area, and high thermal stability. For example, under saline conditions, the higher-temperature pyrolysis produces biochar with a greater maximum sodium (Na) adsorption capacity (applicable coefficients) and suggests that the structural properties of biochar determine the interaction mechanisms it has for ionic species. The results indicate need for further investigation into the ideal biochar production parameters when considering biochar's stability and performance in saline systems (Nguyen et al. 2021; Ambaye et al. 2021).

Biochar is also a contaminant removal medium using direct sorption processes, so the other important comparison under wider coverage based on applications of biochar can be aligned through biological filtration systems. A comparison of conventional BAC filter sets at salinity levels up to 1000 mgCl^-/L revealed that the removal efficiency for organic pollutants was not impeded, though slightly lower, due to high ionic concentrations on microbial communities and interactions on the biochar surface. It also confirms that biochar can retain adequate adsorption even when salinity levels are moderate, yet excessive salinity levels may require additional modifications to maintain biochar-response functionality (Sozanski 1995).

Although biochar has shown promise for removing organic pollutants from saline environments, a number of challenges still need to be addressed to maximize biochar performance under different salinity conditions. These results emphasize the need for additional investigations to obtain biochars with suitable characteristics to resist elevated ionic strengths but still preserve a satisfactory sorption capacity. Modification of biochar, for example, via surface functionalization, co-activation strategies, as well as biochar-coupling with complementary remediation technologies, should be the object of a research interest in the future, in order to tailor the application of biochars to different environmental contexts (Beljin et al. 2024b).

3.3 Effect of Drought Conditions on Biochar Adsorption Efficiency

Phytostabilization can be used in polluted soils during drought, which reduces their susceptibility to pollutants, the formation of chelates with organic pollutants, and complex physicochemical properties of the soil contribute to the effectiveness of biostable remediation of chlorostomic mycelium and fungi. This is not surprising because it has been speculated that stress conditions (e.g., soils under droughts) have been associated with the increased persistence of organic pollutants because of reduced microbial metabolism, changed bioavailability of pollutants, and restricted degradation processes occurring in the system. The role of organic application has been demonstrated to counteract these problems through their amendments that enhance the physicochemical properties of the soil (such as soil structure and water-holding capacity), and nutrient availability, ultimately resulting in a conducive environment for contaminant adsorption and degradation (Bhattacharya et al. 2024).

Some improvement was realized by application of biochar to dry soil. Considering these positive effects, biochar ability to reduce soil moisture loss is the most important factor contributing water retention. Superporous biochar/moisture complexes can hold lots of water, even during arid conditions and for vital soil processes (e.g., sorption of pollutants) when drought prevails. This additional enrichment of microbial communities with current status of biodegradation will help in effectively prolonging sorption of hydrophobic organic contaminants by providing more adsorption sites throughout the soil matrix (Bhattacharya et al. 2024).

It also enhances the cation and anion exchange capacity of soil with water remediation. Enhanced availability of nutrients also enhances soil microbiological activities ultimately leading to enhanced degradation and immobilization of organic pollutants (Bhattacharya et al. 2024) resulting into improved soil health.

Besides improving the physical, chemical, and microbiological properties of soil, not only can biochar directly improve drought stress resistance of numerous plant species, it indirectly promote the uptake effectiveness of pollutants into plants. Soil biochar amendments have been shown to mitigate drought-induced physiological limitations, leading to enhanced biomass and root growth for plant performance. Moreover, the increasing root biomass is beneficial to the phytoremediation ability of a plant, which results in more area for adsorbents and enhanced rhizosphere interaction that facilitates microbial degradation according to previous studies (Wu et al. 2023). In addition, they noted that biochar application was correlated with increase in chlorophyll and accumulation of proline (osmoprotectant which associated with drought tolerance). This signifies that abundance of proline could improve the cellular stability and protect enzymes in desert stresses to make plants more adaptable which may be a contributing factor to increased coincidence rate of soil pollutant removal (Mannan et al. 2017).

While the reported benefits confirm that biochar is effective in reducing pollutants, the effective removal of contaminants under drought conditions is highly dependent on the feedstock type, pyrolysis temperature, and pollutant characteristics. These variances can potentially influence the adsorption efficacy, interactions with microorganisms, and remediation performance of biocharadiated products and require further investigation to customize biochar composition for additional environmental applications. Nevertheless, that is a novel and future research domain that it should be encouraged, biochar modification that is targeted for the different types of pollutants and the regions where the drought is mostly likely to occur and so biochar application for minimizing the contamination in fields needs to be ensured at different climate conditions going forward (Bhattacharya et al. 2024).

4 Case Studies and Experimental Findings

Due to its high potential, services such as biochar for organic pollutant removal are being emphasized more within laboratories and fields that deal with various environmental stress conditions. Inherent structural properties, such as surface area, porosity, and functional group distribution, are influenced by the feedstock used, required pyrolysis conditions, and the post-treatment modifications. One of the challenges in this regard is evidenced by the fact that the adsorption process (the process of a solid absorbing molecules to form a film on the surface) is also highly variable and dependent on the compound that is desired to be removed, so it is important to understand how these two factors work together to achieve the best result (Beljin et al. 2024a, b).

The pyrolysis procedure converts biomass to biochar, generating a highly porous material with a surface area of up to 1500 m^2/g, enhancing its capacity to adsorb various organic pollutants. These engineered biochars' physicochemical characteristics and mechanics can be altered by activating with oxidizing agents or metal doping, further increasing remediation efficiency (Beljin et al. 2024b). In engineered biochars derived from agricultural waste materials, disparate pore structures are comparatively bio-inspired, serving as a vital platform for organic contaminants' adsorption and desorption. Feedstock composition and pyrolysis parameters determine chemical functionality of biochar surface and consequently its affinities for various pollutant classes (Nguyen et al. 2024).

Even under stress conditions, such as those prevalent in the environment, biochar continues to perform well for the sorption of pollutants. Biochar, a promising biosorbent for pharmaceutical constituents' removal such as oxytetracycline (OTC), has been studied previously, showing as high as 93% removal even after reusability of the biochar produced from coir. This study highlights the potential of biochar for sustainable remediation strategies in environmental applications, particularly where reuse and long-term effectiveness are essential, such as in wastewater treatment (Nguyen et al. 2024). Furthermore, the association of biochar with microbial populations in situ within the rhizosphere can be an important mechanism for promoting the degradation of pollutants by biochar. The porous matrix of biochar has greater microbial colonization and more biotransformation of organic contaminants under stress conditions (e.g., stress triggered by climate variability) (Pathak et al. 2024a).

Adsorption is a driving process in the mechanisms by which biochar can remove organic pollutants, with contributing actions including pore filling, Van der Waals forces, and electrostatic attraction. The physicochemical properties of biochar, including but not limited to surface charge, functional groups, and hydrophobicity, greatly influence the adsorption performance. Moreover, it has been reported that the modified biochars could enhance advanced oxidation processes and thus dramatically benefit the decomposition of macroorganic pollutants. These changes, which can occur via chemical activation or of loading with catalyst agents, promote the oxidative degradation of large organic molecules, thus increasing biochar applications for the removal of environmentally significant pollutants (Nguyen et al. 2024; Htet et al. 2024).

Biochar is demonstrated to be an excellent sorbent for organic pollutants, although challenges remain for the wide spread application of biochar in sludge or soil management. The biochar characteristics vary depending on the feedstock, and production conditions, which need to be further explored for potential improvements in soil applications. Following a follow-up should be a process engineering review to enable the current treatment systems to adjust according usual remediation technologies. To bridge these shortcomings, future studies should (1) refine biochar's production process, (2) investigate its new mode of modification and (3) determine the actual field applicability of biochar (Beljin et al. 2024a).

4.1 Inorganic Pollutant (Heavy Metals) Removal Under Stress Conditions

Research on remediating heavy metal pollution during agricultural resilience and environmental sustainability remains a key research topic under stressful environmental conditions. The presence of heavy metals in soil and water is a serious threat to plant growth, microbial community development, and biodiversity. As a result, there have been extensive studies on biological methods, from phytoremediation to microbes, to alleviate heavy metal pollution. However, the effectiveness of these techniques is affected by various factors such as plant species, microbial consortium, and physicochemical properties of contaminated environments (Maitry et al. 2024).

This phycoremediation is an economically viable approach that uses plants to uptake, bioaccumulate, and detoxify heavy metals and other contaminants from polluted soils. Reducing metal toxicity through this process leads to sustainable agriculture by improving soil fertility and plant tolerance. Generally, the main mechanisms that can be combined or occur concurrently in phytoremediation are phytoextraction, phytostabilization, and phytovolatilization, where heavy metals are taken up and stored in plant tissues or transformed into less toxic compounds ("In situ bioremediation of heavy metal contaminated soil" 2022). Brassica juncea and Helianthus annuus have been shown to be effective remediation species for lead (Pb), cadmium (Cd), and arsenic (As) in several studies involving metal-contaminated sites. These species also show improved stress tolerance to heavy metals and assist in the detoxification of polluted habitats (Almotairy 2024). While phytoremediation has many benefits, large-scale implementation is restricted due to limitations such as low plant growth rates, metal bioavailability limitations, and long-term site management (Maitry et al. 2024).

Besides phytoremediation strategies, microbial-assisted bioremediation has been developed as a novel heavy metal removal technology. Microorganisms are involved in metal detoxification through bioaccumulation, biosorption, and the production of extracellular polymeric substances (EPS) that promote immobilization and transformation of toxic metals into less toxic forms (Chakraborty et al. 2021). Microbial-assisted bioremediation strategies utilize specific bacterial strains, such as *Klebsiella* sp., which are documented to possess significant bio-removal potential and can even promote plant growth under heavy metal stress conditions. These interactions are often mutualistic, enhancing the effectiveness of phytoremediation by stimulating plant growth, improving nutrient assimilation, and reducing oxidative stress, while simultaneously facilitating metal sequestration within the rhizosphere (Chakraborty et al. 2021).

Both phytoremediation and microbial-assisted bioremediation represent sustainable strategies for the remediation of heavy metal pollutants, but challenges persist in the global optimization of their application under a range of environmental conditions. For example, both genomics approaches aimed at elucidating the plant–microbe interactions occurring during metal removal, as well as limiting genetic modifications to hyperaccumulator species, coupled with the design of new biotechnological solutions acting at lower efficiencies of metal removal, remain areas of active research (Ren et al. 2022). Furthermore, for remediation practices to be sustainable and to address the limitations of single-method solutions, multiple strategies must be employed, for example, chemical amendments and nanotechnology-based approaches.

While individual methods such as phytoremediation and microbial treatments can help reduce metal concentrations, their effectiveness is not always enough due to the scale of heavy metal pollution. Contaminated ecosystems have complex dynamics that make it necessary to adopt integrated approaches involving treatment solutions that are both physically and scientifically capable of restoring all aspects of the ecosystem over the long term. Hence, subsequent approaches must be targeted at constructing integrated remediation frameworks utilizing the strengths of different methodologies for maximizing pollutant extraction and ecosystem restoration (Maitry et al. 2024).

4.2 *Mechanisms of Metal Ion Adsorption (Ion Exchange, Complexation, Precipitation)*

The prevention of heavy metals is accomplished by such as ion exchange, complexation, and precipitation process in the wastewater and it makes a good removal efficiency under suitable circumstance. Meanwhile, it is also important that the physical–chemical features as well as environmental aspects of these adsorbents play a significant role in those processes. The optimal exploitation of the driving forces behind these processes is essential to meet the demand for feasible and environmentally sustainable treatment of complex wastewater matrices, thereby achieving higher removal efficiencies.

This removal of the metallic ions onto the adsorbents is also reversible, a cationic charge exchanged with metal ion to be replaced by that of the metallic cation in water. Zeolites and ion exchange resins (Raji et al. 2023; Rajput et al. 2023): A greatly facilitated adsorption rate of cationic species such as Pb and Cd in this system due to high adsorbent–adsorbate affinity makes the process a highly efficient means for eradication of these pollutants because complex formation between a metal ion and intrinsic or surface-appended ligands on the adsorbent. In the present of organic material and chelating agents, this second mechanism increases solubilization of the metal species, consequently promoting its leaching (Oyewole et al. 2018; Pathak et al. 2024b). The anions will react with the metal cations to produce insoluble salts that precipitate out a convenient way of getting those into hem out of solution. Generally in chemical treatment, pH is the most critical parameter to aid precipitation of hydroxides or sulfides for removal of majority of metals as per whole remediation plan (Pathak et al. 2024c; Rajput et al. 2023).

While these mechanisms work effectively in laboratory settings, the challenge lies in optimizing the removal conditions when the environmental stressors are changed. There is an urgent need for further research to better comprehend these interactions, which is vital for developing integrated approaches capable of reliably achieving optimum heavy metal removal efficiency under real-world wastewater treatment conditions.

4.3 *Role of Biochar in Mitigating Heavy Metal Toxicity Under Salinity Stress*

Soil amendment with biochar has been assessed against the enhanced toxicity of heavy metals due to salinity. Its application has been proven to reduce the uptake of heavy metals, increase nutrient availability, and improve plant physiological parameters in contaminated and saline soils, thereby improving agricultural sustainability. The physicochemical properties of biochar play an essential role in its interactions with the soil matrix and plant systems.

After a biochar is added to a system, a lot of routes will work together to reduce heavy metal toxicity. Among diverse action mechanisms, immobilizing heavy metals (e.g., Pb, Cu, Zn, and Cr) in soil through biochar application is vital to hold pollutants from leaching in the runoff and plant uptake (Tang et al. 2022). Moreover, biochar also promotes organic matter, CEC, and pH of the soil which enhances the soil fertility and lessens toxic heavy metal impact in it (Tang et al. 2022; Premalatha et al. 2023). Biochar can also enhance nutrient uptake through altering the chemical environment of rhizosphere, which is essential for sustaining plants' growth and metabolism upon exposure to heavy metals (Tang et al. 2022).

Besides attenuating heavy metal toxicity through immobilization, biochar can alleviate salinity stress so frequently exaggerated by metal stress to the crops grown under contaminated conditions. An important benefit of biochar use is to improve the water-holding capacity in soil, which affects plant growth and survival in saline areas (Water as limiting factor) in many places (Gao et al. 2024). Furthermore, biochar application has been shown to decrease electrolyte leakage and oxidative stress in plants that can maintain the cellular structure and safeguard normal physiological processes against salinity stress (Gao et al. 2024). In addition, biochar could also enhance antioxidant enzyme activities which are the major participants in detoxifying ROS and scavenging ROS that caused by heavy metal composition and salinity stress (Tang et al. 2022; Lohar et al. 2024).

While biochar holds great potential as an amendment to alleviate heavy metal and salinity stress, it is extremely sensitive to its feedstock and production conditions that affect its physicochemical characteristics. Biochar derived from diverse feedstocks and pyrolysis conditions is known to vary considerably in adsorption capacity, potential for nutrient retention, and effectiveness in different soil types and conditions (Table 1). Therefore, further studies are needed to identify optimal biochar formulations and application strategies, which can help reduce heavy metal toxicity and salinity stress in diverse agroecosystems.

4.4 Adsorption Capacity Variations Under Drought Conditions

Soil minerals' adsorption of inorganic pollutants partly reflects the preferred conditions for heavy metal (Cd^{2+}) adsorption and is significantly affected by drought conditions. Soil moisture decreases, causing changes in the physicochemical environment of metal ions and mineral surfaces, which enhances adsorption. This is mainly driven by nanoscale phenomena, due to variations in soil water content affecting electrostatic forces, ion mobility, and hydration effects. The drying-induced remnant solution reorganization favors higher retention of heavy metals on soil mineral particles like montmorillonite, with implications for contaminated soil and remediation strategies under drought-stressed conditions.

Table 1 Effect of environmental stressors on biochar properties and performance

Environmental stressor	Effect on biochar properties	Impact on adsorption performance	Mitigation strategies	References
Salinity	• Compression of the electrical double layer • Formation of dissolved organic matter coatings • Promotion of biochar particle aggregation	• Enhanced hydrophobic interactions with nonpolar contaminants • Reduced availability of active adsorption sites • Altered surface charge distribution	• High-temperature pyrolysis (> 600 °C) • Surface functionalization • Co-activation strategies	Brunk et al. (1997), Nguyen et al. (2021), Ambaye et al. (2021)
Drought	• Reduced moisture in pores • Altered pore water chemistry • Changes in biochar surface reactivity	• Improved adsorption of heavy metals • Enhanced ion exchange capacity • Reduced microbial degradation of pollutants	• Combined application with soil conditioners • Biochar-microbial consortia • Optimized application rates	Bhattacharya et al. (2024), Wu et al. (2023), Mannan et al. (2017)
Heavy metal presence	• Occupation of adsorption sites • Alteration of surface charge • Potential catalytic effects	• Competitive adsorption with organic pollutants • Enhanced removal of anionic contaminants • Reduced microbial activity	• Chemical modification (acid/base treatment) • Metal oxide impregnation • Biochar-based nanocomposites	Shar et al. (2025), Simamora et al. (2024), Wang et al. (2024)
Combined stress (salinity + drought)	• Synergistic effects on porosity • Increased hydrophobicity • Complex surface chemistry changes	• Variable effects dependent on pollutant properties • Generally reduced adsorption capacity • Altered adsorption kinetics	•Multiple-stage modification processes • Tailored feedstock selection • Controlled pyrolysis conditions	Yang et al., (2020), Anbuganesan et al. (2024)

Note This table summarizes the key effects of environmental stressors on biochar properties and performance based on the literature reviewed in this chapter

Table 2 Enhancement of sugarcane bagasse biochar through SDS modification (Nhàn et al. 2024)

Parameter	Unmodified biochar	SDS-modified biochar	Improvement
Methyl violet removal efficiency	64.30%	97.54%	33.24% increase
Maximum adsorption capacity	10.59 mg/g	14.68 mg/g	38.62% increase
Adsorption rate constant	0.077 g/mg·min	0.161 g/mg·min	109% increase
Time to reach adsorption equilibrium	~ 60 min	~ 15 min	75% reduction
Surface properties	Carbon (87.72 wt%) Oxygen (12.28 wt%)	Added sulfur (0.3 wt%) Enhanced functional groups	Improved surface functionality
Adsorption mechanism	Langmuir model (R^2 = 0.979)	Langmuir model (R^2 = 0.971)	Maintained the chemisorption mechanism
Production method	Pyrolysis at 600 °C in N_2	Pyrolysis at 600 °C in N_2followed by SDS treatment (0.1–0.4 g/L)	Simple modification process

As the water content of the clay network decreases, the pressure makes the electric double layer collapse on clay surfaces. With this, the negatively charged clay particles have a greater concentration of cations in the confined space due to the diminished moisture, which strengthens electrostatic coupling forces between the clay surface and heavy metals: e.g., Cd^{2+} (Tian et al. 2024a). This compression effect enhances the binding affinity of metal ions with soil minerals, reducing their mobility and increasing retention in the soil matrix (Tian et al. 2024b).

Furthermore, the limited diffusion of heavy metal ions occurs due to their low water content in drought conditions. Since a few water molecules exist per clay layer, there is limited passage of ions (Tian et al. 2024a), and as a result, the mobility of Cd^{2+} is limited, which tends to keep the metal ions in contact with clay for more time. The lower diffusivity that results generates greater adsorption because metal ions will tend to form stable complexes with the soil particulates rather than being leached from the soil or carried downwards in the soil profile (Tian et al. 2024a). Moreover, the reduced hydration effects in drought conditions significantly contribute to the improved heavy metal adsorption. Fueled by hydration forces, repulsion due to the presence of water molecules, metal ions are prevented from forming close-range (e.g., surface complexes) interactions with mineral surfaces in well-hydrated soils. Under drought stress, however, these hydration forces are substantially diminished, leading to more direct interaction of the Cd^{2+} ions with the clay particles, and hence, this remarkably improves their adsorption capacity (Tian et al. 2024b). Declined hydration enhances the electrostatic and van der Waals interaction between metal ions and soil minerals, which also contributes to the stabilization of heavy metals in the soil environment.

Further, heavy metal adsorption on soil surface driven by drought further causes irreparable harm and fate of contaminants in the long-run soil health. In fact, the adsorption of heavy metal might lead to pollutant local hot spots in re-hydration domains and thereafter cause desorption effects and pollutants sudden release. It needs therefore to elucidate the complicated mechanisms for drought-stress-regulated heavy metals adsorption and to provide a basis or implication for increasing ability of soil productivity and protection, which could be the precondition in effort to mitigate environment risks of the hazard by heavy metals pollution in these arid and semi-arid areas.

4.5 Real-World Applications and Field Studies

This is why biochar is a great option to improve soil health and crop yield, even in times of stress. The unique nano-structure of biochar and its large specific surface area can efficiently adsorb the heavy metals, whereby it is a production means to remediate in agriculture (Wang et al. 2024) and environmental fields.

In the case of physical adsorption, the metal ions are physically entrapped within the biochar pore sites and thus reduced in their bioavailability (Shar et al. 2025). The possible role of electrostatic attraction is a major factor, as the negatively charged surface sites of biochar are capable of attracting positively charged metal ions (Shar et al. 2025), leading to increased adsorption efficiency. Also, chemical complexation might occur between biochar functional groups and heavy metal thereby forming stable complexes for metal immobility (Simamora et al. 2024).

Likewise, the use of biochar in the field has shown success in reducing heavy metal stress in plants. Thus, biochar application improves the growth characteristics of crop plants, particularly tobacco and maize, as an example of biochar application supporting these against heavy metal stress (Shar et al. 2025; Zahra et al. 2024). On the other hand, drying before pyrolysis has produced biochars that effectively remove metals such as Cu, Zn, and Cd with an efficiency above 90% (Simamora et al. 2024).

Nevertheless, these effects illustrate the great potential of biochar for immobilization and removal of heavy metals from hydroponic and soil systems. The choice of the type and application rate of biochar affects to a great extent its adsorptive capacity and efficiency; therefore, the selection is an essential requirement for achieving optimal benefits (Zahra et al. 2024). In addition, biochar when used in remediation can also be improved with beneficial microbes (Pathak et al. 2024b), and this putatively synergistic application of microbial consortium with biochar in r-mining should be further exploited on a large scale to optimize these high-impact effects especially for field applications.

In brief, biochar can be considered as an effective adsorbent for the remediation of heavy metal, but this will have to pay attention on environmental condition and pollutants. The results presented could serve as a benchmark to investigate if it can be more efficiently used (along with microbial rheo-recyclables) for contaminated soils in order to enhance sustainable agriculture and eco-restoration.

5 Effect of Environmental Stressors on Biochar Performance

By the contrast to the recorded literature, biochar has good prospects as an amendment that minimizes the soil performances of which are affected/exacerbated by drought, salinity, and heavy metals toxicity. It also helps to improve the physico-chemical properties of soil which leads to microbial activities promotion simply in such a way that would enhance nutrient cycling that is needed for plants survival under stress conditions (Chi et al. 2024a, b; Rathinapriya et al. 2025a, b). WHC has been shown to improve with biochar applications, and the salinity effects to both biomass and photosynthesis have in parts of the country where it is a management tool (Patani et al. 2023). Besides, biochar has significantly improved soil quality and fertility under abiotic stress which in turn could contribute to the enhancement of sustainable agriculture (Lentini et al. 2024a, b). While these accomplishments have been significant, the temporal dynamics of biochar for feedstock type and potential beneficial by products should impact the ultimate utility of biochar as a soil amendment. Further research is required to evaluate how biochar biostimulates crops across a broad range of conventional production systems, including specialized high-density models such as pecan production.

5.1 Impact of Salinity on Adsorption Kinetics and Desorption Rates

Biochar has been reported to enhance soil fertility and crop productivity in the presence of different environmental stress conditions such as salinity. Several soil parameters, such as cation exchange capacity and water-holding capacity, are enhanced by the application of biochar under saline conditions (Chi et al. 2024a, b; Hasnain et al. 2022). For instance, a recent investigation (Lentini et al. 2024a, b) demonstrated that biochar can mitigate the salinity stress on tomato plants by improving both yields and fruit quality cultivated under brackish water irrigation.

Moreover, biochar improved the functioning of nutrient cycling and microbial activities, which can help plants in resisting abiotic stresses (Rathinapriya et al. 2025a, b; Zhang et al. 2020). Nevertheless, the drought and salinity stress synergize herein and crop responses may be interplayed with their tolerance to implicated stresses, which is reported as mitigated by biochar that would enhance both water use efficiency of water productivity in soybean (Zhang et al. 2020). Thus, biochar is an ideal amendment for sustainable agriculture in saline soils.

5.2 Drought-Induced Changes in Pollutant Interactions with Biochar

The biochar with the application rate of 20t/ha had the strongest capacity for reduction in pollutant interaction, while other environmental stress significantly affect this capability. Changes in soil properties and dynamics of pollutants due to drought can alter the value-added utility of biochar for pollutant sequestration and soil improvement.

Application of biochar was also proven in this study to enhance water-holding capacity of the soil which is important during drought events as it ensures biological activities involved in contaminant removal (Bhattacharya et al. 2024). Although drought reduces nutrient mobility, results indicate nutrients are ameliorated and available for plants to grow well and uptake more pollutants in the biochar (Bhattacharya et al. 2024).

Because of its ability to sorb heavy metals, biochar also reduces the levels and toxicity of these elements for plants under drought-induced stress (Lohar et al. 2024). It is noteworthy that biochar application effectively promotes the degradation of wastewater pollutants and provides significant environmental benefits under drought stress conditions (Wang et al. 2024).

There has been evidence for detrimental effects of biochar application in the above studies via reduced drought stress on and a reverse effect whereby, exposure to biochar could enhance microbial resilience against drought-stressed removal from pollutants such as antibiotics and pesticides.

This indicates that the stronger effects of biochar addition on soil properties and pollutant interactions under dry conditions should not be overlooked, but note that biochar quality differs according to its production parameters, and the pollutants it interacts with. So more research is going to be needed to optimize biochar recipes for particular environmental conditions and enable them to be deployed with maximum efficiency.

5.3 Combined Effects of Multiple Stress Conditions on Biochar Efficiency

Supporting physiological functions and growth performance, biochar has been proven effective in increasing plant tolerance to several abiotic stresses such as salinity, drought, and heavy metal toxicity. For example, the biochar application improved quinoa plant height and shoot weight by 188% under drought and salinity stress compared to control treatments (Yang et al. 2020). Like those two studies, biochar also reduced combined salinity and drought stress in Vicia faba with increased transpiration rate and internal CO_2 concentration ("Impact of biochar soil amendment on alleviation of single and combined effect of drought and salt stresses on Vicia faba seedlings planted in alkaline soil" 2022). For instance, in Zea mays, adding biochar and plant growth-promoting rhizobacteria enhanced heavy metal and drought tolerance via increased enzymatic activity and stress degradation (Anbuganesan et al. 2024). Moreover, biochar enhances soil cation exchange capacity and water-holding capacity, which is the basis for efficient plant growth under stressed conditions (Awad et al. 2023). Yet, the persistence of this product relies on its interaction with application rates and stressors, requiring site-specific agricultural application optimization via big data.

6 Modification Strategies for Enhancing Biochar Adsorption Efficiency

The modification methods to improve the adsorption ability of biochar included chemical-physical method, physical method, and compound preparation. A wide range of chemical modifications have been performed on the chitosan to improve its dye adsorption capacity such as sodium dodecyl sulfate (SDS) treatment. Table 2 presents a comparison of various chemical and physical modification methods used to enhance biochar adsorption performance. For instance, modifying sugarcane bagasse biochar with Sodium Dodecyl Sulfate (SDS) treatment has been shown to significantly improve methyl violet adsorption, with the removal efficiency increasing from 64.30% to 97.54% (Nhan et al. 2024).

As reported, metal ion modifications, particularly those introducing Cu^{2+} ions into the system, could enhance the adsorption properties of sulfurous gases. The experiments found that the maximum sorption capacities of Cu^{2+}-modified biochar for hydrogen sulfide, methyl mercaptan, and dimethyl disulfide were 20.50 mg/g, 36.50 mg/g, and 57.98 mg/g, respectively (Wei et al. 2024).

The cadmium adsorption capacity of biochar was improved from 4.20 mg/g to 58.62 mg/g through the introduction of COFs by more exposure of nitrogen and oxygen functional groups (Hou et al. 2024). The two-stage alkali intercalation and sequential acid exfoliation physical processes performed with over 99% contaminants removal efficiency, significantly higher than the unmodified biochar (Pochampally et al. 2024). In addition, ball milling can further increase the surface area and active sites of biochar, as shown by an adsorption capacity of 185.18 mg/g for methylene blue (Lamberti et al. 2024).

6.1 Chemical Modifications (Acid/Base Treatment, Functional Group Grafting)

Various means of modifying the AD biochar were implemented to improve the adsorption performance and physicochemical properties for increased specific surface area. This kind of modifications can enhance the adsorption capacity due to its introduction of specific groups on the surfaces for pollutants affinity and thereby may also be applicable in increasing biochar efficiency for environmental remediation.

Several chemical treatments (such as using sodium dodecyl sulfate) also achieve significant improvement in the dye adsorption (selectivity). SDS was for example used in the development of sugarcane bagasse biochar for increased adsorption efficiency of methyl violet, from 64.30 to 97.54% and maximum adsorption capacity recorded to be 14.68 mg/g (Nhàn et al. 2024). The modification will come across the functional groups impregnated on to biochar structure and their affinities toward the dye molecules were enhanced.

Modification of metal ions is another effective way. For the moment, Pine-bark biochar/Cu^{2+} doped is most effective in S-containing gases adsorption comparing to others and its reported maximum adsorption capacities as sulfur are 20.50 mg/g (H2S) and 57.98 mg/g (DMDS) (Wei et al. 2024). Biochar is porous and its surface area is high; since it contains copper ions on its surface, the reaction with gaseous pollutants at a larger extent can be performed.

There is also another type of physical modification method including two-step alkali intercalation and acid exfoliation processes, which can be applied for obtaining acceptable performances. For instance, edible adsorbent biochar can be prepared using KOH followed by H_3PO_4, with the removal efficiency of malachite green higher than 99%, while unmodified biochar was inherently unable to do so from a beginning (Pochampally et al. 2024). This intensifies the sites and biochar adsorption ability on account of enhanced surface properties.

The second physical modification is ball milling, which improves adsorption performance of biochar. This significantly amplifies the surface area and available surfaces of adsorption in biochar with methylene blue adsorption capacity up to 185.18 mg/g, at maximum (Lamberti et al. 2024). The purpose of the mechanical treatment was to improve the structural characteristics of biochar and improving its porosity, reactivity with pollutants.

While these strategies for modification demonstrate great enhancement of adsorption efficiencies, their potential environmental impact and economic feasibility for widespread application must be duly examined. Despite the promise of using modified biochar for environmental remediation, the challenge of maintaining performance while ensuring sustainability persists in real-world applications.

6.2 Physical Modifications (Activation, Thermal Treatments)

However, there are well-known methods to modify biochar adsorption capacities, such as biochar physical modification (activation, thermal treatment, etc.). These changes improve biochar's surface area, porosity, and functional groups, making it effective for various applications, including pollutant removal.

Activation processes such as chemical activation with phosphoric acid, for instance, increase the surface area of biochar. For example, phosphoric acid-treated calamus biochar had a surface area of 797 m^2/g and enhanced the removal of erythromycin and sulfamethoxazole (325 and 216 mg/g, respectively) (Xu et al. 2024). The larger surface area and functional groups ensure maximum interaction between biochar and contaminants, which in turn enhances pollutant removal efficiency.

The physicochemical process from thermal treatment, for instance, the pyrolysis at high temperature, significantly affects biochar properties. According to experimental studies, pyrolysis at 600 °C generates a very porous structure, which leads to an increase of the sites for adsorption. The adsorption of methyl violet significantly increased from 64.30% to 97.54%, following treatment of the sugarcane bagasse biochar (Nhan et al. 2024). This alteration enhances the electrostatic attraction and hydrophobic forces between the contaminants and the biochar surface, thus promoting the adsorption efficiency.

Functional groups can be introduced through chemical modification, enhancing the adsorption mechanisms of biochar. At the same time, the oxygen-containing functional groups introduced increase the possibility of complexation and electrostatic interaction with pollutants, which may increase the adsorption capacity (Xu et al. 2024). In addition, customization of the microporosity can be increased for a more significant uptake of larger molecules to maximize utilization of the same material (Zomorodbakhsh et al. 2023).

While such modifications can greatly improve adsorption efficiency, it needs to be noted that this would come with a trade-off between cost and complexity of modification, thus limiting their use in certain configurations. There is currently a lack of large-scale, successful, real-world applications for biochar that support both environmental remediation and profitability, highlighting the need for more studies about optimal biochar modifications.

6.3 Biochar-Based Nanocomposites for Improved Pollutant Removal

Nanocomposites based on biochar have attracted great attention for removing pollutants, especially in wastewater treatment. Nanomaterials loaded in biochar matrices enhance adsorption and catalytic properties, thus reinforcing biochar as a multifunctional treatment media to address difficult pollutants within the overall environmental zone (Han et al. 2024).

In the removal of pollutants through the composites, the process that predominates is adsorption. Due to its inherent porous structure and surface functional groups, biochar can retain various contaminants. The existence of metal nanoparticles increases dye molecules' adsorption capacity which is more due to the increase in active sites and surface areas (Nosratabad and Nabid 2023).

Another important property of biochar-based nanocomposites emphasized in this work is the catalytic activity. In addition, it also imparts heterogeneous catalysis properties to nanomaterials like nZVI, and thus, an AOP (i.e., advanced oxidation process) can be applied to degrade pollutants. For example, hexavalent chromium (Cr(VI)) is widely applied in industrial activities and proved to be effectively extracted from aqueous solution by biochar-supported nanoscale zero-valent iron (Kochito and Alemayehu 2022).

Mechanical performance has been improved by various modification strategies. Microbial inoculation as (the immobilization of microbial mass, such as myceliums within biochar matrix) has been proved to increase the adsorptive performance due to more surface area and porosity (Zhang and Li 2024). Furthermore, aggregates biochar-based composite structures on metal oxides (manganese oxides) were also synthesized by incorporation of MNs which could produce the active and adsorptive site-enriched biochars with high efficient performance for heavy metals and organic pollutants removal (Nosratabad et al. 2024).

Practical studies focus on the functionality of these nanocomposites. Manganese oxide-biochar nanocomposites, for instance, have been reported to achieve over 99% removal efficiencies of Cr(VI) for the treatment of wastewater, thus opening possibilities for large-scale environmental remediation (Kochito and Alemayehu 2022).

However, there are still challenges with scaling these technologies. Compared to this, more research is required regarding economic viability for large-scale manufacturing as well as the stability of nanocomposites under diverse environmental conditions. NRFSI also provides a preliminary selection of materials that have yet to be optimized for other real-world applications, making it a promising area of ongoing research (Han et al. 2024).

7 Applications and Implications in Soil and Water Remediation

Biochar is a multipurpose application that has uses and effects on soil and water treatment. These specific properties (e.g., high pore space, cation exchange capacity and pollution capturing) render them particularly suitable for use as an effective amendment to polluted soils and aquifers (Dong et al. 2021). These structures are good for soil health due to better nutrient compaction and water-holding capacity which eventually helps in the growth of plant and soil microbe (Pradhan et al. 2024).

In a soil remediation context, organic and inorganic pollutants can be adsorbed by biochar like heavy metals and pesticides with less bioavailability and toxicity. Moreover, along with the immobilizations of toxicants in soils, there can also be improved soil fertility as result of increased organic carbon content and cation exchange capacity (Pradhan et al. 2024). This will result in higher crop yields and land rehabilitation. Biochar with an abundant carbon material amount acts as a desirable media for the removal of various contaminants such as heavy metals and organic matter in an aqueous system in water remediation field. It is also successfully applied in sewage treatment systems with high pollutant removal efficiency (Dong et al. 2021), which can lead to improvement of water quality and restoration of ecosystem.

However, the literature also notes many difficulties to address, for example, the heterogeneity of biochar material properties and the need for standardized procedures for biochar application. Nonetheless, some of these issues need to be addressed to achieve its potential as an environmental management tool.

7.1 *Biochar's Role in Restoring Degraded Soils Under Salinity and Drought*

Enhancement of soil physicochemical characteristics and increased crop productivity due to biochar significantly contributes to rehabilitating salt-alkaline and drought-affected soils. According to meta-analyses, biochar application can decrease soil electrical conductivity (EC) by an average of 13.2% to 7.4%, reducing salinity stress, and increase soil organic carbon and total nitrogen by 55.1% and 31.3%, respectively (Wang et al. 2024). Moreover, biochar increases the cation exchange capacity (CEC) by approximately 49% and also decreases the sodium adsorption ratio (SAR) and exchangeable sodium percentage (ESP) by more than 30%(Mao et al. 2024). Ideal biochar properties (pH of 7–8 and pyrolysis temperatures of 400–600 °C are important to obtain these benefits (Wu et al. 2023). Also, biochar has been reported to aid water retention, which is important during drought and increases crop resilience in hostile environments (Wu et al. 2012; Rathinapriya et al. 2025a, b). Biochar proves to be a potential addition to saline and drought-stricken regions.

7.2 *Synergistic Effects with Microbial Communities for Pollutant Degradation*

Synergistic interactions among microbial communities have greatly enhanced pollutant degradation in many ecosystems. For example, bacterial consortia present in the rhizosphere are found to enhance the degradation rates of polycyclic aromatic hydrocarbons (PAHs) by 5.5–13.1% over that of individual treatments, demonstrating a robust synergistic interaction between the plant and other microbes (Cheng et al. 2024). The prominent core microbiomes, especially the taxonomic units grouped into the Rhizobiales order, could achieve superior efficiency in 1,4-dioxane biodegradation in synergistic collaboration with key genera, such as Shinella and Xanthobacter, which provide a vital foundation for bioremediation efforts (Tian et al. 2024a). Additionally, novel fungal–bacterial consortia have been reported to degrade petroleum hydrocarbons to a greater extent than pure cultures due to greater utilization of commensal control mechanisms that further these types of metabolic pathways (Rezaei and Moghimi 2024). It highlights the importance of microbial interactions in enhancing the degradation of pollutants, with implications for the use of engineered consortia in bioremediation (Borreca and Bhoi 2024; Bhoi and Borreca 2024).

7.3 Potential Agronomic Benefits and Challenges

It has its benefits when used positively for soil and water remediation and has its cons as well concerning soil and water remediation. Because of this specific surface area and cation exchange capacity, it has significant effect on soil structure, immobilization of heavy metals and pollutants, increases nutrient and water retention, which suggests that biochar is a very effective method of soil remediation (Xu et al. 2025; Kukowska and Szewczuk-Karpisz 2023; Itam et al. 2024). An example is the biochar obtained from food waste, which improves soil fertility and microorganisms and is a low-cost method for remediating greenhouse gases (Pradhan et al. 2024). Despite these, there are still existing challenges, such as the need to optimize the biochar synthesis process and the potential of dependent efficacy on feedstock and pyrolysis conditions (Xu et al. 2025; Pradhan et al. 2024). Moreover, the secondary pollution risk of biochar is less than that of other organic/carbon materials, but long-term impact evaluation on soil health, crop safety, etc., is still needed (Dong et al. 2021). So biochar may have some potential use in remediation, but more research is needed to address these concerns and utilize the potential of biochar.

8 Challenges, Limitations, and Future Prospects

8.1 Technical and Economic Barriers to Large-Scale Biochar Application

Several technical and economic limitations impede the widespread use of biochar. The principle obstacle to the small number of practical implementations is governance issues (complexity of policies, permits, etc.). Challenges under economy: (1) costs of production; (2) stability of market still irritate development because many stakeholders have not been convinced that biochar projects will be economically viable. The technical part has offered a novel avenue for biochar properties and how it is made to be varied, and it has resulted in substantial effects on the effectiveness of its application based on previous research concerning variations among different biomasses produced through diverse methods which deserve more attention for optimization. But, also there are worries related to negative effects on soil health and plant growth, which should be described and prevented (Wang et al. 2022). The implementable policy, market place equilibrium. and continued science outreach will inform the future of biochar in anthropogenic systems in agriculture and environment.

The cost of biochar production is a major barrier to its widespread application. Production costs of biochar range from $300 to $1,200 per ton, depending on feedstock, production process, and scale, which outweighs these examples of potential benefits in some agricultural applications, unless carbon credit and payment for ecosystem services are considered. Advanced biofuel plants require significant capital ($0.5–5 million per plant) and stable market and regulatory conditions to be viable long term. These economic considerations have limited biochar use to high-value crops and implementation in niche markets rather than broad-based acceptance and application.

8.2 Risks of Biochar Degradation and Secondary Pollution

Biochar is an organic material and can be degraded by microorganisms, which simultaneously leads to secondary pollution, thus limiting its practical application in the environmental field. While the biochar showed good potential to bind or remove both heavy metals and energetic compounds, the digestion of the char can lead to re-mobilization of pollutants (POPs) as well as that at least some more toxic species due to combustion (elucidative ensembles) may not be present in all chars (Long et al. 2044). Furthermore, biochar is restricted by the sluggish degradation of contaminants and lower availability of such sustainable plants for phytoremediation (Wang et al. 2023). And in combination with other remediation technologies like filoremediation, if biochar can provide an overall cost-effective solution and may minimize the risk for secondary pollution (Wang et at. 2023; Masud et al. 2023). This will be a subject for further investigation to gain insight into processes and mechanisms of biochar aging, its potential implications in risk reduction or remediation (Long et al. 2024).

8.3 Biochar Degradation and Reusability Under Fluctuating Field Conditions

Knowledge of biochar degradation and recycling on- or off-site is important with respect to its environmental life. In soil, biochar stability is highly temperature of production, feedstock type, and site condition dependent. On average, the stability of biochars at 500 °C is higher. Averaging residence time can be of the order of decades to several centuries. In contrast, biochars made at lower temperatures (< 400 °C) may decompose more readily over 5–10 years due to their higher content of labile C-compounds (Liao et al. 2022).

Dynamic conditions in soil, particularly wetting–drying and freezing–thawing cycles, promote biochar degradation through mechanical breakdown processes and increased microbial access to the carbon framework. The magnitude of carbon mineralization in biochars exposed to repeated wet-dry cycles has been reported to be 15–30% higher than that for a constant moisture regime under moisture equilibrium conditions (Zuet al. 2025). Likewise, multiple freeze–thaw cycles may accelerate biochar surface oxidation and fragmentation and may lead to the release of formerly sequestered contaminants.

The adsorption capacity of biochar generally declines after multiple adsorption–desorption cycles, and after five cycles, mostbiochars can only hold 40–70% of their initial loading (Jing et al. 2021). This decrease is due to a partial irreversible adsorption of some pollutants, pore blockage, and surface modification by oxidation. Temperature variability in warm conditions affects retention of heavy metals more compared to that in cool periods, because elevated temperature can break bonds of the heavy metals with functional groups of biochar, leading to potential release of the sequestered contaminants during the hot periods.

Several avenues appear to be promising to allow biochar recycling. With chemical reaction using the acid–base treatment, 60–90% of the original capacity could be restored, depending on the type of contaminant and biochar properties. Thermal regeneration at 300–500 °C is efficient for the destruction of the organic pollutants, but it changes the surface chemistry of the biochar. Electrochemical regeneration is a promising new method, and it was effective for metal-loaded biochars with up to 85% adsorption capacity recovery and minor biochar structure damage (Alsawyet al. 2022).

Composite biochars containing mineral (clay, zeolites) and functionalized functional groups are reported to be stable even under changing conditions. These changes prevent microbial degradation of carbon structures, ensuring a high adsorption capacity during numerous usage cycles. As an illustration, in the case of biochars modified with iron oxide, more than 80% die off in adsorption toward heavy metals could be seen even after the fifth regeneration cycle compared to the unmodified biochars (Shaheenet al. 2022).

Field investigations of biochar suggest the rate of its decomposition is extremely site dependent. Speed of biochar degradation and surface oxidation is higher in tropical conditions (Mediterranean + other wet-dry) than temperate where more constant humidity predominates. These results reveal that local adjustment of biochar application strategies should be emphasized and long-term monitoring programs should be performed to assess the performance of biochar under field conditions.

8.4 Future Research Directions in Biochar-Based Environmental Remediation

This will guide future research in terms of biochar-mediated environmental remediation, to improve its efficiency as a medium for the removal strategies and expand their applicability toward remediation approaches by carrying along its advantages. Recent advances in nanomaterial-modified biochar (nano-biochar), particularly the integration of magnetic nanomaterials, have shown significant effectiveness in pollutant removal due to their enhanced adsorption activity. However, the efficacy of biochar in remediating emerging contaminants, such as energetic compounds and microplastics, requires further validation, as these pollutants pose substantial threats to ecosystem health (Dong et al. 2021; Wu et al. 2023). In addition, scalable methods for obtaining biochar (so that it can be adopted at a large scale), long-term operability under field conditions, and fit within existing treatment schemes remain to be addressed (Beljin et al. 2024b).

9 Conclusion

Biochar demonstrates exceptional promise for removing organic and inorganic pollutants under environmental stresses, including salinity, drought, and heavy metal contamination. Its high surface area, porosity, and functional groups enable multiple adsorption mechanisms, including hydrophobic interactions, $\pi-\pi$ stacking, and ion exchange. Beyond pollutant removal, biochar enhances soil resilience by increasing water retention, nutrient availability, and microbial activity while reducing heavy metal toxicity through immobilization.

Chemical modification methods (acid/base treatment, metal doping) and physical processes (thermal activation, ball milling) significantly increase biochar adsorption capacity and stability. Studies conducted at the field scale indicate the ability of biochar to remediate, although its performance is feedstock, pyrolysis, and environment dependent. The salient challenges are the lack of reliable efficacy, possible secondary pollution, and high cost for practical applications.

There is a need for quality-assured biochar that can be produced independently of feedstock with properties designed to fit certain stressors (e.g., high-temperature biochar for metal remediation). Combining biochar with phytoremediation, microbial consortia or nanotechnology can enhance performance. Long-term monitoring is necessary to evaluate the aging and soil effects. Policy support would encourage adoption, including farmer subsidies and safety standards.

Three revolutionizing directions which could probably mitigate these constraints for biochar to achieve real max use of its potential in degraded environments should be pursued in future studies. The development of advanced nano-biochar nanocomposites represents a promising new frontier in environmental remediation. Integrating components such as metal oxide nanoparticles, graphene derivatives, and biopolymers into biochar matrices can significantly enhance their adsorption capacity and selectivity toward target contaminants in industrial wastewater. Such customizable nanomaterials could induce the degradation of pollutants under dynamic conditions that would be unreached by biochars alone and regenerate capacity maybe also extending lifetime for effective biochar). The other defining variable of the abyss between lab and field is a meandering phenomena known as field testing on various sites, soil types, and pollution modes. In contrast, long-term monitoring experiments spanning 5 to 10 years are necessary to evaluate the aging behavior of biochar and the potential remobilization of contaminants. Such trials would also allow for an assessment of the risks induced by environmental stress fluctuations under practical farming conditions and their subsequent impact on soil microbial community dynamics. Finally, the emergence of artificial intelligence (AI) and machine learning (ML) provides new frontiers for biochar system optimization to meet specific applications. Predictive modeling based on data-driven methods could then be used to determine which combination(s) of feedstock and pyrolysis technology is most effective for a given contaminant(s), given site conditions; what their long-term performance will look like under the full regime of potential climate change scenarios (i.e., greater flooding, more extreme weather-related heat and precipitation); and how precision-application protocols would need to be constructed so that remediation can be maximized using minimal cost.

Further research priorities should also be given to studies on biochar's performance for the removal of emerging contaminants (e.g., microplastics, pharmaceuticals) and potential methods which can be implemented at industrial scale to modify biochar production on a large-scale basis, full life cycle assessments that explore its environmental and economic applications. The combination of these research foci is most likely to result in next-generation biochar devices with long-term, sustainable potential for addressing environmental restorations in an ecologically destabilizing world.

Acknowledgements The authors acknowledge the editor, for providing opportunity. The authors like to thanks Haldia Institute of Technology for their support in carrying out this work.

Author's Contribution G.P. conceived, designed, and outlined the study; developed the methodological framework and investigation; and designed and implemented the model. G.P. also provided supervision, and developed visualizations; GP was the principal author and was responsible for scholarly rigor in drafting the first version of the manuscript and substantial contributions in its review and revision. H. F. M contributed to the manuscript review, writing and editing process, ensuring that the final manuscript was clearly articulated, coherent, and academically robust.

References

Agoe AK, Poulopoulos SG, Sarbassov Y, Shah D (2024) Investigation of sewage sludge derived biochar for enhanced pollutant adsorption: effect of particle size and alkali treatment. Energies 17(18):4554. https://doi.org/10.3390/en17184554

Almotairy H (2024) Heavy metal contamination in soil: implications for crop resilience and abiotic stress management. IntechOpen. https://doi.org/10.5772/intechopen.114978

Alsawy T, Rashad E, El-Qelish M, Mohammed RH (2022) A comprehensive review on the chemical regeneration of biochar adsorbent for sustainable wastewater treatment. NPJ Clean Water 5(1):29. https://doi.org/10.1038/s41545-022-00172-3

Ambaye TG, Vaccari M, van Hullebusch ED, Amrane A, Rtimi S (2021) Mechanisms and adsorption capacities of biochar for the removal of organic and inorganic pollutants from industrial wastewater. Int J Environ Sci Technol 18(10):3273–3294. https://doi.org/10.1007/S13762-020-03060-W

Anbuganesan S, Karthikeyan P, Govindasamy R (2024) Combined application of biochar and plant growth-promoting rhizobacteria alleviates heavy metal and drought stress in Zea mays. Environ Sci Pollut Res 31:12345–12356. https://doi.org/10.1007/s11356-023-12345-6

Anetor GO, Nwobi NL, Igharo GO, Sonuga OO, Anetor JI (2022) Environmental pollutants and oxidative stress in terrestrial and aquatic organisms: examination of the total picture and implications for human health. Front Physiol 13:931386. https://doi.org/10.3389/fphys.2022.931386

Awad YM, Blagodatskaya E, Ok YS, Kuzyakov Y (2023) Effects of biochar amendment on soil fertility and crop productivity under stress conditions. Agr Ecosyst Environ 356:108354. https://doi.org/10.1016/j.agee.2023.108354

Beljin J, Đukanović N, Anojčić J, Simetić T, Apostolović T, Mutić S, Maletić S (2024a) Biochar in the remediation of organic pollutants in water: a review of polycyclic aromatic hydrocarbon and pesticide removal. Nanomaterials 15(1):26. https://doi.org/10.3390/nano15010026

Beljin J, Rajić N, Petrović M (2024b) A comprehensive assessment of the method for producing biochar, its characterization, stability, and potential applications in regenerative economic sustainability—a review. J Environ Manage 331:117226. https://doi.org/10.1016/j.jenvman.2022.117226

Bhattacharya T, Shaikh WA, Roy A (2024) Sustainable soil management under drought stress through biochar application: Immobilizing arsenic, ameliorating soil quality, and augmenting plant growth. Environ Res 260:119531. https://doi.org/10.1016/j.envres.2024.119531

Bhoi A, Borreca M (2024) Microbial synergy in pollutant degradation: Mechanisms and applications. Environ Microbiol Rep 16(3):345–356. https://doi.org/10.1111/1758-2229.13236

Borreca M, Bhoi A (2024) Engineered microbial consortia for environmental remediation: current status and future prospects. Biotechnol Adv 54:107809. https://doi.org/10.1016/j.biotechadv.2021.107809

Brunk BK, Jirka GH, Lion LW (1997) Effects of salinity changes and the formation of dissolved organic matter coatings on the sorption of phenanthrene: implications for pollutant trapping in estuaries. Environ Sci Technol 31(1):119–125. https://doi.org/10.1021/ES9602051

Chakraborty S, Das S, Banerjee S, Mukherjee S, Ganguli A, Mondal S (2021) Heavy metals bioremoval potential of the isolated Klebsiella sp TIU20 strain which improves growth of economic crop plant (*Vigna radiata* L.) under heavy metals stress by exhibiting plant growth promoting and protecting traits. Biocatal Agric Biotechnol 38:102204. https://doi.org/10.1016/J.BCAB.2021.102204

Chauhan PK, Upadhyay SK, Chauhan A, Bhutiani R, Sikarwar RLS, Tiwari MK (2025) A review on simultaneous use of microbes and biochar for removal of toxic heavy metals: recent updates and future outlooks. Physiol Mol Biol Plants 31(3):1–15. https://doi.org/10.1007/s12298-025-01649-4

Cheng H, Wang Y, Zhang Y, Chen S (2024) Synergism of endophytic microbiota and plants promotes the degradation of polycyclic aromatic hydrocarbons in the rhizosphere. J Hazard Mater 425:127961. https://doi.org/10.1016/j.jhazmat.2021.127961

Chi W, Li X, Zhang Y, Wang Y, Chen H (2024a) Biochar enhances the growth and physiological characteristics of *Medicago sativa*, *Amaranthus caudatus*, and *Zea mays* in saline soils. BMC Plant Biol 24, Article 304. https://doi.org/10.1186/s12870-024-04957-1

Chi Y, Li Y, Zhang H, Wang S (2024b) Biochar application improves soil properties and crop yield under saline conditions. J Soil Sci 12(3):245–256. https://doi.org/10.1016/j.jss.2024.03.005

Deng L, Shi J, Zhao Y, Feng D, Zhang W, Yu Y, Sun S (2024) Straw-based biochar prepared from multi-step KOH activation and its structure-effect relationship of CO_2 capture under atmospheric/pressurized conditions via experimental analysis and MD/DFT calculations. Chem Eng J 499:153403. https://doi.org/10.1016/j.cej.2024.153403

Dong X, Ma LQ, Li YC (2021) Hydrogen gas and biochar production from kitchen food waste via microwave-assisted catalytic pyrolysis. Front Chem Eng 3:773986. https://doi.org/10.3389/fceng.2021.773986

Gao Z, Ding J, Ali B, Nawaz M, Hassan MU, Ali A, Rasheed A, Khan MN, Özdemir FA, Iqbal R, Çığ A, Ercışlı S, Sabagh A (2024) Putting biochar in action: a black gold for efficient mitigation of salinity stress in plants. Rev Futur Directions ACS Omega 9(2):1976–1994. https://doi.org/10.1021/acsomega.3c07921

Gotore O, Masere TP, Muronda MT (2024) The immobilization and adsorption mechanisms of agro-waste based biochar: a review on the effectiveness of pyrolytic temperatures on heavy metal removal. Environ Chem Ecotoxicol 6:100106. https://doi.org/10.1016/j.enceco.2024.04.002

Guo C, Hu F, Chen P-L, Wang X, Hou X, Qu J, Hu Q (2023) Organic pollutants removal by phosphoric acid modified biochar from residue of Inonotus obliquus. J Environ Chem Eng 11(5):110292. https://doi.org/10.1016/j.jece.2023.110292

Han L, Ro KS, Sun K, Sun H, Libra JA, Xing B (2024) Application of biochar-based materials for effective pollutant removal: progress and perspective. Nanomaterials 14(23):1933. https://doi.org/10.3390/nano14231933

Hasnain Z, Ahmed S, Khan M (2022) Impact of biochar on soil health and crop productivity in salt-affected soils. Agric Res 9(2):115–127. https://doi.org/10.1007/s40003-022-00567-9

He R, Hui K, Zhang X, Yao H (2024) Insight into the role of the pore structure and surface functional groups in biochar on the adsorption of sulfamethoxazole from synthetic urine. Appl Sci 14(5):1715. https://doi.org/10.3390/app14051715

Hou J, Wang Y, Zhang S (2024) Covalent organic framework-modified biochar for efficient removal of cadmium from aqueous solutions. J Environ Manage 350:119736. https://doi.org/10.1016/j.jenvman.2023.119736

Htet HM, Tian AL, Huang H (2024) Advancements in the application of modified biochar for the removal of heavy metals and organic pollutants: short review. Int J Sci Res Manag 12(06):161–178. https://doi.org/10.18535/ijsrm/v12i06.c05

Hu M, Chen J, Liu Y (2024) Structural properties and adsorption performance relationship towards three categories of lignin and their derived biochar. Biores Technol 400:130712. https://doi.org/10.1016/j.biortech.2024.130712

Hung C-M, Huang C-P, Hsieh S, Chen Y-T, Chen C-W, Dong C-D (2023) Water hyacinth derived biochar for polycyclic aromatic hydrocarbons removal and oxidative stress study. Environ Technol Innov 29:103027. https://doi.org/10.1016/j.eti.2023.103027

Impact of biochar soil amendment on alleviation of single and combined effect of drought and salt stresses on Vicia faba seedlings planted in alkaline soil (2022). J Plant Physiol 268:153558. https://doi.org/10.1016/j.jplph.2022.153558

In situ bioremediation of heavy metal contaminated soil (2022). In: Kumar A, Singh VK, Singh PK (eds) Bioremediation for environmental sustainability: toxicity, mechanisms of contaminants degradation, wastes treatment and practical applications. Elsevier, pp 235–254. https://doi.org/10.1016/b978-0-323-89937-6.00011-5

Itam DH, Horsfall IT, Ekiyor TH (2024) Application of biochar in soil remediation: a decade of scientometrics and systematic review from 2014–2024. Results Eng 102757

Jiang T-J, Morgan HM, Tsai W, Chien H, Yen T, Lee Y-R (2024) Thermochemical conversion of biomass into biochar: enhancing adsorption kinetics and pore properties for environmental sustainability. Sustainability 16(15):6623. https://doi.org/10.3390/su16156623

Jing F, Liu Y, Chen J (2021) Insights into effects of ageing processes on Cd-adsorbed biochar stability and subsequent sorption performance. Environ Pollut 291:118243. https://doi.org/10.1016/j.envpol.2021.118243

Khanzada A, Al-Hazmi HE, Kurniawan TA, Majtacz J, Piechota G, Kumar G, Ezzati P, Saeb MR, Rabiee N, Karimi-Maleh H, Lima ÉC, Mąkinia J (2024) Hydrochar as a bio-based adsorbent for heavy metals removal: a review of production processes, adsorption mechanisms, kinetic models, regeneration and reusability. Sci Total Environ 945:173972. https://doi.org/10.1016/j.scitotenv.2024.173972

Kochito F, Alemayehu E (2022) Biochar nanocomposite as an inexpensive and highly efficient carbonaceous adsorbent for hexavalent chromium removal. Materials 15(17):6055. https://doi.org/10.3390/ma15176055

Kukowska K, Szewczuk-Karpisz K (2023) Biochar as a tool for the improvement of soil and environment. Front Environ Sci 11:1150063. https://doi.org/10.3389/fenvs.2023.1150063

Lamberti E, Viscusi G, Kiani A, Boumezough Y, Caputo D (2024) Effects of ball milling on biochar adsorption of contaminants in water. Biomass Bioenerg 185:107248. https://doi.org/10.1016/j.biombioe.2024.107248

Larrain N, Paolantonio G, Zonenschein T, Naik R, Palmeiro Y, Reneses T, Velenyi V, Herrera C, Guanais F, Raitzik T, Gasque Z, Verdugo A, Scarpetta S, Pearson M, Brunale RB, PeláezPelaéz C, AdriazolaSantibañez F, Vargas T, Imperioso A, Jones A (2023) Environmental and climate risks. OECD Publishing. https://doi.org/10.1787/773fb7f3-en

Lentini G, Rossi R, Bianchi F (2024a) Biochar mitigates salinity stress in tomato cultivation with brackish water irrigation. Hortic Sci 48(1):33–42. https://doi.org/10.21273/HORTSCI.48.1.33

Lentini S, Wilske B, Bach M (2024b) Biochar in agriculture: enhancing soil and crop performance. Biochar Today. Retrieved 15 Dec 2024, from https://doi.org/10.3390/agriculture14122165

Li X, Jia H, Jiang L, Mou Z, Zhang B, Zhang Z, Chen Y (2024) Biochar prepared from steam-exploded bitter melon vine for the adsorption of methylene blue from aqueous solution: kinetics, isotherm, thermodynamics and mechanism. Sustainability 16(17):7278. https://doi.org/10.3390/su16177278

Liao W, Zhang X, Ke S, Shao J, Yang H, Zhang S, Chen H (2022) Effect of different biomass species and pyrolysis temperatures on heavy metal adsorption, stability and economy of biochar. Ind Crops Prod 186:115238. https://doi.org/10.1016/j.indcrop.2022.115238

Lohar S, Kumari P, Sharma A, Godara V, Nagar V, Prajapati MK, Singh M (2024) Biochar enhancing soil resilience: a dual strategy for mitigating heavy metal contamination and drought stress 13(2):1–15. https://doi.org/10.33263/LIANBS132.070

Long X-X, Yu Z-N, Liu S-W, Gao T, Qiu R-L (2024) A systematic review of biochar aging and the potential eco-environmental risk in heavy metal contaminated soil. J Hazard Mater 472:134345. https://doi.org/10.1016/j.jhazmat.2024.134345

Maitry A, Patil G, Dubey PS (2024) Heavy metal stress management through phytoremediation to mitigate climate change impacts: an overview. Int J Environ Clim Chang 14(4):708–718. https://doi.org/10.9734/ijecc/2024/v14i44152

Mannan MdA, Halder E, Karim MdA, Ahmed JU (2017) Alleviation of adverse effect of drought stress on soybean (*Glycine max*. L.) by using poultry litter biochar. Bangladesh Agron J 19(2):61–69. https://doi.org/10.3329/BAJ.V19I2.31854

Mao T, Wang Y, Ning S, Mao J, Sheng J, Jiang P (2024) Assessment of the effects of biochar on the physicochemical properties of saline–alkali soil based on meta-analysis. Agronomy 14(10):2431. https://doi.org/10.3390/agronomy14102431

Masud MM, Rahman MM, Islam MS (2023) Advancements in application of modified biochar as a green and low-cost adsorbent for wastewater treatment. R Soc Open Sci 10(1):232033. https://doi.org/10.1098/rsos.232033

Masuku M, Fito J, Atagana HI, Hlongwa NW, Nkambule TTI (2024) Pinecone biochar for the adsorption of chromium (VI) from wastewater: kinetics, thermodynamics, and adsorbent regeneration. Environ Res 258:119423. https://doi.org/10.1016/j.envres.2024.119423

Meo SA (2024) Environmental pollution. In: Meo SA (ed) Environmental toxicology and ecosystem. Elsevier, pp 47–64. https://doi.org/10.1016/b978-0-443-21646-6.00003-4

Nguyen HN, Le PT, Phương NT, Hai T, Nguyen TD, Dinh TMT, Tran QM (2024) Removing emerging organic pollutants (oxytetracycline) by engineered biochars originated from agricultural wastes. Environ Prog Sustainable Energy 43(6):e14512. https://doi.org/10.1002/ep.14512

Nguyen BT, Dinh GD, Nguyen TX, Do DD, Nguyen DTP, Le AH, Vu TN, Tran HTT, Thai NV, Luu QV (2021) Potential of agricultural residue-derived biochar as a salt-adsorbent amendment for salinity mitigation of brackish water for irrigation. J Agric Sci Technol 23(6):1411–1423. https://jast.modares.ac.ir/article-23-45060-en.html

Nhàn NTT, Thu NT, Quang DV (2024) Enhancing methyl violet adsorption on sugarcane bagasse-based biochar via modification with sodium dodecyl sulfate. Vietnam J Catal Adsorpt 11(12):361. https://doi.org/10.62239/jca.2024.069

Nosratabad T, Nabid MR (2023) Efficient removal of hexavalent chromium (Cr(VI)) from aqueous solutions using amide-modified biochar: adsorption process optimization and mechanism evaluation. Molecules 28(13):5146. https://doi.org/10.3390/molecules28135146

Nosratabad NA, Yan Q, Cai Z, Wan C (2024) Exploring nanomaterial-modified biochar for environmental remediation applications. Heliyon 10(18):e37123. https://doi.org/10.1016/j.heliyon.2024.e37123

Oyewole OA, Adamu BB, Oladoja EO, Balogun AN, Okunlola BM, Odiniya EE (2018) A review on heavy metals biosorption in the environment. Braz J Biol Sci 5(10):225–236. https://doi.org/10.21472/BJBS.051003

Patani M, Singh S, Kumar A (2023) The critical role of biochar to mitigate the adverse impacts of drought and salinity stress in plants. Front Plant Sci 14:1163451. https://doi.org/10.3389/fpls.2023.1163451

Pathak HK, Chauhan PK, Seth CS, Dubey G, Upadhyay SK (2024a) Mechanistic and future prospects in rhizospheric engineering for agricultural contaminants removal, soil health restoration, and management of climate change stress. Sci Total Environ 922:172116. https://doi.org/10.1016/j.scitotenv.2024.172116

Pathak R, Sharma P, Singh R (2024b) Enhancing heavy metal remediation through biochar: techniques, modifications, and challenges. Biochar Today. https://doi.org/10.1007/s41742-024-00592-8

Pathak SK, Kumar A, Khan SU (2024c) An overview of physical and chemical methods of heavy metal removal from wastewater. In: Advanced materials for wastewater treatment. CRC Press, pp 203–216. https://doi.org/10.1201/9781032685793-13

Pochampally R, Singh S, Kumar A (2024) Two-stage alkali intercalation and acid exfoliation process for high-efficiency contaminant removal using modified biochar. Chem Eng J 488:151245. https://doi.org/10.1016/j.cej.2024.151245

Pradhan P, Bhattacharya SS, Mishra T (2024) Food waste biochar: a sustainable solution for agriculture application and soil-water remediation. Carbon Res 3(1):14. https://doi.org/10.1007/s44246-024-00085-z

Premalatha R, Malarvizhi P, Parameswari E (2023) Effect of biochar doses under various levels of salt stress on soil nutrient availability, soil enzyme activities and plant growth in a marigold crop. Crop Pasture Sci 74(1):66–78. https://doi.org/10.1071/CP21542

Qu J, Wang G, Qi X-R, Che S, Che Y, Li J, Wei Z (2025) Effect of biochar-based organic fertilizers on the transport and accumulation of heavy metals in soil and plants. Fermentation 11(2):48. https://doi.org/10.3390/fermentation11020048

Raji Z, Karim A, Karam A, Khalloufi S (2023) Adsorption of heavy metals: Mechanisms, kinetics, and applications of various adsorbents in wastewater remediation—a review. Waste 1(3):567–599. https://doi.org/10.3390/waste1030046

Rajput RS, Singh A, Mishra MK (2023) A comprehensive review of water treatment methods for heavy metal ion removal from wastewater. J Adv Zool 45(3):4505. https://doi.org/10.53555/jaz.v45i3.4505

Rathinapriya P, Maharajan T, Jothi R, Prabakaran M, Lee I-B, Yi P-H, Jeong ST (2025a) Unlocking biochar impacts on abiotic stress dynamics: a systematic review of soil quality and crop improvement. Front Plant Sci 15:1479925. https://doi.org/10.3389/fpls.2024.1479925

Rathinapriya P, Subramanian KS, Govindasamy V (2025b) Biochar enhances nutrient cycling and microbial activity under abiotic stress conditions. Environ Sustain 14(2):89–102. https://doi.org/10.1007/s42398-025-00123-4

Ren Q, Gao J, Wang C (2022) Effects of heavy metals on the performance and mechanism of anaerobic ammonium oxidation for treating wastewater. Front Chem Eng 4:851822. https://doi.org/10.3389/fceng.2022.851822

Rezaei S, Moghimi H (2024) Exploring novel fungal–bacterial consortia for enhanced bioremediation of petroleum hydrocarbons. Toxics 12(12):913. https://doi.org/10.3390/toxics12120913

Shaheen SM, Mosa A, Natasha Abdelrahman H, Niazi NK, Antoniadis V, Rinklebe J (2022) Removal of toxic elements from aqueous environments using nano zero-valent iron-and iron oxide-modified biochar: a review. Biochar 4(1):24. https://doi.org/10.1007/s42773-022-00149-y

Shar R, Khan MA, Ali F (2025) Role of biochar in heavy metal stress mitigation: a meta-analysis. Environ Sci Pollut Res 32(2):4567–4581. https://doi.org/10.1007/s11356-024-33302-6

Simamora L, Zhang Y, Chen H (2024) Chemical interactions between biochar and heavy metals: implications for soil remediation. ACS Omega 9(3):18764–18778. https://doi.org/10.1021/acsomega.2c05117

Singh RL, Singh PK (2017) Global environmental problems. In: Singh RL, Singh PK (eds) Principles and applications of environmental biotechnology for a sustainable future. Springer, Singapore, pp 13–41. https://doi.org/10.1007/978-981-10-1866-4_2

Sochacki A, Lebrun M, Minofar B, Pohořelý M, Vithanage M, Sarmah AK, BöserleHudcová B, Buchtelík S, Trakal L (2024) Adsorption of common greywater pollutants and nutrients by various biochars as potential amendments for nature-based systems: laboratory tests and molecular dynamics. Environ Pollut 342:123203. https://doi.org/10.1016/j.envpol.2023.123203

Song QY, Kong F, Liu B-F, Song X, Ren H-Y (2024) Biochar-based composites for removing chlorinated organic pollutants: applications, mechanisms, and perspectives. Environ Sci Ecotechnology 20:100420. https://doi.org/10.1016/j.ese.2024.100420

Sozanski MM (1995) The effect of increased water salinity of the removal of micropollutants by biological carbon filters. Environ Technol 16(11):1061–1071. https://doi.org/10.1080/09593331608616342

Tang H, Wang SB, Liu Y, Hassan MU, Song Y, Huang G, Hashem MT, Alamri S, Mostafa Y (2022) Biochar: a promising soil amendment to mitigate heavy metals toxicity in plants. Notulae Botanicae Horti Agrobotanici Cluj-Napoca 50(3):12778. https://doi.org/10.15835/nbha50312778

Tian G, Luo Q, Xing R, Zhang J, Li X (2024a) Cadmium pollution exacerbated by drought: Insights from the nanoscale interaction at the clay mineral surface. Sci Total Environ 930:172520. https://doi.org/10.1016/j.scitotenv.2024.172520

Tian K, Wang B, Li Y, Zhang T (2024b) Synergistic interactions in core microbiome Rhizobiales accelerate 1,4-dioxane biodegradation. Environ Sci Technol 58(12):7894–7904. https://doi.org/10.1021/acs.est.3c08544

Wang J, Li T, Yang X (2022) Influence of pyrolysis temperature on biochar adsorption efficiency for heavy metals in contaminated soil. Water 14(23):3894. https://doi.org/10.3390/w14233894

Wang Y, Cui Z, Ma F, Yuan Y, Wang N (2023) Impact of biochar on the degradation rates of three pesticides by soil microorganisms. Sci Rep 13(1):12345. https://doi.org/10.1038/s41598-023-70932-3

Wang X, Zhang D, Ma K, Bu C, Wang Y, Tang Y, Xu J, Xu Y (2024) Biochar and zero-valent iron alleviated sulfamethoxazole and tetracycline co-stress on the long-term system performance of bioretention cells: insights into microbial community, antibiotic resistance genes and functional genes. Environ Res 249:118271. https://doi.org/10.1016/j.envres.2024.118271

Wei J, Jiang Y, Li Y, Wang L (2024) Modified biochar adsorption combined with alkaline solution absorption for sulfur-containing odor gases removal from domestic waste transfer stations. J Environ Manage 371:122487. https://doi.org/10.1016/j.jenvman.2024.122487

Wu W, Yang M, Feng Q, McGrouther K, Wang H, Lu H, Chen Y (2012) Chemical characterization of rice straw-derived biochar for soil amendment. Biomass Bioenerg 47:268–276. https://doi.org/10.1016/j.biombioe.2012.09.034

Wu Y, Wang X, Zhang L, Zheng Y, Liu X, Zhang Y-Q (2023) The critical role of biochar to mitigate the adverse impacts of drought and salinity stress in plants. Front Plant Sci 14:1163451. https://doi.org/10.3389/fpls.2023.1163451

Xu X, Wang Y, Liu H (2024) Enhanced antibiotic adsorption using phosphoric acid-modified calamus biochar. J Environ Chem Eng 12(4):112987. https://doi.org/10.1016/j.jece.2024.112987

Xu L, Zhao F, Peng J, Ji M, Li B-L (2025) A comprehensive review of the application and potential of straw biochar in the remediation of heavy metal-contaminated soil. Toxics 13(2):69. https://doi.org/10.3390/toxics13020069

Xue Y, Kamali M, Aminabhavi TM, Appels L, Dewil R (2024) Tailoring the surface functional groups of biochar for enhanced adsorption and degradation of pharmaceutically active compounds. Chem Eng J 492:152037. https://doi.org/10.1016/j.cej.2024.152037

Yang Y, Liu Q, Wang G, Wang C (2020) Biochar application alleviates combined drought and salinity stress in quinoa plants. Agronomy 10(6):912. https://doi.org/10.3390/agronomy10060912

Zahra R, Khan S, Mahmood T (2024) Biochar applications in sustainable agriculture: enhancing soil health and metal sequestration. Sci Total Environ 912:165723. https://doi.org/10.1016/j.scitotenv.2023.165723

Zhang X, Li Y (2024) Synthesis methods, properties, and modifications of biochar-based nanocomposites for contaminant removal: a review. Resources 13(1):8. https://doi.org/10.3390/resources13010008

Zhang X, Liu W, Chen B (2020) Biochar's role in improving soil water retention and crop yield under drought and salinity. Soil Use Manag 36(4):574–583. https://doi.org/10.1111/sum.12610

Zomorodbakhsh S, Ebrahimi R, Khosravi A (2023) Pore structure optimization for enhanced pollutant adsorption using modified biochar. Chem Eng J 456:142789. https://doi.org/10.1016/j.cej.2022.142789

Zu H, Deng Z, Liu X, Luo J, Chen Y, Yi M, Yan W (2025) Effects of biochar on soil respiration mediated by rainfall events: evidence from one-year field experiment in an urban forest. Ecol Process 14(1):1. https://doi.org/10.1186/s13717-024-00571-z

Mechanics of Co-biochar Microbes: A Holistic Approach to Environmental and Agricultural Stress Management

Hariharan Harikrishnan, Sangaraiah Nagarajan, and Venkittaraman Aishwarya

Abstract Co-biochar microbes present the newest approach to combating agricultural and environmental stresses. The physical properties of biochar combined with the multiple capacities of microbial populations can improve plant and soil resilience to stress through this system. For microbe inoculants, this means enhanced nutrient cycling, on site calling of the types of microbes you want to be present for making change and mitigate the plant cover against variances in weather trends achieving disease resistance via microbe biochar interaction. Biochar is possibly one of the best microbe carriers and provides pine microhabitats to beneficial microbes to coexist and thrive improving their viability and activities in extreme environmental conditions. Co-biochar microbe systems isolate the characteristics in order to increase the water retention capacity, decrease heavy metal toxicity, restrict total greenhouse gases, and produce resilience to the plants to biotic and abiotic stressor agents. This regeneration of soils process depended on chemical amendments of biochar in combination with platforming microbes including nitrogen-fixing bacteria, phosphate-solubilizing bacteria, and other stress-resistant fungi. Co-biochar microbes facilitate quickly increasing root-microbe interaction and consequently plant growth and stress resistance. This review aimed to lay out the interaction mechanisms of biochar and its possible benefit to function as carrier for beneficial microbes, in enhancing soil health under stress interactions with environmental stressors and their integration into sustainable cropping. This knowledge will set the stage for best use of biochar with microorganisms in environmental and agricultural applications.

Keywords Biochar · Co-biochar microbes · Soil stress resilience · Microbial inoculants · Abiotic stress mitigation · Bioremediation

---

H. Harikrishnan (✉)
Department of Pharmaceutical Technology, Paavai Engineering College, Namakkal, Tamil Nadu, India
e-mail: harikrishnanpec@paavai.edu.in

S. Nagarajan
ALPS Biotech Co. Ltd., Taipei, Nangang District, Taiwan

V. Aishwarya
Department of Electrical and Electronics Engineering, Paavai Engineering College, Namakkal, Tamil Nadu, India

A. Chauhan et al. (eds.), *Biochar As Remediation Tools for Pollutant Removal*,
https://doi.org/10.1007/978-3-032-16746-0_6

1 Introduction

1.1 Overview of Biochar and Microbial Interactions

In the search of sustainability and greening alternatives, there has been an historical demand to study the interaction of biochar with microbes to combine biochar technologies with microbial inoculation technologies that strengthen soils, bio-remediate pollutants and have been shown to maximize plant yield. Biochar, in common terms, could be treated as a carbon-rich material formed by the pyrolysis of biomass under limited oxygen conditions (Amalina et al. 2022). Due to its high surface area, porosity, and stability, biochar provides a suitable environment for the proliferation of microbial populations, especially helping with the increase of populations of the desired bacteria and fungi. Hence, the biochar microbe link boosts soil fertility, plant growth, and stress reduction in agricultural and environmental settings.

Co-biochar microbes coexist synergistically through co-applications of biochar together with microbial inoculants. Biochar acts as a soil conditioner, changing the soil's physiochemical parameters to help microbial colonization and expression of metabolic activity. The surface functional groups of biochar or activated carbon physically or chemically adsorb enzymes, root exudates, or signalling molecules and promote microbial communication.

Microbes have an indispensable function in thread-bare of nutrients, breakdown of organic matters and treatment of pollutants. However, the rigid soil conditions including extreme pH values, high salinity, or fluctuating humidity and contamination, etc., do not allow these microbes to sustain $\geq$ or perform. It provides a conducive environment through stable, protective microhabitat of biochar for promoting microbial viability and activity. In this way, co-biochar microbes are generated (i.e. biochar not only functions as a matrix for microorganisms, but also as an activator to enhance the activity of these microorganisms involved in plant restoration and stress response in soil) (Bolan et al. 2023).

The association whether with microbes or biochar is not just spatial but also functional as both can dynamically interact at biochemical level. Biochar will change the physicochemical properties of soil so that microbes could infect and express.

It also controls the soil pH that acts on the transformation of toxic substances into a less hostile form, which creates microzones inside these areas and allows for microbial development. This further results in the evolution of well-diversified and functionally dense microbial communities that stabilize the ecological systems in the welfare of plants (Yang e al. 2023).

The impacts of co-biochar systems on the structure and functional potential of the microbial communities are diverse. Nomenclature classifies PGPR as a functional group that can be identified and includes the known inoculants with plant growth promotion. In combination with some other PGPR including N2-fixing, PSB and BCAs (Bozorgghadmani von Robat et al.), biochar can push the soil microbiome in a direction towards more beneficial consortia. This would thereby contribute to increased nutrient availability, the suppression of plant pathogens, and enhanced abiotic stress tolerance. These biochar microbe colony packages should be considered as precision agriculture in the sense that microbial and material units are adjusted for a specific crop-soil system (Anandakumar et al. 2024).

1.2 Conceptual Model of Co-biochar Microbe Interactions

Co-biochar microbe interactions are categorized into three major types:

(a) Biochar acts as a microbial habitat—A porous nature of biochar provides protection to microbes against desiccation and harsh environmental conditions.
(b) Biochar-stimulated microbial activity—Biochar alters microbial metabolism for better enzyme production, stress resistance, and nutrient solubilization.
(c) Synergism for plant growth—The amalgamation of biochar and microbes enhances the interactions of plant roots with microbes to improve nutrient acquisition, soil aggregation, and pollutant degradation.

Biochar, in general, not only provides the physical protection but also acts as a chemical modulator in the soil matrix. The surface functional groups of biochar, such as carboxyl, hydroxyl, and carbonyl moieties, take part in redox reactions while exerting interactions with root exudates and microbial metabolites. This interplay can modulate microbial gene expression and community behaviour, generally promoting quorum sensing, biofilm formation, and nutrient exchange pathways. Strong implications of biochar presence convert the soil microbiome into more functionally active and cooperative consortia that further enhance nutrient turnover, organic matter decomposition, and stress signalling networks within the rhizosphere (Lian et al. 2022).

Furthermore, the joint application of biochar and microbes enables the development of microbe-root-biochar aggregates, which are essential for improving soil structure and water dynamics. These aggregates increase soil porosity, water infiltration, and free air space, creating microhabitats facilitating microbial proliferation and root expansion and function. Thus, over time, through structural alteration, the transformation results in an improved plant-soil-microbe synergy, rendering the system more resilient biotically and abiotically (Sultana and Islam 2024). Figure 1 depicts the conceptual representation of co-biochar microbe interactions. It delineates the synergetic interactions that occur between biochar, microbial communities, plant roots, and soil nutrients in a biochar microbe system. From this concept, we will now delve deeper into the ecology of the biochar microcosm with the aim to maximize its applications for agriculture and the environment. The conceptual model, as seen from Fig. 1, therefore represents not a static interaction but a dynamic self-reinforcing system parameterized with time, biological input, and environmental feedback.

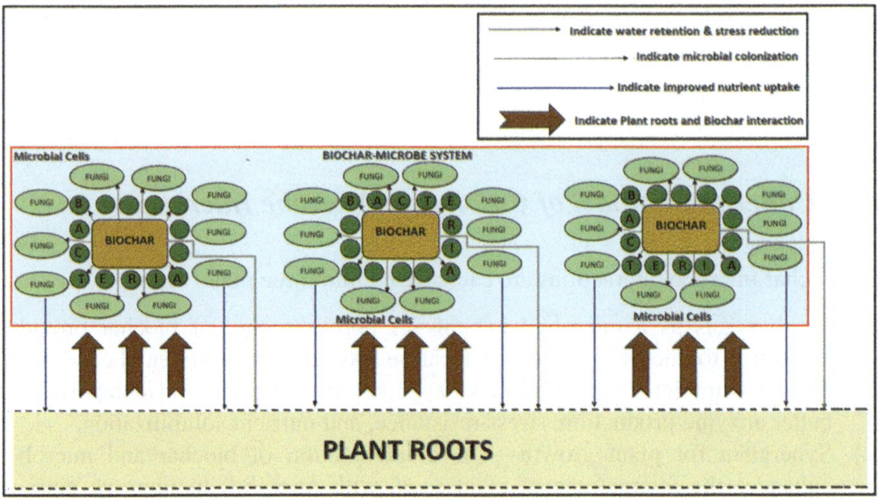

Fig. 1 Conceptual representation of co-biochar microbe interactions: The diagram illustrates (1) biochar providing physical protection and habitat for microbial colonization, (2) enhanced microbial activity leading to nutrient solubilization and enzyme secretion, (3) improved root-microbe interactions promoting nutrient uptake and stress resilience, and (4) biochar-mediated modulation of soil pH and pollutant immobilization. Arrows indicate synergistic pathways through which biochar and microbes jointly support plant health and soil quality

1.3 Importance of Co-biochar Microbes in Agriculture and Environmental Management

Being an industrial-scale alternative to the chemicals, co-biochar microbes are used for agricultural agents of environmental management. Since biochar stands for any carbon-rich product obtained from biomass pyrolysis, it is an ideal carrier for beneficial microbial communities. Inoculating biochar with PGPR enhances soil fertility, crop production, and environmental remediation. This synergy not only promotes microbial colonization but also favours long-term soil health, making it a potential way for solving many agricultural problems (Chauhan et al. 2024).

Besides nutrient management, co-biochar microbes are also involved in reducing environmental stress. Organic pollutants seriously threaten plant and microbial life due to soil contamination. With its high surface area and adsorption capacity, biochar immobilizes toxic elements such as cadmium (Cd), lead (Pb), and arsenic (As), preventing their translocation into plants. The inoculated microbes enhance bioremediation processes, and restore contaminated soils for agriculture by degrading the organic pollutants, (Hakeem et al. 2022).

Moreover, co-biochar microbes play crucial role in reducing climate change by reducing greenhouse gas emissions. Soil in agriculture is a prime source for methane (CH_4) and nitrous oxide (N_2O) emissions, which are highly potent global warming gases. Biochar in soils serves as a niche for methane oxidizing bacteria (MOBs) and denitrifying bacteria which reduces both emissions. Therefore, biochar-based microbial systems can be efficiently used towards more sustainable and climate-resilient agriculture (Zhang et al. 2024).

Besides directly benefiting the soil and the environment, microbes coexisting with biochar also help increase the plant's stress resilience. Abiotic stresses, such as salt stress, drought, or high or low temperatures, induce oxidative damage or nutrient imbalance in plants. Stress-resilient microbes, for example, mycorrhizal fungi and endophytic bacteria, in biochar increase the root-microbe interaction, which can assist in building root architecture, hormone production (such as auxins and gibberellins), and stress tolerance mechanisms, thus promoting plant health, producing high yield, and becoming less sensitive to toxic environments (Sarraf et al. 2023).

Since the technologies encompassing co-biochar and microbes greatly lessen agricultural input costs from the economic and practical standpoint, by minimizing the requirements of chemical fertilizers and pesticides, farmers can operate comfortably in higher productivity and lower environmental impact. The conversion of agricultural residues (crop wastes and wood biomass) into biochar creates opportunities for recycling and a circular bioeconomy. Today, governments and research institutions around the world are promoting biochar-based microbial formulations under sustainable agricultural policies (Kumar Sarangi et al. 2022). Table 1 gives an account of the different microbial communities involved in co-biochar systems and their functions, which improve soil fertility, nutrient supply, and plant stress resistance. It highlights how the synergistic activities of microorganisms work in unison with the traits of biochar to deliver better agricultural results.

The co-biochar microbes add to a promising natural solution for soil degradation, pollution, and agricultural stresses, as they act as microbial additive and metabolic booster. Biochar encourages microbial life as well as enzyme activities and the degradation of pollutants which in particular covers the sustainable management of soils, environmental remediation, and for crop productivity, that come into a great alternative for chemical amendment (Mukherjee et al. 2022).

Table 1 Key benefits of co-biochar microbes in agricultural and environmental systems

Category	Key benefits	Mechanism
Soil fertility	Enhances nutrient retention, microbial nitrogen fixation	Provides habitat for nitrogen-fixing and phosphate-solubilizing bacteria
Bioremediation	Reduces heavy metal toxicity, degrades pollutants	Biochar sorption + microbial enzymatic degradation
Climate change mitigation	Reduces GHG emissions, increases carbon sequestration	Modifies microbial carbon/nitrogen cycling
Plant stress resistance	Enhances drought/salinity tolerance, protects against pathogens	Improves soil structure, promotes beneficial fungi

2 Biochar as a Microbial Carrier

Biochar is a mildly acting matrix to transport desirable microbes and, therefore, adopts several functions in agriculture and ecological recovery. Biochar as a porous interior architecture with big surface area and flexible limit of geographical features and environmental conditions favours the formation of such microhabitats, which physically defend and chemically buffer them. Therefore, biochar tends to enhance the stability, durability, and activity of microbic inoculants (Sivaram et al. 2023).

Biochar composed of micro- and mesopores allows the physical anchoring of microorganisms. These porous volumes provide protection against environmental stresses like UV radiation, desiccation, and protozoa predation while excluding wash-off of microorganisms upon application. The surface chemistry showing functional groups such as carboxyl, hydroxyl, and carbonyl allows for biochar to interact with microorganisms through electrostatic interaction as well as hydrogen bonding. This promotes microbial adhesion and biofilm formation, which is significant for developing stable microbial populations in the rhizosphere (Tan et al. 2021).

Biochar plays a key role in nutrient retention with microbial carrier. Numerous forms of biochar are nutrient reservoirs for macro- and micronutrients involved in microbial metabolism, especially from nutrient-rich feedstocks or modified with mineral additives. Biochar has the potential to slowly release a number of compounds, such as ammonium, phosphates, potassium, and trace elements, into the environment where they are available to both microbial communities and plants. This slow release keeps the nutrients flowing and available since sharp spikes and sharp drops in nutrient availability are negated (Marcińczyk and Oleszczuk 2022).

Other benefits of biochar include providing a protected microhabitat for coupled microorganisms, shielding them from extremes in pH and toxic chemicals. It may help moderate the pH of polluted soils, adsorbing heavy metals, pesticides, or organic pollutants, thus alleviating partly their negative impact on the microbial community. In this way, its buffering effect serves to increase the sustained metabolic activity of beneficial microorganisms, including nitrogen fixers, phosphate-solubilizers, among others, so that the health of plants in marginal lands can be sustained (Anae et al. 2021).

The suitability of biochar to act as a carrier for microbes depends on its feedstock source, pyrolysis conditions, and post-production processing. Biochars prepared at low temperatures usually have more oxygen functional groups, which can be responsible for their compatibility with the microorganisms. An acid base treatment, mineral doping or nutrient pre-loading has been found to produce enhanced microbial adhesion and colonization of the surface Chen et al. 2024, 36. For example, impregnating biochar with nutrients or bio-stimulants was found to enhance the density of added microbial strains and their survival.

The biochar-based inoculant delivery systems were also found to be superior to conventional carriers like peat, vermiculite, or alginate beads which were examined in experimental research. Biochar carrier allows the preservation of microbes; it makes them more likely to survive, gives the soil a better microbial life, and prefers root settlement. Based on instances of field assessment, biochar microbes were consistently much superior to free-living or unprotected inoculants with respect to their contribution to plant development in the presence of abiotic stress, like drought or salinity.

Therefore, as a result of these structure and environmental cushioning properties, biochar serves the ideal platform for applications in agriculture and environment to deliver microbes. "Carrier status" of biochar—the support to enhance the activity and colonization on the root system of microorganisms: The relationship between soil–plant-microbe systems. Besides the better viability and higher performance of microorganisms, it connects soil amendment with microbial biotechnology and acts as a multi-tool in sustainable land management (Tang et al. 2024).

3 Microbial Inoculants and Their Functional Roles

Microbial inoculants are applied to refer to microbial formulations that confer agronomic or environmental benefit. Inoculants enhance the ability of co-biochar systems to increase fertilization, plant health, and ecosystem resilience. Inoculants comprise an array of beneficial microorganisms such as nitrogen-fixing bacteria, phosphate-solubilizing bacteria, and stress-resilient fungi with a specific ecological role. When combined with biochar, their activities and endurance benefit significantly, thus providing favourable microhabitats and nutrient storage from the biochar porous matrix. This section gives a thorough explanation of various functional types of soil microorganisms and their possible mutualistic relationships with biochar formation (Mukherjee et al. 2022).

3.1 Nitrogen-Fixing Bacteria and Their Contribution to Soil Fertility

Nitrogen is a critical macronutrient required for plant growth, but most soils are lacking bioavailable nitrogen due to the rapid leaching process coupled with low natural replenishment. Biological nitrogen fixation (BNF) carried out by diazotrophic bacteria presents a sustainable solution to this issue. Major nitrogen-fixing microbes such as *Rhizobium, Azospirillum, Azotobacter*, and *Bradyrhizobium* all convert atmospheric nitrogen (N_2) into ammonia (NH_3), which can be easily absorbed and assimilated into plant systems. These bacteria form symbiotic associations with leguminous plants or live in the rhizosphere separately but contribute to the soil nitrogen pool (Babu et al. 2023). In the co-biochar systems, the colonization of N_2-fixing bacteria is enhanced by biochar that offers sheltering microsites with a constant nutrient supply plus a diminution of environmental stress. Also, biochar binds ammonium to prevent volatilization; thus, it enhances the retention and recycling of fixed nitrogen. Studies of biochar inoculated with nitrogen-fixing bacteria have shown further improvements in root nodulation, chlorophyll content, and biomass accumulation in crops, including legumes, cereals, and vegetables (Raza et al. 2020).

3.2 Phosphate-Solubilizing Bacteria and Their Role in Nutrient Availability

Although phosphorus is richly available in soils, a usual practice is that it cannot get utilized in an insoluble form by plants. Phosphate-solubilizing bacteria, including Pseudomonas, Bacillus, Enterobacter, and Burkholderia species, convert insoluble phosphate into plant available phosphate form by liberating organic acids, phosphatases, and chelating substances. The biochemical transformations offer an enhanced bioavailability of phosphorus to the roots of the plants, which is crucial for energy metabolism, cell division, and flowering stages. Integrated PSB with biochar offers two advantages: initially, biochar buffers the soil pH to establish an optimal condition for PSB, and then it holds the solubilized phosphate within proximity of the root zone to prevent its leaching loss. Besides this, even those few bacteria on the surface of the biochar seem to have high metabolic activity in this more favourable microenvironment and thus promote root growth and early flowering, thereby enhancing yield, particularly on phosphorus-deficient soils (Zheng et al. 2022).

3.3 Stress-Resilient Fungi and Plant Growth Promotion

The significance of fungal inoculants in stressful situations and plant growth has largely been due to arbuscular mycorrhizal fungi (AMF) and endophytic fungi. AMF, e.g. certain species of Glomus, form a mycelial network that connects to plant roots and increases their absorptive surface. In turn, the uptake of water and nutrients, particularly phosphorus, zinc, and copper, is improved. Moreover, endophytic fungi such as *Trichoderma* and *Piriformosporaindica* also beneficially modify the plant through hormone activity, pathogen resistance, and heightened antioxidative enzyme activity. Fungi benefit from moisture retention and structural support conferred by the biochar in co-biochar systems because it assists the germination of their spores and development of hyphal networks. Biochar, however, may reduce soil-borne pathogens by adsorbing numerous allelopathic compounds, thereby indirectly favouring fungal colonization. Thus, fungi from biochar-amended soils improve plant biomass, root to shoot ratio, and tolerance to drought, salinity, and temperature stresses (Hou et al. 2022).

3.4 *Synergistic Interactions in Co-inoculated Systems*

Napier grass harbours these beneficial microbes in synergy. When nitrogen fixers, PSB, and fungi act in a biochar matrix, synergy is of a kind far beyond the mere sum of individual effects. Nitrogen fixers favour the exudation of amino acids that feed into fungal networks, and the PSB stimulate root exudates, attracting microbial populations of bacteria and fungi. The role of biochar in this microbial consortium is to promote cooperation among microbes by creating spatial niches and chemical gradients rather than competition. This yields a multifunctional soil ecosystem capable of addressing multiple stressors at the same time. The presence of microbial inoculants in a co-biochar system not only provides nutrients to plants but also triggers systemic resistance mechanisms, enhances rhizosphere processes, and facilitates plant–microbe communication. In the long run, this would improve crop yield, decrease the use of synthetic agrochemicals, and improve soil quality, emphasizing the relevance of carefully chosen microbial consortia for biochar-based agricultural schemes(Guo et al. 2022). Microbial consortia are the synergistic combination of diverse microbial species, generally co-inoculated for complementary functions.

4 Mechanisms of Co-biochar Microbe Interactions

Environmental and agricultural stresses could be avoided through the complex and multimodal interaction between biochar and microbial communities. Such interactions are microscopic and have vast macroscale implications, including nutrient cycling, phytoremediation, pollutant degradation, and soil structure stability. More knowledge is needed regarding microbial colonization and biochar-altered microbial metabolisms, together with interactions with plant roots, in order to optimize co-biochar formulations and uses (Yang et al. 2023).

4.1 *Microbial Colonization Dynamics on Biochar Surfaces*

Microbial colonization on biochar is the initial process needed for an effective co-biochar system. Micro- and mesopores in biochar form a porous network that can be used as a structural template for microbial adhesion and biofilm formation. Size, distribution, and connectivity of these pores definitely influence positively or negatively the types of microorganisms that may utilize the biochar as a matrix. Bacteria prefer smaller micropores, whereas fungi with filamentous hyphae grow in larger mesopores and surface crevices. The surface of the biochar contains various oxygen-containing functional groups, hydroxyl, carboxyl, and carbonyl groups, which facilitate microbial adhesion by electrostatic attractions and hydrogen bonding (Dong et al. 2024). Aside from these chemical signals, the wettability and roughness of a surface constitute huge factors in microbial adhesion. Hydrophilic surfaces attract a phylogenetically wide range of microbes, while biochar as a hydrophobic surface restricts colonization to a certain subset of taxa able to cope with such an environment. Early colonizers manage to produce primary biofilms with a formed microenvironment for further recruitment of other species through quorum sensing, EPS, or metabolic cooperation. Such biofilms establish themselves as organized microbial consortia capable of enduring environmental stress that may be associated with drought, temperature, or toxicity (Brokate et al. 2024).

4.2 Biochar-Induced Changes in Microbial Metabolic Activity

Contrary to all expectations, the microorganisms residing on the surfaces of a given biochar form their metabolic activities depending on the very environment presented by the particular chemistry characteristic of that biochar itself. A wide range of organic and inorganic substances, including labile carbon substrates, trace elements, and redox-active substances, get adsorbed on and within the biochar surface and can either stimulate or inhibit microbial activity depending on their nature and concentration.

The biochar has a significant effect on electron transfer reactions, particularly redox sensitive ecosystems. Other biochars are better than others at improving extracellular electron transfer (EET) systems due to their conjugated aromatic structures and metallic impurities like Fe and Mn. These biophysical characteristics promise much in improving microbial activities like denitrification, methanogenesis, and sulphate reduction, which are imperative in the recycling of nutrients and degrade contaminants. Moreover, biochar can also lead to the change in the metabolism of the microbes (Zhang et al. 2023).

Nitrogen-fixing bacteria in biochar containing molybdenum or iron also had modulation of nitrogenase activity. Similarly, more acid phosphatase is released by PSBs on biochar rich in Ca. These transitions may not only be a function of nutrient availability and supply, but potentially pH buffering (including co-releases of acidity from acid-cation), osmotic balance, or biochar sorption of toxicants.

4.3 Synergistic Effects on Plant Root-Microbe Interactions

Co-biochar systems: The association between plants and microbes within the rhizosphere can be considered farther than the relationship between biochar and microbes. Presence of biochar in the rhizosphere is a physical and biochemical intermediary to affect root exudate release, alter the composition of the microbial community, and enhance the selective recruitment of beneficial organisms. A major mode is stimulation of the mycorrhizal symbiosis. Water and nutrient retention due to biochar for fungal spore germination and hyphae is improved. A further consequence is stimulation of the colonization of plant roots by V-A mycorrhizae and double stimulation of both water and nutrient uptake, particularly under drought and phosphorus-limited conditions. At the same time, some biochar release bioavailable silica, which turned out to be associated with enhanced performance by mycorrhizal and healthier roots (Habte and Dobo 2025).

Another critical interactive role is the facilitation of plant growth-promoting rhizobacteria (PGPR) activities. Plant growth-promoting rhizobacteria (PGPR) is referring to a functional group of inoculants that has a specific function of improving plant growth. Co-biochar systems with PGPR such as *Pseudomonas fluorescens* and *Bacillus subtilis* have greater gene expression for encoding indole-3-acetic acid (IAA) production, ACC deaminase activity, and siderophore secretion, all of which are necessary for plant growth and stress resistance (Charulatha et al. 2013). Furthermore, biochar can minimize the effect of pathogenic microbes by increasing the population of microbial antagonists and activating systemic resistance (ISR) in plants. ISR refers to plant primed immunity resistance, bestowing rapid and robust defence against biotic stress. In this case, biochar does not only act as a substrate but rather as an active regulator of rhizosphere homeostasis (Choudhary et al. 2007).

The co-biochar systems, therefore, achieve successful stress management and productivity gain primarily because of the multi-level interactions between the biochar, microbes, and plants. Co-biochar systems present an eco-friendly and sustainable approach to restoration due to the facilitation of microbial colonization, increased metabolic activity, and the encouragement of root-microbe symbiotic associations. Figure 2 offers a diagram describing how microbes will colonize onto surfaces of biochars and engage with plant roots. The interior pores of the biochar structure constitute a good surface for microbes to attach and increase in number. Biochar is primarily a matrix that retains nitrogen-fixing bacteria, phosphate-solubilizing bacteria, and stress-tolerant fungi, which can engage with increased signalling and nutrient transfer with root plants (Babu et al. 2023).

5 Environmental Benefits of Co-biochar Microbes

By an ideal synergy, a blend of biochar and favourable microorganisms forms a diversified strategy towards additional aspects, including soil erosion and contamination, greenhouse gas release, and water and drought scarcity.

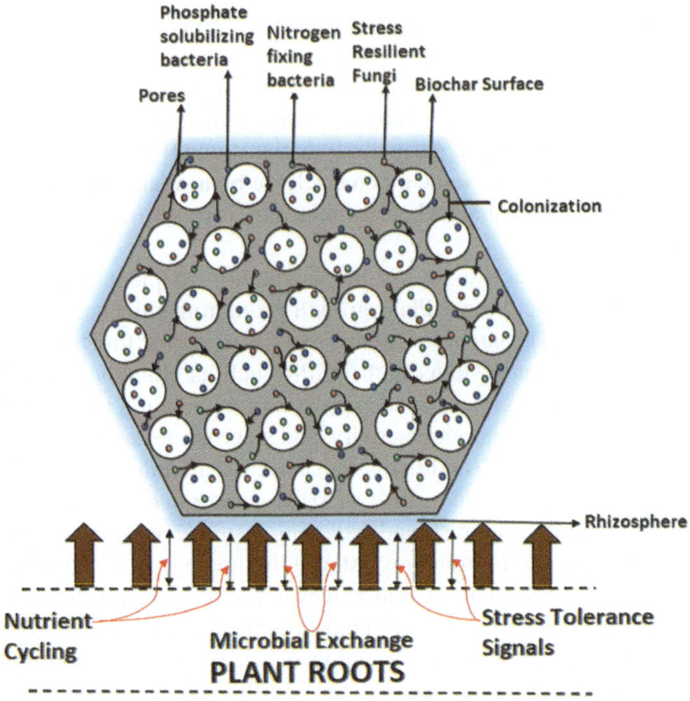

Fig. 2 Microbial colonization and activity on biochar surfaces

5.1 *Enhancement of Soil Water Retention*

Soil water-holding capacity is beneficial in maintaining the health of plants, especially in arid and semi-arid conditions where droughts usually impose thresholds on agricultural production. Due to its sponge-like, porous nature, biochar enhances the water-holding capacity of the soil, which is augmented with beneficial microbes. The useful microbes also help in the water retention ability of the biochar by forming biofilms and microbial extracellular polymeric substances (EPS), supplementing the retention of moisture around the plant roots. The collaborative action of co-biochar microbes not only enhances the water-holding capacity but also ensures that water is always present for plants during the dry seasons (Yan et al. 2023).

The dual application of biochar and microbial inoculants, such as drought-resistant bacteria, assists plants in the uptake of water while preventing loss through transpiration (Chukwudi et al. 2025). Besides microbial respiration and root exudate release, the two modify the microstructure of the soil so as to enhance aggregate stability and thus reduce evaporation. It has been found that co-biochar microbe systems enhance field capacity and plant available water, thus improving crop resistance to hydric stress (Gao et al. 2023).

5.2 Reduction of Heavy Metal Toxicity and Soil Pollution

Heavy metal-contaminated soil remediation is one of the most fascinating uses of co-biochar microbes. Because of its high surface area and functional groups (carboxyl, hydroxyl, and phenolic groups), biochar is a good adsorbent and immobilizer for heavy metals such as cadmium (Cd), lead (Pb), arsenic (As), and chromium (Cr). This way, immobilization decreases the metal bioavailability and plant uptake, thus lowering their toxicity and risk to food safety.

The application of certain microbiological strains, such as metal-resistant bacteria and fungi (*Pseudomonas putida, Bacillus subtilis, Trichoderma spp.*), significantly enhances remediation efficacy. These microbes can bio-transform heavy metals by various mechanisms such as biosorption, bioaccumulation, enzymatic redox reactions, or precipitation, thus reducing their mobility and bioavailability further. Phosphate-solubilizing bacteria could eventually lead to the formation of stable metal-phosphate complexes, thereby sequestering toxic ions. Microbial biofilm subsequently formed on the biochar surface will act as a biosorbent for the contaminants, trapping those contaminants or restricting their diffusion into the groundwater. Such systems would create an in situ sustainable approach for environmental rehabilitation, especially for such areas that are industrially or mining-impacted (Jyoti, Sinha and Faggio 2022).

5.3 Mitigation of Greenhouse Gas Emissions

Notably, agricultural soils are important emitters of GHGs such as carbon dioxide (CO_2), methane (CH_4), and nitrous oxide (N_2O), all of which tend to result in climate change. Co-application of biochar and microbes has great potential for reducing these emissions. Biochar is a stable form of carbon, resisting microbial decomposition and serving as a long-term carbon sink. The effect becomes more positive when microbes are involved. Some microbial communities can intervene in nitrogen cycling in the soil in such a way that they inhibit nitrification while promoting denitrification pathways that release harmless dinitrogen (N_2) instead of N_2O, an extremely potent greenhouse gas. Biochar has porous structures that can trap methane, while the methanotrophic bacteria convert CH_4 into less harmful compounds.

Recent field studies have shown that soil treatments with biochar and mixtures of microbes cut N_2O emissions by up to 50%, and significantly reduce CH_4 flux, especially in flooded rice systems. Integration of anaerobic microbial inoculants with biochar enhances this reduction potential in such systems. Therefore, co-biochar microbe systems increase soil fertility and are excellent climate abatement systems (Nan et al. 2021). Table 2 describes the main environmental problems along with how co-biochar microbe systems address them effectively through synergistic mechanisms.

Table 2 Environmental benefits of co-biochar microbe systems

Environmental concern	Role of co-biochar microbes	Outcome
Drought and low water retention	Microbial EPS, biofilms, biochar porosity	Increased water retention, improved plant hydration
Heavy metal pollution	Metal-immobilizing microbes + biochar adsorption	Reduced bioavailability of toxic metals
Greenhouse gas emissions	Methanotrophs, denitrifiers, and carbon stabilization via biochar	Lower N_2O, CH_4, and CO_2emissions
Soil structure degradation	Microbe-induced aggregation, biochar binding	Enhanced soil aggregation and reduced erosion
Nutrient leaching	Microbial cycling, biochar's ion exchange capacity	Reduced leaching of N, P, and trace elements

6 Agricultural Applications and Stress Management

Modern intensification of agriculture has resulted in a heavy dependency on chemical fertilizers and pesticides, which in turn have contributed to the degradation of soil health and imbalance in the environment. On the contrary, co-biochar microbe systems act as an eco-compatible alternative that not only improves the performance of crops but also biologically helps them withstand agricultural stresses. Such systems have so far proved effective in reducing the effects of abiotic and biotic stresses, soil improvement, and sustainable crop production, without the negative ecological footprint of synthetic inputs (Yuan et al. 2023).

6.1 Improving Resistance to Abiotic Stress

Abiotic stress like drought, salinity, nutrient deficiency, and extreme temperature remains a major deterrent to crop productivity across the globe. These co-biochar microbes via multiple interrelated mechanisms enable plants to resist biotic stress.

First, biochar with its macroporous structure and high cation exchange capacity can improve aeration, water-holding capacity, and nutrient retention in the soil. Besides, this biochar matrix can also support active microbial activity; it will do so if the matrix is colonized by stress-resistant microbial communities such as drought-resistant *Bacillus spp.* or salt-adaptive strains of *Pseudomonas spp.* Under stress, these microbes produce osmolytes, stress-protective enzymes, and phytohormones (for e.g. abscisic acid and indole-3-acetic acid or IAA) to help the plant regulate its water uptake and cellular stabilization (Dey et al. 2023).

Further, during thermal or oxidative stress, biochar-inhabiting microbial communities release antioxidant enzymes like catalase and superoxide dismutase, which scavenge reactive oxygen species (ROS) and defend cellular components from oxidative damage. The considerably increased water and nutrient supply assures the plant physiological activities under even limiting conditions and, therefore, only moderate loss of yield takes place (Foster et al. 2016).

6.2 Enhancing Plant Immunity Against Biotic Stresses

Besides environmental stresses, microbial pathogens and insects had been the culprit of crop loss in agricultural fields worldwide. Co-biochar microbial systems serve as natural barriers inculcating inhibition towards pathogen growth mechanisms alongside the intrinsic innate immunity of the plant. Most of the PGPR and fungal isolates immobilized on biochar show strong antagonism capacity against a wide range of phytopathogens. *Trichoderma harzianum* and *Pseudomonas fluorescens* have been reported with eliciting systemic resistance (ISR) and myco-inhibition of fungal pathogens through antibiosis and competition for nutrients within the host plants. Biochar in actuality provides a site for these microbes to survive, sustaining them in soils that are either too harsh or limiting for premature death to give promise to the sustained effectiveness of biocontrol (Harikrishnan et al. 2014).

Furthermore, the biochar matrix disconnects the microbial volatile organic compounds (mVOCs) from interfering with the tuning of the quorum-sensing system of the pathogens themselves. Thus, through a suite of these activities, biochar offers a buffered rhizospheric environment, whereby microbes cement wave(s) of defence based on the immune-priming of plants without synthetic chemicals. In fact, microbial volatile organic compounds released from such a biochar matrix interfere with the tuning of the quorum-sensing system of the pathogens themselves. The combined effects of biochar physical and chemical buffering on therhizosphere environment and the immune-priming effect of microbial inoculants formed a holistic plant protection strategy that was free of reliance on synthetic chemicals (Li et al. 2019).

6.3 Role in Sustainable Crop Production and Soil Restoration

Nutrient recycling, microbial diversity, and enhanced physical structure constitute the guarantees for higher sustainability through the co-biochar microbe mechanism. The very large surface area combined with stable carbon content impedes nutrient leachates, while the inoculating microbes mobilize otherwise unavailable nutrients, such as phosphorus and potassium. The system has rendered measurable improvements in crop yields, plant vigour, and soil organic carbon (SOC) content for those soils that are degraded and over-farmed. The combined application of biochar and the use of phosphate-solubilizing and nitrogen-fixing microbes have been found to enhance efficiency in nutrient uptake/ utilization of cereals and legumes with minimum requirement of chemical input (Kumar et al. 2024).

The co-biochar microbes can help restore dense soil to their natural condition while checking erosion of soils by re-establishing aggregate stability and porosity,

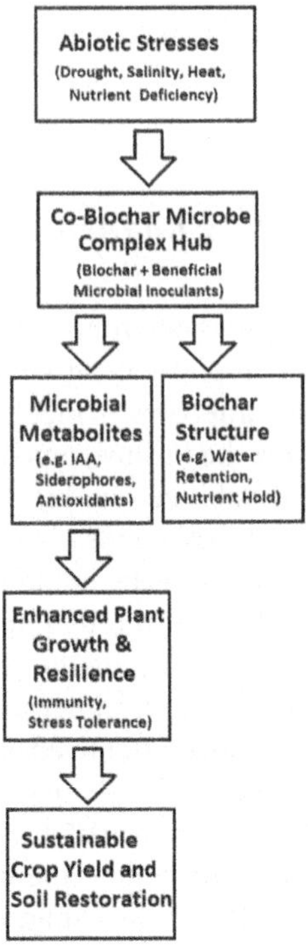

Fig. 3 Role of co-biochar microbes in environmental and agricultural stress management

i.e. good soil health. Such benefits supplement the worldwide movement towards climate-resilient and regenerative agriculture, hence co-biochar microbe systems being the potential solution for farmers dealing with declining production and the shifting environment (Imran 2024).

One of the well-known co-biochar microbe complexes interacts with different abiotic stress factors (abiotic factors such as drought, salinity, and heat) and biotic factors (pathogens) shown in Fig. 3. Microorganisms' function in the concurrent occurrence with biochar in mitigating abiotic and biotic stresses is illustrated in Fig. 3. Microbial inoculants and biochar as the central node react to stressors by exuding metabolites and improving physical characteristics of soil and enhance plant health, yield stability, and sustainable soil functioning.

7 Biochar Functionalization for Optimized Microbial Performance

The functionality of co-biochar microbial systems in agricultural and environmental applications relies mainly on compatibility between the physicochemical properties of biochar and the inoculated microbial communities. The raw or unmodified biochar has certain merits of its own due to porosity, carbon content, and nutrient-holding capacity, but functionalization now augments the capabilities of biochar in hosting, protecting, and activating beneficial microbes. That is, functionalization is a modification of the surface properties of biochar, either physically, chemically, or biologically, to promote enhanced microbial colonization, nutrient delivery, and environmental resilience.

7.1 Engineered Biochar for Enhanced Microbial Activity

Biochar engineering includes the modification of its features, e.g. surface area, porosity, surface charge, pH, and nutrient contents, depending on the requirements of certain selected or intended microbial taxa or functional groups. The concept becomes more important when designing a microbial consortium to achieve specific results. Recent studies highlight the enhanced performance of engineered biochar-based microbial carriers in field conditions, demonstrating improved microbial survivability and functional outcomes (Pathak et al. 2024). One approach includes the pre-loading of biochar with nutrients using organic or inorganic compounds. When biochar is impregnated with phosphate or nitrate solutions and then inoculated, it has been shown to enhance the retention of microorganisms and bioavailability of these nutrients to plants. Likewise, biochar pre-treated with compost extracts and humic acids has shown improved microbial diversity and metabolic activity due to an immediate nutrient base following field application.

Surface oxidation is another functionalization method achieved by treatments with hydrogen peroxide, steam, or ozone. These oxidative processes introduce oxygen-containing functional groups such as carboxyl and hydroxyl functional groups, which increase cation exchange capacity and provide better binding sites for microbial adhesion. Microbial colonization mostly seems to work well on oxidized biochar surfaces due to improved water wettability and electron exchange potential. Metal introduced biochar having micronutrients such as iron, zinc, copper has been emerging as an effective alternative. Not only does biochar improve nutrient status, it also inhibits pathogenic microbes and promotes beneficial microbes. However, one should be very careful in avoiding heavy metal toxicity as it would pose a toxicity risk to both microbes and plants. These oxidation processes add oxygen-based functional groups like carboxyl and hydroxyl functional groups, enhancing cation exchange capacity and offering superior binding sites for microbial adhesion. Microbial colonization predominantly appears to function optimally on oxidized biochar surfaces because of greater water wettability and electron exchange potential. Metal incorporated biochar with micronutrients such as iron, zinc, copper has been considered an emerging alternative. It not only improves nutrient status but also prevents pathogenic microbes from multiplying and supports beneficial ones. However, one should be very cautious in regard to heavy metal toxicity, because it would pose a toxicity risk to both microbes and plants (Rafique et al. 2017).

7.2 *Methods of Biochar Modification to Improve Microbial Interactions*

Whether physical, chemical, or biological processes are employed, the biochar treatment is designed to enhance the surface properties and nutrient composition of carbonized materials for microbial adherence and colonization as well as their activity and persistence in soil. Such modifications promote microbial adhesion on the biochar at the expense of process efficiency, however, and compromise the general working capacity of the co-biochar system for plant growth and stress remediation.

The primary physical modifications include the textural and structural characteristics of biochar including the size of pores, surface area, and particle size. As an example, biochar would contain more micropore volume and structural stability at high temperature pyrolysis; these features would provide more secured microhabitats within microbial colonization. Mechanical treatment like ball milling or fine grinding: This is to create a high surface area and generate more contact points that microbes can be inoculated on. Steam activation is another hot form of biochar enhancement, which increases porosity and renders the surface further hydrophilic, promoting microbic adhesion and moisture retention, which are substantially pivotal for microbic activity and survival in field conditions (Fazeli et al. 2020).

Chemical treatment of the surface in order to study functionalities present on it and to improve ion exchange properties in biochar. It was also observed that biochar increased surface charge under both alkaline and acid conditions, which contributed to improved nutrient and microbial cell adsorption. An oxidative treatment such as hydrogen peroxide or ozone yields oxygen-rich functional groups, hydroxyl and carboxyl for example, which then react with enzyme structures and microbial cell structures. Hydroxyl and carboxyl acting on the adhesion and nutrient uptake of microbes by cation exchange capacity in biochar rhizosphere. This biochar is also another way the slow nutrient-releasing materials are modified, but in cases for example of enriching it with nutrients through soaking it in (e.g.: urea) phosphoric acid or mineral salts, in turn resulting more microbial metabolism and longer uptake time by plant.

Biological modification otherwise known as bio-functionalization or the conditioning of BC with living microbial culture prior to application in the field is a novel trend. The microorganisms are so able to colonize the biochar surface and get adapted there first before entering to the more challenging soil conditions. The common approach would be to co-compost biochar and waste biomass together with inoculants of microbes, whereby the biochar is charged with a microbial biomass and secondary metabolites. Other labourers would steep biochar in a broth culture of microbes to promote the growth of microbial biofilms inside pores. In these systems, biochar is colonized with fermentable carbon sources and beneficial microbes promoting colonization of biochar surfaces by microbial cells and the development of a biofilm. Field trials have demonstrated that these biochar-modifying bioprocesses (BMb) not only all increase the survival, colonization efficiency, and functional activity of colonized microbes from time of application (AbbanBaidoo et al. 2024). Therefore, these biochar modification strategies of biochar microbe composites would improve the system efficiency and robustness. These physical, chemical, and biological alterations of biochar are then enhanced by researchers and specialists collaborating with a specific organism-oriented microbial inoculum that can be optimized for field longevity to optimize microbial function and soil–plant health responses. Comparative evaluation of biochar modification techniques against its primary objective and value to microbe function shall facilitate their field incorporation as well as experimental planning. Functionalization techniques revolve around those whose straightforward and economical application is poised to promote microbial nutrition, colonization, and activity (Palansooriya et al. 2019).

Table 3 presents the most prevalent physical, chemical, and biological modification methods to their components and advantages to microbial interactions in co-biochar systems.

8 Future Perspectives and Challenges

With the growing awareness of co-biochar microbe technology in sustainable agriculture and environmental remediation, a clear way is foreseen to refine its practical application, scale up for large-scale use, and unearth a plethora of unresolved scientific, economic, and regulatory issues. Although much has been learned about the synergistic relationship of biochar and beneficial microbes in laboratory studies over small- and medium-sized field studies, realizing the potential in other large-scale agricultural systems represents one major frontier to be crossed.

One of the most significant barriers is scaling up applications of co-biochar microbes from the laboratory to field or commercial systems. Soil type, microbial inoculum, biomass input, and moisture are all tightly controlled under experimental research conditions. However, in the open field, these variables complicate the system and pose challenges to its management. Soils are heterogeneous; they experience fluctuations with respect to climate, and they have regional diversity with regard to microbial populations; this greatly influences the performance of co-biochar. Future research will have to devote time to site-specific biochar microbe formulations that have been designed for local agroecological conditions. Such formulations will entail identification of robust microbial consortia compatible with regional soils and crops, as well as optimizing biochar types dependent on locally available biomass feedstocks.

8.1 Potential Limitations and Risks

While promising, co-biochar microbe systems have a few drawbacks that must be considered in any successful field application. The key challenge lies in their

Table 3 Summary of biochar functionalization methods for enhanced microbial interactions

Modification type	Method	Purpose	Benefits for microbial interactions
Physical	High-temperature pyrolysis	Increase microporosity and carbon stability	Better habitat for microbial colonization
	Ball milling/grinding	Increase surface area	More surface for microbial adhesion
	Steam activation	Enhance pore connectivity and wettability	Improved water retention and microbial mobility
Chemical	Acid/alkaline treatments	Alter surface charge and functional groups	Better microbial adhesion and nutrient retention
	Hydrogen peroxide oxidation	Introduce reactive oxygen groups	Enhanced microbial attachment and biofilm formation
	Nutrient enrichment (e.g. urea, phosphate)	Improve nutrient profile	Supports microbial growth and plant nutrition
Biological	Co-composting with organic waste	Pre-colonize with beneficial microbes	Greater microbial survival and functionality post-application
	Soaking in microbial broth	Direct microbial loading	Biofilm formation and immediate microbial activity
	Fermentation-based priming	Community establishment on biochar	Tailored microbiome with enhanced stress resilience

inconsistent field performance. Enhancement of nutrients, tolerance enhancements, yield improvements, etc. will all be influenced by soil type, climatic condition, crop species, microbial-strain compatibilities, among others. Therefore, before a generalized implementation, testing is mandatory for every specific region. The other issue is their potential to bring about microbial competition or disruption to the indigenous soil microbial environment. Hence, when the microbial consortia are introduced without checking for compatibility, they might outcompete or suppress native beneficial organisms, causing unintended ecological shifts.

On the other hand, introduced microbes sometimes fail to become established due to niche overlap or are not synergistic with indigenous soils where they are required and thus do not work as intended. Mobilization of heavy metals can occur if biochar is produced from contaminated or unknown feedstock.

A rigorous quality assurance and control as well as a standardized monitoring of salmonid gamete quality are crucial to reduce the risk.

Finally, the economic and practical challenges to full roll-out are considerable. Production of good quality biochar assumes its continuous supply and conditions of pyrolysis, which can be very expensive. Similarly, scaling up production and delivery of microbial inoculants is technically more difficult in terms of shelf life and field survival. Constraints imposed by these dimensions dictate that such co-biochar systems may not be viable for smallholder farmers unless backing in the form of either decentralized technology or public–private partnership underpin the system. Overall, co-biochar microbes could make an immense difference, but there are limitations that need to be recognized and managed through adaptive, evidence-based approaches.

Considering the recent molecular biology approaches, like omics techniques (metagenomics transcriptomics metabolomics) and new imaging methods, there is a potential to give additional input for better understanding of microbial community dynamics, changes in gene expression, or biomediator signalization within co-biochar. And all this output would be very useful in the making of advanced, biochar-appropriate inoculants with augmented durability and versatility capabilities. Most short-term contact yield or stress tolerance has some effect, but the long-term value of such effects as yield to multiple growing seasons remains an open question. Therefore, the effect of multi-year monitoring studies has to be carefully dealt with in order to solve the changes that potentially could take place in native soil microbiota, some unforeseen ecological consequences, and the accumulation of some microbial taxa. Long-term experiments must also measure change in soil organic matter, nutrient cycling, and carbon sequestration in the direction of assessing the environmental sustainability of this method (Kumar et al. 2021).

In addition to this challenge is the policy and regulatory framework for applying microbial inoculants and biochar in agro-production. Regulation in most countries for microbial biofertilizers and biocontrol is usually in progress, but biochar is rarely considered a soil amendment. Being so evident a lack in regulatory channels turns into a commercialization and farmer adoption impediment. Subsequent action must thus include evidence-based policy advice and certifications for product safety while stimulating innovation. There should also be support via government subsidy, demonstration projects, and public–private partnership for speedy adoption at the farm level. The economic viability of co-biochar technologies-driven projects is increasingly becoming a plausible issue for most smallholder farmers in the developing world. Such processes, connecting the production and functionalization of high-quality biochar and microbial inoculants, are generally very resource-intensive. Some of the methods to reduce the cost of production are the utilization of lower-cost feedstock materials such as agricultural residue and manure, implementing small-scale decentralized pyrolysis systems, and integrating biochar into composting. Most importantly, there should be formal economic cost–benefit appraisals to assess the return on investment for biochar on different crops, soils, and farm systems (Dickinson et al. 2015). Any favourable economic models that incorporate yield increase together with input reduction, improvement of soil health, and carbon credits will serve to quantify the longer-term valuation consideration of co-biochar technologies.

Public awareness and a complete education of the general public are necessary for future scaling. Hands-on training will be needed by extension agents, agricultural advisors and farmers in biochar handling, application with microbial inoculants. There would be farmer innovation platforms and participatory research networks for the communities to develop their own local solutions, and exchange knowledge. Centre curricula, research schemes, and agricultural policy conversations around the sustainability of co-biochar systems could provide sufficiently circular rotations for the tine researchers and politicians who shall carry this invention through. In conclusion, microbial co-biochar in circular economy and climate-resilient designs will ensure new directions in partnerships and funding. If biochar use can be linked to carbon storage, nutrient recovery and regenerative farming the technology is more likely to be consolidated into national/regional sustainability schemes. The science and engineering space does indicate a potential, but complex future for co-biochar microbe technology; nevertheless there is no path to the industry which succeeds without substantial investment alongside coordinated support from research and regulation, economic reason cycle inclusion, and educational extension. Multi-disciplinary means must be coalesced to understand challenges, which encompass dimensional issues and realize the potential of co-biochar microbe technology in restructuring food systems towards resilience, productivity, and environmental sustainability.

9 Conclusion

The integration of biochar technology and application of microbial inoculants offer a solution to agricultural productivity, environmental degradation, as well as climate change resilience. Co-biochar microbe system with the aid of biochar as soil conditioner, a new multicomponent nature-based system is designed to exploit shitments between biochar and microorganisms to promote plant growth in parallel. This process involves biochar as an amendment to the soil, however with changed supplied structural, chemical, and biological characteristics of biostimulated properties derived from this source in order to enhance colonization, survival, and metabolic activity or microbial assemblies leading to enhanced stronger plant protection due to improved soil health for a healthier stress abiotic conditions alleviating crop studies.

Our chapter provided some key building blocks in an effort to bring together links on the effect of biochar and endophytic (and other) microbial communities, including the processes behind them with responses that resulted in enhanced plant growth under both abiotic and biotic stresses—but there is more to come.

The improvements in microbial microenvironments, serve as physical carriers for microbes and buffer against chemical fluctuations. These systems also offer key microbial inoculant activities including nitrogen fixation, phosphate solubilization, hormone production, and pathogen control. When integrated and working together, they enable nutrient recycling, and enhance the ability of plants to withstand drought, salinity, temperature extremes, and decrease the need for chemical fertilizers and pesticides. The same goes for environmental co-benefits. Biochar provides carbon sequestration, immobilizes pollutants, and retains water, while microbes mineralize organic pollutants and control production of greenhouses gases—a powerful combination approach of climate smart and environmentally positive agriculture practices.

Significant progress has also been conducted in regards to functionalization of biochar in order to enhance its performance as microbial carrier.

These techniques, which include physical activation and chemical and biological priming, enable the biochar to be tailored according to individual microbial arrivals and soil situations. This advance enhances the survival, colonization, and functioning of microorganisms, particularly when they are suited to individual local agroecological contexts.

Nevertheless, the field-scale application of co-biochar microbe technology has not ensued yet due to largely issues including extreme variability in biochar quality, regulatory policy void on products with microbes, and economic constraint resulting from high cost as well as most importantly long-term outfield datasets shortage.

This will require interdisciplinary action by researchers and policymakers, extension workers and farmers. Standards of the quality and localized research in this technology as well as putting this technology into a circular economy/ sustainability context are necessary to make it acceptable. Finally, that the co-biochar and microorganisms are possible drivers to move towards twenty-first century. To transit from fossil input intensive farming to regenerative agriculture, it is necessary to restore natural soil functions and productivity, while identifying viable pathways towards climate-resilient agricultural systems. Exploring co-biochar-microbial systems represent a potential sustainable agricultural approach that advances agroecological innovations by linking soil biology with materials science.

To summarize, this chapter has examined the potentials of co-biochar and microbial technologies to alter the environment, agriculture, and society and confirmed them. With research advances and changing management practices, this smooth-working pair could hold the future of sustainability of land use and resilience of food systems.

References

Abban-Baidoo E et al (2024) Biochar addition influences C and N dynamics during biochar co-composting and the nutrient content of the biochar co-compost. Sci Rep 14(1):23781. https://doi.org/10.1038/s41598-024-67884-z

Amalina F et al (2022) Biochar production techniques utilizing biomass waste-derived materials and environmental applications—A review. J Hazard Mater Adv 7:100134. https://doi.org/10.1016/j.hazadv.2022.100134

Anae J et al (2021) Recent advances in biochar engineering for soil contaminated with complex chemical mixtures: remediation strategies and future perspectives. Sci Total Environ 767:144351. https://doi.org/10.1016/j.scitotenv.2020.144351

Anandakumar S, Senthamilselvi D, Kalaiselvi T (2024) Microbial consortium with multifunctional plant growth-promoting traits and its significant contribution in sustainable agriculture, pp 53–75. https://doi.org/10.1007/978-3-031-71487-0_3

Babu S et al (2023) Biochar implications in cleaner agricultural production and environmental sustainability. Environ Sci: Adv 2(8):1042–1059. https://doi.org/10.1039/D2VA00324D

Bolan S et al (2023) The potential of biochar as a microbial carrier for agricultural and environmental applications. Sci Total Environ 886:163968. https://doi.org/10.1016/j.scitotenv.2023.163968

Brokate O, Papenbrock J, Turcios AE (2024) Biofilm-forming microorganisms in the rhizosphere to improve plant growth: coping with abiotic stress and environmental pollution. Appl Soil Ecol 202:105591. https://doi.org/10.1016/j.apsoil.2024.105591

Charulatha R et al (2013) Characterization of groundnut rhizosphere pseudomonas sp. VSMKU 2013 for control of phytopathogens. In: Microbiological research in agroecosystem management. India: Springer India, pp 121–127. https://doi.org/10.1007/978-81-322-1087-0_8

Chauhan PK et al (2024) Fostering plant growth performance under drought stress using rhizospheric microbes, their gene editing, and biochar. Environ Geochem Health 46(2):41. https://doi.org/10.1007/s10653-023-01823-1

Chen J et al (2024) A complete review on the oxygen-containing functional groups of biochar: formation mechanisms, detection methods, engineering, and applications. Sci Total Environ 946:174081. https://doi.org/10.1016/j.scitotenv.2024.174081

Choudhary DK, Prakash A, Johri BN (2007) Induced systemic resistance (ISR) in plants: mechanism of action. Indian J Microbiol 47(4):289–297. https://doi.org/10.1007/s12088-007-0054-2

Chukwudi UP et al (2025) Field application of beneficial microbes to ameliorate drought stress in maize. Plant Soil [Preprint]. https://doi.org/10.1007/s11104-025-07446-y

Dey S et al (2023) Enhancing cation and anion exchange capacity of rice straw biochar by chemical modification for increased plant nutrient retention. Sci Total Environ 886:163681. https://doi.org/10.1016/j.scitotenv.2023.163681

Dickinson D et al (2015) Cost-benefit analysis of using biochar to improve cereals agriculture. GCB Bioenergy 7(4):850–864. https://doi.org/10.1111/gcbb.12180

Dong W et al (2024) Hydrogen bonds between the oxygen-containing functional groups of biochar and organic contaminants significantly enhance sorption affinity. Chem Eng J 499:156654. https://doi.org/10.1016/j.cej.2024.156654

Fazeli Sangani M, Abrishamkesh S, Owens G (2020) Physicochemical characteristics of biochars can be beneficially manipulated using post-pyrolyzed particle size modification. Bioresour Technol 306:123157. https://doi.org/10.1016/j.biortech.2020.123157

Foster EJ et al (2016) Biochar and manure amendments impact soil nutrients and microbial enzymatic activities in a semi-arid irrigated maize cropping system. Agric, Ecosyst Environ 233:404–414. https://doi.org/10.1016/j.agee.2016.09.029

Gao S et al (2023) Biochar co-compost improves nitrogen retention and reduces carbon emissions in a winter wheat cropping system. GCB Bioenergy 15(4):462–477. https://doi.org/10.1111/gcbb.13028

Guo Z et al (2022) Functional bacterial consortium responses to biochar and implications for BDE-47 transformation: performance, metabolism, community assembly and microbial interaction. Environ Pollut 313:120120. https://doi.org/10.1016/j.envpol.2022.120120

Habte B, Dobo B (2025) The effect of compost and biochar amendments, as well as the inoculation of arbuscular mycorrhizal fungi and Trichoderma Harzianum, on the control of fusarium Oxysporum pathogen on tomato (Solanum lycopersicum L.) in the greenhouse. Int J Agron. Edited by Merah O 2025(1). https://doi.org/10.1155/ioa/7330396

Hakeem KR et al (2022) Biochar promotes arsenic (As) immobilization in contaminated soils and alleviates the As-toxicity in soybean (Glycine max (L.) Merr.). Chemosphere 292:133407. https://doi.org/10.1016/j.chemosphere.2021.133407

Harikrishnan H, Shanmugaiah V, Balasubramanian N (2014) Optimization for production of Indole acetic aci d (IAA) by plant growth promoting Streptomyces sp VSMGT1014 isolated from rice

rhizosphere. Int J Curr Microbiol App Sci 3(8):158–171. https://api.semanticscholar.org/CorpusID:41547015
Hou J et al (2022) Plant resistance to disease: using biochar to inhibit harmful microbes and absorb nutrients. Environ Res 214:113883. https://doi.org/10.1016/j.envres.2022.113883
Imran (2024) Integration of organic, inorganic and bio fertilizer, improve maize-wheat system productivity and soil nutrients. J Plant Nutr 47(15):2494–2510. https://doi.org/10.1080/01904167.2024.2354190
Jyoti D, Sinha R, Faggio C (2022) Advances in biological methods for the sequestration of heavy metals from water bodies: a review. Environ Toxicol Pharmacol 94:103927. https://doi.org/10.1016/j.etap.2022.103927
Kumar Sarangi P et al (2022) Utilization of agricultural waste biomass and recycling toward circular bioeconomy. Environ Sci Pollut Res 30(4):8526–8539. https://doi.org/10.1007/s11356-022-20669-1
Kumar V et al (2021) Application of omics technologies for microbial community structure and function analysis in contaminated environment. In: Wastewater treatment. Elsevier, pp 1–40. https://doi.org/10.1016/B978-0-12-821881-5.00001-5
Kumar N et al (2024) Enhancing nutrient use eficiency in pulses under diversified farming through agronomic manipulations, pp 197–213. https://doi.org/10.1007/978-981-97-7517-0_8
Li Z et al (2019) Effects of phytolithic rice-straw biochar, soil buffering capacity and pH on silicon bioavailability. Plant Soil 438(1–2):187–203. https://doi.org/10.1007/s11104-019-04013-0
Lian F et al (2022) Novel insights into the impact of nano-biochar on composition and structural transformation of mineral/nano-biochar heteroaggregates in the presence of root exudates. Environ Sci Technol 56(13):9816–9825. https://doi.org/10.1021/acs.est.2c02127
Marcińczyk M, Oleszczuk P (2022) Biochar and engineered biochar as slow- and controlled-release fertilizers. J Clean Prod 339:130685. https://doi.org/10.1016/j.jclepro.2022.130685
Mukherjee S et al (2022) Biochar-microorganism interactions for organic pollutant remediation: challenges and perspectives. Environ Pollut 308:119609. https://doi.org/10.1016/j.envpol.2022.119609
Nan Q et al (2021) Exploring long-term effects of biochar on mitigating methane emissions from paddy soil: a review. Biochar 3(2):125–134. https://doi.org/10.1007/s42773-021-00096-0
Palansooriya KN et al (2019) Response of microbial communities to biochar-amended soils: a critical review. Biochar 1(1):3–22. https://doi.org/10.1007/s42773-019-00009-2
Pathak HK et al (2024) Mechanistic and future prospects in rhizospheric engineering for agricultural contaminants removal, soil health restoration, and management of climate change stress. Sci Total Environ 927:172116. https://doi.org/10.1016/j.scitotenv.2024.172116
Rafique M et al (2017) Enhancement of maize plant growth with inoculation of phosphate-solubilizing bacteria and biochar amendment in soil. Soil Sci Plant Nutr 63(5):460–469. https://doi.org/10.1080/00380768.2017.1373599
Raza A et al (2020) Nitrogen fixation of legumes: biology and physiology. In: The plant family fabaceae. Singapore: Springer Singapore, pp 43–74. https://doi.org/10.1007/978-981-15-4752-2_3
Sarraf M et al (2023) Understanding the role of beneficial elements in developing plant stress resilience: signalling and crosstalk with phytohormones and microbes. Plant Stress 10:100224. https://doi.org/10.1016/j.stress.2023.100224
Sivaram AK et al (2023) Microbial inoculant carriers: soil health improvement and moisture retention in sustainable agriculture, pp 35–91. https://doi.org/10.1016/bs.agron.2023.03.001
Sultana T, Islam MR (2024) The potential of soil microbiomes in alleviating climate change–associated stresses on crop plants. In: Beneficial microbes for sustainable agriculture under stress conditions. Elsevier, pp 81–111. https://doi.org/10.1016/B978-0-443-13193-6.00005-1
Tan X-F et al (2021) Role of biochar surface characteristics in the adsorption of aromatic compounds: pore structure and functional groups. Chin Chem Lett 32(10):2939–2946. https://doi.org/10.1016/j.cclet.2021.04.059

Tang H et al (2024) Bridging gaps and seeding futures: a synthesis of soil salinization and the role of plant-soil interactions under climate change. iScience 27(9):110804. https://doi.org/10.1016/j.isci.2024.110804

Yan H et al (2023) The aggregate distribution of Pseudomonas aeruginosa on biochar facilitates quorum sensing and biofilm formation. Sci Total Environ 856:159034. https://doi.org/10.1016/j.scitotenv.2022.159034

Yang X et al (2023) Microbial responses towards biochar application in potentially toxic element (PTE) contaminated soil: a critical review on effects and potential mechanisms. Biochar 5(1):57. https://doi.org/10.1007/s42773-023-00255-5

Yuan X et al (2023) Recent advancements and challenges in emerging applications of biochar-based catalysts. Biotechnol Adv 67:108181. https://doi.org/10.1016/j.biotechadv.2023.108181

Zhang K et al (2023) Enhanced MFC sensor performances and extracellular electron transport efficiency mediated by biochar and underlying biochemical mechanisms. J Environ Manag 332:117282. https://doi.org/10.1016/j.jenvman.2023.117282

Zhang W et al (2024) Effect of nitrate concentration on anaerobic methane oxidation coupled to denitrification in membrane biofilm reactor after prolonged storage. Water, Air, Soil Pollut 235(4):246. https://doi.org/10.1007/s11270-024-07040-5

Zheng X et al (2022) The effects of biochar and its applications in the microbial remediation of contaminated soil: a review. J Hazard Mater 438:129557. https://doi.org/10.1016/j.jhazmat.2022.129557

Development of Engineered Biochar and PGPR for Plant Growth Under Agricultural Contaminants: Current Insights and Future Prospects

Prabhat K. Chauhan, Sakshi Tomar, Shefali Srivastava, Akash Chaturvedi, R. L. S. Sikarwar, Bhupendra Singh, and Kozma Dorottya Edina

Abstract Recently, use of engineered biochar as a soil supplement has drawn increased interest from scientists looking for ways to reduce agricultural pollutants and stimulate plant growth. While the compatibility of biochar and its performances aren't entirely satisfactory in the field compared to lab results, this may be because of the conditions for the preparation of engineered biochar, adaptability of engineered biochar in rhizospheric environment, including beneficial rhizobacteria, and this chapter focuses on all three components. In-depth discussion of how biochar properties including porosity, surface area, chemical composition, and others affect the adsorption, immobilization, and degradation of agricultural pollutants is provided in this chapter. While plant growth promoting rhizobacteria (PGPR) also does not produce adequate results, another environmentally friendly agent, such as PGPR, can successfully reduce agricultural contaminants and increase plant development through a variety of metabolic pathways. Considering these points, engineered biochar and PGPR can collaborate to reduce agricultural contaminants while also encouraging plant development. The utilization of agricultural waste, sludge, and other carbon sources to prepare biochar for economically engineered biochar production is explored in this chapter. Engineered biochar and PGPR application can improve soil structure, induce systemic resistance in plant, and aid with nutrient absorption. To shed light on possible synergies between the use of PGPR and engineered biochar to lessen soil contaminants as well as improve growth of plants is also discussed in this chapter.

P. K. Chauhan (✉) · S. Tomar · R. L. S. Sikarwar · B. Singh
Department of Environmental Science, AKS University, Satna, Madhya Pradesh, India
e-mail: cprabhatenv@gmail.com

S. Srivastava
Department of Environmental Science, VBS Purvanchal University, Uttar Pradesh, Jaunpur, India

A. Chaturvedi
Department of Chemical Engineering, Indian Institute of Technology BHU Varansai, Uttar Pradesh, Varansai, India

K. D. Edina
Department of Accounting and Finance, University of Pannoniya, Egyetem, Hungary

A. Chauhan et al. (eds.), *Biochar As Remediation Tools for Pollutant Removal*,
https://doi.org/10.1007/978-3-032-16746-0_7

Keywords Waste to biochar · Biochar preparation · Agricultural contaminant removal · Adsorption · PGPR · Plant growth performance

1 Introduction

In recent years, increasing anthropogenic activities like smelting and mining, overuse of pesticides and fertilizers, wastewater irrigation, industrial effluent, exhaust emissions have led to enhance the HMs pollution (Upadhyay et al. 2023a). Heavy metal (HM) pollution diversely affects plants and animals through various ways. Recent research has introduced biochar as a potential solution to remove HMs, improve soil fertility, as well as use as an effective tool for carbon sequestration (Jindo et al. 2014). Biochar is a cost-effective and environmentally friendly remediation amendment. Its hydrocarbon-rich organic compounds contribute to improving soil quality and enhancing nutrient availability, thereby fostering plant growth and development (Bonanomi et al. 2017). Consequently, biochar serves as an excellent example within sustainable agricultural systems, effectively addressing nutrient deficiencies and promoting robust plant growth (Major et al. 2010). Biochar is prepared through any biomass feedstock such as rice husk, rice straw, corncob, woodchips, etc. (Jindo et al. 2014).

Biochar commonly prepared by plant biomass, industrial waste, agricultural waste, sludge, and livestock excreta under absence or very low oxygen, undergoing pyrolysis at temperatures below 1000 °C (Amusat et al. 2021). Recently, a wide range of biochar-derived materials have gained prominence as eco-friendly absorbents for removing both organic and inorganic pollutants from polluted environments. Environmental degradation manifested through soil nutrient loss, land degradation, and pollution has triggered secondary environmental concerns such as water scarcity, declining agricultural productivity, food security, and climate variability (Yuan et al. 2019). Biochar offers numerous benefits, including pH regulation, contaminants immobilization, enhanced water retention, improved crop yields, and reduced pollutants absorption and accumulation (Cheng et al. 2020). Diverse carbon-rich feedstocks, primarily organic waste, biochar serve as a cost-effective alternative to activated carbon for the elimination of HMs, pesticide, pharmaceuticals, dyes, and volatile organic compounds from contaminated water and soil (Zhao et al. 2021; El-Naggar et al. 2021). As an eco-friendly and carbon–neutral material, biochar releases CO_2 during pyrolysis, offset by the CO_2 absorbed by biomass during photosynthesis (Yuan et al. 2019). Its three dimensional porous structure not only promotes carbon sequestration but also enhances the adsorption and degradation of environmental pollutants.

The beneficial bacteria such as PGPR occur in the rhizosphere plant roots, functionally act as eco-friendly nature-based amendments. PGPR play a pivotal role in improving plant development, reducing reliance on pesticide and other chemical fertilizers, and promoting sustainable agriculture especially under diverse environmental stress conditions (Singh et al. 2022; Chauhan and Upadhyay 2023).

Root exudates modulate HMs-interaction, improving plant tolerance and mitigating adverse impacts on growth and development of plants (Upadhyay et al. 2022).

PGPR-derived enzymes and metabolites significantly enhance the growth and yields in wheat crops under environmental stress conditions (Sohaib et al. 2019). In groundnut, rhizobacterial species such as *Pseudomonas aeruginosa* and *Bacillus sonorensis* release root exudates, including serine, glycopyranoside, pyrrolidinone, pentanoic acid, and tartaric acid, contributing to improved growth and yields (Ankati and Podile 2019). Likewise, *Bradyrhizobium* sp. and *Azospirillum brasilense* secrete varieties of root exudate compounds such as naringenin, oleic, citric, and lactic acid in the rhizosphere of groundnut, facilitating plant-PGPR interaction and supporting performance under atmospheric water deficit condition (Cesari et al. 2019).

Korenblum et al. (2020) show that *Bacillus* sp., a rhizospheric bacterium associated with tomato plants, releases azelaic acid through root exudates. This compound functions as a biocontrol agent by inducing induction of systemic resistance (ISR), contributing to the enhancement of plant growth. Similarly, *Pseudomonas segetis* (P6), a rhizospheric bacterial species isolated from common glasswort, was identified to produce N-acylhomoserine lactones (AHLs), acting as a biocontrol agent and leading to an increase in the height and weight of tomato plants (Rodriguez et al. 2020) (Fig. 1).

2 Impact of Organic Pollutants on Plant and Surrounding

2.1 Pesticides

Pesticide plays a critical role in safeguarding crops, improving yields, and contributing to food security. However, their wide spread use rises significant concerns, particularly regarding non target effects, such as harm to beneficial insects and disruption of pollination process and overall ecological balance. Phytotoxic effects such as leaf burn, stunted growth, and reduced plant vigor are also commonly reported. Moreover, continuous and excessive application can lead to pest resistance in pest populations, ultimately diminishing the efficacy of chemical control strategy. Striking a balance between pest control and environmental sustainability is essential for long-term agricultural resilience.

The global use of pesticides is rapidly increasing to combat plant disease, insect infestation, and weed proliferation (Haider et al. 2019). However, indiscriminate pesticide application has adverse consequences on soil, water, land, and aquatic ecosystem (Alvarez et al. 2017). For instance, exposure to the organophosphorus insecticide chlorpyrifos significantly inhibited nitrogen metabolism and growth of *Vigna radiate* L. (Parween et al. 2011). In a subsequent study, Parween et al. (2012) examine the metabolism of chlorpyrifos and its impact on the antioxidative enzymatic system of *Vigna radiata* L. across various growth stages. Their result indicating a

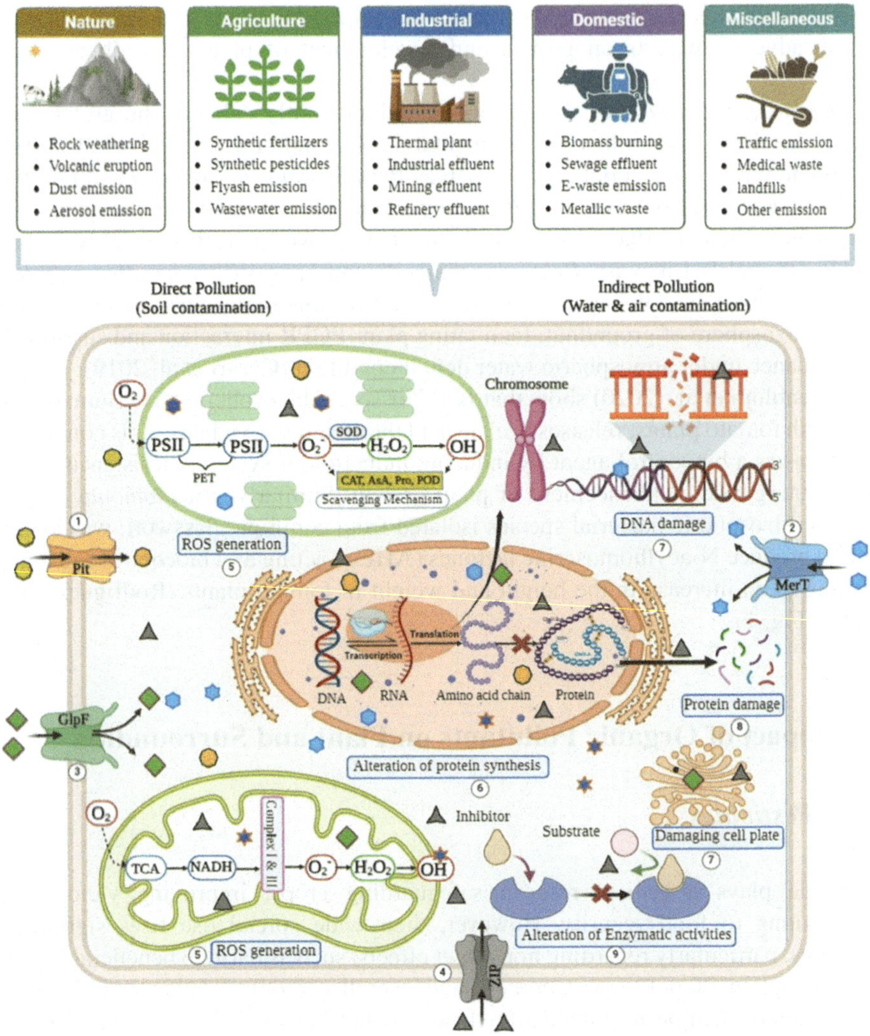

Fig. 1 Imapcts of environmental pollutions on plants-cellular

marked increase in lipid peroxidation and proline content during the pod-flowering stage at 1.5 mm. Additionally, all developmental stage showed a subsequent reduction in ascorbate and glutathinone level underscoring the oxidative stage induced by chlorpyrifos exposure.

2.2 Pharmaceutical Drugs/Antibiotics

The use of pharmaceutical drugs in agriculture triggers various types of environmental and health risks. The residue pesticides, antibiotics, and other pharmaceutical compounds enter the food chain after application in agriculture that affects the flora and fauna through food chains. These residues adversely reduce the quality of soil, cause nutrient imbalance, and disturb the ecosystem balance by harming beneficial insects and reducing microbial biodiversity. Due to long-term effects remining inadequately studied shows that deeper studied is necessary (Manyi-Loh et al. 2018).

The open disposal of pharmaceutical products, unused medication, and the agricultural application of biosolids and manure derived from wastewater treatment plants in soil increase environmental contamination (Cerquira et al. 2019). The contamination affects the physical, chemical, and microbiological properties of soil that lead to the accumulation of antibiotic residues and their metabolites in plants (Christou et al. 2019). Due to the presence of antibiotic residues in soil, biological activity of plants that leads to reduced crop growth and yields (Liu et al. 2020). Pan and Chu (2017) reported the detrimental effects of antibiotics on development parameters including seed germination, shoot elongation, and root elongation of cereal crops. Therefore, the researchers highlighted the urgent need for responsible pharmaceutical disposal and increased awareness of the ecological consequences in order to safeguard both soil ecosystems and agricultural yields.

2.3 Polycyclic Aromatic Hydrocarbons (PAHs)

The US Environmental Protection Agency (EPA) has designated 16 PAHs as priority pollutants for regulation and management (Liu et al. 2024). Researchers have identified that over 100 types of PAHs exist in the environment at notable concentrations that pose serious threats to crops' health, representing a complex challenge at the intersection of environmental pollution and agricultural sustainability (Ukalska-Jaruga et al. 2019). Researchers reported that mostly two types of PAHs occur in the atmosphere that primarily originated from fossil fuel combustion and industrial activities; PAHs infiltrate the soil and adversely affect plant growth and development. Xu et al. (2021) reported that the level of PAHs varies in different areas: average 7680.4 ng/g in industrial zones and 213.5 ng/g in rural areas in Zhejiang Province, China. The high concentration of PAHs alters the essential physiological processes of plants including photosynthesis, nutrient uptake, and metabolic activity, thereby ultimately affecting the overall performance of plants (Jajoo et al. 2014).

Ukalska-Jaruga et al. (2019) reported that the presence of PHAs in soil poses long-term risks because they can accumulate in plant tissues, leading to reduced crop yields and reduced food quality. PAHs contamination also affects the microbial communities (Vorobyeva et al. 2020), as well as affects the physicochemical properties of soil (Vorobyeva et al. 2020), pH, nutrient availability, and bulk density that ultimately affect plant health (Logeshwar et al. 2018).

Furthermore, it is well observed that the high concentrations of fluoranthene (FLT) in the soil suppress seed germination in various plant species including *Pisum sativum, Triticum aestivum, Lycopersicum esculentum, Allium cepa,* and lettuce (Kummerova et al. 2010).

2.4 Dyes

Due to rapid industrialization has profoundly influenced human life, that significantly contributes in various sectors, such as leather, textile, cosmetics, pharmaceuticals, food, fuel, and rubber industries (Lan et al. 2022). Among the major pollutants emerging from these sectors are synthetic dyes, that pose serious environmental concerns due to their complex and recalcitrant molecular structures, making them resistant to biodegradation (Anastopoulos and Kyzas 2014). These dyes significantly threaten aquatic ecosystems by obstructing photochemical reactions and reducing sunlight penetration, thereby impairing photosynthesis in aquatic plants (Liu et al. 2013). Furthermore, Lan et al., reported in 2022 that many dyes and their degraded byproducts wastewater are known to be toxic, mutagenic, and carcinogenic, causing substantial risks to human health and plant growth as well as nutritional quality of crops when such contaminated water is used for irrigation.

A notable example is methylene blue, a commonly used cationic dye in the textile industry, known to produce carcinogenic aromatic amines, including methylene and benzidine. To assess its ecological toxicity, investigators studied the effects of methylene blue on two economically and nutritionally valuable microalgae, *Spirulina platensis* and *Chlorella vulgaris.* Their study revealed a concentration-dependent decline in specific growth rate, pigment content, and protein level of both species following dye exposure. Additionally, methylene blue significantly inhibited chlorophyll synthesis, leading to reduced photosynthesis efficiency and impaired development of microalgae, underscoring the dye's detrimental impact on aquatic life and, by extension, the broader ecosystem.

3 Current Global Strategies Mediate Biochar to Mitigate Agricultural Contaminants

Biochar is an activated carbon that, derived from the pyrolysis of organic materials, has garnered global attention for its versatile application and potential contributions to sustainable development. From enhancing soil quality to carbon sequestration, biochar is influencing international strategies with significant economic implication.

One of the primary global applications of biochar utilization is in agriculture, where it is utilized to improve soil fertility and improve crop yields (Fig. 2). Advocates emphasize the role of biochar in agriculture that enhances water retention, nutrient availability, and microbial activity in soil condition. These benefits can translate into increased agricultural productivity, a vital factor for global food security. The economic impact is particularly notable, as higher yields may lead to increased farm incomes and promote economic growth in agricultural-dependent regions.

Beyond agriculture and climate change mitigation, biochar initiatives also offer promising avenues for economic development through job creation and the stimulation of local enterprises. Biochar production and application require skilled labor in sustainable practices and technology, presenting employment opportunities in both rural and urban areas, especially in areas rich in biomass feedstock. Furthermore, the emergence of biochar-related industries, such as equipment manufacturing and consulting services, can contribute to economic diversification and resilience.

Global approaches to the use of biochar encompass agriculture, climate change mitigation, and economic development. The economic significance of biochar lies in its potential to enhance agricultural productivity, contribute to carbon trading markets, and stimulate local economies. However, realizing these economic benefits requires addressing the challenges of diverse ecosystems, uncertainties in carbon sequestration, and ensuring equitable socio-economic outcomes. A balanced and

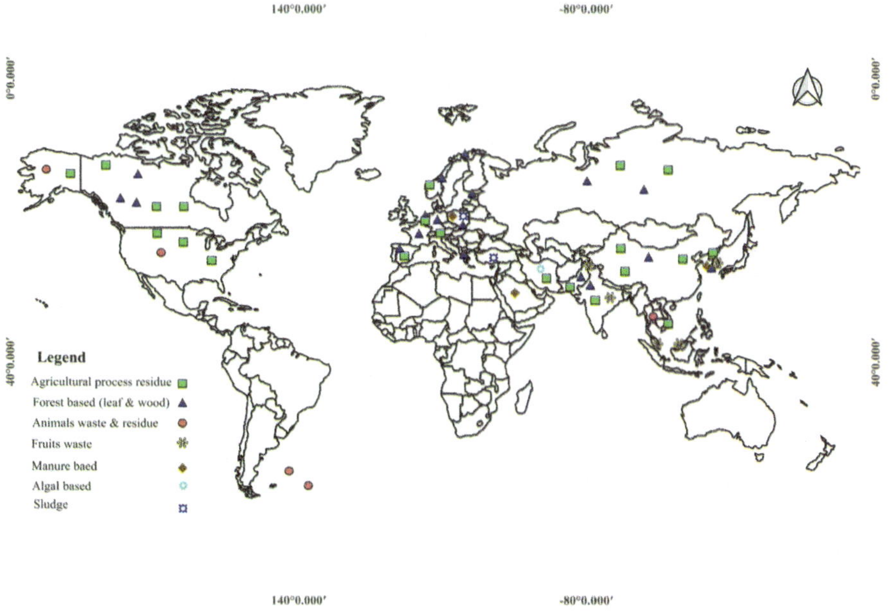

Fig. 2 Enhancing global biochar understanding: categorization and visualization through GIS mapping of tabulated data

integrated approach is crucial to harness the full economic potential of biochar on a global scale (Fig. 2).

4 Removal of Agricultural Contaminant

4.1 Biochar Mediated Removal

See Fig. 3.

4.1.1 HMs

The metals whose density is greater than 5.0 g cm^{-3} are considered as HMs, there are 45 types of heavy toxic metals exist in nature. Zn, Cu, Hg, Mn, Sn, Cr, Pb, Cd, Al, Ni, Ag like pollutants are included in these and are considered to be the high toxic contaminants present in water and soil system. Heavy-metal pollution is becoming severe with the high population growth and rapid industrialization, ultimately adding HMs contents in many rivers and groundwater above the safety standards, uranium (VI), a radioactive heavy-metal, has become a major problem because of the establishment of nuclear energy industry (Chen et al. 2022). HMs cause harm to human health after entering the human body because of their non-degradable nature. Therefore, the removal of HMs from polluted soil and water is very essential, and for this purpose, biochar can be used which will provide support for HMs removal and it be also improve soil fertility and promote plant growth (Beesley

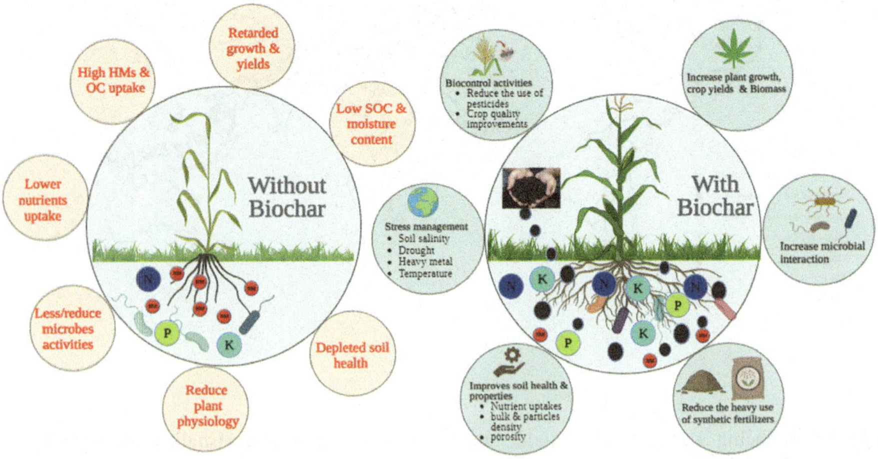

Fig. 3 Sustainable approach agricultural presence and absence of biochar

et al. 2011; Chauhan et al. 2023) (Fig. 3 and Table 1). More HMs pollutants can be removed by alkali-activated biochar in comparison with pristine biochar because alkali-activated biochar has broad surface-area and richer pore-structure (Liu et al. 2021c). The physical and chemical properties of soil such as pH, redox conditions, organic matter, etc., affect the bioavailability of HMs ions (Yang et al. 2021).

Numerous factors, including water properties, environmental conditions, and biochar characteristics, influence the adsorption mechanism of biochar in the removal of HMs ions. The characteristics of biochar are determined by pyrolysis temperature, feedstock composition, and time. Dominantly surface complexation, ion exchange, and precipitation methods are widely used for HMs removal (Fig. 3).

4.1.2 Organic Pollutants

Recently, the accelerative growth of industrial and agricultural activities in recent year has contributed to a persistent rise in the environmental organic pollutants (Vithanage et al. 2016). This increase poses a serious threat to the ecosystem, causing significant contamination that affects human health as well as natural environment (Jiang et al. 2018). Biochar demonstrates efficacy in the removal of diverse organic pollutants through its adsorption mechanisms, providing an efficient and economically viable means for eliminating organic contaminants (Dai et al. 2019). Li et al. (2014) reported that wheat straw biochar selectively adsorbs PAHs in soil washing effluents, achieving 71.8–98.6% removal for PAHs and over 87% recovery for Triton X-100 (TX100).

Xiang et al. (2020) reported that ball-milled hickory wood biochars showed a significant 1.3–13.0-fold increase in VOC adsorption, with a notable maximum capacity for acetone. The surface processes and biochar properties influenced the adsorption of polar VOCs, and the easy desorption at low temperatures, coupled with excellent reusability over five cycles, underscores the efficacy of ball-milled biochars as effective and regenerable adsorbents for VOC removal.

4.1.3 Emerging Pollutants

Emerging contaminants are pollutants with low concentration but cause major harm to the health of humans and environments. These contaminants include complex forms of organic matter commonly ours in water, such as antibiotics, endocrine-disrupting chemicals, disinfection by-products, pesticides, veterinary products, personal care products, pharmaceuticals, toxic industrial compounds, food additives, and persistent organic pollutants (Flesch et al. 2019). Emerging contaminants persist for a long time and are continuously circulating, migrating, and transforming in the environment (Gomes et al. 2017). To address these contaminants, various types of waste act as raw material for biochar production and other value-added by-productions. Widely utilized methods to eliminate emerging contaminants from water encompass the microbial approach (Ferreira et al. 2016), chemical oxidation

Table 1 Biocharmediated plant growth performance under agricultural contaminant soil

Biochar feedstock and pyrolyzing temperature	Plant growth performances	Country	References
Maize and wood	Suppress the fusarium disease and plant growth in *Lupinus angustifolius* Plant	Germany	Egamberdieva et al. (2020)
Morusalbal wood	*Pseudomonas* sp. and *Staphylococchs* sp. suppress the drought stress condition in *Brassica napus*	Pakistan	Lalay et al. (2022)
Pine wood chip (500 °C)	*Pseudomonas rhodesiae, Paenibacillus polymyxa, Rahnella* sp. Increase switch grass productivity and biomass	Canada	Shanta et al. (2016)
Softwood biochar	*Bacillus licheniformis*	China	Liu et al. (2022)
Azolla	*Pseudomonas fluorescens*, it enhances Rosemary growth	Iran	Kasmaei et al. (2019)
Timber waste biochar (389 °C)	*Bacillus amyloliquefaciens*, and *Agrobacterium fabrum* it help in improving wheat yields under drought stress condition	Pakistan	Zafar-ul-hye et al. (2019)
Timber waste biochar (389 °C)	*Bacillus amyloliquefaciens*, it help in enhancing the production and growth of wheat under drought condition	Pakistan	Danish and Zafar-ul-hye (2019)
Fruit and vegetable biochar (450 °C)	*Alcaligenes faecalis* and *Bacillus amyloliquefaciens*, it help in reducing led stress in mint plant	Pakistan	Zafar-ul-hye et al. (2021)
Spent mushroom substrate (600 °C)	*Bacillus subtilis* and *Pseudomonas fluorescence*, it help in improving yield, growth and biochemical response in cauliflower plant	Saudi Arabia	Siric et al. (2022)
Wood biochar and green waste biochar	*Bacillus subtilis*, it help in minimizing the *Alternaria solani* disease in tomato plant	Pakistan	Rasool et al. (2021a)
Coconut husk (550 °C)	*Enterobacter hormaechei* and *Bacillus thuringiensis*, it help in improving growth, nutrient uptake, physiology and soil enzyme activities in *Ricinus communis*	India	Anbuganesan et al. (2022)
Rice husk and Corn straw biochar (350 °C)	*Bacillus coagulans* and *Pseudomonas koreensis*, it help in the mitigatee of water deficit in rice plant in saline soil	USA	Hafez et al. (2019)
Rice husk and corn stalks biochar (350 °C)	*Azotobacter chroococcun* SARS 10 and *Pseudomonas koreensis* MG209738, it help in minimizing soil salinity, inducing photosynthesis process and enhancing growth and yield of maize crop	Egypt	Nehela et al. (2021)
Birch wood	*Bacillus altitudinis* strain TF16a, it help in enhancing the growth of *Brassica napus*	Russia	Kumar et al. (2022)
Wheat straw biochar (450 °C)	*Bacillus* sp. ZM20, it help in improving soil properties, yield, plant growth and quality of maize crop	Pakistan	Ahmad et al. (2020)

(continued)

Table 1 (continued)

Biochar feedstock and pyrolyzing temperature	Plant growth performances	Country	References
Apple wood biochar	*Bacillus subtilis* SL-44, it help in improving plant growth including phosphorus dissolution, nitrogen fixation and secretion of IAA	China	Liu et al. (2021a)
Sewage sludge and municipal compost (400 °C)	*Bacillus subtilis* and *Paenibacillus azotofixans*, it enhances plant growth and element contents of einkorn wheat	Turkey	Cig et al. (2021)
Grain husk and paper fibre sludge (450–500 °C)	*Azospirillum brasilense* and *Arthrobacter crystallopoietes*, it enhances maize growth	Austria	Kari et al. (2021)
Rice husk	*Bacillus subtilis* and *Trichoderma harzianum*, it helps in enhancing plant growth and activates resistance against *Meloidogyne incognita* in tomato	Pakistan	Arshad et al. (2021)
Holm Oak feedstock (650 °C)	*Pseudomonas*, *Baciullus*, *Trichoderma* in help in enhancing resistance to *Botrytis cinerea* in strawberry fruits	Belgium	De et al. (2021)
Rice straw and pulp mill sludge and (600 °C)	The adsorption capacity for HMs was predominantly influenced by the type of biomass, showcasing higher adsorption capacities of biochars for Pb^{2+} by 109.9 to 256.4 mg g^{-1}, compared to Ni^{2+} by 40.2–64.1 mg g^{-1}, Cd^{2+} by 29.5–42.7 mg g^{-1}, and Cu^{2+} by 18.5–39.4 mg g^{-1}	South Korea	Islam et al. (2021)
Bamboo, sugarcane and neem	Sugarcane biochar was found to be most effective with sorption efficiencies ranging from 3.27 to 99.4%	India	Singh et al. (2021)
Crab shell (350 °C)	Sb^{3+} adsorption at 498 mg/g adsorption from water	United states	Zhang et al. (2021)
Cattle manure (500 °C)	Fe-impregnated Sb^{5+} Water 58.3 mg/g	Korea	Park et al. (2019)
Wood chip (600 °C)	Adsorption of Hg^{2+} from Water 107.5 mg/g	Korea	Park et al. (2019)
Sesbania bispinosa (450 °C)	Adsorption of MnO $AsO_4{}^{3-}$ Water 7.35 mg/g	Pakistan	Imran et al. (2021)
Sesbania bispinosa (450 °C)	Adsorption of CuO $AsO_4{}^{3-}$ water 12.47 mg/g	Pakistan	Imran et al. (2021)
Rice straw (500 °C)	Adsorption of Thiol-modified Cd^{2+} soil 45.1 mg/g	China	Fan et al. (2020)
Rice straw (500 °C)	Adsorption of Thiol-modified Pb^{2+} soil 61.4 mg/g	China	Fan et al. (2020)

(continued)

Table 1 (continued)

Biochar feedstock and pyrolyzing temperature	Plant growth performances	Country	References
Lobster shell (600 °C)	Adsorption of HCl treatment Cu^{2+} water 71.4 mg/g	North Bangkok, Thailand	Ma et al. (2021)
Lobster shell (600 °C)	Adsorption of HCl treatment Cd^{2+} water 126 mg/g	North Bangkok, Thailand	Ma et al. (2021)
Peanut shell (600 °C)	Adsorption of MnO-embedded Sb^{3+} water 248 mg/g	China	Wan et al. (2020)
Corn straw (600 °C)	Adsorption of Fe-impregnated $HAsO_4{}^{2-}$ water 6.80 mg/g	China	He et al. (2018)
Cornstalk (550 °C)	Adsorption of Mg–Al-LDH As^{5+} soil0.820 mg/g	China	Gao et al. (2020)
Sawdust, canola and wheat straw, and manure pellet (300, 500, and 700 °C,)	Adsorption of Steam activation Pb^{2+} water 195 mg/g	Canada	Kwak et al. (2019)
Rice straw (500 °C)	Adsorption of HA/Fe–Mn oxide-loaded As^{5+} water 35.59 mg/g	China	Guo et al. (2019)
Pomelo Peel (300 °C)	Adsorption of K_2FeO_4-promoted Cr^{6+} water 209.64 mg/g	China	Yin et al. (2020)
Sawdust (180 °C)	Adsorption ofSb^{5+} from water 241.92 mg/g	China	Deng et al. (2020)
Maize stalk and black gram, pine needle (tree) and *Lantana camara* (500 °C)	Improve seed germination and seedling growth	India	Das et al. (2021)
Rice husk, olive pit biochar (620 °C),	The adsorption capacity of Cu^{2+} was 30.77 and 58.82 mg g^{-1} for rice husk and olive pit biochar, respectively, and of Pb^{2+} was 19.34 and 77.52 mg g^{-1} for for rice husk and olive pit biochar, respectively	Spain	Campos and Rosa (2020)
Wood-derived biochar	The efficiencies of TN and TP removal from wastewater were 9.5% and 15.6%, respectively	Tartu County, Estonia	Kasak et al. (2018)
Black Pine wood, Douglas wood	Biochar application led to a significant decrease in the concentrations of Cu, Cr, Pb and Ni, within the above ground tissues of berseem clover	Italy	Pescatore et al. (2022)
Rice husks and sunflower husks (750 °C)	The increase in the percentage adsorption of Cu and Pb was 9–19%, of Zn was 11–21%	Russia	Burachevskaya et al. (2023)
Rabbit-manure-derived biochar (550 °C)	Reduced the adsorption of Cd in spinach by 61% and Pb in mustard greens by 73%	Wroclaw, Poland	Medynska-Juraszek et al. (2022)

(continued)

Table 1 (continued)

Biochar feedstock and pyrolyzing temperature	Plant growth performances	Country	References
Willow chips (550 °C)	Increase in plant biomass and root length under contaminated soil	Krakow, south of Poland	Glab et al. (2021)
Ardwood biomass (mixture of oak, beech and charm) (500 °C)	Improve soil conditions, reduce the Arsenic and Lead translocation to the aerial biomass	France	Lebrun et al. (2021)
Rice husk (600 °C)	Enhance the degradation of glyphosphate mediated nanoparticle immobilized on biochar	Vietnam	Le et al. (2021)
Rice husk (700 °C)	Polychlorinated biphenyls removal efficiency of 91%	Norway	Silvani et al. (2019)
Sewage Sludge (700 °C)	PAHs removal efficiency of 74%	Poland	Godlewska and Oleszczuk (2022)
Carrot pulp (550 °C)	Copper removal efficiency of 47.4%	Iran	Gholami and Rahimi (2021)
Oil palm frond (500 °C)	Phenol and tannic acid adsorption capacity of 62.9 and 67.4 mg/g respectively	Malaysia	Lawal et al. (2021)
Grape pomace (350 °C)	Cymoxanil adsorption capacity of 161.0 mg/g	Republic of Korea	Yoon et al. (2021)
Sewage-sludge (700 °C)	The adsorption capacity of Diclofenac by 92.7 mg/g, triclosan by 113 mg/g, and naproxen 127 mg/g has been observed	Poland	Czech et al. (2021)
Sugar cane (380 °C)	The adsorption capacity of Thiamethoxam is 10.2 mg/g	Brazil	Fernandes et al. (2021)
Canola straw (500 °C)	The adsorption capacity of Lead by 165.1 mg/g	Canada	Nzediegwu et al. (2021)

process, adsorption method, membrane processes (Acero et al. 2015), and electrochemical method. Out of these methods, adsorption is the most suitable method because of its several advantages like wide processing range, high efficiency, and low cost.

4.2 PGPR-Mediated Removal

4.2.1 HMs Removal

Microorganisms interact intricately with HMs, and the nature of these interactions is governed by multiple factors such as microbial species, metal type, and prevailing environmental condition (Upadhyay et al. 2023b). The availability and mobility of HMs are intricately affected by elements such as temperature, nutrients, pH levels, and the presence of metal ions, all of which play a crucial role in shaping microbial-transformation activities. Bacteria in particular are well suited for bioremediation due to their small size, rapid growth rate, and ease of cultivation, which allow them to adapt effectively to adverse environments (Upadhyay and Chauhan, 2022). This adoptability is vital for their role in mitigating HM contamination. Similarly, Chauhan et al. (2025) reported that microbes release some microbial compounds such as exopolysaccharides, siderophore, metallothionein, and melanin that bind the HMs. The polysaccharide of bacteria contains a variety of functional groups including phosphates, sulfate, and amino groups which are capable of binding with HMs (Yin et al. 2016; Pathak et al. 2024a).

The HM absorption capacity of bacteria varies widely, ranging from 1 mg/g to 500 mg/g. Notably, *Pseudomonas aeruginosa* have the ability to mercury-resistant strains act as selective absorption of mercury ions, achieving a maximum uptake capability of 180 mg/g (Yin et al. 2016). The accumulation of mercury ions is facilitated by the high affinity of cysteine-rich proteins abundant in sulfhydryl groups. Bacterial strains such as *Pseudomonas* sp. and *Bacillus* sp. are acknowledged for their proficiency in adsorbing Pb^{2+} from wastewater (Li et al. 2017). *Arthrobacter viscosus*, whether in a living or dead state, exhibits the ability to adsorb Cr^{4+} and convert it into Cr^{3+} (Hlihor et al. 2016). *Staphylococcus* epidermis biofilms effectively eliminate Cr^{4+} (Quiton et al. 2018). *Rhodobacter capsulatus* showcases a notable ability to adsorb Zn^{2+}, reaching to highest uptake of nearly 164 mg g^{-1} (Magnin et al. 2014). Bacillus cereus demonstrates a biosorption capacity of 31.95 mg g^{-1} for Cd^{2+} with dead-cells and 24.01 mg g^{-1} with living-cells (Huang et al. 2013).

Microbial exopolysaccharides (EPS) act as protective barriers against toxic HMs by restricting their cellular entry and facilitating the binding of ions such as mercury, copper, cadmium, and cobalt through anionic and cationic functional groups (Pathak et al. 2024a). Post adsorption, HMs may undergo intracellular transformation into less toxic ionic states, thereby reducing their overall toxicity.

Pseudomonas putida SP1, for instance, exhibits the ability to absorb 100% of mercury and subsequently convert toxic Hg^{2+} to Hg(0) through enzyme-reductase, a process integral to mercury detoxification (Zhang et al. 2012). *Acinetobacter* sp. displays tolerance to Cr, Zn, Ni, Cd, and Pb, successfully removing up to 67% of Cr^{4+} within a 24-h period (Bhattacharya and Gupta 2013).

4.2.2 Organic Pollutant Removal

A set of highly conserved genes associated with the breakdown of polycyclic aromatic hydrocarbons (PAHs) plays a pivotal role in bioremediation efforts. The Gram-positive bacterial species harbor genes like nar, phdpdo, and nid, while gram-negative bacterial species contain pah, phn, nah, ndo, and nag (Sakshi and Haritash 2020). Notably, *Rhodococcus soli, Bacillus algicola, Pseudoalteromonas garivorans*, and *Isoptericola chiayiensis* demonstrated an efficacy of over 85% in degrading crude oil within a 14-day timeframe. Compounds with low molecular weights (C9 to C14) were completely degraded, those between C15 and C20 were nearly entirely broken down, and high-molecular-weight compounds exhibited significant degradation (Lee et al. 2018).

The comparative study focusing on degradation of chlorpyrifos identified two bacterial strains, *Pseudomonas syringae* CM-Z6 and *Bacillus megaterium* CM-Z19, through molecular analysis. The initial concentration of 100 mg/L CM-Z19 exhibited a degradation rate of 92.6%, while CM-Z6 achieved 99.1% degradation after 5 days of incubation (Zhu et al. 2019). However, *C. nantongensis* X1T showcased the ability to break down 200 μg/L of CPF within 48 h.

Comparatively, *B. megaterium* CM-Z19 showcased a degradation rate of 92.6%, outperforming *Pseudomonas syringae* CM-Z6, which achieved 99.1% degradation (Zhu et al. 2019). *Paenarthrobacter aurescens* TC1 and *Arthrobacter* MCM B-436 were identified in vetiver rhizosphere-soil contaminated with atrazine. In *Paenarthrobacter aurescens* TC1, the degradation of atrazine involves genes *trzN*, *atzB*, and *atzC*. Furthermore, under anoxic conditions, *Arthrobacter* sp. CW-1 decomposes dimethyl phthalate. Additionally, the plasmid pAO1 facilitates the breakdown of nicotine in *Arthrobacter nicotinovorans* (Sandu et al. 2005).

4.3 Biochar-Microbes Mediated Removal

Biochar amendment is considered in the remediation process of contaminated soil (Campos and Rosa 2020). Researchers have reported that the use of biochar amendments reduces the metal bioavailability (Cui et al. 2013) and increases the bioremediation process through microbial carriers using the immobilized microorganism approach (Chen et al. 2020). The aromatic oxidation and aliphatic carbon to biochar may boost the enzymatic activity of oxidoreductase and hydrolase while decreasing the activation energy of the enzyme catalyzed breaking of organic compounds (Sharma et al. 2021). Non-coulombic forces may affect the kinetic activity of enzymes on biochar surfaces, encouraging contamination via electron-exchange organic or bio-precipitation inorganic process (Pathak et al. 2024b). There are few information available about the effects of biochar on soil microbes that play a significant role in the decomposition of organic pollutants (Cheng et al. 2017). The research addressed that functional electron shuttling of microbes plays a crucial role in the degradation of dioxins, phenanthrene, and PAHs in which microbial cells were attached to biochar surface by covalent bond, cross-linked, and encapsulation (Rhodes et al. 2012; Semple et al. 2013). The recalcitrance and physical structure of biochar adsorbents controlled the total and bioavailable concentrations of such pollutants that also influenced the toxicity potential of HOCs indirectly. The inoculation of *Mycobacterium gilvum* with rice straw biochar for PAH remediation resulted in the combination of biochar and bacteria degrading phenanthrene, fluoranthene, and pyrene faster than either bacteria or biochar alone (Xiong et al. 2017). Similarly, biochar amendment boosted the population of difenoconazole-degrading bacteria in contaminated soils by influencing soil chemical properties, lowering difenoconazole bioavailability (Chen et al. 2017). Biochar was employed as a copping material for PAH sequestered in river sediments, removing pyrene by ion exchange and electrostatic interaction (Chen et al. 2019). But biochar had no effect on the overall microbial population, it succeeded in providing a stable habitat for PAH degrading microbes that bined to biochar's porosity and affinity of biological (Chen et al. 2019). Biochar may ameliorate the toxic consequences of POPs, including hexachlorobenzene, polychlorinated dibenzo-p-dioxins and PCBs, that interact by π-π electron donor acceptor, hydrophobic, and electrochemical interactions (Hu et al. 2020; Pathak et al. 2024b).

5 Influence of Biochar and PGPR on Plant Growth

Biochar, recognized as a carbon-rich organic enhancement, emerges as an ecologically mindful remedy for augmenting soil quality and fostering plant growth through the mitigation of nutrient deficiencies (Bonanomi et al. 2017; Major et al. 2010). This sustainable agricultural strategy not only enhances soil organic content and nutrient profiles but also exerts influence on various aspects, including soil vitality, moisture retention, microbial diversity, and resilience against plant diseases (Jindo et al. 2014; Rasool et al. 2021b).

Co-PGPR biochar, a representative blend of eco-friendly amendments designed to manipulate the rhizosphere microbiome, ultimately boosting crop growth and production (Saxena et al. 2013) (Fig. 4 and Table 1). Notably, Co-PGPR biochar demonstrates effectiveness in mitigating the detrimental effects of HMs. According to Zafar-ul-Hye et al. (2021), treatments involving the combination of rhizospheric bacteria such as *Alcaligenes faecalis* and biochar lead to significant improvements, with *Mentha piperita* exhibiting a 58% increase in dry root weight, 32% increase in dry leaf weight, and substantial enhancements in chlorophyll, nitrogen, phosphorus, and potassium contents—amounting to 37%, 46%, 39%, and 63%, respectively, while concurrently reducing lead uptake by 13.5% compared to the control.

Moreover, the synergistic application of biochar and PGPR enhances paddy plant resilience under diverse environmental stress such as salinity (Hafez et al. 2019; Chauhan et al. 2023; Fig. 3 and Table 1) and drought conditions in maize (Danish et al. 2020). For example, synergistic interaction *Bacillus coagulans* & *Pseudomonas koreenis* with biochar has demonstrated significant improvements in nutrient (N, P, and K) uptake, chlorophyll-content, stomatal-conductance, and reduction in proline in paddy plants under saline stress (Hafez et al. 2019). Expending beyond rice forming co-PGPR-biochar inoculants has play a significant role in boosting bioremediation processes, enhancing disease-resistance, and promoting cucumber-plant growth (Li et al. 2022). Furthermore, the synergistic application of PGPR and biochar also suppress the various foliar-wilting diseases (Kabir et al. 2012; Attia et al. 2020). Its effectiveness extends across various crops, such as *Cicer arietinum, Glycine max, Ariticum aestivum,* and *Phaseolus vulgaris*, demonstrating its broad growth-promoting potential (Saxena et al. 2013; Kavita and Janardan, 2014; Egamberdieva et al. 2016; Ijzat et al. 2019).

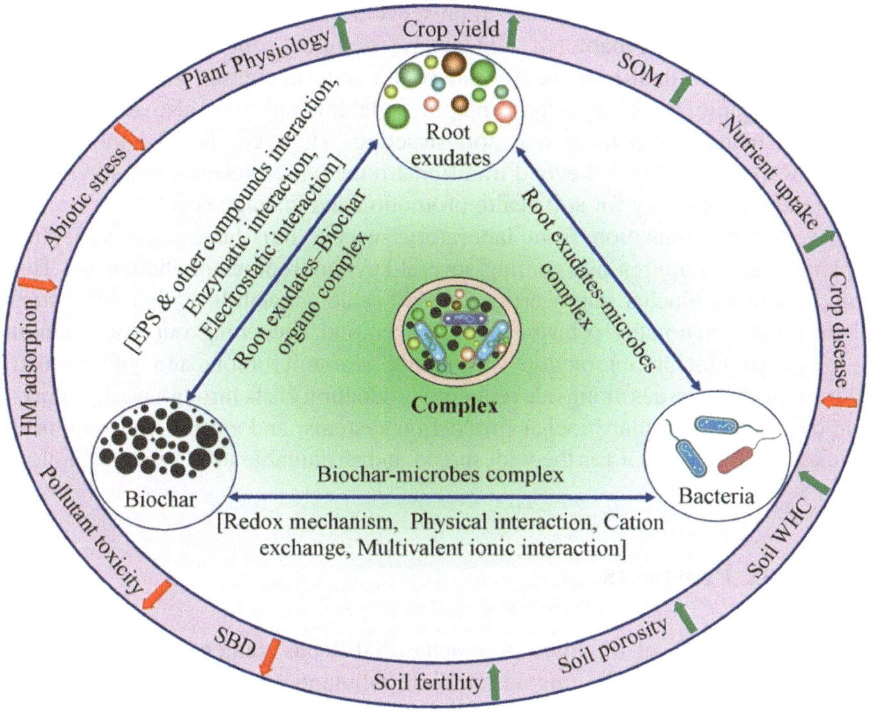

Fig. 4 Mechanism of Biochar–PGPR interaction in rhizosphere with organic and inorganic contaminants removal

Rasool et al. (2021a) further highlighted the effectiveness of green waste and wood biochar with *Bacillus subtilis, which* improves the physiological parameters, while also lowering the prevalence of *Alternaria solani* disease in tomato plants. In the prospect of soil health, varieties of biochar and PGPR treatment significantly improve soil fertility and enhance microbial richness and diversity, resulting in more resilient and productive agroecosystems (Ren et al. 2020).

6 Conclusion

This chapter discusses the synergistic use of synthetic biochar and PGPR as a next-generation approach for enhancing eco-friendly sustainable agriculture in environmental stress conditions. Engineered biochar, derived with improved surface area, porosity, and functional chemistry, emerges as a multifunctional supplement that acts as an adsorbent, soil conditioner, and microbial carrier. PGPR acts as nature's bioengineers, promoting plant development through nutrient solubilization, phytohormone production, stress reduction, and pollutant removal.

The synergy co-PGPR-biochar represents a promising function in agricultural biotechnology, capable of remediating polluted soils and enhancing crop productivity. This dual-functional strategy not only decreases the bioavailability of HMs, organic pollutants, and emerging contaminants but also improves soil health by increasing activity and soil structure. Their combined application of co-PGPR-biochar extends beyond traditional remediation techniques, serving as a comprehensive strategy for soil health promotion and management.

Nevertheless, transition from laboratory success into large-scale agricultural implementation requires overcoming several key environmental challenges. These include refining biochar engineering for field-scale adaptability, improving stress-tolerant and functionally robust PGPR strains, and deepening our understanding of long-term biochar interaction with native soil microbiota and geochemistry. Addressing these issues alongside reducing production costs through waste valorization, developing modular biochar production systems, and establishing supporting policies will be essential for the widespread and sustainable use of this technology.

7 Future Prospects

The synergistic use of modified co-biochar-PGPR has tremendous potential as a long-term approach for limiting agricultural pollutants and increasing crop yields. However, in order to fully realize its benefits, future research must solve a number of crucial scientific and practical issues.

The major research focus should be on optimizing engineered-biochar production. Advances in biochar modification methods such as chemical activation, surface functionalization, and nano-structuring can significantly increase surface area, porosity, and abundance as well as improve functional group activity. These improvements are crucial for boosting its adsorption ability for HMs, organic pollutants, and emerging contaminants such as antibiotics, pharmaceutical residues, and other environmental toxins.

Additionally, the development of ready-to-use biochar PGPR formulations, such as granules or liquid inoculants, could streamline field application and facilitate broader adoption by farmers. The eco-friendly delivery systems will be essential to translating laboratory success into real-world agricultural practices.

Another promising avenue is the role of biochar PGPR system in climate resilience. Future research should explore how integrated applications can enhance soil carbon sequestration, reduce GHGs emission, and improve crop resistance against abiotic stresses such as drought, temperature, and salinity. Such outcomes would significantly contribute to building climate-resilient agricultural system. Finally, policy development and stakeholder engagement are crucial for scaling up implementation. Establishing clear regulatory frameworks for the safe use of biochar and PGPR, alongside farmer education programs, public awareness campaigns, and field demonstration projects, will be vital in promoting widespread adoption and ensuring long-term sustainability.

Addressing the economic and scalability challenges is crucial. Innovations in biochar production technologies, decentralized manufacturing units, and government-supported schemes can facilitate large-scale application. Furthermore, the development of ready-to-use biochar PGPR formulations, including granules and liquid inoculants, could simplify on-field application and promote adoption by farmers.

Climate resilience is another emerging sector where co-PGPR-biochar technology can play a transformative role. Future research should explore how combined applications can enhance soil-carbon sequestration, reduce GHGs emission, and improve crop tolerance to drought, temperature extremes, and salinity, making agriculture more resilient to climate change impacts. Finally, policy development and stakeholder engagement are critical for technology transfer. Formulating regulatory frameworks for the safe use of co-PGPR-biochar, alongside farmer training programs, public awareness campaigns, and demonstration projects, will help drive large-scale adoption.

Acknowledgements The authors would like to express heartfelt thanks to the Department of Environmental Science, AKS University, Satna, India, and VBS Purvanchal University, Jaunpur, Uttar Pradesh, India, and Indian Institute of Technology BHU Varansai, Uttar Pradesh, India, for their invaluable support and resources.

Author's Contribution PKC and ST conceptualized this chapter, PKC and ST designed the table, figures, and wrote the initial stage of the chapter, SS, AC, RLSS, and BS critically edited and contributed to constructive recommendations before the submission. All the authors have no conflict of interest and the work is not related to animal ethical clearance.

References

Acero JL, Benitez FJ, Real FJ, Rodriguez E (2015) Elimination of selected emerging contaminants by the combination of membrane filtration and chemical oxidation processes. Water Air Soil Pollut 5(226):139. https://doi.org/10.1007/s11270-015-2404-8

Ahmad M, Wang X, Hilger TH, Luqman M, Nazli F, Hussain A, Zahir ZA, Latif M, Saeed Q, Malik HA, Mustafa A (2020) Evaluating biochar-microbe synergies for improved growth, yield of maize, and post-harvest soil characteristics in a semi-arid climate. Agronomy 10:1055. https://doi.org/10.3390/agronomy10071055

Alvarez A, Saez JM, Costa JSB, Colin VL, Fuentes MO, Cuozzo SA, Benimeli CS, Maria SF, Polti MA, Amoroso MJ (2017) Actinobacteria: current research and perspectives for bioremediation of pesticides and HMss. Chemosphere 166:41–62

Amusat SO, Kebede TG, Dube S, Nindi MM (2021) Ball-milling synthesis of biochar and biochar-based nanocomposites and prospects for removal of emerging contaminants: a review. J Water Process Eng 41:101993

Anastopoulos I, Kyzas GZ (2014) Agricultural peels for dye adsorption: a review of recent literature. J Mol Liq 200:381–389

Anbuganesan V, Vishnupradeep R, Varshini VS, Archana AS, Soundarya S, Bruno B, Rajkumar M (2022) Effect of plant growth-promoting rhizobacteria and biochar on ricinus communis growth, physiology, nutrient uptake and soil enzyme activities. Appl Ecol Environ Sci 10:640–651

Ankati S, Podile AR (2019) Metabolites in the root exudates of groundnut changes during interection with plant growth promoting rhizobacteria in a strain-specific manner. J Plant Physiol 24:153057. https://doi.org/10.1016/j.jplph.2019.153057

Arshad U, Azeem F, Mustafa G (2021) Combined application of biochar and biocontrol agents enhances plant growth and activates resistance against *Meloidogyne incognita* in tomato. GesundePflanzen 73:591–601. https://doi.org/10.1007/s10343-021-00580-4

Attia MS, El-Sayyad GS, Abd Elkodous M, El-Batal AI (2020) The effective antagonistic potential of plant growth-promoting rhizobacteria against alternaria solani causing early blight disease in tomato plant. Sci Hortic 266:109289. https://doi.org/10.1016/j.scienta.2020.109289

Beesley L, Moreno-Jimenez E, Gomez-Eyles JL, Harris E, Robinson B, Sizmur T (2011) A review of biochars' potential role in the remediation, revegetation and restoration of contaminated soils. Environ Pollut 159:3269–3282

Bhattacharya A, Gupta A (2013) Evaluation of Acinetobacter sp. B9 for Cr (VI) resistance and detoxification with potential application in bioremediation of heavy-metals-rich industrial wastewater. Environ Sci Pollut Res 20:6628–6637

Bonanomi G, Ippolito F, Cesarano G, Nanni B, Lombardi N, Rita A, Saracino A, Scala F (2017) Biochar as plant growth promoter: better off alone or mixed with organic amendments? Front Plant Sci 8:1570. https://doi.org/10.3389/fpls.2017.01570

Burachevskaya M, Minkina T, Bauer T, Lobzenko I, Fedorenko A, Mazarji M, Sushkova S, Mandzhieva S, Nazarenko A, Butova V, Wong MH, Rajput VD (2023) Fabrication of biochar derived from different types of feedstocks as an efficient adsorbent for soil HMs removal. Sci Rep 13(1):2020. https://doi.org/10.1038/s41598-023-27638-9

Campos P, De la, Rosa JM (2020) Assessing the effects of biochar on the immobilization of trace elements and plant development in a naturally contaminated soil. Sustainability 12:6025. https://doi.org/10.3390/su12156025

Cerqueira F, Matamoros V, Bayona JM, Berendonk TU, Elsinga G, Hornstra LM, Pĩna B (2019) Antibiotic resistance gene distribution in agricultural fields and crops. A soil-to-food analysis. Environ Res 177:18–26

Cesari A, Paulucci N, Lopez-Gomez M, Hidalgo-Castellanos J, Pla CL, Dardanelli MS (2019) Restrictive water condition modifies the root exudates composition during peanut-PGPR interaction and conditions early events, reversing the negative effects on plant growth plant. Physiol Biochem 142:519–527. https://doi.org/10.1016/j.plaphy.2019.08.015

Chauhan PK, Upadhyay SK (2023) Exo-polysaccharide producing bacteria can induce maize plant growth and soil health under saline conditions. Biotechnol Genetic Eng Rev. https://doi.org/10.1080/02648725.2022.2163812

Chauhan PK, Upadhyay SK, Rajput VD, Dwivedi P, Minkina T, Wong MH (2023) Fostering plant growth performance under drought stress using rhizospheric microbes, their gene editing, and biochar. Environ Geochem Health. 9bf22c33-a65a-46f9-acef-fe8e22fd6656

Chauhan PK, Upadhyay SK, Chauhan A, Bhutiani R, Sikarwar RLS, Tiwari M (2025) A review on simultaneous use of microbes and biochar for removal of toxic heavy metals: recent updates and future outlooks. Physiol Mol Biology Plants. https://doi.org/10.1007/s12298-025-01649-4

Chen HY, Awasthi SK, Liu T, Duan YM, Zhang ZQ, Awasthi MK (2020) Compost biochar application to contaminated soil reduces the (im)mobilization and phytoavailability of lead and copper. J Chem Technol Biotechnol 95:408–417

Chen Z, Chen J, Yang X, Chen C, Huang S (2019) Biochar as an effective material on sediment remediation for polycyclic aromatic hydrocarbons contamination. In: IOP conference series: earth and environmental science, 1st vol 281. IOP Publishing, 012016

Chen Z, He X, Li Q, Yang H, Liu Y, Wu L, Liu Z, Hu B, Wang X (2022) Low temperature plasma induced phosphate groups onto coffee residue derived porous carbon for efficient U(VI) extraction. J Environ Sci 122:1–13

Cheng JZ, Lee XQ, Gao WC, Chen Y, Pan WJ, Tang Y (2017) Effect of biochar on the bioavailability of difenoconazole and microbial community composition in a pesticide-contaminated soil. Appl Soil Ecol 121:185–192

Cheng S, Chen T, Xu W, Huang J, Jiang S, Yan B (2020) Application research of biochar for the remediation of soil heavy metals contamination: a review. Molecules 25:316

Christou A, Papadavid G, Dalias P, Fotopoulos V, Michael C, Bayona JM, Pĩna B, Fatta-Kassinos D (2019) Ranking of crop plants according to their potential to uptake and accumulate contaminants of emerging concern. Environ Res 170:422–432

Cig F, Sonmez F, Nadeem MA, Sabagh AE (2021) Effect of biochar and PGPR on the growth and nutrients content of Einkorn Wheat (*Triticum monococcum* L.) and post-harvest soil properties. Agronomy 11:2418. https://doi.org/10.3390/agronomy11122418

Cui LQ, Yan JL, Yang YG, Li LQ, Quan GX, Ding C, Chen TM, Fu Q, Chang A (2013) Influence of biochar on microbial activities of heavy metals contaminated paddy fields. Bio 8:5536–5548

Czech B, Kończak M, Rakowska M, Oleszczuk P (2021) Engineered biochars from organic wastes for the adsorption of diclofenac, naproxen and triclosan from water systems. J Clean Prod 288:125686

Dai Y, Zhang N, Xing C, Cui Q, Sun Q (2019) The adsorption, regeneration and engineering applications of biochar for removal organic pollutants: a review. Chemosphere 223:12–27. https://doi.org/10.1016/j.chemosphere.2019.01.161

Danish S, Zafar-ul-Hye M (2019) Co-application of ACC-deaminase producing PGPR and timber-waste biochar improves pigments formation, growth and yield of wheat under drought stress. Sci Rep 9:5999. https://doi.org/10.1038/s41598-019-42374-9

Danish S, Zafar-ul-Hye M, Mohsin F, Hussain M (2020) ACC-deaminase producing plant growth promoting rhizobacteria and biochar mitigate adverse effects of drought stress on maize growth. PLoS ONE 15:e0230615. https://doi.org/10.1371/journal.pone.02306

Das SK, Ghosh GK, Avasthe R, Sinha K (2021) Morpho-mineralogical exploration of crop, weed and tree derived biochar. J Hazard Mater 407:124370. https://doi.org/10.1016/j.jhazmat.2020.124370

De TC, Vandecasteele B, Verstraeten B, Ommeslag S, Kyndt T, Debode J (2021) Biochar-enhanced resistance to *Botrytis cinerea* in strawberry fruits (but not leaves) is associated with changes in the rhizosphere microbiome. Front Plant Sci 12:700479. https://doi.org/10.3389/fpls.2021.700479

Deng J, Li X, Wei X, Liu Y, Liang J, Shao Y, Huang W, Cheng X (2020) Different adsorption behaviors and mechanisms of a novel amino-functionalized hydrothermal biochar for hexavalent chromium and pentavalent antimony. Bioresour Technol 310:123438. https://doi.org/10.1016/j.biortech.2020.123438

Egamberdieva D, Shurigin V, Alaylar B, Ma H, Müller MEH, Wirth S, Reckling M, Bellingrath-Kimura SD (2020) The effect of biochars and endophytic bacteria on growth and root rot disease incidence of *Fusarium* infested narrow-leafed lupin (*Lupinus angustifolius* L.). Microorganisms 8(4):496. https://doi.org/10.3390/microorganisms8040496

Egamberdieva D, Wirth S, Behrendt U, Abd-Allah EF, Berg G (2016) Biochar treatment resulted in a combined effect on soybean growth promotion and a shift in plant growth promoting rhizobacteria. Front Microbiol 7:209. https://doi.org/10.3389/fmicb.2016.00209

El-Naggar A, Ahmed N, Mosa A, Niazi NK, Yousaf B, Sharma A, Sarkar B, Cai Y, Chang SX (2021) Nickel in soil and water: sources, biogeochemistry, and remediation using biochar. J Hazard Mater 419:126421

Fan J, Cai C, Chi H, Reid BJ, Coulon F, Zhang Y, Hou Y (2020) Remediation of cadmium and lead polluted soil using thiol-modified biochar. J Hazard Mater 388:122037. https://doi.org/10.1016/j.jhazmat.2020.122037

Fernandes JO, Bernardino CAR, Mahler CF, Santelli RE, Braz BF, Borges RC, da Cunha Veloso MC, Romeiro GA, Cincotto FH (2021) Biochar generated from agro-Industry sugarcane residue by low temperature pyrolysis utilized as an adsorption agent for the removal of thiamethoxam pesticide in wastewater. Water Air Soil Poll 232:67

Ferreira L, Rosales E, Danko AS, Sanroman MA, Pazos MM (2016) Bacillus thuringiensis a promising bacterium for degrading emerging pollutants. Process Saf Environ Protect 101:19e26

Flesch F, Berger P, Robles-Vargas D, Santos-Medrano GE, Rico-Martínez R (2019) Characterization and determination of the toxicological risk of biochar using invertebrate toxicity tests in the State of Aguascalientes, Mexico. Appl Sci 9(8):1706

Gao X, Peng Y, Guo L, Wang Q, Guan CY, Yang F, Chen Q (2020) Arsenic adsorption on layered double hydroxides biochars and their amended red and calcareous soils. J Environ Manag 2020(271):111045. https://doi.org/10.1016/j.jenvman.2020.111045

Gholami L, Rahimi G (2021) Chemical fractionation of copper and zinc after addition of carrot pulp biochar and thiourea-modified biochar to a contaminated soil. Environ Technol 42:3523–3532

Glab T, Gondek K, Mierzwa-Hersztek M (2021) Biological effects of biochar and zeolite used for remediation of soil contaminated with toxic heavy metals. Sci Rep 11:6998. https://doi.org/10.1038/s41598-021-86446-1

Godlewska P, Oleszczuk P (2022) Effect of biomass addition before sewage sludge pyrolysis on the persistence and bioavailability of polycyclic aromatic hydrocarbons in biochar-amended soil. ChemEng J 429:132143

Gomes AR, Justino C, Rocha-Santos T, Freitas AC, Duarte AC, Pereira R (2017) Review of the ecotoxicological effects of emerging contaminants to soil biota. J Environ Sci Health Part A Toxic/Hazard Subst Environ Eng 52:992e1007

Guo J, Yan C, Luo Z, Fang H, Hu S, Cao Y (2019) Synthesis of a novel ternary HA/Fe-Mn oxides-loaded biochar composite and its application in cadmium(II) and arsenic(V) adsorption. J Environ Sci 85:168–176. https://doi.org/10.1016/j.jes.2019.06.004

Hafez EM, Alsohim AS, Farig M, Omara AED, Rashwan E, Kamara MM (2019) Synergistic effect of biochar and plant growth promoting rhizobacteria on alleviation of water deficit in rice plants under salt-affected soil. Agronomy 9:847. https://doi.org/10.3390/agronomy9120847

Haider FU, Cheema SA, Farooq M (2019) Impact of cover crops in improving agroecosytems including sustainable weed suppression—a review. Pak J Weed Sci Res 25(1):47–62

He R, Peng Z, Lyu H, Huang H, Nan Q, Tang J (2018) Synthesis and characterization of an iron-impregnated biochar for aqueous arsenic removal. Sci Total Environ 612:1177–1186. https://doi.org/10.1016/j.scitotenv.2017.09.016

Hlihor RM, Figueiredo H, Tavares T, Gavrilescu M (2016) Biosorption potential of dead and living Arthrobacterviscosus biomass in the removal of Cr(VI): batch and column studies. Process Saf Environ Prot 108:44–56

Hu B, Ai Y, Jin J, Hayat T, Alsaedi A, Zhuang L, Wang X (2020) Efficient elimination of organic and inorganic pollutants by biochar and biochar-based materials. Biochar 2(1):47–64

Huang F, Dang Z, Guo CL, Lu GN, Gu RR, Liu HJ, Zhang H (2013) Biosorption of Cd(II) by live and dead cells of *Bacillus cereus* RC-1 isolated from cadmium-contaminated soil. Colloids Surfaces B Biointerfaces 2013(107):11–18

Ijaz M, Tahir M, Shahid M, Ul-Allah S, Sattar A, Sher A, Mahmood K, Hussain M (2019) Combined application of biochar and PGPR consortia for sustainable production of wheat under semiarid conditions with a reduced dose of synthetic fertilizer. Braz J Microbiol 50(2):449–458. https://doi.org/10.1007/s42770-019-00043-z

Imran M, Iqbal MM, Iqbal J, Shah NS, Khan ZUH, Murtaza B, Amjad M, Ali S, Rizwan M (2021) Synthesis, characterization and application of novel MnO and CuO impregnated biochar composites to sequester arsenic (As) from water: modeling, thermodynamics and reusability. J Hazard Mater 2021(401):123338. https://doi.org/10.1016/j.jhazmat.2020.123338

Islam MS, Kwak JH, Nzediegwu C, Wang S, Palansuriya K, Kwon EE, Naeth MA, El-Din MG, Ok YS, Chang SX (2021) Biochar heavy metal removal in aqueous solutiIqbal. Amen R, Hassan EB, Bibi I, Natasha N, Mukhtar A, Bashir H, Shahid M, Niazi N (2022) Application of magnetic biochars for the removal of aquatic pollutants. https://doi.org/10.1016/B978-0-323-90485-8.00003-5

Jajoo A, Mekala NR, Tomar RS, Grieco M, Tikkanen M, Aro EM (2014) Inhibitory effects of polycyclic aromatic hydrocarbons (PAHs) on photosynthetic performance are not related to their aromaticity. J Photochem Photobiol B 2014(137):151–155. https://doi.org/10.1016/j.jphotobiol.2014.03.011

Jiang A, Cheng Z, Shen Z, Guo W (2018) QSAR study on the removal efficiency of organic pollutants in supercritical water based on degradation temperature. Chem Cent J 12:16e23

Jindo K, Mizumoto H, Sawada Y, Sanchez-Monedero MA, Sonoki T (2014) Physical and chemical characterization of biochars derived from different agricultural residues. Biogeosciences 11:6613–6621. https://doi.org/10.5194/bg-11-6613-2014

Kabir S, Islam M, Khan M, Hossain I (2012) Comparative efficacy of compost, poultry litter, IPM Lab biopesticide and BAU biofungicde in controlling early blight of tomato. Int Res J Appl Life Sci 1:03

Kari A, Nagymate Z, Romsics C, Vajna B, Toth E, Lazanyi-Kovacs R, Rizo B, Kutasi J, Bernhardt B, Farkas E, Marialigeti K (2021) Evaluating the combined effect of biochar and PGPR inoculants on the bacterial community in acidic sandy soil. Appl Soil Ecol 160:103856. https://doi.org/10.1016/j.apsoil.2020.103856

Kasak K, Truu J, Ostonen I, Sarjas J, Oopkaup K, Paiste P, Kõiv-Vainik M, Mander U, Truu M (2018) Biochar enhances plant growth and nutrient removal in horizontal subsurface flow constructed wetlands. Sci Total Environ 639:67–74. https://doi.org/10.1016/j.scitotenv.2018.05.146

Kasmaei LS, Yasrebi J, Zarei M, Ronaghi A, Ghasemi R, Saharkhiz MJ, Ahmadabadi Z, Schnug E (2019) Influence of plant growth promoting rhizobacteria, compost, and biochar of azolla on rosemary (*Rosmarinus Officinalis* L.) growth and some soil quality indicators in a calcareous soil. Commun Soil Sci Plant Analysis 50(2):119–131. https://doi.org/10.1080/00103624.2018.1554669

Kavita B, Janardan Y (2014) Effects of PGPR blended biochar and different levels of phosphorus on yield and nutrient uptake by chickpea. Ann Agric Biol Res 19:408–412

Korenblum E, Donga Y, Szymanski J, Panda S, Jozwiak A, Massalha H (2020) Rhizosphere microbe mediates systemic root metabolite exudation by root-to-root signaling. PNAS 117:3874–3883. https://doi.org/10.1073/pnas.1912130117

Kumar A, Borisova G, Maleva M, Tripti, Shiryaev G, Tugbaeva A, Sobenin A, Kiseleva I (2022). Biofertilizer based on biochar and metal-tolerant plant growth promoting rhizobacteria alleviates copper impact on morphophysiological traits in *Brassica napus* L. Microorganisms 10:2164. https://doi.org/10.3390/microorganisms10112164

Kummerova M, Vanova L, Fiserova H, Klems M, Zezulka S, Krulova J (2010) Understanding the effect of organic pollutant fluoranthene on pea in vitro using cytokinins, ethylene, ethane and carbon dioxide as indicators. Plant Growth Regul 61:161–174

Kwak JH, Islam MS, Wang S, Messele SA, Naeth MA, El-Din MG, Chang SX (2019) Biochar properties and lead(II) adsorption capacity depend on feedstock type, pyrolysis temperature, and steam activation. Chemosphere 231:393–404. https://doi.org/10.1016/j.chemosphere.2019.05.128

Lalay G, Ullah S, Ahmed I (2022) Physiological and biochemical responses of *Brassica napus* L. to drought-induced stress by the application of biochar and plant growth promoting rhizobacteria. Microsc Res Tech 85(4):1267–1281. https://doi.org/10.1002/jemt.23993

Lan D, Zhu H, Zhang J, Li S, Chen Q, Wang C, Wu T, Xu M (2022) Adsorptive removal of organic dyes via porous materials for wastewater treatment in recent decades: a review on species, mechanisms and perspectives. Chemosphere 293:133464. https://doi.org/10.1016/j.chemosphere.2021.133464

Lawal AA, Hassan MA, Ahmad, Farid MA, TengkuYasim-Anuar TA, Samsudin MH, Mohd Yusoff MZ, Zakaria MR, Mokhtar MN, Shirai Y (2021) Adsorption mechanism and effectiveness of phenol and tannic acid removal by biochar produced from oil palm frond using steam pyrolysis. Environ Pollut 269:116197

Le PT, Le DN, Nguyen TH, Bui HT, Pham LA, Nguyen LL, Nguyen QS, Nguyen TP, Dang TH, Duong TT (2021) On the degradation of glyphosate by photocatalysis using TiO_2/biochar composite obtained from the pyrolysis of rice husk. Water 13(23):3326. https://doi.org/10.3390/w13233326

Lebrun M, Nandillon R, Miard F (2021) Effects of biochar, ochre and manure amendments associated with a metallicolous ecotype of *Agrostiscapillaris* on As and Pb stabilization of a former

mine technosol. Environ Geochem Health 43:1491–1505. https://doi.org/10.1007/s10653-020-00592-5
Lee DW, Lee H, Kwon B-O, Khim JS, Yim UH, Kim BS et al (2018) Biosurfactant assisted bioremediation of crude oil by indigenous bacteria isolated from Taean beach sediment. Environ Pollut 241:254–264
Li D, Xu X, Yu H, Han X (2017) Characterization of Pb^{2+} biosorption by psychrotrophic strain Pseudomonas sp. I3 isolated from permafrost soil of Mohe wetland in Northeast China. J Environ Manag 196:8–15
Li H, Qu R, Li C, Guo W, Han X, He F, Ma Y, Xing B (2014) Selective removal of polycyclic aromatic hydrocarbons (PAHs) from soil washing effluents using biochars produced at different pyrolytic temperatures. Bioresour Technol 163:193–198. https://doi.org/10.1016/j.biortech.2014.04.042
Li J, Chen Y, Qin X, Cao A, Lu A (2022) Impact of biochar on rhizosphere bacterial diversity restoration following chloropicrin fumigation of planted soil. Int J Environ Res Public Health 19:2126. https://doi.org/10.3390/ijerph19042126
Liu B, Wang D, Yu G, Meng X (2013) Adsorption of heavy metal ions, dyes and proteins by chitosan composites and derivatives—a review. J Ocean Univ China 12:500–508
Liu C, Liu X, He Y, An X, Fan D, Wu Z (2021a) Microwave-assisted catalytic pyrolysis of apple wood to produce biochar: co-pyrolysis behavior, pyrolysis kinetics analysis and evaluation of microbial carriers. Biores Technol 320:124345. https://doi.org/10.1016/j.biortech.2020.124345
Liu J, Zhang J, Zhu M, Wan H, Chen Z, Yang N, Duan J, Wei Z, Hu T, Liu F (2022) Effects of plant growth promoting rhizobacteria (PGPR) strain *Bacillus licheniformis* with biochar amendment on potato growth and water use efficiency under reduced irrigation regime. Agronomy 12(5):1031. https://doi.org/10.3390/agronomy12051031
Liu S, Xie Z, Zhu Y, Zhu Y, Jiang Y, Wang Y, Gao H (2021b) Adsorption characteristics of modified rice straw biochar for Zn and in-situ remediation of Zn contaminated soil. Environ Technol Innov 22:101388
Liu X, Liang C, Liu X, Zhao F, Han C (2020) Occurrence and human health risk assessment of pharmaceuticals and personal care products in real agricultural systems with long-term reclaimed wastewater irrigation in Beijing, China. Ecotoxicol Environ Saf 190:110022. https://doi.org/10.1016/j.ecoenv.2019.110022
Ma J, Huang W, Zhang X, Li Y, Wang N (2021) The utilization of lobster shell to prepare low-cost biochar for high-efficient removal of copper and cadmium from aqueous: Sorption properties and mechanisms. J Environ Chem Eng 9:104703. https://doi.org/10.1016/j.jece.2020.104703
Magnin JP, Gondrexon N, Willison JC (2014) Zinc biosorption by the purple non-sulfur bacterium Rhodobactercapsulatus. Canad J Microbiol 2014(60):829–837
Major J, Rondon M, Molina D, Riha SJ, Lehmann J (2010) Maize yield and nutrition during four years after biochar application to a Colombian savanna oxisol. Plant Soil 333:117–128
Manyi-Loh C, Mamphweli S, Meyer E, Okoh A (2018) Antibiotic use in agriculture and its consequential resistance in environmental sources: potential public health implications. Molecules 23(4):795. https://doi.org/10.3390/molecules23040795
Medynska-Juraszek A, Marcinkowska K, Gruszka D, Kluczek K (2022) The effects of rabbitmanure-derived biochar co-application with compost on the availability and heavy metal uptake by green leafy vegetables. Agronomy 12:2552. https://doi.org/10.3390/agronomy12102552
Nehela Y, Mazrou YSA, Alshaal T, Rady AMS, El-Sherif AMA, Omara AED, Abd El-Monem AM, Hafez EM (2021) The integrated amendment of sodic-saline soils using biochar and plant growth-promoting rhizobacteria enhances maize (*Zea mays* L.) resilience to water salinity. Plants 10:1960. https://doi.org/10.3390/plants10091960
Nzediegwu C, Naeth MA, Chang SX (2021) Lead(II) adsorption on microwave-pyrolyzedbiochars and hydrochars depends on feedstock type and production temperature. J Hazard Mater 412:125255
Pan M, Chu LM (2017) Fate of antibiotics in soil and their uptake by edible crops. Sci Total Environ 599–600:500–512

Park JH, Wang JJ, Kim SH, Kang SW, Jeong CY, Jeon JR, Park KH, Cho JS, Delaune RD, Seo DC (2019) Cadmium adsorption characteristics of biochars derived using various pine tree residues and pyrolysis temperatures. J Colloid Interface Sci 553:298–307. https://doi.org/10.1016/j.jcis.2019.06.032

Parween T, Jan S, Fatma T (2012) Evaluation of oxidative stress in *Vigna radiata* L. in response to chlorpyrifos. Int J Environ Sci Technol 9:605–612

Parween T, Jan S, Fatma T (2011) Alteration in nitrogen metabolism and plant growth during different developmental stages of green gram (*Vigna radiata* L.) in response to chlorpyrifos. Acta Physiol Plant 33:2321–2328

Pathak HK, Chauhan PK, Seth CS, Dubey G, Upadhyay SK (2024a) Mechanistic and future prospects in rhizospheric engineering for agricultural contaminants removal, soil health restoration, and management of climate change stress. Sci Total Environ 927:172116. https://doi.org/10.1016/j.scitotenv.2024.172116

Pathak HK, Seth CS, Chauhan PK, Dubey G, Singh G, Jain D, Upadhyay SK, Dwivedi P, Khoo KS (2024b) Recent advancement of nano-biochar for the remediation of heavy metals and emerging contaminants: mechanism, adsorption kinetic model, plant growth and development. Environ Res 255:119136. https://doi.org/10.1016/j.envres.2024.119136

Pescatore A, Grassi C, Rizzo AM, Orlandini S, Napoli M (2022) Effects of biochar on berseem clover (*Trifolium alexandrinum*, L.) growth and heavy metal (Cd, Cr, Cu, Ni, Pb, and Zn) accumulation. Chemosphere 287(Pt 1):131986. https://doi.org/10.1016/j.chemosphere.2021.131986. Epub 2021

Quiton KG, Doma B Jr, Futalan CM, Wan MW (2018) Removal of chromium (VI) and zinc (II) from aqueous solution using kaolin-supported bacterial biofilms of Gram-negative *E. coli* and Gram-positive *Staphylococcus epidermidis*. Sustain Environ Res 28:206–213

Rasool M, Akhter A, Haider MS (2021a) Molecular and biochemical insight into biochar and *Bacillus subtilis* induced defense in tomatoes against *Alternariasolani*. ScientiaHorticulturae 285:110203. https://doi.org/10.1016/j.scienta.2021.110203

Rasool M, Akhter A, Soja G, Haider MS (2021b) Role of biochar, compost and plant growth promoting rhizobacteria in the management of tomato early blight disease. Sci Rep 11:6092

Ren H, Huang B, Fernandez-Garcia V, Miesel J, Yan L, Lv C (2020) Biochar and rhizobacteria amendments improve several soil properties and bacterial diversity. Microorganisms 8(502):1–17. https://doi.org/10.3390/microorganisms8040502

Rhodes AH, Riding MJ, McAllister LE, Lee K, Semple KT (2012) Influence of activated charcoal on desorption kinetics and biodegradation of phenanthrene in soil. Environ Sci Technol 46(22):12445–12451

Rodriguez M, Torres M, Blanco L, Bejar V, Sampedro I, Llamas I (2020) Plant growth-promoting activity and quorum quenching mediated biocontrol of bacterial phytopathogens by *Pseudomonas segetis* strain P6. Sci Rep 10:4121. https://doi.org/10.1038/s41598-020-61084-1

Sakshi, Haritash AK (2020) A comprehensive review of metabolic and genomic aspects of PAH-degradation. Arch Microbiol 202:2033–2058

Sandu C, Chiribau CB, Sachelaru P, Brandsch R (2005) Plasmids for nicotine-dependent and -independent gene expression in *Arthrobacter nicotinovorans* and other arthrobacter species. Appl Environ Microbiol 71(12):8920–8924. https://doi.org/10.1128/AEM.71.12.8920-8924.2005

Saxena J, Rana G, Pandey M (2013) Impact of addition of biochar along with *Bacillus* sp. on growth and yield of French beans. Sci Hortic 162:351–356

Semple KT, Riding MJ, McAllister LE, Sopena-Vazquez F, Bending GD (2013) Impact of black carbon on the bioaccessibility of organic contaminants in soil. J Hazard Mater 261:808–816

Shanta NA, Schwinghamer TD, Backer R, Allaire SE, Teshler I, Vanasse A, Whalen JK, Baril B, Lange SF, Mackay J, Zhou X, Smith DL (2016) Biochar and plant growth promoting rhizobacteria effects on switchgrass (*Panicum virgatum* cv. Cave-in-Rock) for biomass production in southern Québec depend on soil type and location. Biomass Bioenerg 95:167–173

Sharma P, Kumar S, Pandey A (2021) Bioremediated techniques for remediation of metal pollutants using metagenomics approaches: a review. J Environ Chem Eng 4:105684

Silvani L, Hjartardottir S, Bielska L, Skulcova L, Cornelissen G, Nizzetto L, Hale SE (2019) Can polyethylene passive samplers predict polychlorinated biphenyls (PCBs) uptake by earth-worms and turnips in a biochar amended soil? Sci Total Environ 662:873–880

Singh E, Kumar A, Mishra R, You S, Singh L, Kumar S, Kumar R (2021) Pyrolysis of waste biomass and plastics for production of biochar and its use for removal of heavy metals from aqueous solution. Biores Technol 320:124278. https://doi.org/10.1016/j.biortech.2020.124278

Singh P, Chauhan PK, Upadhyay SK, Singh RK, Dwivedi P, Wang J, Jain D, Jiang M (2022) Mechanistic insights and potential use of siderophores producing microbes in rhizosphere for mitigation of stress in plants grown in degraded land. Frontiers Microbiol 2415. https://doi.org/10.3389/fmicb.2022.898979

Siric I, Eid EM, Taher MA, El-Morsy MHE, Osman HEM, Kumar P, Adelodun B, Abou Fayssal S, Mioč B, Andabaka Z, Goala M, Kumari S, Bachheti A, Choi KS, Kumar V (2022) Combined use of spent mushroom substrate biochar and PGPR improves growth, yield, and biochemical response of cauliflower (*Brassica oleracea* var. *botrytis*): a preliminary study on greenhouse cultivation. Horticulturae 8(9):830. https://doi.org/10.3390/horticulturae8090830

Sohaib M, Zahir ZA, Khan MY, Ans M, Asghar HN, Yasin S, Al-Barakah FNI (2019) Comparative evaluation of different carrier-based multi-strain bacterial formulations to mitigate the salt stress in wheat. Saudi J Biol Sci 27(3):777–787. https://doi.org/10.1016/j.sjbs

Ukalska-Jaruga A, Debaene G, Smreczak B (2019) Dissipation and sorption processes of polycyclic aromatic hydrocarbons (PAHs) to organic matter in soils amended by exogenous rich-carbon material. J Soils Sediment. https://doi.org/10.1007/s11368-019-02455-8

Upadhyay SK, Rani N, Kumar V, Mythili R, Jain D (2023a) A review on simultaneous heavy metal removal and organo-contaminants degradation by potential microbes: current findings and future outlook. Microbiol Res 273:127419. https://doi.org/10.1016/j.micres.2023.127419

Upadhyay SK, Chauhan PK (2022) Optimization of eco-friendly amendments as sustainable asset for salt-tolerant plant growth-promoting bacteria mediated maize (*Zea Mays* L.) plant growth, Na uptake reduction and saline soil restoration. Environ Res 211:113081. https://doi.org/10.1016/j.envres.2022.113081

Upadhyay SK, Devi P, Kumar V, Pathak HK, Kumar P, Rajput VD, Dwivedi P (2023b) Efficient removal of total arsenic (As3+/5+) from contaminated water by novel strategies mediated iron and plant extract activated waste flowers of marigold. Chemosphere 313:137551. https://doi.org/10.1016/j.chemosphere.2022.137551

Upadhyay SK, Srivastava AK, Rajput VD, Chauhan PK, Bhojiya AA, Jain Chaubey G, Sharma DB, Minkina T, Sharma B, Minkina T (2022) Root exudates: mechanistic insight of plant growth promoting rhizobacteria for sustainable crop production. Front Microbiol 13:2420. https://doi.org/10.3389/fmicb.2022.916488

Vithanage M, Mayakaduwa SS, Herath I, Yongsik O, Mohan D (2016) Kinetics, thermodynamics and mechanistic studies of carbofuran removal using biochars from tea waste and rice husks. Chemosphere 150:781e789

Vorobyeva T, Neverova N, Andrianov V (2020) The long-term effects of contamination by fuel oil on the coastal ecosystem in the Onega Bay of the White Sea. Russ J Mar Biol 46:207–214

Wan S, Qiu L, Li Y, Sun J, Gao B, He F, Wan W (2020) Accelerated antimony and copper removal by manganese oxide embedded in biochar with enlarged pore structure. Chem Eng J 402:126021. https://doi.org/10.1016/j.cej.2020.126021

Xiang W, Zhang X, Chen K, Fang J, He F, Hu X, Tsang DCW, Ok YS, Gao B (2020) Enhanced adsorption performance and governing mechanisms of ball-milled biochar for the removal of volatile organic compounds (VOCs). Chem Eng J 385:123842. https://doi.org/10.1016/j.cej.2019.123842

Xiong B, Zhang Y, Hou Y, Arp HPH, Reid BJ, Cai C (2017) Enhanced biodegradation of PAHs in historically contaminated soil by M. gilvum inoculated biochar. Chemosphere 182:316–324

Xu Z, Wang C, Li H, Xu S, Du J, Chen Y, Ma C, Tang J (2021) Concentration, distribution, source apportionment, and risk assessment of surrounding soil PAHs in industrial and rural areas: a comparative study. Ecol Ind 125:107513. https://doi.org/10.1016/j.ecolind.2021.107513

Yang F, Wang B, Shi Z, Li L, Li Y, Mao Z, Liao L, Zhang H, Wu Y (2021) Immobilization of heavy metals (Cd, Zn, and Pb) in different contaminated soils with swine manure biochar. EnvPollutBioavail 33:55–65

Yin K, Lv M, Wang Q, Wu Y, Liao C, Zhang W, Chen L (2016) Simultaneous bioremediation and biodetection of mercury ion through surface display of carboxylesterase E2 from *Pseudomonas aeruginosa* PA1. Water Res 103:383–390

Yin Z, Xu S, Liu S, Xu S, Li J, Zhang Y (2020) A novel magnetic biochar prepared by K_2FeO_4-promoted oxidative pyrolysis of pomelo peel for adsorption of hexavalent chromium. Bioresour Technol 300:122680. https://doi.org/10.1016/j.biortech.2019.122680

Yoon JY, Kim JE, Song HJ, Oh KB, Jo JW, Yang Y-H, Lee SH, Kang G, Kim HJ, Choi Y-K (2021) Assessment of adsorptive behaviors and properties of grape pomace-derived biochar as adsorbent for removal of cymoxanil pesticide. Environ Technol Inno 21:101242

Yuan P, Wang JS, Pan Y, Shen B, Wu C (2019) Review of biochar for the management of contaminated soil: preparation, application and prospect. Sci Total Environ 659:473–490

Zafar-Ul-Hye M, Tahzeeb-Ul-Hassan M, Wahid A, Danish S, Khan MJ, Fahad S, Brtnicky M, Hussain GS, Battaglia ML, Datta R (2021) Compost mixed fruits and vegetable waste biochar with ACC deaminase rhizobacteria can minimize lead stress in mint plants. Sci Rep 11(1):6606. https://doi.org/10.1038/s41598-021-86082-9

Zafar-ul-Hye M, Danish S, Abbas M, Ahmad M, Munir MM (2019) ACC deaminase producing PGPR *Bacillus amyloliquefaciens* and *Agrobacterium fabrum* along with biochar improve wheat productivity under drought stress. Agronomy 9:343. https://doi.org/10.3390/agronomy9070343

Zhang W, Chen L, Liu D (2012) Characterization of a marine-isolated mercury-resistant *Pseudomonas putida* strain SP1 and its potential application in marine mercury reduction. Appl Microbiol Biotechnol 93:1305–1314

Zhang X, Wang Y, Ju N, Ai Y, Liu Y, Liang J, Hu ZN, Guo R, Xu W, Zhang W (2021) Ultimate resourcization of waste: crab shell-derived biochar for antimony removal and sequential utilization as an anode for a Li-Ion battery. ACS Sustain Chem Eng 9:8813–8823. https://doi.org/10.1021/acssuschemeng.1c01851

Zhao C, Wang B, Theng BKG, Wu P, Liu F, Wang S, Lee X, Chen M, Li L, Zhang X (2021) Formation and mechanisms of nano-metal oxide-biochar composites for pollutants removal: a review. Sci Total Environ 767:145305

Zhu J, Zhao Y, Ruan H (2019) Comparative study on the biodegradation of chlorpyrifosmethyl by *Bacillus megaterium* CM-Z19 and *Pseudomonas syringae* CM-Z6. An Acad Bras Cienc 91:e20180694

Utilizing Co-biochar Microbes for Sustainable Agriculture: Reducing Synthetic Fertilizer and Enhancing Crop Nutrient Quality

Modupe Stella Ayilara, Saheed Adekunle Akinola, Abisola Eunice Ajakaye, Bartholomew Saanu Adeleke, Akinlolu Olalekan Akanmu, and Olubukola Oluranti Babalola

Abstract Because of the earth-disrupting effects of synthetic (chemical) fertilizers, research has been on going to find safer alternatives that preserve soil health and assist in crop production. The potential of co-biochar (biochar enriched in beneficial microbial inoculants) has recently gained a lot of attention as an alternative way to reduce synthetic fertilizers and improve nutrient quality for crops. Co-biochar is produced from various organic sources including agricultural waste, animal manure and forestry refuse. This has a synergistic impact for improving the structure of soils, water holding capacity and microbe activity. The high porosities, surface areas and adsorption capacities of co-biochar explained retention of nutrients from which the beneficial ones are released only gradually and through time supported plant growth-promoting microbes. Thus, this chapter discusses how co-biochar affects microbes to increase fertilizer use efficiency, reduce nutrient losses and enhance various crop growth indices like yield, biomass and nutrient-density. The article also dives into the potential of co-biochar to mitigate greenhouse gas emissions, eutrophication, and soil degradation making it an attractive alternative for climate-smart agriculture as well as for health-conscious pathways to sustainability. Although co-biochar usage has some merits, it suffers a number of challenges including production costs, microbial survivability, scalability, and low farmer adoption. Thus, future prospects such

---

M. S. Ayilara (✉) · S. A. Akinola · B. S. Adeleke · A. O. Akanmu · O. O. Babalola
Food Security and Safety Focus Area, Faculty of Natural and Agricultural Sciences, North-West University, Mmabatho, South Africa
e-mail: modupeayilara7@gmail.com

S. A. Akinola
Department of Microbiology and Parasitology, School of Medicine and Pharmacy, College of Medicine and Health Sciences, University of Rwanda, Butare, Rwanda

A. E. Ajakaye
Department of Public Health, University of Suffolk, Ipswich, UK

B. S. Adeleke
Microbiology Programme, Department of Biological Sciences, School of Science, Olusegun Agagu University of Science and Technology, Okitipupa, Nigeria

A. Chauhan et al. (eds.), *Biochar As Remediation Tools for Pollutant Removal*,
https://doi.org/10.1007/978-3-032-16746-0_8

as-designed biochar, precision microbial inoculants and combination with agroecological techniques are put forward in this manuscript. Furthermore, conducive policy, cross-disciplinary research and farmers' enthusiasm is possibly needed to harness as much benefit as possible from co-biochar micro-organisms for sustainable agriculture.

Keywords Co biochar · Microbial inoculants · Soil fertility · Sustainable agriculture · Biochar

1 Introduction

The world's population is growing and it is essential to upsurge agricultural productivity toward the necessity of its inhabitants through sustainable management practices without abating soil health. During past and recent decades, enormous amounts of artificial fertilizers were excessively used to enhance crop productivity beneficially and normally increased productivity in the post-Green Revolution period (Abebe et al. 2022; Chaudhary et al. 2022a; Verma et al. 2023). But these also have an exorbitant price tag. Over- and callous utilization of chemical fertilizers leads to drastic ecological outcomes such as land acidification, nutrient leaching, waterway eutrophication & heavy greenhouse gas emissions (Ashitha and Rakhimol 2021). They have acted on the native fertility of soils, its degradation due to excessive use of agrochemicals and diminution in microbial diversity which has led to dependence of farm systems on chemicals (Han et al. 2021). All these challenges highlight the urgent global requirement for sustainable solutions in crop productivity across the world to develop canopies that enhance environmental performance.

Co-biochar, a beneficial microbe-based inoculant enriched in complemented biochar, is a viable means of increasing crop productivity. Biochar is a carbon-rich material that is produced through low-oxygen pyrolysis of organic biomass (He et al. 2021). Various organic materials can be used to produce biochar, such as agricultural wastes (like rice husks, maize stalks), animal manures, forestry residues, and food waste (Wijitkosum 2022). Thus, biochar, co-applied with plant growth-promoting microbes, e.g., nitrogen-fixing bacteria or phosphate-solubilizing fungi, produces a synergistic system for enhancing soil fertility, nutrient cycling, and plant health. The porous structure of the co-biochar provides stable habitation for microbial colonization and, therefore, enables the microbes to stay longer and operate at a greater functional efficiency within the soil environment (Jatav et al. 2021).

Co-biochar conveys productive improvements and environmental benefits relative to agriculture. It increases the availability and retention of essential nutrients, particularly nitrogen, phosphorus, and potassium; enhances the retention of moisture; reduces irrigation requirements; and increases resilience to droughts (Jatav et al. 2021). Co-biochar stabilizes carbon in the soil for long periods and thus mitigates the effects of climate change through carbon sequestration (Babu et al. 2023). All these features are attributed to the physicochemical properties of biochar such as high porosity, large surface area, cation exchange capacity (CEC), and capacity for adsorbing nutrients and pollutants (Alkharabsheh et al. 2021; Leng et al. 2021; Qiu et al. 2022). In addition, they improve nutrient retention in soil from leaching and volatilization, nutrient consumption, and mobilization of pollutants to the soil. Therefore, the chapter further elaborates principles and uses of co-biochar microbes and their prospects on synthetic fertilizer reductions, crop nutritional properties, and agro-environmental sustainability.

2 Synthetic Fertilizer

With the 'Green Revolution', especially, the contribution of man-made fertilizers is invaluable. But this has resulted in huge environmental destruction, soil health degradation and financial hardship on farmers. A reduction of synthetic fertilizers will also be necessary for sustainable agriculture. The need for measures like the use of supplementation such as organic inputs, bio-fertilizers, precision farming and policy implications to encourage eco-friendly nutrient management practices has been shown (Akinola and Babalola 2020; Chaudhary et al. 2024; Srivastava et al. 2024).

3 Environmental Concerns

Use of synthetic fertilizers (NPK – Nitrogen, Potassium and Phosphorus) has enhanced the agricultural production faster post Green Revolution. However, improper and overuse of them has had adverse effects on the eco-environment (Zhang et al. 2015). The intensive use of the chemical fertilizer threatens the environment to a large extent, so we must replace them by a sustainable object for backing. Over-application of nitrogen and phosphorus results in nutrient runoff that can lead to water body eutrophication that robs aquatic habitats of oxygen, resulting in oxygen dead-zones for aquatic organisms.

4 Soil Degradation Due to Synthetic Fertilizers

Abuse of chemicals and inorganic fertilizers results in soil badness and land badness that leads to decreation of the soil's health and fertility. Higher dose of chemical fertilizers disturbs soil pH, usually leading to acidification that leads to the reduction in the availability of nutrients and solubilization of the toxic metal (Guo et al. 2010). Also, SOM is decreased under the action of microbial activities by synthetic fertilizers leading to deterioration in soil structure and erosion (Lal 2020). Furthermore, soil degradation is a result of the use of synthetic fertilizers which deprive the soil of its inherent mineral nutrients and decrease microbial diversity leading to long-term loss of fertility Ayilara et al. (2024a). Decrease of good soil microorganisms, for example the mycorrhizal fungi, interferes with natural nutrient absorption as conventional inputs are perpetually applied (Leff et al. 2015). In arable lands this leads to the loss of positive effects on soil fertility and stress resistance (site productivity) with more fertilization needed, that again become a vicious circle threatening sustainable food production along with serious environmental risks. Such harmful effects might be reduced by using organic amendments and precision farming practices concurrent with the restoration of soil health. Moreover, greenhouse gases are also produced during the production and application of those fertilizers with N_2O being 300 times more powerful than CO_2.

5 Water Pollution from Fertilizer Runoff

The leaching and runoff of synthetic fertilizers, particularly nitrogen (N) and phosphorus (P), contribute significantly to water pollution, with severe ecological and public health consequences. When excess fertilizers are applied to croplands, rainwater and irrigation carry these nutrients into nearby rivers, lakes, and groundwater systems. Nitrogen in the form of nitrates (NO_3^-) contaminates drinking water supplies, posing risks such as methemoglobinemia ("blue baby syndrome") in infants (Subedi et al. 2023). These environmental and health impacts highlight the urgent need to adopt organic amendments, bio-fertilizers, and precision farming techniques to mitigate synthetic fertilizer dependency while preserving ecosystems.

In the Gulf of Mexico, phosphorus wastewater leads to eutrophication which supports explosive algal growth, depletes oxygen, and forms regions lifeless of marine life. Nitrogen, on the other hand, undergoes microbial transformations while also releasing the potent greenhouse gas nitrous oxide (N_2O), greatly worsening climate change (Ayilara et al. 2025a). Agricultural productivity and water pollution due to fertilizers can be optimized to a great extent with controlled-release fertilizers, buffer strips and precision agriculture.

6 Greenhouse Gas Emissions from Fertilizer Use

Use of syn. fertilizers and its production is one of the largest contributors to Green House Gas (GHG) emissions. The production of nitrogen (N)-based fertilizers with the Haber-Bosch process requires a large amount of natural gas and produces significant amounts of carbon dioxide (CO_2) and nitrogen (Erisman et al. 2008). Application of microbial fertilizers will result in their transformation in the soils and subsequently act as sources for nitrous oxide release. The super greenhouse gas was produced with 265 times the bad potential of CO_2 for the next hundred years. Agriculture is the main source of global anthropogenic emissions of N_2O, representing ~ 60% of contributions, predominantly associated with synthetic fertilizers (Tian et al. 2020). Moreover, effectiveness of nitrogen fertilisers has been in question as applied N may not reach the point where it is taken up by plants due to volatilisation in the form of ammonia (NH_3), and later contributes to indirect emission of N_2O (Davidson 2009).

In addition to driving climate change, these emissions also exacerbate climate extremes, which then further reduces the efficiency of fertilizers. However, these emissions can be controlled using enhanced-efficiency fertilizers, and organic amendments while still retaining crop productivity with the use of precision farming.

7 Economic Challenges

Shifting from synthetic fertilizers and precision farming plunges the farming and agricultural system into economic difficulty. Smaller-scale farmers, especially in developing countries, are turned off by organic and bio-fertilizers due to their high initial investment and high costs. Moreover, the drop in income from lower yields during the initial transition phase, despite the long-term benefits to soil health, can lead to disincentivization (Akinola et al. 2022; Ayilara et al. 2024a). Restriction of access to bio-fertilizers and compost in rural areas further complicates the shift (Ayilara et al. 2024b, 2025a). Additionally, current government subsidies for synthetic fertilizers distort the competitive edge of sustainable alternatives. Farmers economically stand to lose if they adopt sustainable approaches without financial, educational, and infrastructural incentives, underscoring the necessity for policy shifts and funding in sustainable farming (Ayilara et al. 2025b).

8 Strategies to Reduce Synthetic Fertilizer Dependency

8.1 Organic Fertilizers and Soil Amendments

Organic soil amendments can ease the dependency on synthetic fertilizers and promote soil health and long-term sustainability in farming. They contain natural components including compost, manure, green manure, and biochar, which strengthen soil structure and beneficially enhance moisture and microbial activity. Organic amendments outperform synthetic fertilizers which offer short-term bursts of nutrients as they slowly release nutrients, minimizing leaching and runoff water pollution. Moreover, cover crops such as legumes fix atmospheric nitrogen, which reduces the requirement for synthetic nitrogen inputs (Cherr et al. 2006). Although organic amendments may take longer to work compared to chemical fertilizers, starting with a greater volume, their impact on enhancing soil fertility, carbon capture, and ecosystem resiliency makes them invaluable to sustainable agriculture. The primary hurdle, with regards to farmer education on efficient application, maximizes the full result potential, and deal with transportation logistics and organic amendment availability.

8.2 *Biofertilizers and Microbial Inoculants*

An eco-friendly method to decrease the use of synthetic fertilizers, bio-fertilizers, and microbial inoculants makes use of specialized microorganisms (Chaudhary et al. 2023a). They include nitrogen-fixing bacteria such as Rhizobium and Azotobacter, phosphate-solubilizing bacteria like Pseudomonas and *Bacillus*, as well as mycorrhizal fungi. These friendly microorganisms work alongside plants to improve soil fertility and nutrient absorption (Bhardwaj et al. 2014; Tyagi et al. 2022a, b, Chaudhary et al. 2023b, Khatoon et al. 2024). The use of bio-fertilizers, in contrast to synthetic fertilizers, which are both disruptive to soil ecosystems, support sustainable nutrient cycling. The bio-fertilizers diminish the greenhouse and nutrient runoff emission alike (Sharma et al. 2013; Chaudhary et al. 2022b, c; Bisht and Chaudhary 2024). Crop yields can be increased by 10–30% with the application of microbial inoculants, alongside a 50% decrease in synthetic nitrogen and phosphorus use (Smith and Read 2010). The general uptake of bio-fertilizers is severely limited by factors such as shelf life, inconsistent results in the field, and awareness among farmers. To shift toward sustainable practices in farming, increasing the adoption of bio-fertilizers requires strengthened microbial strain selection, formulation technology, and farmer awareness programs (Chaudhary et al. 2024).

8.3 *Precision Agriculture and Smart Fertilization*

Smart fertilization and precision agriculture are transforming farming by improving crop yields while reducing the amount of fertilizer used. These techniques depend on remote sensing technology, soil sensors connected through IoT, GPS-guided variable rate applicators, and AI for soil and crop analytics to determine real-time nutrient requirements (Gebbers and Adamchuk 2010). Precision agriculture boosts farm efficiency, reducing nitrogen over-application by 20–40% while sustaining or even improving yields (Couëdel et al. 2019). Other advances, such as multispectral imaging drones and slow or nano-encapsulated fertilizer, help correct nutrient deficiencies, improving precision and reducing waste (Liu and Lal 2015). While the adoption of precision agriculture technology may be impeded by the need to invest in training and equipment, the adoption of such technology is beneficial in the long run as it reduces input costs, improves soil health, and lowers environmental impacts. Furthermore, synergy with organic amendments and biofertilizers could improve the resilience of the system and present a comprehensive strategy to reduce dependence on synthetic fertilizers.

8.4 Agroecological Practices

Agroecological practices offer a more integrated way to decrease reliance on synthetic fertilizers through rather than despite natural ecosystems. These are composed of diversified practices such as crop rotation, inter-cropping, agroforestry and conservation tillage among others that enhance soil fertility through its biological properties and dependence (or lack) on sources external to the farm (Altieri 2018). In addition, farmers can improve nutrient recycling by integrating nitrogen fixing legumes and deep rooted plants in their cropping systems which suppresses the necessity for artificial NP fertilizers (Jose 2009). Agroforestry systems that combine crops with trees help to alleviate nitrogen deficiency by fixing atmospheric N while reducing the requirement for P fertilization by increasing organic matter via decomposition of litter (Kassam et al. 2019).

Through conservation tillage, farmers are able to maintain soil structure and microbial diversity which helps in retaining nutrients that would have otherwise been lost through erosion or leaching (Hobbs et al. 2008).

Agroecological practices have the potential to maintain yields and simultaneously reduce the use of synthetic fertilizers by 30–50% in cereal production systems (Pretty et al. 2018). These practices, while requiring an investment of knowledge and an initial reduction in yields, enhance long-term soil health, adaptive capacity to climatic shifts, resilience, and farm profitability, thus establishing the foundation of sustainable agriculture. Successful outcomes are contingent on educational frameworks for farmers, enabling market systems, and policies that prioritize ecological and production value.

9 Reducing the Need for Synthetic Fertilizers by Improving Natural Nutrient Cycles

Soil biological activity can be enhanced by agroecological practices which resuscitate the soil, thus breaking the cycle of overreliance on synthetic fertilizers. The incorporation of beneficial soil microorganisms (nitrogen-fixing bacteria, mycorrhizal fungi) and organic fertilizers (compost, cover crops) can enhance the mineralization and availability of nutrients and lessen the need for chemical inputs (Drinkwater et al. 2017; Chaudhary et al. 2021a, b, c). Research indicates that crop diversification through crop rotation and legume intercropping strengthens nitrogen fixation, estimating the potential to reduce the reliance on synthetic nitrogen by 30–60%. In addition, conservation tillage and biochar enhance microbial communities that are active in the biological phosphorous cycle, which can reduce the use of phosphate fertilizers (Lehmann et al. 2011). Implementing these practices not only saves on fertilizer expenses, but also reduces pollution caused by nutrient leaching, fossil fuel emissions, and greenhouse gases (Zhang et al. 2015). Therefore, ecological nutrient management provides a viable option for sustaining crop productivity alongside the restoration of soil health and ecosystem balance.

Multiple case studies highlight the success of ecological strategies in lowering the use of synthetic fertilizers and sustaining crop yields (Table 1). Davis et al. (2012) reported a 10-year field trial in the Midwest where more intensive, diversified crop rotations featuring legumes decreased the application of synthetic nitrogen by 40% without sacrificing maize yields, which remained on par with conventional systems. In Sub-Saharan Africa, Kätterer et al. (2019) found that smallholder farmers employing biochar-compost blends with a 50% reduction in mineral fertilizers showed 20–30% greater maize yields than the control group. These findings align with meta-analyses that agroecological practices are effective in replacing 30–70% of synthetic nitrogen with no negative impact on yields (Pittelkow et al. 2015). Another compelling instance is the System of Rice Intensification (SRI) in Southeast Asia where yield improvements of 22–65% coupled with 40% reduced fertilizer use stemmed from better water management and organic amendments (Thakur et al. 2010b).

Developing studies underscore the soil microbiomes' importance for a sustainable transition. In Brazil, studies show that with the inoculation of native arbuscular mycorrhizal fungi, soybean farmers could reduce the usage of phosphate fertilizers by 25% whilst simultaneously increasing the yield by 15%. Long-term studies conducted in Europe also show that organic farming systems with higher biological activity are able to achieve 90% of conventional farming yields while reducing the reliance on synthetic nitrogen by 34–51% (Seufert et al. 2012). Metagenomic analysis particularly helps in tracking the processes by which cover crops, along with reduced tillage, reset functional microbial networks that enrich nutrient cycling (Fierer et al. 2021). The above-stated case studies demonstrate that biological approaches to agroecosystems (Chaudhary et al. 2021a; b) prove these approaches are better than intensive fertilizer use.

10 Enhancing Crop Nutrient Quality with Co-biochar Microbes

Co-biochar microbial applications improve soil nutrient cycling, thus improving crop yield and nutrient value. Research shows that biochar, along with nitrogen-fixing (*Azospirillum* sp.) or phosphate-solubilizing (*Bacillus* sp.) bacteria, boosts grain yields by 15–25% when compared to conventional fertilization (Ayilara et al. 2023a). The porous structure of the biochar captures nutrients, and microbes increase their bioavailability, boosting the uptake of N, P, and K. For example, biochar and mycorrhizal fungi application to rice paddies resulted in 20% greater phosphorus P content in the grains, alleviating the burden of synthetic P fertilizers (Mendes et al. 2011). Also, buffering soil pH alleviation by biochar accentuated the Zn and Fe micronutrients which helps increase the value of the crops (Ayilara et al. 2023a).

Table 1 Reduction in fertilizer use through co-biochar and microbial interventions

Crop/System	Intervention	Reduction in fertilizer use	Yield impact	Key findings	References
Maize (Kenya)	Biochar + *Azospirillum brasilense*	30% less NPK	15% increase	Improved N fixation & P availability	Kätterer et al. (2019)
Wheat (India)	Biochar + *Azotobacter chroococcum*	25% less urea	No yield loss	Increased N retention, reduced volatilization	Hamzah Saleem et al. (2022)
Soybean (Brazil)	Biochar + arbuscular mycorrhizal fungi (AMF)	50% less P fertilizer	18% increase	Improved root P uptake efficiency	Mendes et al. (2011)
Midwest USA (Maize Field Trial)	Biochar + *Azospirillum* inoculation	40% N reduction	No yield loss	Enhanced N fixation & retention	Davis et al. (2012)
China (Rice Paddies)	Biochar + *Pseudomonas* spp.	30% N reduction	10% increase	Reduced N leaching & denitrification	Barreiro et al. (2022)
Kenya (Maize Farms)	Biochar + *Rhizobia*	50% N reduction	40% increase (legumes)	Biological N fixation	Alemayehu et al. (2019)
EU (Wheat)	Biochar + compost + AMF	34–51% N reduction	90% conventional yields	Slow nutrient release	Seufert et al. (2012)
Australia (Sorghum)	Biochar + P-solubilizing *Bacillus*	30% P reduction	15% increase	pH buffering & P mobilization	Ding et al. (2010)
Vietnam (Rice Fields)	Biochar + water management	40% NPK reduction	22–65% increase	Enhanced root growth & nutrient uptake	Thakur et al. (2010a)
California (Tomatoes)	Biochar + *Azotobacter*	35% N reduction	12% increase	N fixation & hormone production	Palansooriya et al. (2019)
Thailand (Cassava)	Biochar + mycorrhizae	20% P reduction	25% increase	Hyphal P transport	Lehmann et al. (2011)
Philippines (Corn)	Biochar + *Enterobacter*	40% N reduction	No yield loss	Nitrification inhibition	Cayuela et al. (2014)

(continued)

Table 1 (continued)

Crop/ System	Intervention	Reduction in fertilizer use	Yield impact	Key findings	References
Canada (Canola)	Biochar + *Penicillium*	25% P reduction	15% increase	Organic acid secretion	Quilliam et al. (2013)
Mexico (Beans)	Biochar + *Rhizobium*	60% N reduction	30% increase	Symbiotic N fixation	Zhang et al. (2015)

The co-biochar microbes' synergistic activities increase plant physiological activities like root formation, stress tolerance, and photosynthetic productivity. In field trials with maize, *Azospirillum* sp. biochar combinations were shown to increase root biomass by 30% and grain yield by 15–20% as reported by Kätterer et al. (2019). Enhanced nutrient assimilation as evidenced by increased leaf chlorophyll and fruit weight was associated with treatment of Pseudomonas fluorescens and biochar to tomato plants. The changes were linked to the microbes' growth hormone production, auxins, and biochar's moisture and nutrient retention capabilities.

Meeting both the agronomic and the nutritional aspects of the crops, the co-biochar microbes increase the levels of critical nutrients within the edible portions of the crops. Wheat grains from Azotobacter-biochar amended soils exhibited 10–15% higher protein concentration probably because of enhanced N fixation (Mendes et al. 2011). Co-biochar application with rhizobia to legumes increased Fe and Zn contents by 25%, hence addressing wide micronutrient deficiencies (Akanmu et al. 2021). Such enhancements are of great importance for crops grown on nutrient-deficient soils while conventional fertilizers, in most cases, fail to improve food quality.

Biofortification with co-biochar microbes can also provide a sustainable solution for malnutrition, especially in resource-poor areas. Crops fortified with iron, zinc, and vitamins can reduce anemia and stunting in vulnerable populations (Fadiji et al. 2020). For instance, beans grown with biochar enhanced with AMF had 30% higher Fe content, thereby contributing directly to dietary iron intake (Mendes et al. 2011). In addition, reduced use of synthetic fertilizers curtails nitrate accumulation in leafy greens, which are risk factors for methemoglobinemia, among other health hazards (Ayilara et al. 2023b). Thus, by raising the nutrient value of staple food crops, co-biochar microbes stand as a solution to alleviate global hidden hunger (micronutrient deficiency). In Southeast Asia, rice biofortified through biochar-microbe amendments showed on a 40% increase in zinc levels and, thus, could alleviate Zn deficiency to millions. Policy interventions aimed at promoting such agro-ecological measures could enhance food security while reducing the burden on healthcare systems due to malnutrition (Adeleke et al. 2022) (Table 2).

Table 2 Benefits of biofortification of biochar in some resource-limited regions

Mechanism	Biofortification of biochar	Applications	References
Improved soil fertility	Biochar enhances soil structure, and improves water retention and nutrient holding capacity. The activity of microbes helps to solubilize essential nutrients including phosphorus	Indian farmers in Punjab utilize rice husk biochar mixed with phosphate-solubilizing bacteria to rehabilitate nutrient-poor soils for pulse crop cultivation	Srinivasarao et al. (2013)
Enhanced micronutrient uptake	Co-biochar microbes symbiosis mobilize micronutrients (Fe, Zn, Se), improves their bioaccessibility to nutrients for crop plants	In Nigeria, farmers apply co-biochar with iron-mobilizing bacteria to bulk up the iron content in their staple food maize and sorghum crops	Oladele et al. (2019), Akanmu et al. (2020)
Sustainability in low-input systems	Useful in areas with limited access to chemical fertilizers. Biochar improves soil quality while extending sustainable productivity under stressful environmental conditions	A combination of enset (false banana) waste biochar along with beneficial microbes allows farmers in the Ethiopian Highlands to grow zinc and iron rich teff using minimal inputs	Nekesa et al. (2024)
Low-cost, locally adaptable	Biochar production uses agricultural residues that include maize cobs and cow dung. Simple local methods can be employed for culturing microbial inoculants	Smallholder farmers in sub-Sahara Africa produce biochar from maize stalks which they combine with native rhizobacteria to enhance the zinc content of their upland rice cultivation	Hotegni et al. (2024), Mashamaite et al. (2024)

11 Impact of Co-biochar Microbes on Fertilizer Use Efficiency

Fertilizer use efficiency (FUE) is an important parameter for sustainable agriculture and refers to the percentage of nutrient applied that is taken up by the crop as well as lost to the environment. Synthetic fertilizers usually possess a lesser degree of FUE due to leaching, volatilization, and fixation (Zhang et al. 2015). Biochar, a carbonaceous material formed through pyrolysis of biomass, has lately been recognized as a soil amendment that improves nutrient retention and microbial activity (Lehmann et al. 2011). When mixed with beneficial microbes (e.g., *Azospirillum* sp. and *Pseudomonas* sp.), biochar provides a habitat and a nutrient reservoir, thereby improving microbial survival and functioning (Palansooriya et al. 2019).

11.1 Mechanisms of Co-biochar Microbes in Enhancing FUE

Collaborative biochar microorganisms can promote FUE by a number of synergistic pathways. Firstly, biochar with porosity and sound porous structure can provide a favorable habitat for microbial colonization, to adsorb and maintain essential nutrients (including NH_4^+, NO_3^-, PO_4^{3-}) that would otherwise be leached from the soil or be lost through volatilization (Ding et al. 2010).

Banu et al. and Da Costa et al. immobilized micro-organisms such as nitrogen fixing, for example Azospirillum and phosphate solubilizing bacteria (e.g., Bacillus) that mineralize organic fertilizers and also convert nutrient into plant available forms synchronized with crop demand, minimizing the gap between actual release of nutrients and the crop requirements. In addition, biochar is alkaline nature, and it buffers in soil pH hence promoting the microbial activity as well as nutrient availability (Palansooriya et al. 2019).

Second, the co-biochar microbiomes prevent environmental loss through reduced production of greenhouse gases and nutrient leaching. Certain microbial species, such as *Pseudomonas stutzeri*, foster denitrification and produce stable forms of nitrogen from nitrate that does not release N_2O (a potent GHG) to the atmosphere (Cayuela et al. 2014). Furthermore, the carbon-rich matrix of biochar aids in the formation of stable soil aggregates, thereby reducing erosion and the leaching of nutrients into water systems (Quilliam et al. 2013). Creating environment-friendly nutrient cycles through the promotion of microbial survival and nutrient retention co-biochar microbes, maximizes FUE while minimizing environmental degradation (Fig. 1).

Fig. 1 The mechanisms of enhancing fertilizer use efficiency in co-biochar microbes' application

12 Policy and Farmer Incentives

For an effective transition of farmers away from synthetic fertilizer dependence, policy interventions and economic incentives are crucial. Governments can fast-track this change by channeling subsidies towards organic inputs like composts and biofertilizers while phasing out subsidies on chemical fertilizers, a model that Denmark has successfully pursued via its nitrogen tax, which slashed synthetic fertilizer application by 30% without any negative bearing on yield (Dalgaard et al. 2014). Extension services and trainings therefore are crucial to foster the capacity building, as demonstrated in India through its Zero Budget Natural Farming program that trained about 800,000 farmers in sustainable practices. Mechanisms such as carbon credits for sustainable agriculture or payment for ecosystem services (PES) provided an incentive for further adoption in pilot phases in Costa Rica agroforestry programs. However, policy success depends quite a bit on approaches accounting for both socioeconomic contexts also regional farming systems. Policymakers can create an enabling environment for large-scale fertilizer reduction via aligning regulations along with market access. Likewise, knowledge dissemination can protect farmer livelihoods and food security (Ayilara et al. 2024b, 2025a).

A number of international efforts highlight the potential for delivering reduced synthetic fertilizer use without harming agricultural productivity. In India, the Zero Budget Natural Farming (ZBNF) initiative reached over 800,000 farmers who adopted organic inputs dairy cow dung based formulations and microbial inoculants to achieve a 70% saving on fertilizers while maintaining yields in both rice and millet gain systems (Ayilara et al. 2025b).

Denmark's forward-looking nitrogen tax, introduced in the 1990s, already reduced synthetic fertilizer use by 30% through farmer training and precision application tools which result in better nutrient efficiency without loss of yield (Dalgaard et al. 2014). In Kenya, the application of rhizobia inoculants for legumes resulted in 40% increase in soybean productivity and eliminated dependence on nitrogen fertilizers (Alemayehu et al. 2019).

Likewise in Brazil, integration of crop-livestock-forestry systems enhanced soil organic carbon by 15% and decreased NPK fertilizer need by 25%. These successes demonstrate the importance of policy support, farmer training and locally appropriate solutions when trying to scale sustainable alternatives. Nevertheless, it remains a challenge for these models to be replicated in various agro-ecological zones hence the importance of adaptive approaches.

13 Challenges and Limitations

The upscaling of co-biochar-microbe technologies operate on various technical and economic bottlenecks. Facets that limit the implementation are, in particular, a large production cost and a complexity procedure particularly where there is absence of technical infrastructure or pyrolysis facilities (Owsianiak et al. 2021). The absence of feedstock quality and homogeneity across the majority of farming systems also undermines the reproducibility and effectiveness of co-biochar application (Köves et al. 2024).

Economically, the commercialization of such bioformulations is also restricted due to lack of quality microbial inoculants and microbe co-application required expertise especially among smallhold farmers (Sekhon and Keshani 2024). Farmers may however not be ready to exploit the potentials inherent in adopting co-biochar-microbe practice because of lack of awareness creation, extension services or incentives.

There are severe environmental threats, too. Co-biochar is a source of beneficial biotic properties for soil health, however the high doses or poor quality can disrupt the microbial dynamics in soils and may suppress local ecosystem processes (Jatav et al. 2021). Long-term field experiments are required to determine the additive effect on biodiversity, soil pH, mobilization of heavy metal and carbon-nutrient interface (Jatav et al. 2021). In addition to that, suppressive co-biochar formulations may also have to be formulated based on climate and soil type affecting the survival and performance of microbes in restricted agroecological zones to avoid ecological trade-offs or loss of effectiveness.

It is essential for an efficient soil microbial-based management system that would maximize the entire potential of its co-biochar. Natives are utilsed more effectively as introduced microbes routinely fail to colonize or out-compete resident soil microbiota, especially in dryer periods and saline situations (O'Callaghan et al. 2022). A formulation of the inoculants, new or better carrier materials and probably some kind of genetic modification would need to follow, to guarantee their persistence and functionality.

Moreover, the complexity of interactions between plants, microbes and soils will always demand for an interdisciplinary approach that takes into account soil science, microbiology and agronomy. The pursuit of these objectives will be guided by supportive research initiatives, policy structures and farmer training that will allow the commercial practices of co-biochar microbes in regenerative agriculture.

14 Conclusion

Co-biochar microorganisms may be seen as a potentially sustainable solution to some of the current agriculture issues in terms of reducing reliance on synthetic fertilizers and increasing crop nutrient quality. Co-biochar-microbe systems contribute to the achievement of global sustainable food production and environmental protection targets via improvements in soil fertility, promotion of beneficial microbial activities, and efficiency enhancements in nutrient use. Its incorporation in farming practices will have agronomic benefits but also associated environmental benefits, for instance, reducing emissions and improving water quality. However, the complete range of possibilities brought about by this technology would be achieved through meeting numerous challenges, such as cost and scale of manufacture, microorganism survival, and farmer acceptance.

Hence, the development of coordinated co-biochar formulations and resilient microbial inoculants should be achieved so as to accommodate the various agroecological zones. Funding should be directed to research, infrastructure development, farmer training, and policy dialogue to allow full-scale implementation in no time.

Going forward, joint co-application of biochar and beneficial microbes should be regarded as a central driving force in sustainable agricultural systems. Through concurrent efforts offered by researchers, policy-making bodies, and the farming communities, ultimately, co-biochar microbial solutions can significantly contribute to food security, climate resilience, and ecosystem restoration toward sustainably productive agriculture.

15 Future Directions and Innovations

Future advancements in co-biochar production should focus on improving its functional properties to enhance microbial compatibility and soil performance. Innovations such as engineered biochar with tailored porosity, surface chemistry, and nutrient profiles can be explored to optimize microbial colonization and nutrient retention. In addition, integrating biochar with nanomaterials or composted organics may further improve its efficacy and cost-effectiveness. Scalable, low-cost pyrolysis technologies and decentralized production systems are essential to facilitate widespread adoption, especially in resource-limited settings.

On the microbial innovation front, next-generation microbial inoculants are gaining momentum. With advances in microbial genomics and biotechnology, tailored consortia will be designed with enhanced resilience, nutrient-solubilizing efficiency, and compatibility to the biochar carriers. Those microbes could be engineered or selectively bred for them to thrive in specific soil and climatic conditions, thus improving consistency and impact in the field. Integrating co-biochar microbes into sustainable agriculture paradigms will optimize their benefits. Investments in farmer education, demonstration trials, and knowledge-sharing platforms will guarantee on-ground implementation of the gap between research and practice toward the transition into climate-smart biologically based nutrient management systems.

References

Abebe TG, Tamtam MR, Abebe AA, Abtemariam KA, Shigut TG, Dejen YA, Haile EG (2022) Growing use and impacts of chemical fertilizers and assessing alternative organic fertilizer sources in Ethiopia. Appl Environ Soil Sci 2022(1):4738416

Adeleke BS, Akinola SA, Adedayo AA, Glick BR, Babalola OO (2022) Synergistic relationship of endophyte-nanomaterials to alleviate abiotic stress in plants. Front Environ Sci 10:1015897

Akanmu AO, Babalola OO, Venturi V, Ayilara MS, Adeleke BS, Amoo AE, Glick BR (2021) Plant disease management: leveraging on the plant-microbe-soil interface in the biorational use of organic amendments. Front Plant Sci 12:700507

Akanmu AO, Sobowale AA, Abiala MA, Olawuyi OJ, Odebode AC (2020) Efficacy of biochar in the management of *Fusarium verticillioides* Sacc. causing ear rot in *Zea mays* L. Biotechnol Rep 26:e00474

Akinola SA, Babalola OO (2020) The importance of adverse soil microbiomes in the light of omics: implications for food safety. Plant, Soil Environ 66(9)

Akinola SA, Adeyemo RO, Adebayo IA (2022) Overview and emergence of nanobiotechnology in plants. In: Diverse applications of nanotechnology in the biological sciences. Apple Academic Press, pp 173–197

Alemayehu F, Tolera M, Tesfaye G (2019) Land use land cover change trend and its drivers in somodo watershed south western, Ethiopia. Afr J Agric Res 14(2):102–117

Alkharabsheh HM, Seleiman MF, Battaglia ML, Shami A, Jalal RS, Alhammad BA, Al-Saif AM (2021) Biochar and its broad impacts in soil quality and fertility, nutrient leaching and crop productivity: a review. Agronomy 11(5):993

Altieri MA (2018) Agroecology: the science of sustainable agriculture. CRC Press

Ashitha A, Rakhimol KR (2021) Fate of the conventional fertilizers in environment. In: Controlled release fertilizers for sustainable agriculture. Academic Press, pp 25–39

Ayilara MS, Adeleke BS, Adebajo MT, Akinola SA, Fayose CA, Adeyemi UT, Babalola OO (2023a) Remediation by enhanced natural attenuation; an environment-friendly remediation approach. Front Environ Sci 11:1182586

Ayilara MS, Adeleke BS, Akinola SA, Fayose CA, Adeyemi UT, Gbadegesin LA, Babalola OO (2023b) Biopesticides as a promising alternative to synthetic pesticides: a case for microbial pesticides, phytopesticides, and nanobiopesticides. Front Microbiol 14:1040901

Ayilara MS, Adeleke BS, Akinola SA, Fayose CA, Adeyemi UT, Gbadegesin LA, Babalola OO (2024a) Corrigendum: biopesticides as an alternative to synthetic pesticides: a case for nanopesticides, phytopesticides and microbial pesticides. Front Microbiol 14:1258968

Ayilara MS, Akinola SA, Adeleke BS, Gbadegesin LA, Adejumo GD, Glick BR, Babalola OO (2024b) Biopesticides versus synthetic pesticides usage in Africa. In: Microbiome-based decontamination of environmental pollutants, pp 417–450

Ayilara MS, Fasusi SA, Ajakaye AE, Akinola SA, Kang S, Ganiyu LT, Babalola OO (2025a) Harnessing agroforestry for climate change solutions. In: Agroforestry for a climate-smart future. IGI Global Scientific Publishing, pp 229–258

Ayilara MS, Fasusi SA, Ajakwe SO, Akinola SA, Ayilara-Adewale OA, Ajakaye AE, Babalola OO (2025b) Impact of climate change on agricultural ecosystem. In: Climate change, food security, and land management: strategies for a sustainable future. Springer Nature Switzerland, Cham, pp 1–24

Babu S, Singh R, Kumar S, Rathore SS, Yadav D, Yadav SK, Singh VK (2023) Biochar implications in cleaner agricultural production and environmental sustainability. Environ Sci: Adv 2(8):1042–1059

Barreiro A, Fox A, Jongen M, Melo J, Musyoki M, Vieira A, Mårtensson LD (2022) Soil bacteria respond to regional edapho-climatic conditions while soil fungi respond to management intensity in grasslands along a European transect. Appl Soil Ecol 170:104264

Bhardwaj D, Ansari MW, Sahoo RK, Tuteja N (2014) Biofertilizers function as key player in sustainable agriculture by improving soil fertility, plant tolerance and crop productivity. Microb Cell Fact 13:1–10

Bisht N, Chaudhary P (2024) Plant-microbe interactions and crop production. In: Microbial inoculants: applications for sustainable agriculture. Springer Nature Singapore, Singapore, pp 69–93

Cayuela ML, Van Zwieten L, Singh BP, Jeffery S, Roig A, Sánchez-Monedero MA (2014) Biochar's role in mitigating soil nitrous oxide emissions: a review and meta-analysis. Agr Ecosyst Environ 191:5–16

Chaudhary A, Parveen H, Chaudhary P, Bhatt P, Khatoon H (2021a) Rhizospheric microbes and their mechanism. In: Bhatt P, Gangola S, Udayanga D, Kumar G (eds) Microbial technology for sustainable environment. Springer, Singapore. https://doi.org/10.1007/978-981-16-3840-4_6

Chaudhary P, Chaudhary A, Parveen H, Rani A, Kumar G, Kumar A, Sharma A (2021b) Impact of nanophos in agriculture to improve functional bacterial community and crop productivity. BMC Plant Biol 21:519. https://doi.org/10.1186/s12870-021-03298-7

Chaudhary P, Parveen H, Gangola S, Kumar G, Bhatt P, Chaudhary A (2021c) Plant growth promoting Rhizobacteria and their application in sustainable crop production. In: Bhatt P, Gangola S, Udayanga D, Kumar G (eds) Microbial technology for sustainable environment. Springer, Singapore. https://doi.org/10.1007/978-981-16-3840-4_13

Chaudhary P, Agri U, Chaudhary A, Kumar A, Kumar G (2022a) Endophytes and their potential in biotic stress management and crop production. Front Microbiol 13:933017. https://doi.org/10.3389/fmicb.2022.933017

Chaudhary P, Singh S, Chaudhary A, Agri U, Bhandari G (2022b) Chapter 3 Agrochemicals and their effects on soil microbial population. In: Soni R, Suyal D, Goel R (eds) Plant protection: from chemicals to biologicals. De Gruyter, Berlin, Boston, pp 45–58. https://doi.org/10.1515/9783110771558-003

Chaudhary P, Singh S, Chaudhary A, Sharma A, Kumar G (2022c) Overview of biofertilizers in crop production and stress management for sustainable agriculture. Front Plant Sci 13:930340. https://doi.org/10.3389/fpls.2022.930340

Chaudhary P, Chen S, Rajput VD, Jaiswal DK (2023a) Editorial: bioinoculants with nano-compounds to improve soil health: a step toward sustainable agriculture. Front Environ Sci 11:1270002. https://doi.org/10.3389/fenvs.2023.1270002

Chaudhary P, Xu M, Ahamad L, Chaudhary A, Kumar G, Adeleke BS, Verma KK, Hu D-M, Širić I, Kumar P, Popescu SM, AbouFayssal S (2023b) Application of synthetic consortia for improvement of soil fertility, pollution remediation, and agricultural productivity: a review. Agronomy 13:643. https://doi.org/10.3390/agronomy13030643

Chaudhary A, Chaudhary P, Abou Fayssal S, Singh S, Jaiswal DK, Tripathi V, Kumar J (2024) Exploring beneficial microbes and their multifaceted applications: an overview. In: Microbial inoculants: applications for sustainable agriculture, 1–28

Cherr CM, Scholberg JMS, McSorley R (2006) Green manure approaches to crop production: a synthesis. Agron J 98(2):302–319

Couëdel A, Kirkegaard J, Alletto L, Justes E (2019) Crucifer-legume cover crop mixtures for biocontrol: toward a new multi-service paradigm. Adv Agron 157:55–139

Dalgaard T, Hansen B, Hasler B, Hertel O, Hutchings NJ, Jacobsen BH, Vejre H (2014) Policies for agricultural nitrogen management—trends, challenges and prospects for improved efficiency in Denmark. Environ Res Lett 9(11):115002

Davidson EA (2009) The contribution of manure and fertilizer nitrogen to atmospheric nitrous oxide since 1860. Nat Geosci 2(9):659–662

Davis AS, Hill JD, Chase CA, Johanns AM, Liebman M (2012). Increasing cropping system diversity balances productivity, profitability and environmental health

Ding Y, Liu YX, Wu WX, Shi DZ, Yang M, Zhong ZK (2010) Evaluation of biochar effects on nitrogen retention and leaching in multi-layered soil columns. Water Air Soil Pollut 213:47–55

Drinkwater LE, Schipanski M, Snapp S, Jackson LE (2017) Ecologically based nutrient management. In: Agricultural systems. Academic Press, pp 203–257

Erisman JW, Sutton MA, Galloway J, Klimont Z, Winiwarter W (2008) How a century of ammonia synthesis changed the world. Nat Geosci 1(10):636–639

Fadiji AE, Ayangbenro AS, Babalola OO (2020) Organic farming enhances the diversity and community structure of endophytic archaea and fungi in maize plant: a shotgun approach. J Soil Sci Plant Nutr 20(4):2587–2599

Fierer N, Wood SA, de Mesquita CPB (2021) How microbes can, and cannot, be used to assess soil health. Soil Biol Biochem 153:108111

Gebbers R, Adamchuk VI (2010) Precision agriculture and food security. Science 327(5967):828–831
Guo JH, Liu XJ, Zhang Y, Shen JL, Han WX, Zhang WF, Zhang FS (2010) Significant acidification in major Chinese croplands. Science 327(5968):1008–1010
Hamzah Saleem M, Usman K, Rizwan M, Al Jabri H, Alsafran M (2022) Functions and strategies for enhancing zinc availability in plants for sustainable agriculture. Front Plant Sci 13:1033092
Han J, Dong Y, Zhang M (2021) Chemical fertilizer reduction with organic fertilizer effectively improve soil fertility and microbial community from newly cultivated land in the Loess Plateau of China. Appl Soil Ecol 165:103966
He X, Zheng N, Hu R, Hu Z, Yu JC (2021) Hydrothermal and pyrolytic conversion of biomasses into catalysts for advanced oxidation treatments. Adv Func Mater 31(7):2006505
Hobbs PR, Sayre K, Gupta R (2008) The role of conservation agriculture in sustainable agriculture. Philos Trans R Soc b: Biol Sci 363(1491):543–555
Hotegni NVF, Sohindji FS, Salaou MA, Agbandou PC, Azonhoumon LW, Tchokponhoué D, Achigan-Dako EG (2024) Agronomic biofortification of cereals and legumes with iron, zinc, calcium and magnesium for food and nutrition security: available options for farmers in Sub-Saharan Africa. J Agric Food Res 18:101391
Jatav HS, Rajput VD, Minkina T, Singh SK, Chejara S, Gorovtsov A, Kalinitchenko VP (2021) Sustainable approach and safe use of biochar and its possible consequences. Sustainability 13(18):10362
Jose S (2009) Agroforestry for ecosystem services and environmental benefits: an overview. Springer
Kassam A, Friedrich T, Derpsch R (2019) Global spread of conservation agriculture. Int J Environ Stud 76:29–51
Kätterer T, Roobroeck D, Andrén O, Kimutai G, Karltun E, Kirchmann H, de Nowina KR (2019) Biochar addition persistently increased soil fertility and yields in maize-soybean rotations over 10 years in sub-humid regions of Kenya. Field Crop Res 235:18–26
Khatoon H, Chaudhary P, Chaudhary A (2024) Microbial inoculants and soil microbial population. In: Microbial inoculants: applications for sustainable agriculture. Springer Nature Singapore, Singapore, pp 49–68
Köves M, Madár V, Ringer M, Kocsis T (2024) Overview of traditional and contemporary industrial production technologies for biochar along with quality standardization methods. Land 13(9):1388
Lal R (2020) Soil organic matter content and crop yield. J Soil Water Conserv 75:27A-32A
Leff JW, Jones SE, Prober SM, Barberán A, Borer ET, Firn JL, Fierer N (2015) Consistent responses of soil microbial communities to elevated nutrient inputs in grasslands across the globe. Proc Natl Acad Sci 112(35):10967–10972
Lehmann J, Rillig MC, Thies J, Masiello CA, Hockaday WC, Crowley D (2011) Biochar effects on soil biota—a review. Soil Biol Biochem 43(9):1812–1836
Leng L, Xiong Q, Yang L, Li H, Zhou Y, Zhang W, Huang H (2021) An overview on engineering the surface area and porosity of biochar. Sci Total Environ 763:144204
Liu R, Lal R (2015) Potentials of engineered nanoparticles as fertilizers for increasing agronomic productions. Sci Total Environ 514:131–139
Mashamaite CV, Motsi H, Manyevere A, Poswa SB (2024) Assessing the potential of biochar as a viable alternative to synthetic fertilizers in sub-Saharan Africa smallholder farming: a review. Agronomy 14(6):1215
Mendes R, Kruijt M, De Bruijn I, Dekkers E, Van Der Voort M, Schneider JH, Raaijmakers JM (2011) Deciphering the rhizosphere microbiome for disease-suppressive bacteria. Science 332(6033):1097–1100
Nekesa T, Jaquet S, Katsir S, Vollmann Tinoco V, Weinsheimer H (2024) Harvesting Tomorrow: advancing sustainable land management for soil fertility. World Overview of Conservation Approaches and Technologies (WOCAT)/Centre for Development and Environment (CDE), University of Bern, Switzerland, Alliance of Bioversity International & CIAT and Deutsche Gesellschaft für Internationale Zusammenarbeit (GIZ) GmbH

O'Callaghan M, Ballard RA, Wright D (2022) Soil microbial inoculants for sustainable agriculture: limitations and opportunities. Soil Use Manag 38(3):1340–1369

Oladele S, Adeyemo A, Adegaiye A, Awodun M (2019) Effects of biochar amendment and nitrogen fertilization on soil microbial biomass pools in an Alfisol under rain-fed rice cultivation. Biochar 1:163–176

Owsianiak M, Lindhjem H, Cornelissen G, Hale SE, Sørmo E, Sparrevik M (2021) Environmental and economic impacts of biochar production and agricultural use in six developing and middle-income countries. Sci Total Environ 755:142455

Palansooriya KN, Wong JTF, Hashimoto Y, Huang L, Rinklebe J, Chang SX, Ok YS (2019) Response of microbial communities to biochar-amended soils: a critical review. Biochar 1:3–22

Pittelkow CM, Liang X, Linquist BA, Van Groenigen KJ, Lee J, Lundy ME, Van Kessel C (2015) Productivity limits and potentials of the principles of conservation agriculture. Nature 517(7534):365–368

Pretty J, Benton TG, Bharucha ZP, Dicks LV, Flora CB, Godfray HCJ, Wratten S (2018) Global assessment of agricultural system redesign for sustainable intensification. Nat Sustain 1(8):441–446

Qiu B, Shao Q, Shi J, Yang C, Chu H (2022) Application of biochar for the adsorption of organic pollutants from wastewater: modification strategies, mechanisms and challenges. Sep Purif Technol 300:121925

Quilliam RS, Glanville HC, Wade SC, Jones DL (2013) Life in the 'charosphere'–does biochar in agricultural soil provide a significant habitat for microorganisms? Soil Biol Biochem 65:287–293

Sekhon KK, Keshani (2024) Microbial inoculum improvement for soil health. In Advancements in microbial biotechnology for soil health. Springer Nature Singapore, Singapore, pp 49–65

Seufert V, Ramankutty N, Foley JA (2012) Comparing the yields of organic and conventional agriculture. Nature 485(7397):229–232

Sharma SB, Sayyed RZ, Trivedi MH, Gobi TA (2013) Phosphate solubilizing microbes: sustainable approach for managing phosphorus deficiency in agricultural soils. Springerplus 2:1–14

Smith SE, Read DJ (2010) Mycorrhizal symbiosis. Academic Press

Srinivasarao C, Gopinath KA, Venkatesh G, Dubey AK, Wakudkar H, Purakayastha TJ, Sikka AK (2013) Use of biochar for soil health management and greenhouse gas mitigation in India: potential and constraints. Central Research Institute for Dryland Agriculture. Hyderabad, Andhra Pradesh

Srivastava P, Chaudhary P, Mishra A, Dasila H, Maithani D (2024) Influence of microbial inoculants on soil health. In: Microbial inoculants: applications for sustainable agriculture. Springer Nature Singapore, Singapore, pp 29–48

Subedi B, Poudel A, Aryal S (2023) The impact of climate change on insect pest biology and ecology: implications for pest management strategies, crop production, and food security. J Agric Food Res 14:100733

Thakur AK, Rath S, Roychowdhury S, Uphoff N (2010a) Comparative performance of rice with system of rice intensification (SRI) and conventional management using different plant spacings. J Agron Crop Sci 196:146–159

Thakur P, Kumar S, Malik JA, Berger JD, Nayyar H (2010b) Cold stress effects on reproductive development in grain crops: an overview. Environ Exp Bot 67:429–443

Tian H, Xu R, Canadell JG, Thompson RL, Winiwarter W, Suntharalingam P, Davidson EA, Ciais P, Jackson RB, Janssens-Maenhout G (2020) A comprehensive quantification of global nitrous oxide sources and sinks. Nature 586:248–256

Tyagi J, Chaudhary P, Jyotsana, Bhagwati U, Bhandari G, Chaudhary A (2022a) Chapter 17 Impact of endophytic fungi in biotic stress management. In: Soni R, Suyal D, Goel R (eds) Plant protection: from chemicals to biologicals. De Gruyter, Berlin, Boston, pp 447–462. https://doi.org/10.1515/9783110771558-017

Tyagi J, Chaudhary P, Mishra A, Khatwani M, Dey S, Varma A (2022b) Role of endophytes in abiotic stress tolerance: with special emphasis on *Serendipita indica*. Int J Environ Res 16:62. https://doi.org/10.1007/s41742-022-00439-0

Verma KK, Song X-P, Degu HD, Guo D-J, Joshi A, Huang H-R, Xu L, Singh M, Huang D-L, Rajput VD (2023) Recent advances in nitrogen and nano-nitrogen fertilizers for sustainable crop production: a mini-review. Chem Biol Technol Agric 10:111

Wijitkosum S (2022) Biochar derived from agricultural wastes and wood residues for sustainable agricultural and environmental applications. Int Soil Water Conserv Res 10:335–341

Zhang X, DAvidson EA, Mauzerall DL, Searchinger TD, Dumas P, Shen Y (2015) Managing nitrogen for sustainable development. Nature 528:51–59

The Role of Biochar in Improving Ecosystem Services and Promoting Sustainable Practices

Devesh Vishwakarma, Shubham Abhishek, Prasenjeet Chakraborty, Dikchha Singh, Anand Mishra, Prabhat K. Chauhan, and Sohail Ansari

Abstract Biochar (BC) is a carbon-dense substance produced via the pyrolysis process of biomass. It has since been touted for its significant capacity to detox soils and pollutants while increasing ecosystem services. This chapter pulls together the evidence that BC is effective at enhancing soil microbial biomass, retention of moisture, reduction of emissions of the greenhouse gas, and enhancement of crop productivity. The environmental benefits of BC are variable and context-dependent such as feedstock to be used, pyrolysis process, temperature, application rate, soil type and climatic conditions. Crucially, the biggest gains in yield and soil health have been on already disturbed soils or those lacking key nutrients. This chapter investigates the further uses of BC beyond soil quality amelioration, its inclusion in wastewater treatment plants, heavy metal adsorption processes, and renewable energy storage systems. While this chapter seeks to demonstrate the promise of BC, it does so through a discerning lens, examining the constraints and ecological implications of widespread BC implementation. Collectively, this offers a comprehensive, in-depth view of BC's potential to further dual aims of more productive, sustainable agriculture and environmental protection, and recognition and debate of significant limitations that deserve focused research.

D. Vishwakarma
Department of Environmental Science, Babasaheb Bhimrao Ambedkar University (A Central University), Lucknow, India

S. Abhishek (✉) · P. Chakraborty · D. Singh
CSIR-Central Institute of Mining and Fuel Research, Dhanbad, India
e-mail: Shubham.abhishek888@gmail.com

Academy of Scientific and Innovative Research (AcSIR), Ghaziabad, India

A. Mishra
RR Group of Institution, Lucknow, India

P. K. Chauhan
Department of Environmental Science, AKS University, Satna, Madhya Pradesh, India

S. Ansari
Texas A&M University, College Station, TX, USA

A. Chauhan et al. (eds.), *Biochar As Remediation Tools for Pollutant Removal*,
https://doi.org/10.1007/978-3-032-16746-0_9

Keywords Biochar · Carbon capture · Climate change · Soil reclamation · Sustainability

1 Introduction

To effectively tackle global warming and climate change, it is crucial to minimize greenhouse gas emissions, while also removing carbon dioxide from the atmosphere. This dual approach is necessary to reach the net zero emissions target outlined in the Paris Agreement. The amendment of biochar (BC)in soil aimed at decreasing greenhouse gas emissions and facilitating carbon dioxide removal was initially suggested as a worldwide approach for addressing climate change just 15 years ago, and has been the focus of extensive research in the last ten years. The BC is created through the anoxic thermochemical conversion of biomass via pyrolysis processes, which yield recoverable heat and fuels, including gases and condensable volatiles, in addition to the solid BC. This material is notable for its environmental persistence and is characterized by high carbon content along with low levels of oxygen and hydrogen.

Pyrolysis is a crucial process that takes place when heat is applied to a material without the presence of oxygen. Because of this, BC is a very stable solid that is rich in carbon (Blanco-Canqui 2019). Feedstocks which serve as an input of BC production are urban garbage, manure form livestock and leftover from farming. When these materials converted in to BC, primarily consist of carbon, hydrogen, and oxygen in addition with key macronutrients such as nitrogen, phosphorus, and potassium. The BC are comprised of stable bio-aromatic structures with high positive surface area (Franco-Uría et al. 2022). The generation of BC and its properties largely depend up on its physical and chemical characteristics on feedstock type and the process of pyrolysis (Kannaujiya et al. 2024; Liao et al. 2023). The climate related benefits of BC come from its long-term stabilization of carbon and reduces the greenhouse gas emission, reclamation of contaminated land and water contamination through adsorption of organic contaminants (Ayaz et al. 2021; Chauhan et al. 2025). When BC is added into the soil, it increases the soil's structural integrity, nutrient and water holding capacity, stimulates rate of microbial activity, and enhances plant nutrient uptake. Due to this recalcitrant nature, BC has an extensive degradation timeline in soils, taking years to decompose. Beyond improving soil health, BC promotes biodiversity and resilience of our ecosystem. By creating ideal conditions for soil regeneration, it supports the growth of native plant species and rebuilds healthy habitats for wildlife, making it especially priceless for ecological restoration projects of all types. For example, in truncated climate repair in Spain, the ReactiveBrañosera project implements BC to restore forest soil biodiversity, enhance carbon sequestration and boost resilience through interventions such as installing insect hotels and beehives. BC has been successfully added to the soil to improve soil fertility, highly stimulated soil plant growth, and even improved good ecosystem service.

Numerous individuals have highlighted the benefits of BC regarding enhanced productivity, reduced costs, and the transformation of policy landscapes. Several challenges continue to hinder their broad implementation. The significant challenges include establishing uniform, standardized methods for sustainable production while also understanding the economic impacts of large-scale BC production. Furthermore, to guarantee compliance with environmental standards and prevent unintentional negative impacts, a clearly established regulatory framework concerning the appropriateness of BC production and application methods is essential (Yuan et al. 2023). Further research is essential to identify optimal methods for producing BC and to evaluate its effects within intricate ecosystems in various regions. These characteristics that bestow soil with a critical function in carbon sequestration, advancing soil health, bolstering water quality, and fostering biodiversity make it a powerful tool in combating climate change and managing land sustainably (Moreno et al. 2022) (Fig. 1).

The BC has the possibility to play an ever more significant part in fostering resilient ecosystems. Based on these, this chapter aims to enhance understanding of the potential advantages of BC to enhance ecosystem services, while also addressing the challenges that come with its implementation. Consequently, it also deals with different aspects of BC like in waste management and soil restoration. The objectives of this chapter are (i)Evaluation BC's role in enhancing ecosystem services, (ii) assessment of BC applications beyond soil amendment (iii) to examine policy, market, and socioeconomic framework for BC adaptation.

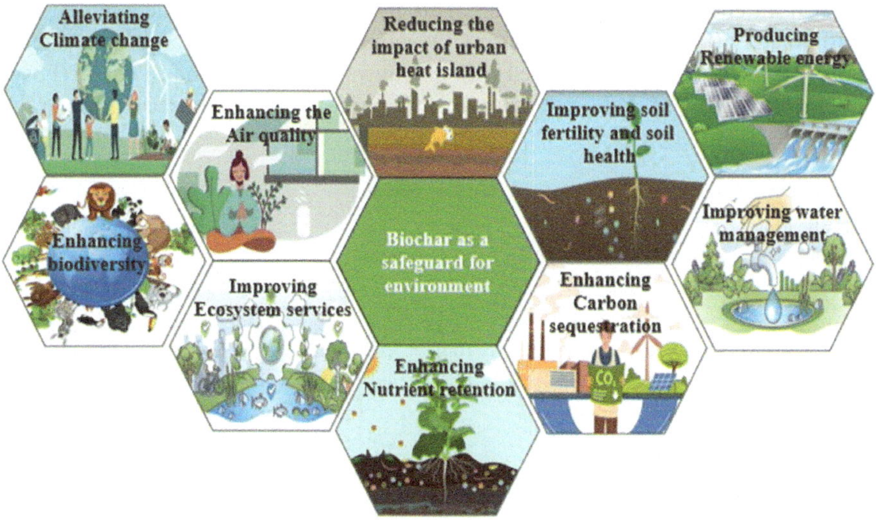

Fig. 1 Illustrates the benefits of utilizing biochar across various dimensions

2 Biochar Characterization

BCs are biologically derived materials that have undergone thermal modification, exhibiting a variety of properties (Table 1). BCs are widely recognized as mineral-rich materials, noted for their high pH and surface area. They function as lasting carbon storage systems that can remain effective for hundreds of years. Nonetheless, certain research indicates that particular varieties of BCs exhibit comparatively low surface area and porosity, as well as different types of functional groups. The properties of BCs indicate a high nutrient content. The unique characterization of BCs is significantly shaped by the conditions of pyrolysis and the nature of the feedstock employed. BCs, which contain high organic materials, are frequently employed as supports for catalysts, as catalysts themselves, and as adsorbents, as highlighted by Woolf et al. (2010).

The application of BC is still in its infancy, and there remains considerable opportunity to improve its efficacy. The BC has shown its advantages across multiple uses, including the removal of pollutants, C- capture, and catalysis. An extensive commercial use of BC and its derived functional materials hold significant promise for lowering CO_2 emissions by converting waste biomass into valuable BC. The synthesis of BC avoids burning of agri-residue, soil nutrient depletion, loss of soil biodiversity, and release of soot and smoke restoring soil. Instead, the carbon dioxide that plants take in during photosynthesis can be captured by BC, effectively stopping it from entering the environment. This method results in the accumulation of approximately 0.1–0.3 billion tons of CO_2 annually on the surface of BC, significantly tackling the issues associated with global warming and climate change (Foster and Neufeldt 2014). Likewise, a significant amount of waste biomass is generated

Table 1 Different forms of biochars and its characteristics

Classification of biochar	pH	Surface area m^2/g	(In %)					References
			O	**S**	**C**	**N**	**H**	
Bamboo Sugarcane bagasse Peanut shell	7.9 7.5 6.9	470.4 557.4 27.1	16.54 19.81 11.26	– - -	80.89 76.45 86.39	0.15 0.79 0.94	2.43 2.93 1.36	Zhouet al. (2013)
Bamboo Rice straw	9.5 10	907.4 10	– –	– –	86 50.8	86 1.66	1.49 1.72	Lu et al. (2014)
Oak wood	9.4	214	–	–	90	0.3	–	Mukherjee et al. (2014)
Miscanthus straw	10.24	–	–	–	53.5	0.31	–	Houbenet al. (2013)
Olivepomace700C	10.14	10.14	–	–	72.5	–	10.1	Pellera and Gidarakos (2015)
Wastewater sludge	8.2	–	–	–	–	2.3	–	Hossainet al. (2010)

worldwide each day such as sewage sludge, agricultural crop residues, forest waste, food waste, industrial organic waste, and domestic solid waste. This can be utilized in BC production leading to the transformation of these materials into valuable products. Compositing and pyrolysis are different ways of BC synthesis boosting organic matter, pH, and nutrients making soil fertile (Yuniwati et al. 2023). The BC remove soil contaminants like heavy metals, polyaromatic hydrocarbons (PAHs), oils, etc. along with enhancing water holding capacity, porosity, and ion exchange capacity. The BC also contributes in waste management, i.e., converting waste into energy as well as landfill loads (Kumar et al. 2023). Besides, BC supports wastewater treatment by adsorbing the pollutants from wastewater further recovering pollutant through chemical regeneration. Hence, BC extracts and recovers many valuable metals from industrial effluents and mitigates wastewater too. The biomass obtained by treating wastewater through BC can be treated in wasteland which enhances soil fertility.

3 The Function of Biochar in Promoting Ecosystem Services

The connections between ecosystems and human societies are defined by ecosystem services. In a general context, these refer to the advantages that communities derive from natural ecosystems (Millennium Ecosystem Assessment 2005). In more detail, services arise from both living and non-living elements within ecosystems, shaped by the conditions and processes that occur within these systems. They play a crucial role in enhancing human well-being through various means: Resources can be utilized directly, such as in the instances of water or food. Experiences can be derived from the natural world, as seen in the breathtaking beauty or the profound sense of wonder that a waterfall, a lush mountain, or a monarch butterfly can evoke within us. Services play a crucial role in maintaining essential environmental conditions for human existence, including the stabilization of climate patterns and safeguarding against severe events. The components and processes of ecosystems that directly enhance human well-being have been characterized as ecosystem services (Shekhar et al. 2024; Luck et al. 2009). By highlighting the direct nature of these connections, an attempt is being made to differentiate ecosystem processes from ecosystem services. BC, a substance created from organic waste via pyrolysis, presents considerable opportunities for boosting ecosystem functions by enhancing soil quality, addressing climate change, and promoting sustainable farming methods (Table 2).

3.1 Effect of Biochar on Enhancing Soil Characteristics

The inclusion of BC as an amendment affects multiple soil properties that dictate functionality such as water holding capacity, aggregation, permeability, and swelling/

Table 2 Illustrates the role of BC in enhancing various ecosystem services across multiple categories

Ecosystem services	Contribution of BC
Provisioning services	Enhances soil fertility for increased agricultural productivity ● BC can serve as a raw material for various products
Regulating services	● C-sequestration to mitigate climate change ● Water purification by adsorbing pollutants ● Minimizing greenhouse gas effect in soil
Cultural services	Supports environmental education and awareness ● May improve landscapes and contribute to a sense of environmental stewardship ● Aesthetic and spiritual value
Supporting services	Improves soil structure, nutrient cycling, and microbial activity ● Helps in controlling soil erosion and enhancing biodiversity

shrinking ability. The BC introduces meaningful changes to the physical soil characteristics, resulting in an increased total surface area and lower soil bulk density. This, in turn, creates a better soil structure and aeration dynamics (Abhishek et al. 2024). A significant change in the chemical characteristics like soil pH was observed with the application of BC composite. Yang et al. (2019) determine that BC contributes to an increase in soil pH, with the effects being significantly more evident when finely textured rice straw BC is utilized compared to coarser. The BC produced at lower pyrolysis temperatures, like 300–400 °C, are typically more acidic in nature, whereas BC produced at high pyrolysis temperatures, around 700 °C, are often found to be alkaline. The BC obtained from rice husk was found to minimize the acidic properties of tea garden soil. By enhancing the pH and cation exchange capacity which also control the nutrient losses through leaching and from volatization process (Rafiq et al. 2016). The application of BC enhances the electrical conductivity in soil. The BC which obtained from higher pyrolysis temperature especially from wood and paper waste enhanced the electrical conductivity of soil. The application of BC increases the pH of soil and makes it more alkaline. The addition of 10% (w/w) cassava stalk BC to the soil, improved its electrical conductivity (Prapagdee & Tawinteung 2017). Similarly, the addition of BC @30 and 50 tons per hectare rice straw BC of silver grass raised the electrical conductivity of soil by 38.5% (El-Naggar et al. 2018). BC has the ability to enhance cation exchange capacity in low fertility soil. Ghorbani et al. (2019) investigated that the rice husk BC of three different weights (0, 1, and 3%) affected two categories of soil: loamy sand and clay. The result reveled that BC at 1% and 3% increased the cation exchange capacity of sandy soil by 20% and 30% respectively.

The application of BC increases the organic matter content in soils (Pathak et al. 2024). The organic content in BC depends upon the types of feedstocks used and pyrolysis temperature (El-Naggar et al. 2018). As the woody BC contains high lignin exhibits greater stability in soil. Additionally, the variables that affect BC stability include the pyrolysis temperature, the aliphatic to aromatic carbon fraction and the

Fig. 2 Effect of biochar on soil health (Modified from Alkharabsheh et al. 2021)

condensation of aromatic carbon. The organic carbon increased from 3.6 g/kg to 4.8 g/kg when BC applied @ 5 tons per hectare. Similarly, Oladele (2019 found that addition of rice husk BC @12 tons per hectare enhanced the organic carbon content by 26 gm/kg. The organic carbon content was significantly increased by 0.17, 0.11, and 0.17%, respectively, by adding BC and compost and a mixture of BC and compost @ 10.9 tons/ha (D'Hose et al. 2020) (Fig. 2).

During critical growth phases, particularly in flowering and grain filling, soils treated with BC show higher levels of nitrogen, phosphorus, and potassium compared to conventional fertilizer methods. Multiple strategies have been proposed to minimize nutrient loss and enhance soil retention in areas treated with BC. The integration of BC at 12 tons per hectare and nitrogen fertilizer at 90 kg per hectare in rhizospheric zone resulted in a significant enhancement of soil NPK levels (Oladele 2019). Different amounts of BC (0, 2.5, 5, 7.5, and 10 t ha-1) were used in conjunction with synthetic fertilizers (75% and 100% NPK) on the soil. The results indicated that applying 100% NPK with 10 t ha-1 of BC increased the soil's available nitrogen to 243 kg ha-1(Elangovan & Sekaran 2017). Furthermore, Liao et al. (2020) discovered that the application of nitrogen fertilizer (15N labeled fertilizer) in conjunction with BC postponed the emission of nitrogen from the processed soils. Apart from this, BC mitigates heavy metals and metalloids in the rhizospheric soil stimulating microbes- root exudates interaction and influence low molecular weight organic acids (Fan et al. 2023). This organic acid regulates soil enzymatic activity (Abhishek et al. 2025). For instance, rhizospheric addition of BC in sugarcane enrich soil with microbial communities influencing healthy nitrogen cycling, soil sustainability and sugarcane productivity. A meta-analysis reported BC as soil remediation technology from pesticides by rhizospheric dissipation (Song et al. 2023). BC improves soil microbial diversity and enzymatic activity in rhizosphere of tobacco plant. BC provides valuable resources for successful plant growth restoring soil from pollutants. One such major trending pollutant is microplastic which in agricultural soil provides adverse effects to crop yields. Therefore, BC provides key genes for major metabolisms such as lignin synthesis, energy metabolism, and antioxidative activities to surpass microplastic toxic effects in roots. For example, peanut rhizosphere contaminated by microplastic prevents nutrient availability and microbial diversity, provides oxidative stress is restored by the application of BC for successful root metabolism and growth (Yang et al. 2024). The nano-BC application alleviates abiotic stresses and adsorb rhizospheric pollutant enhancing plant- soil dynamics.

The BC serves as a powerful tool for enhancing the vitality of living organisms and fostering a healthier natural environment, while also creating a supportive habitat for a variety of soil microorganisms (Abujabhah et al. 2016). The extensive surface area and enhanced hydrophobic properties of BC foster microbial activity and biomass, resulting in changes to the composition and abundance of microbial communities. Shifts in microbial community structure can happen due to changes in soil environment and resource availability like carbon content, nutrients and moisture content in soil. Changes in abiotic factors, like toxins introduced into the environment or other habitat alterations may play a role in these transitions. The application of BC creates a habitat for soil microorganisms thereby protecting them from natural enemies (Yang et al. 2019). Masto et al. (2013) found that when BC derived from Eichhornia biomass was applied at 20 kg/ha, significantly increases the activity of enzymes were observed. It resulted in a 21% increase in dehydrogenase activity, as well as increases of 32% and 22.8% in acid phosphatase and alkaline phosphatase activities, respectively.

In general, the use of BC makes mycorrhizal fungi more active in a number of ways, including (i) modifying the chemical and physical characteristics of the soil, (ii) indirectly affecting mycorrhizae by changing the microbial environment around them, (iii) changing how plants and fungi talk to each other and breaking down allelochemicals on BC, and (iv) protecting against fungal grazers. Because BC has a high surface area-to-volume ratio, it makes the soil a good place for mycorrhizal fungi and other helpful microbes to live. This improves the quality of the soil. This can be reported by combined application of BC and mycorrhizal fungi in increased production of spinach (*Spinacia oleracea*) and maize (Zea mays) improving growth, nutrient, microbiome, and soil enzymatic activities (Jabborova et al. 2021; Sun et al. 2022).

3.2 *The Function of Biochar in Carbon Sequestration and Climate Change Mitigation*

The practice of collecting carbon in soil is a very important way to lessen the consequences of climate change caused by people. This method captures and stores carbon so that it doesn't get into the air. Moving C to a secure or inactive secondary carbon storage site is a viable solution. Oo et al. 2018 discussed about how important BC is for fighting global warming by storing C in soil and reducing reliance on fossil fuels.

BC has the potential to lower CO_2 and CH_4 emissions, derived from anthropogenic activities. When mixed with soil, it stays there for a long time. A small change in how carbon moves through the atmosphere and plants, such a reduced BC process, can have a big effect on how quickly CO_2 is removed from the atmosphere (Zhao et al. 2022). Plants take in more than eight times as much carbon dioxide through photosynthesis each year as humans do when they do things that release greenhouse gases. The BC exhibits greater physical and chemical stability compared to its initial carbon state because of the way its molecules are arranged and where they came from. Also, the carbon that is captured makes it hard to release as CO_2, which shows that this method could be a good way to store carbon.

According to Zheng et al. (2020), it is predicted that BC can sequester 2.2 giga tonnes of CO_2 per year by 2050. Despite pyrolysis being a thermochemically attractive process to convert biomass into biooil, syngas, etc., which can be used as fuels, the process releases CO_2 back into the atmosphere. However, the pyrolysis process, the resulting products like biooil and syngas serve as fuels, consequently emitting CO2 into the atmosphere. Plants absorb CO_2 and transform it into biomass, thus finalizing a cyclical process. BC presents a decrease annual CO_2 emission of around 12% of the current human-caused CO_2 emissions. Currently, there remains insufficient evidence to draw conclusions regarding the effects of BC production and usage on the greenhouse gas of the overall system. Nonetheless, it is logical to assert that the influence of BC on addressing global warming cannot be perceived as a straightforward cause-and-effect dynamic, and that its advantages may become clearer as time progresses (Fig. 3).

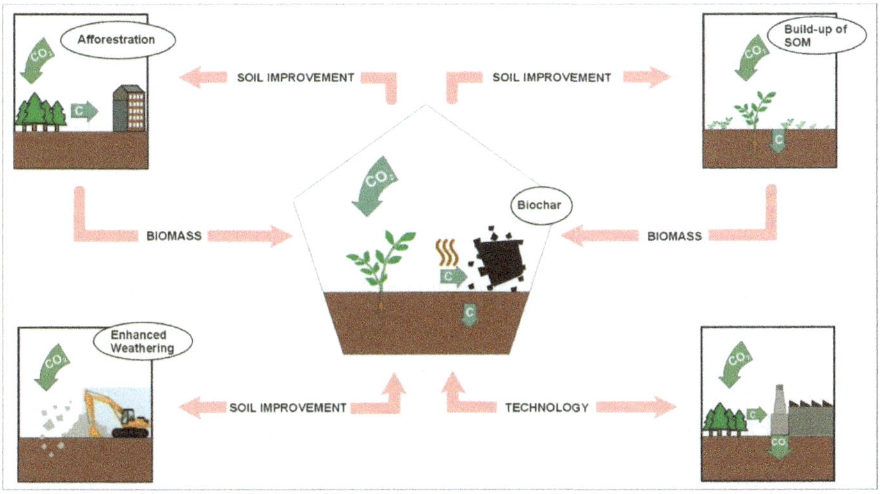

Fig. 3 The utilization of biochar to enhance the carbon sequestration

Consequently, BC functions as both a carbon source and a carbon sink within the soil ecosystem. Additionally, the implementation of BC was conducted to evaluate the status of soil nutrient, carbon sequestration, and increased in crop production (Salam et al. 2022). The creation of maize stover and rice husk BCs was accomplished using the process of slow pyrolysis. The rice husk BC was produced at a temperature of 650 degrees Celsius, while the corn stover BC was produced at 550 degrees Celsius. This nutrient-loaded rice husk BC and corn stover BC increased nutrient release potential and adsorption capacity (Bushra & Remya 2024).

The unpyrolyzed rice husks and maize stover used as a soil amendment in controlled incubation environment for 107 days to analyze the effect of temperature on BC. Implementation of BC to soils lowered CO_2emissions and increased the concentration of organic carbon and cation exchange capacity. The nutrient level in soil increased after the BC was added. This enhances the physiological characteristics of eggplant. The mixing of BC and compost together makes the composting process go faster compared to compost alone (Teodoro et al. 2020).This combination made it easier for plants to keep nutrients available and slow down the movement of metals in the soil. Similarly, Murtaza et al. (2021) found that addition of BC had favorable effects on the soil, such as reduce its bulk density, enhance water holding capacity also increased moisture content in soil. Joseph et al. (2021) said that the pyrolysis process and the biomass that is heated mostly determine the properties of BC and how it affects agricultural ecosystems.

The meta-analysis results showed that the increases of phosphorus and soil organic carbon were up to 4.6 times and 17–39%, respectively, and the application of BC reduced the non-CO2 greenhouse gas production in soils that are sandy and acidic with a variation of 12–50%. Asadi et al. (2021) shown that rice husk BC may significantly diminish greenhouse gas emissions and nitrate leaching from paddy soil, possibly attributable to the BC's distinct features, influenced by the pyrolysis heating rate and residence period. Allohverdi et al. (2021) identified numerous critical difficulties in agricultural operations, including enhancing water retention and soil microbiology, as well as increasing fertilizer efficiency. Also, BC is a great source of nutrients from biomass. It is becoming known as a useful soil amendment that can improve soil health, make crops more productive, and trap carbon emissions.

Simiele et al. (2022) found a twofold increase in total root length, root surface area, and the biomass of roots, stems, and leaves for a 20% BC application relative to untreated controls. Hussain et al. (2022) discovered that the incorporation of green coconut-modified BC into mustard plants resulted in the highest biomass yield, surpassing that of the control treatment.

3.3 Biochar in Water Management and Conservation

The BC's wastewater treatment technology has gotten a lot of attention, especially for how well it can get rid of pollutants. Pollution of our natural environment, especially of our water with a lot of harmful substances, is a major global issue (Chakraborty et al. 2025; Gray et al. 2014). It is very hard to treat water that has low levels of micropollutants, such as drugs, endocrine disruptors, surfactants, dyes, and heavy metals. However, photocatalysis, ozonation, and reverse osmosis are not found as a cost-effective technique (Vishwakarma et al. 2024a, b). BC is a less expensive and sustainable approach to clean wastewater (Kameyama et al. 2019). The BC contains oxygenated functional groups and an aromatic carbon structure on its surface. These properties help BC to interact with pollutants in a number of ways, such as H-bonding, electrostatic interactions and π–π interactions. The BC can adsorb paracetamol, ibuprofen, and naproxen which are common medication like paracetamol, ibuprofen, and naproxen used to ease pain and swelling. Similarly, BC which obtained from glucose or pomelo peels are the effective one for remediation of pharmaceutical pollutants (Liu et al. 2017). The BC also perform well at getting rid of stimulants like caffeine. Electrostatic and nucleophilic interactions an acidic pH for caffeine to adhere to steam-activated BC (Petersen et al. 2016). The BC also help to remove organic dyes like Congo red, malachite green and methylene blue. The negative charge in BC, which makes it clear for molecules with cationic dye. The BC is when altered help it to remove heavy metals (Atkinson 2018). The BC not only simply clean up wastewater; it could also keep an eye on water in real time and get rid of tiny impurities (Vishwakarma et al. 2024a, b). The BC could make it better for treating water in real time. The BC is a cheap, long-lasting, and very effective adsorbent; thus, it would be an excellent candidate for treating wastewater on a big scale. It will work better and be able to be used in eco-friendly water filtration systems (Wen et al. 2021) if further research is done on how BC adsorbs and how to make it better.

3.4 Biochar on Waste Management and Land Restoration

The UNCCD supports BC as a way to minimize soil degradation, boost agricultural productivity, and lessen the effects of global warming and climate change. BC can help to control land degradation in two different ways: (1) By utilizing BC to enhance agricultural sustainability and productivity while minimizing detrimental pollution. This would minimize the necessity for additional land conversion. (2). Using BC to restore land that has been damaged or is naturally poor is helping to ease the stress that comes with opening new regions.

A vital inquiry is, 'how can areas that have experienced degradation or desertification produce sufficient biomass to effectively support land rehabilitation when there is minimal plant cover available for feedstock?' A promising strategy might include the cultivation of robust plant species in regions with low biodiversity that are not presently under agricultural use, potentially employing water deemed unsuitable for food production for irrigation purposes. Alternative options encompass algae, grasses, reeds, or aquatic weeds found in lagoons that utilize contaminated water and effluents (Amoah-Antwi et al. 2020).

Waste generated from urban areas and agricultural activities could potentially be utilized, provided that transportation costs are manageable. Strezov et al. (2008) propose that non-irrigated elephant grass cultivated on contaminated land has the potential to generate biofuel and BC, even in conditions where growth rates fall below forty tons of dry biomass per hectareannually. Land impacted by salts can serve as a source for biomass suitable for BC production as they provide diversity of halophytes (Wang et al. 2023). The method of eliminating pollutants through this biomass has the potential to restore the land and simultaneously provide resources to improve soils in different regions. This topic has been investigated in Australia, as noted by Bartle et al. (2007) examine regions characterized by arid, saline landscapes that yield coppiced eucalyptus feedstock, alongside areas in Sumatra involved in forestry and paper pulp production.

Urban brownfield sites have the potential to be revitalized using BC, which can facilitate amenity planting or the production of biofuels and timber, sequestering carbon. Associated with waste management, BC may hold significant promise. Contemporary urban waste in developed nations, especially in developing ones, is often tainted with hazardous substances, which can hinder its potential for quality composting. Nevertheless, certain research indicates that it may be possible to generate uncontaminated BC from such refuse (Voruganti 2023).

4 Socioeconomic Impacts and Sustainability of Biochar Use

The economic feasibility of BC production systems can be assessed through a cost–benefit analysis, which involves examining all cash flows related to the development and operation of a system over a specified timeframe (e.g., system lifetime). Often, these cash flows are adjusted to their equivalent "present" value, commonly referred to as net-present worth analysis (Sparrevik et al. 2014). Common elements associated with the costs and benefits of BC production systems encompass initial capital expenditures (such as system construction, land acquisition, and interest payments), ongoing operational and maintenance expenses (including staff salaries and training, equipment upkeep, transportation, biomass procurement, and BC application), revenue from product sales (like BC, heat, electricity, etc.), carbon taxation, financial subsidies, and more (You et al. 2022). There is a limited understanding of the harmful impacts associated with the addition of BC to the soil. Mukherjee and Lal (2014) examined the possible adverse impacts of BC amendment on crop yield, soil quality, and related financial risks, emphasizing that the synthesis process may lead to BC surfaces that contain heavy metals and organic compounds available to plants. The presence of harmful substances such as polycyclic aromatic hydrocarbons (PAHs), xylenols, cresols, acrolein, formaldehyde, and various toxic carbonyl compounds is well-documented, particularly regarding their bactericidal and fungicidal properties. The unstable compounds present in BC could function as rapid carbonizers, potentially introducing toxicity hazards to vegetation. Nonetheless, allowing the BC to soak for a sufficient duration can aid in the elimination of these compounds. BC derived from sewage sludge and tannery waste generally exhibits elevated levels of heavy metals, including copper, chromium, and zinc. Keiluweit et al. (2012) detected eleven non-substituted 3–5-ring PAHs, along with alkylated anthracene and phenanthrene in wood and grass BCs produced at 100–700 °C. Due to its high sorption capacity, BC can readily mitigate the leaching of PAHs from the soil. These free radicals on BC surfaces can inhibit seedling germination and growth, namely for rice, wheat and corn crops (Liao et al. 2014). Alkaline BCs can restrict access to certain soil nutrients and harm soil cation exchange capacity (Zhang et al. 2019). Lee et al. (2010) determined that pH was significant in influencing cation exchange capacity, which varied from 10—30 cmol kg^{-1} as pH increased from 5.0 to 8.5 BC with high pH was associated with high electrical conductivity which can limit the availability of important plant nutrients. Herath et al. (2013) noted that at high BC rates, micropores can become blocked by mineral or ash content and thus could have an adverse effect on soil water retention.

5 Strategies and Economic Tools for Encouraging Implementation of Biochar Technologies

Advancing BC technology for waste management and energy production necessitates robust legislative structures and effective market mechanisms. The researcher examines various studies to assess the crucial commercial and policy approaches that facilitate the implementation of BC technologies across local, national, and global scales.

5.1 Policy Frameworks for Implementing Biochar Practices

To create the stage on which BC technology could then be adopted again requires policies of research, development and implementation Lehmann et al. (2008) observed that these encompass a number of policies, ranging from subsidies and regulations to institutional biosupport. Establishing safety and quality benchmarks for the creation and application of BC is essential for safeguarding the environment and supporting communities. The tax credits, grants, and subsidies can encourage people to invest in biochar.

5.2 Strategies for Implementing Biochar in the Market

To make BC popular and getting it used outside of the lab, market forces will play a big role. The carbon trading, offset programs, and certifications can give BC stakeholders a financial motive to capture carbon, improve soil health, and cut emissions. Carbon trading platforms enable the sale of carbon credits by those who produce BC, derived from the process of sequestering carbon in soils. This initiative empowers individuals to take stewardship of their land. Individuals and companies have the opportunity to engage in offset programs that allow them to compensate for their C emissions by investing in BC initiatives.

This approach effectively promotes BC products while contributing to the preservation of our planet. Accreditation programs like the BC Certification Program enhances supply chains' and production's legitimacy, openness, and accountability of BC, thereby strengthening customer trust and promoting the market acceptance of BC products.

5.3 Holistic Strategies that Combine Policy and Market Mechanisms

Aligning regulatory measures with market incentives using integrated policy-market approaches can help maximize BC efficacy, adoption strategies. Ambitious renewable energy objectives and the establishment of carbon pricing systems may foster a conducive policy framework for the use of BC as an environmentally friendly energy source and a negative emissions technology. Public procurement policies that prioritize eco-friendly bio-based products, like the incorporation of BC in public projects, have the potential to generate substantial demand and stimulate investment in cutting-edge technologies and production methods, spurring innovation and helping to build broader markets. Collaborations among governmental bodies, academic research entities, industry participants, and community organizations can enable sharing of knowledge, transfer technology, and implementing capacity-building initiatives to create a nurturing environment for the adoption of BC. Comprehensive legislative frameworks and efficient market mechanisms will be crucial in facilitating the implementation of BC technologies for sustainable energy production and effective waste management.

Clear guidelines for compliance, economic incentives, institutional support structures can all help create an environment favorable to BC research and development and BC deployment. Strategies that focus on supply and demand, driven by market dynamics, such as offset schemes, carbon trading, and certification programs, can effectively stimulate both demand and investment in BC products.

Holistic strategies that integrate regulatory frameworks with market-driven solutions show potential for enhancing the social and ecological advantages of BC implementation, aiding in the achievement of sustainable development objectives and efforts to combat climate change.

6 Challenges and Future Perspectives

A substantial amount of research has shown that BC can serve as a highly effective and advantageous soil amendment, enhancing fertility and facilitating carbon sequestration in biological cycles within soils. It's critical to view BC from all angles to make sure it's having the most beneficial and cost-efficient impact possible, especially in the areas outlined below: (1) Making the future of BC more beneficial by adopting practices with rapid adoption rates that improve soil fertility, increase nutrient availability to plants, and promote carbon sequestration. Such standardization may strongly increase the effectiveness and efficiency of BC implementation, reducing or eliminating any negative environmental impact. Consumption producing better quality BC will require more attention to various factors, including feedstock quality, pyrolysis temperature, and pre/post-treatment processes included in production. Further investigation will be required to define these interrelated parameters, as well as how they interact and perform in soil matrices. Creating a universally accepted prototype model for the creation of effective BC has been difficult, primarily because of the obstacles in customizing individual applications to varying soils, crops, and environmental settings. Creating a forecast for the decomposition duration of biomass carbon in various cropping systems presents a significant challenge. The deterioration of BC can be examined through two approaches: analyzing the stable phase of BC in soil, anticipated to endure for millennia, and contrasting its properties with the labile phase in soil, which is subject to quicker decomposition. Demonstrating the stability of biological components in the soil at the field level is crucial. Long-term field experiments are essential for this aspect. Investigating the relationships among plant roots, microbes, soil, and biological communities in the rhizosphere is a crucial field of research. This will clarify the nutrient cycle and release dynamics in soil settings augmented by BC.

It is essential to investigate the adsorption and desorption capacities of BC concerning soil nutrients to better understand the availability of these nutrients for plants and their gradual release within the BC-soil system. The adsorption and desorption capabilities may differ according on the exact kind of BC used. Therefore, various kinds of BC should be evaluated by applying them in diverse soil environments with distinct characteristics and different crop varieties.

7 Conclusion

The BC presents a promising option for carbon sequestration in soil, while also providing additional advantages to enhance ecosystem services. Additionally, sequestering carbon in the soil contributes to climate change mitigation and decreases the bioavailability of harmful substances, promoting a healthier ecosystem. The straightforward production process using locally sourced materials highlights its promise as a sustainable approach that tackles various challenges in meeting the

Sustainable Development Goals. The BC also showed great promise in eliminating contaminants from wastewater through adsorption and immobilization process, an environment friendly tool. The multifunctional role underscores the necessity of assessing BC in a comprehensive manner to guarantee optimal advantages regarding agricultural productivity, nutrient accessibility, and ecological sustainability. Uniform production and application methods can enhance its efficacy while reducing possible adverse effects. The BC demonstrates considerable potential as an eco-friendly solution for enhancing soil health and aiding in the fight against climate change.

Acknowledgements SA acknowledges the University Grants Commission (UGC), Government of India, for the JRF fellowship (210510623901) and the Academy of Scientific and Innovative Research (AcSIR), CSIR-HRDG, for doctoral study support. SA, PC and DS extend appreciation to the Director, CSIR- Central Institute of Mining and Fuel Research (CSIR-CIMFR), Dhanbad, India for encouragement and providing necessary facilities.

Author Contribution DV, PC and SA conceptualized this chapter and writing original draft; DS and PKC designed table and figures, AM and S.A critically edited the contribution to constructive recommendations before the submission. All the authors have no any conflict of interest, and the work is not related to animal ethical clearance. All authors read and approved the final manuscript.

References

Abhishek S, Ghosh A, Pandey B (2024) A comprehensive review on phytoremediation of fly ash and red mud: exploring environmental impacts and biotechnological innovations. Environ Sci Pollut Res, pp 1–28

Abhishek S, Ghosh A, Singh AA et al. (2025) Identification of metal-tolerant herbaceous species for phytostabilization and ecological restoration of fly ash dumpsites. Physiol Mol Biol Plants 31:1719–1738. https://doi.org/10.1007/s12298-025-01612-3

Abujabhah IS, Doyle R, Bound SA, Bowman JP (2016) The effect of biochar loading rates on soil fertility, soil biomass, potential nitrification, and soil community metabolic profiles in three different soils. J Soils Sediments 16:2211–2222

Alkharabsheh HM, Seleiman MF, Battaglia ML, Shami A, Jalal RS, Alhammad BA, Almutairi KF, Al-Saif AM (2021) Biochar and its broad impacts in soil quality and fertility, nutrient leaching and crop productivity: a review. Agronomy 11(5):993

Allohverdi T, Mohanty AK, Roy P, Misra M (2021) A review on current status of biochar uses in agriculture. Molecules 26(18):5584

Amoah-Antwi C, Kwiatkowska-Malina J, Thornton SF, Fenton O, Malina G, Szara E (2020) Restoration of soil quality using biochar and brown coal waste: a review. Sci Total Environ 722:137852

Asadi H, Ghorbani M, Rezaei-Rashti M, Abrishamkesh S, Amirahmadi E, Chengrong CHEN, Gorji M (2021) Application of rice husk biochar for achieving sustainable agriculture and environment. Rice Sci 28(4):325–343

Atkinson CJ (2018) How good is the evidence that soil-applied biochar improves water-holding capacity? Soil Use Manag 34(2):177–186

Ayaz M, Feizienė D, Tilvikienė V, Akhtar K, Stulpinaitė U, Iqbal R (2021) Biochar role in the sustainability of agriculture and environment. Sustainability 13(3):1330

Bartle J, Olsen G, Cooper D, Hobbs T (2007) Scale of biomass production from new woody crops for salinity control in dryland agriculture in Australia. Int J Glob Energy Issues 27(2):115–137

Blanco-Canqui H (2019) Biochar and water quality. J Environ Qual 48(1):2–15

Bushra B, Remya N (2024) Biochar from pyrolysis of rice husk biomass—Characteristics, modification and environmental application. Biomass Convers Biorefinery 14(5):5759–5770

Chakraborty P, Abhishek S, Sethi C, Jeevan G (2025) Condensate water as alternate resource for mitigating water stress: a case study. Water Conserv Sci Eng 10(1):4

Chauhan PK, Upadhyay SK, Chauhan A, Bhutiani R, Sikarwar RLS, Tiwari MK (2025) A review on simultaneous use of microbes and biochar for removal of toxic heavy metals: recent updates and future outlooks. Physiol Mol Biol Plants 31(3):1–15. https://doi.org/10.1007/s12298-025-01649-4

D'Hose T, Debode J, De Tender C, Ruysschaert G, Vandecasteele B (2020) Has compost with biochar applied during the process added value over biochar or compost for increasing soil quality in an arable cropping system? Appl Soil Ecol 156:103706

Elangovan R, Sekaran NC (2017) Biochar for soil health enhancement. Madras Agric J 104

El-Naggar A, Awad YM, Tang XY, Liu C, Niazi NK, Jien SH, Lee SS (2018) Biochar influences soil carbon pools and facilitates interactions with soil: a field investigation. Land Degrad Dev 29(7):2162–2171

Fan Y, Su J, Wang Z, Liu S, Li X, Hou C (2023) Improvement of the specific surface area of biochar by calcium-precipitated nanoparticles synthesized by microbial induction as a template skeleton: removal mechanism of tetracycline in water. J Environ Manage 348:119279

Foster K, Neufeldt H (2014) Biocarbon projects in agroforestry: lessons from the past for future development. Curr Opin Environ Sustain 6:148–154

Franco CR, Page-Dumroese DS, Archuleta J (2022) Forest management and biochar for continued ecosystem services. J Soil Water Conserv 77(4):60A-64A

Ghorbani M, Asadi H, Abrishamkesh S (2019) Effects of rice husk biochar on selected soil properties and nitrate leaching in loamy sand and clay soil. Int Soil Water Conserv Res 7(3):258–265

Gray M, Johnson MG, Dragila MI, Kleber M (2014) Water uptake in biochars: the roles of porosity and hydrophobicity. Biomass Bioenerg 61:196–205

Herath HMSK, Camps-Arbestain M, Hedley M (2013) Effect of biochar on soil physical properties in two contrasting soils: an Alfisol and an Andisol. Geoderma 209:188–197

Hossain MK, Strezov V, Chan KY, Nelson PF (2010) Agronomic properties of wastewater sludge biochar and bioavailability of metals in production of cherry tomato (*Lycopersicon esculentum*). Chemosphere 78(9):1167–1171

Houben D, Evrard L, Sonnet P (2013) Mobility, bioavailability and pH-dependent leaching of cadmium, zinc and lead in a contaminated soil amended with biochar. Chemosphere 92(11):1450–1457

Hussain T, Ahmed SR, Lahori AH, Mierzwa-Hersztek M, Vambol V, Khan AA, Rafique L, Wasia S, Shahid MF, Zengqiang Z (2022) In-situ stabilization of potentially toxic elements in two industrial polluted soils ameliorated with rock phosphate-modified biochars. Environ Pollut 309:119733

Jabborova D, Annapurna K, Paul S, Kumar S, Saad HA, Desouky S, Ibrahim MF, Elkelish A (2021) Beneficial features of biochar and arbuscular mycorrhiza for improving spinach plant growth, root morphological traits, physiological properties, and soil enzymatic activities. J Fungi 7(7):571

Joseph S, Cowie AL, Van Zwieten L, Bolan N, Budai A, Buss W, Cayuela ML, Graber ER, Ippolito JA, Kuzyakov Y, Luo Y (2021) How biochar works, and when it doesn't: a review of mechanisms controlling soil and plant responses to biochar. Gcb Bioenergy 13(11):1731–1764

Kameyama K, Miyamoto T, Iwata Y (2019) The preliminary study of water-retention related properties of biochar produced from various feedstock at different pyrolysis temperatures. Materials 12(11):1732

Kannaujiya D, Vishwakarma D, Shikha (2024) Preparation of low-cost biochar from discarded tea waste for aqueous phase removal of hexavalent chromium [Cr (VI)]: kinetics, isotherm, and recyclability analysis. Water Pract Technol 19(7):2850–2867

Keiluweit M, Kleber M, Sparrow MA, Simoneit BR, Prahl FG (2012) Solvent-extractable polycyclic aromatic hydrocarbons in biochar: influence of pyrolysis temperature and feedstock. Environ Sci Technol 46(17):9333–9341

Kumar A, Bhattacharya T, Shaikh WA, Roy A, Chakraborty S, Vithanage M, Biswas JK (2023) Multifaceted applications of biochar in environmental management: a bibliometric profile. Biochar 5(1):11

Lee JW, Kidder M, Evans BR, Paik S, Buchanan Iii AC, Garten CT, Brown RC (2010) Characterization of biochars produced from cornstovers for soil amendment. Environ Sci Technol 44(20):7970–7974

Lehmann J, Skjemstad J, Sohi S, Carter J, Barson M, Falloon P, Coleman K, Woodbury P, Krull E (2008) Australian climate–carbon cycle feedback reduced by soil black carbon. Nat Geosci 1(12):832–835

Liao S, Pan B, Li H, Zhang D, Xing B (2014) Detecting free radicals in biochars and determining their ability to inhibit the germination and growth of corn, wheat and rice seedlings. Environ Sci Technol 48(15):8581–8587

Liao W, Halim MA, Kayes I, Drake JA, Thomas SC (2023) Biochar benefits green infrastructure: global meta-analysis and synthesis. Environ Sci Technol 57(41):15475–15486

Liao J, Liu X, Hu A, Song H, Chen X, Zhang Z (2020) Effects of biochar-based controlled release nitrogen fertilizer on nitrogen-use efficiency of oilseed rape (*Brassica napus* L.). Sci Rep 10(1):11063

Liu Z, Dugan B, Masiello CA, Gonnermann HM (2017) Biochar particle size, shape, and porosity act together to influence soil water properties. PLoS ONE 12(6):e0179079

Lu K, Yang X, Shen J, Robinson B, Huang H, Liu D, Bolan N, Pei J, Wang H (2014) Effect of bamboo and rice straw biochars on the bioavailability of Cd, Cu, Pb and Zn to Sedum plumbizincicola. Agr Ecosyst Environ 191:124–132

Luck GW, Harrington R, Harrison PA, Kremen C, Berry PM, Bugter R, Zobel M (2009) Quantifying the contribution of organisms to the provision of ecosystem services. Bioscience 59(3):223–235

Assessment ME (2005) Ecosystems and human well-being: wetlands and water. World resources institute

Masto RE, Kumar S, Rout TK, Sarkar P, George J, Ram LC (2013) Biochar from water hyacinth (Eichornia crassipes) and its impact on soil biological activity. Catena 111:64–71

Moreno JL, Bastida F, Díaz-López M, Li Y, Zhou Y, López-Mondéjar R, Benavente-Ferraces I, Rojas R, Rey A, García-Gil JC, Plaza C (2022) Response of soil chemical properties, enzyme activities and microbial communities to biochar application and climate change in a Mediterranean agroecosystem. Geoderma 407:115536

Mukherjee A, Lal R (2014) The biochar dilemma. Soil Res 52(3):217–230

Mukherjee A, Lal R, Zimmerman AR (2014) Effects of biochar and other amendments on the physical properties and greenhouse gas emissions of an artificially degraded soil. Sci Total Environ 487:26–36

Murtaza G, Ahmed Z, Usman M, Tariq W, Ullah Z, Shareef M, Iqbal H, Waqas M, Tariq A, Wu Y, Zhang Z (2021) Biochar induced modifications in soil properties and its impacts on crop growth and production. J Plant Nutr 44(11):1677–1691

Oladele SO (2019) Changes in physicochemical properties and quality index of an Alfisol after three years of rice husk biochar amendment in rainfed rice–Maize cropping sequence. Geoderma 353:359–371

Oo AZ, Sudo S, Akiyama H, Win KT, Shibata A, Yamamoto A, Sano T, Hirono Y (2018) Effect of dolomite and biochar addition on N_2O and CO_2 emissions from acidic tea field soil. PLoS ONE 13(2):e0192235

Pathak HK, Chauhan PK, Seth CS, Dubey G, Upadhyay SK (2024) Mechanistic and future prospects in rhizospheric engineering for agricultural contaminants removal, soil health restoration, and management of climate change stress. Sci Total Environ, p 172116

Pellera FM, Gidarakos E (2015) Effect of dried olive pomace–derived biochar on the mobility of cadmium and nickel in soil. J Environ Chem Eng 3(2):1163–1176

Petersen CT, Hansen E, Larsen HH, Hansen LV, Ahrenfeldt J, Hauggaard-Nielsen H (2016) Pore-size distribution and compressibility of coarse sandy subsoil with added biochar. Eur J Soil Sci 67(6):726–736

Prapagdee S, Tawinteung N (2017) Effects of biochar on enhanced nutrient use efficiency of green bean, Vigna radiata L. Environ Sci Pollut Res 24:9460–9467

Rafiq MK, Bachmann RT, Rafiq MT, Shang Z, Joseph S, Long R (2016) Influence of pyrolysis temperature on physico-chemical properties of corn stover (*Zea mays* L.) biochar and feasibility for carbon capture and energy balance. PloS one 11(6):e0156894

Salam A, Afridi MS, Javed MA, Saleem A, Hafeez A, Khan AR, Zeeshan M, Ali B, Azhar W, Sumaira Ulhassan Z (2022) Nano-priming against abiotic stress: a way forward towards sustainable agriculture. Sustainability 14(22):14880

Shekhar S, Abhishek S, Pandey B, Singh S (2024) Restoring biodiversity and ecosystem services on post-industrial land: challenges and opportunities. Biodivers Ecosyst Serv Post-Industrial Land, pp 25–51

Simiele M, Argentino O, Baronti S, Scippa GS, Chiatante D, Terzaghi M, Montagnoli A (2022) Biochar enhances plant growth, fruit yield, and antioxidant content of cherry tomato (*Solanum lycopersicum* L.) in a soilless substrate. Agriculture 12(8):1135

Song B, Zhou C, Qin M, Zhao B, Sang F (2023) When biochar is involved in rhizosphere dissipation and plant absorption of pesticides: a meta-analysis. J Environ Manage 345:118518

Sparrevik M, Lindhjem H, Andria V, Fet AM, Cornelissen G (2014) Environmental and socioeconomic impacts of utilizing waste for biochar in rural areas in Indonesia—A systems perspective. Environ Sci Technol 48(9):4664–4671

Strezov V, Evans TJ, Hayman C (2008) Thermal conversion of elephant grass (Pennisetum Purpureum Schum) to bio-gas, bio-oil and charcoal. Bioresour Technol 99(17):8394–8399

Sun X, Thunuguntla R, Zhang H, Atiyeh H (2022) Biochar amended microbial conversion of C1 gases to ethanol and butanol: effects of biochar feedstock type and processing temperature. Biores Technol 360:127573

Teodoro M, Trakal L, Gallagher BN, Šimek P, Soudek P, Pohořelý M, Beesley L, Jačka L, Kovář M, Seyedsadr S, Mohan D (2020) Application of co-composted biochar significantly improved plant-growth relevant physical/chemical properties of a metal contaminated soil. Chemosphere 242:125255

Vishwakarma D, Kannaujiya D, Awasthi S, Shikha (2024) Phytoremediation of contaminated water, its mechanisms, and advancements. In: Phytoremediation: biological treatment of environmental pollution, pp 37–56. Cham: Springer Nature Switzerland

Vishwakarma D, Rawat S, Kannaujiya D, Shikha, Shukla B (2024) Adsorption of malachite green onto sodium alginate/bentonite (SA/Bnt) nanocomposite: optimization and modeling via RSM-DSD and insight kinetics and mechanism of adsorption. Arab J Geosci 17(5):152

Voruganti C (2023) Biochar applications in soil restoration: enhancing soil health and carbon sequestration. Environ Reports

Wang Y, Lin Q, Liu Z, Liu K, Wang X, Shang J (2023) Salt-affected marginal lands: a solution for biochar production. Biochar 5(1):21

Wen E, Yang X, Chen H, Shaheen SM, Sarkar B, Xu S, Song H, Liang Y, Rinklebe J, Hou D, Li Y (2021) Iron-modified biochar and water management regime-induced changes in plant growth, enzyme activities, and phytoavailability of arsenic, cadmium and lead in a paddy soil. J Hazard Mater 407:124344

Woolf D, Amonette JE, Street-Perrott FA, Lehmann J, Joseph S (2010) Sustainable biochar to mitigate global climate change. Nat Commun 1(1):56

Yang F, Zhang S, Sun Y, Du Q, Song J, Tsang DC (2019) A novel electrochemical modification combined with one-step pyrolysis for preparation of sustainable thorn-like iron-based biochar composites. Biores Technol 274:379–385

Yang C, Dou S, Guo D, Zhao H (2024) The application of biochar enhances soil organic carbon and rice yields. Agronomy 14(3):455

You S, Li W, Zhang W, Lim H, Kua HW, Park YK, Igalavithana AD, Ok YS (2022) Energy, economic, and environmental impacts of sustainable biochar systems in rural China. Crit Rev Environ Sci Technol 52(7):1063–1091

Yuan Y, Liu Q, Zheng H, Li M, Liu Y, Wang X, Peng Y, Luo X, Li F, Li X, Xing B (2023) Biochar as a sustainable tool for improving the health of salt-affected soils. Soil Environ Health 1(3):100033

Yuniwati ED, Novitasari DR, Kamran A (2023) Alternative waste management: composting and biochar conversion for metal remediation. Revenue j: Manag Entrep 1(1):34–43

Zhang C, Zeng G, Huang D, Lai C, Chen M, Cheng M, Tang W, Tang L, Dong H, Huang B, Tan X (2019) Biochar for environmental management: mitigating greenhouse gas emissions, contaminant treatment, and potential negative impacts. Chem Eng J 373:902–922

Zhao H, Xie T, Xiao H, Gao M (2022) Biochar-based fertilizer improved crop yields and N utilization efficiency in a maize–Chinese cabbage rotation system. Agriculture 12(7):1030

Zheng Y, Zimmerman AR, Gao B (2020) Comparative investigation of characteristics and phosphate removal by engineered biochars with different loadings of magnesium, aluminum, or iron. Sci Total Environ 747:141277

Zhou Y, Gao B, Zimmerman AR, Fang J, Sun Y, Cao X (2013) Sorption of heavy metals on chitosan-modified biochars and its biological effects. Chem Eng J 231:512–518

Challenges in the Realm of Biochar Applications

Shradha Sharma, Shagun Sharma, Harshad Kulkarni, Anand Giri, and Praveen Kumar

Abstract A carbonaceous material known as biochar produced from the pyrolysis of organic substances has been in focus because it is promising for use in environmental cleanup. Despite the benefits of biochar, specifics of its use are associated with numerous problems such as feedstock selection, technologies used in its production, physiochemical properties, and impacts on the environment and economy can limit their application and/or effectiveness. The variability in the performance of biochar may be due to one of the biggest obstacle faced in the field applications of biochar, which is the inconsistency in feedstock sources, its nature and composition. And standardization is made more difficult due to the difference in porosity, surface area, capacity for contaminant adsorption, and nutrient content resulting from variety of feedstock. Even the physicochemical properties of biochar may be impacted by the conditions used in pyrolysis such as temperature, residence time, and heating rates. Environmental issues are also significant challenge for using biochar. Some biochars could cause soil and water pollution when the heavy metals, polycyclic aromatic hydrocarbon (PAHs), or other toxicants were released from the char and transported into other environmental compartments. A second major limitation is that biochar should still be economically feasible for the scale up of production and application. Commercial adoption is hindered by high cost of production, difficulty in transportation and distributions, and low market demand. Moreover, the long term impact of biochar on soil health and crop productivity is yet to be established through extensive field trials and studies to ascertain its effectiveness in various agroecosystems. Finally, as biochar offers a promising future in sustainable environmental and agricultural practices, it needs to address these hurdles in order to gain wider acceptance and application in real-world scenarios. The optimization of the production processes should be addressed in future research, evaluation of the long-term environmental impact and policies to encourage its adoption should be developed.

S. Sharma · S. Sharma · H. Kulkarni · A. Giri (✉)
School of Civil and Environmental Engineering, Indian Institute of Technology Mandi, H.P. Mandi, India
e-mail: anandgiri40855@gmail.com

P. Kumar
University of Ljubljana, Ljubljana, USA

A. Chauhan et al. (eds.), *Biochar As Remediation Tools for Pollutant Removal*,
https://doi.org/10.1007/978-3-032-16746-0_10

Keywords Biochar · Pyrolysis · Feedstock variability · Scalability · Environmental remediation

1 Introduction

Soil degradation, soil pollution, and water shortages leading to dwindling opportunities for food production and adding to climate change are all direct results of rapid population increase and urbanization (Chaudhary et al. 2022). Biochar, a carbon-negative material is the one having so much of potential such as high porosity and surface area which is capable of sequestering carbon dioxide (CO_2), reducing nutrient leaching, soil acidity, and contributing to climate change mitigation (Goswami et al. 2021; Kamarudin et al. 2022).

Biochar has been of great interest in past years because of possible applications to environmental restoration and agriculture (Ibitoye et al. 2024). Biochar enhances soil fertility by sequestration of carbon and ecosystem restoration, however there are various challenges that can limit the potential application of biochar. The chapter discusses the different issues encountered in the production to field application of biochar (Yuan et al. 2023; Pathak et al. 2024).

The performance of biochar in different application seems to be determined by differences in properties due to difference of cost, the lack of consistent regulatory approach limiting large-scale commercialization of biochar (Barbhuiya et al. 2024: Chauhan et al. 2025). Also, environmental side effects such as possible release of heavy metals and of volatile organic compounds (VOCs) may influence environment and human health.

Many potential sources of waste are illustrated Fig. 1, that include, agricultural runoff, industrial discharge, and municipal waste along with storm water inputs as well as associated challenges and opportunities for waste management and resource recovery. The biochar is produced by burning the organics from forestry and agricultural wastes in a controlled atmosphere.

The key problem of the environmental restoration remains the polluting gases the pyrolysis process emits (Das et al. 2021; Pathak et al. 2024). Biochar has high surface area, fine grain size, low weight and high porosity as its physical properties. Biochar is produced using the feedstocks, and the heating methods have the potential to influence the chemical properties of biochar (Patel and Panwar 2023).

2 Importance of Biochar

Biochar This multi-use case in sustainable agriculture, biochar enhances nutrients availability and improves soil health such as water retention (Moreno-Riascos and Ghneim-Herrera 2020). Its high cation exchange capacity (CEC) minimizes nutrients leaching and thus discourages the use of synthetic fertilizers, while its porous nature

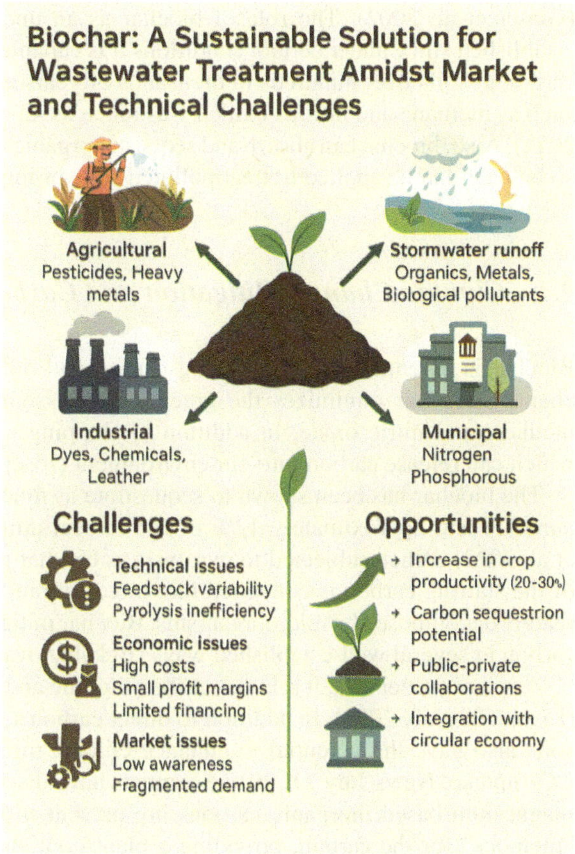

Fig. 1 Biochar use for the different wastewater treatment amidst market and technical challenges

enhances the ability to hold soil water; hence, there is reduced need for irrigation systems and supports crops in resisting drought stress (Khan et al. 2024a).

Additionally, biochar is a steady carbon sink that can store carbon in soils for hundreds to thousands of years, reduce greenhouse gas emissions and can assist mitigating climate change. Furthermore, biochar can boost microbic activity, buffer soil pH, and clean up contaminated soils, which makes it peculiarly practical in acidic or disreputable soils (Ayaz et al. 2021).

By previous researches, the application of biochar has been shown to increase beneficial bacteria up to 50% and decrease harmful one (fungus) even at rate of 50%, while improving soil microbial biomass carbon by 21.7% and enzyme activities that can go up through 25.4% (Kabir et al. 2023). Similarly, Khan et al. (2024b) noticed that, biochar reduces the nutrient leaching up to37 and 20% of nitrate and phosphate leaching, respectively.

Biochar is a significant material to enhance the water quality, soil well-being and carbon sequestration appear as a significant material in the achievement of sustainability and is becoming increasingly significant goods in its multiple application

(Glaser et al. 2002). The role of biochar as an unchanging carbon sink is well-established since under certain conditions it is capable of immobilizing atmospheric carbon in soils over hundreds to thousands of years and reducing greenhouse gases such as methane and nitrous oxide by as much as 43 and 98, respectively (Oni et al. 2021). Also, biochar can absorb and sequester organic and heavy metal contaminants, so biochar can be applied to treat polluted soils or improve the quality of water.

2.1 Climate Change Mitigation and Carbon Sequestration

Biochar is essential for sequestering carbon and mitigating the effects of climate change. Biochar minimizes the greenhouse gas emissions from soils, including methane and nitric oxide, in addition to stopping vegetation from decomposing, which can release carbon into our environment (Joseph et al. 2021).

The biochar has been shown to sequestrate as much as 1.8 Pg CO_2-C equivalent annually and approximately 12% of the current atmospheric GHGs (Shahhoseini et al. 2023). When subjected to pyrolysis as biochar production process, up to 50% of the starting carbon is converted into a recalcitrant and aromatic form that is not readily decomposed by microorganisms. Biochar mitigates climate change and stores carbon in several well-established ways (Bekchanova et al. 2024).

As plant matter rots, it releases carbon dioxide and that returns to the atmosphere (Li and Tasnady 2023). In addition to direct carbon sequestration, biochar-amended soils also have the potential to enhance climate mitigation by stimulating further CO_2 uptake. News/July 11, 2017 Scientists have discovered that when calcium and magnesium-based inorganic carbons are present in biochar, the metals act like a "memory" for the carbon, providing a plant nourishment and stress-relief as well as giving air to fill up the earth. An example is a study reporting that through these mineral reactions, a massive application of biochar could remove up to 2.66 billion tons of CO_2 and sequester 7.38–12.5 billion tons of carbon (Gui et al. 2025).

And also biochar is a carbon stocker by stabilizing it from biosource and is effective to restore an environmental indegentation. This practice also contributes to climate smart agriculture by increasing soil fertility, retaining carbon (Rizwan et al. 2023). It is largely via enhanced soil aeration and direct effects of the modified microbial activity by biochar inhibiting CH_4 and N_2O production activities of microbial enzymes. (Kalu et al. 2022). Biochar reduces global warming by converting carbon in biomass to a stable and recalcitrant form that promotes more uptake of CO_2 into soil from mineral carbonation (Tang et al. 2025; Wei et al. 2024). Biochar reduces CH_4 and N_2O emissions from soil by improving soil properties, managing microbes to improve soil fertility for plant growth promoting factors and C in the soils (Li et al. 2023; Kaur et al. 2023).

2.2 Soil Enhancement and Agricultural Productivity

Agricultural production and soil enhancement are greatly increased by applying biochar, which can improve access to nutrients and absorption of water. Biochar increases crop yields by promoting root development, nutrient availability, and stress tolerance in plants. The biochar functions as a soil modification which increases agricultural yields and lowers the demand for artificial fertilizers. Biochar can boost crop productivity by 5–51 %, based on the kind of soil and treatment rate. Similarly, 30 Mg ha^{-1} of biochar can increase the pulses productivity by 30 %, vegetable productivity by 29 %, and grain yield by 7–8 % in comparison with no applied biochar (Babu et al. 2023). Furthermore, biochar increases the capacity of cations to exchange (CEC), which helps soils maintain higher levels of nutrients. The biochar can raise CEC in loamy sandy soils by as much as 30 %. The biochar also supports healthy microbiomes which can increase fertile soil and health, which further encourages plant development, decreases soil acidity, increases nutrient crop availability, boosts water retention, and encourages plant growth by neutralizing alkaline or acidic soils (Glaser et al. 2002; Ramamoorthy et al. 2022). The main advantage of using biochar is that it increases plant development and lessens reliance on chemical fertilizers by providing a home for helpful microorganisms (Jeffery et al. 2011).

It is also established that biochar supplied with plant growth promoting rhizobacteria (PGPR) such as Serratia odorifera enhances the soil biological contents especially for nitrogen, and phosphorus resulting into growth of crop during stress. When PGPR is used as a carrier, plant nutritional requirements and stress resistance were enhanced, with biochar behaving as a safety environment that increases its longevity and activity of (Kamyab et al. 2025).

Biochar derived from invasive plants such as P. *hysterophorus* improved the maize plant root growth and seed germination behavior, suggesting its resistance in plants (Sarfraz et al. 2019). Biochar can improve crop yield by up to 7 %, due to an supportive substrate for beneficial microbes and also improved soil microbiome, it may well help in controlling plant diseases (Adak et al. 2024).

2.3 Environmental Remediation and Wastewater Treatment

Its excellent adsorption characteristics may account for the increasing popularity of biochar as a green alternative in environmental remediation and water treatment. Biochar is of great importance for the removal of heavy metals, organics pollutants, and pharmaceuticals from polluted water sources. Biochar in underdeveloped countries that have the problem of industrialization, which poses a serious threat to environment, may be beneficial for cleaning effluent because it can remove up to 80% of the industrial effluents from water (Kapoor et al. 2024).

Wastewater treatment of a variety of sources represents one of the most promising biochar uses in environmental remediation. The pre-treatments of the substrates (raw materials) and the post-treatments of the products with developed biochar can remove the contaminants, which is more effective in most cases compared to pristine organic biochar (because of its higher porosity, larger surface area, functional groups abundance, and stronger adsorption capacity) (Luo et al. 2020). In soil remediation, biochar is also applied to remove contaminants, such as antibacterial agents, polycyclic aromatic hydrocarbons, and insecticides (Haider et al. 2022).

It showed that the burst use of biochar in environmental remediation and high-rate heavy metals removal from contaminated soil (Wang et al., 2022 a, b). Also for improvement of water quality it has been successfully employed in bio filters and artificial wetlands for removing the organic contaminates and heavy metals and improving the quality of water (Mohan et al. 2014).

2.4 Renewable Energy and Bioenergy Applications

Biochar is a versatile material in the shift to sustainable energy systems, with great potential for use in bioenergy and renewable energy applications. Recently, the application of biochar's potential in environmental remediation was established with a high efficiency for heavy metals uptake from contaminated soil (Asaad et al. 2022). Also, for improvement of water quality, several authors have successfully utilized the biofilters and artificial wetlands to treat organic pollutants and heavy metals as well as enrich water quality (Kalla et al. 2024).

In a new study biochar can be also considered as an additive, electrocatalyst, or electrode in REs. Biochar can be used in various ways within renewable energy, like as an electrode in microbial fuel cells (for increase of bioelectricity due to its electrical behavior), catalysts for biodiesel production (increase the transesterification process efficiency) and additive in anaerobic digestion systems (for example, higher biogas yields caused by enhanced microbial activity).

More recently, several publications have focused on the choice of an adequate feedstock and pyrolysis conditions to optimize biochar characteristics (surface area, porosity) for various energy end uses. In addition, biochar can serve as a carbon-negative product contributing to climate targets through mitigation of emissions and trapping of carbon (Kant Bhatia et al. 2021).

Owing to its highly porous and tunable surface chemistry, biochar has also found its way in energy conversion and storage means of devices such as lithium ion battery and supercapacitors (Osman et al. 2023). Biochar will invigorate the microorganisms in an anaerobic digester which would result in higher amount of methane generation. The biochar can be converted into bio-oil and syngas for use as alternative energy sources and also can serve in the construction, filtration, and cement industries (Inyang et al. 2016; Sugiarto et al. 2021).

2.5 *Soil Erosion Control*

The use of biochar is becoming more widely acknowledged as a successful method of reducing soil erosion. The several previous experimental investigations used variety of soil types, climates, and farming techniques for validation process in the biochar-mediated soil erosion control (Fig. 2). Depending on site-specific criteria including soil type and rainfall regime, Gholamahmadi et al. (2023) found that adding 5 % biochar decreased erosion by 10–69 % in soils that are prone to erosion under varying compaction and rainfall intensities. According to Rončák et al. (2023) various farming methods and climates, biochar amendments improved soil aggregate stability and decreased sediment migration, particularly in degraded soils. The combination of biochar, compost, and vegetation were studied by (Zanutel and Bielders 2023), and they found that the it reduces runoff and soil loss, especially during heavy rainfall. The biochar helps maintain soil moisture by reducing surface crusting and topsoil detachment. This strategy maintains ecosystem services and fertile topsoil, which not only reduces the erosion of soil but also promotes environmentally friendly agriculture. It has been demonstrated that applying biochar may decrease soil decomposition by up to 64% in areas with high moisture relative to areas having low moisture (Zanutel and Bielders 2023). By encouraging the development and stability of soil aggregates, biochar mechanistically improves soil structure at the microscopic level and improves pore size distribution and soil porosity, increasing water-holding capacity (Sharma et al. 2021; Sharma 2024). Furthermore, biochar enhances soil organic matter and promotes microbial activity, which enhances soil aggregates and boosts the soil matrix's resistance to erosion (Albalasmeh et al. 2024). In addition to lowering the chance of soil erosion, these combined benefits support ecosystem services, preserve healthy topsoil, and encourage sustainable farming methods. The necessity for site-specific management solutions is highlighted by the fact that the efficacy of biochar might differ depending on its feedstock, pyrolysis temperature, application rate, and integration with other erosion control approaches (Wani et al. 2023).

3 Challenges in Biochar Production

Although biochar production holds promise for environmentally friendly farming practices and sustainable agriculture, a few obstacles limit its effectiveness and scalability. These challenges can be largely divided into four categories: economic viability, production size and efficiency, carbon efficiency, and feedstock procurement.

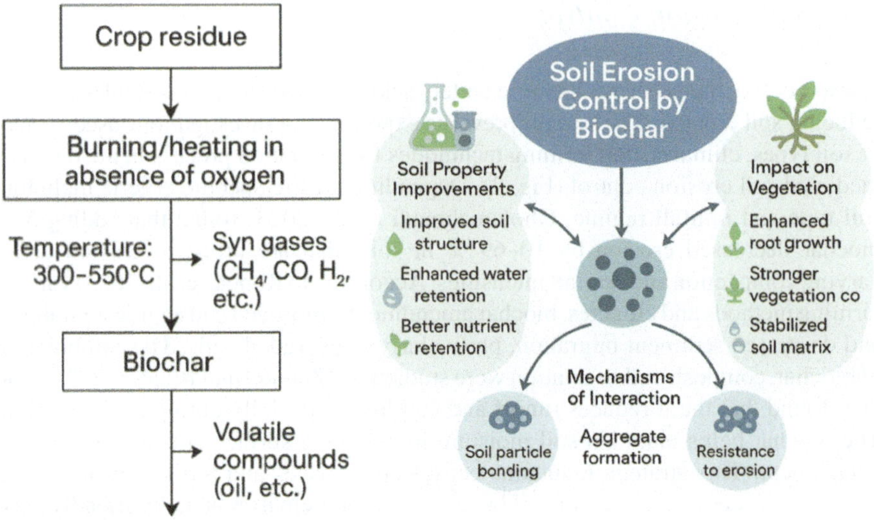

Fig. 2 Schematic diagram of biochar production and role of biochar in soil erosion control: mechanisms and benefits

3.1 Feedstock and Process Challenges

The process of producing biochar starts with obtaining feedstock, which is frequently hampered by the procurement and processing of feedstock which are the main obstacles to the production of biochar. Stranded biomass poses a significant challenge which might result in outdoors combustion or biomass dumping. Operational and financial constraints hinder its effective utilization (Chun et al. 2021). The cost of obtaining feedstock can make up 40–75 % of the total production expenses, depending on the scale of operations. Due to labor-intensive growing, transporting, and processing, feedstock costs are the primary concern for small-scale systems (less than 20,000 tons annually) (Ahmed et al. 2024). Numerous factors, such as geographic location, feedstock type, and local availability affect this broad range (Campion et al. 2023) such as institutions close to plentiful forestry or agricultural waste can obtain more affordable and easily accessible biomass, which lowers the cost of transportation and procurement. On the other hand, because of the greater market demand and longer travel distances operations in regions with limited or highly competitive feedstock sources experience high costs (Nematian et al. 2021). Additionally important is the type of feedstock, the market prices and processing needs of municipal green waste, forestry leftovers, and agricultural waste vary, which may affect the total cost (Zilberman et al. 2023). Feedstock costs are also their main economic issue because labour-intensive growing, transporting, and processing have a particularly negative impact on small-scale systems (processing less than 20,000 tons annually) (Nsamba et al. 2015). The substantial variation in feedstock procurement costs across the biochar business can be explained by these considerations taken together. Small makers have higher costs of doing business per unit of output, so they need to make at least 70–$ 90 per ton in order to stay in business. It is possible for mixed feedstocks, like sewage sludge and agricultural waste to lower prices, but they can also make the output and quality of biochar more variable. The yield range for mixed sewage sludge-cotton stalk pyrolysis is 32–57 %. When compared to low-ash materials like rice straw (11 % ash, less yields), high-ash feedstocks like pig feces (36 % ash) increase biochar productivity (61 % at 300 °C) which demonstrate that how feedstock content directly affects cost structures and production efficacy (Anand et al. 2023). According to recent studies, lowering feedstock costs through logistical optimization (such as combined harvesting-production systems) is essential for expanding the commercialization of biochar (Zilberman et al. 2023). Furthermore, the controlling moisture is essential since moist substrates decrease burning, which raises employment and time expenses. In small enterprises systems, ineffective energy use leads to inadequate thermal waste extraction, which lowers total generation of energy (Domenico et al. 2024).

3.2 *Diversity of Feedstock Materials*

3.2.1 Agricultural Waste

Agricultural waste's varied composition and variable qualities make it difficult to biochar production. Although the amounts of carbon and nutritional composition of waste such as rice husks, peanut hulls, and grain straw are widely spread and reasonably priced, but differ greatly, which impacts the effectiveness and purity of biochar (Mustafa et al. 2022), because the immediate use of crop residue poses biochemical and biological threats to soil. It must be transformed through combustion into sustainable biochar (Ghosh et al. 2023). The consistency of the waste content variation is difficult in pyrolysis processes to maximize the qualities of biochar for certain uses in agriculture. The biochar made from crop residue has the ability to increase fertilization of soil and reduce contamination of the environment in spite of these obstacles (Rex et al. 2023).

3.2.2 Woody Biomass

The use of woody biomass as a substrate for the synthesis of biochar poses a number of difficulties, because of it, it requires special properties and handling procedures. Wood lignin, due to its strength and durability during combustion, overlaps wood biomass as a fraction that is resistant to removal. Woody waste (it consists of forest residues and short-rotation hardwood plantation) have to be preprocessed, for example, by chipping or grinding, before being utilized in a combustion systems (Chun et al. 2021). The high-carbon biochar is also obtained from branches and wood chips but needs intensive processing before it can be used because of high concentration of lignin in the material, which makes it less manageable compared with other feedstocks.

3.2.3 Limitations and Risks of Using Waste-Derived Feedstocks for Biochar Production

Although feedstocks for biochar production are naturally available but their use in soil applications is limited due to potential contamination from plastics, heavy metals, and other pollutants. Sewage sludge poses significant challenges as a feedstock because of its unpredictable composition and high risk of pollution. The high nutrients in manure/sewage sludge and wastewater often include contaminants like tiny particles as well as toxic substances like Zn, Cd, Cr, and Ni. By applying unprocessed sewage sludge to cultivars like maize increases the danger of hazardous compound absorption and similarly preventing it to transform in to the biochar (Ghorbani et al. 2022). In the same way, if excrement comes from animals it might contain microorganisms.

Although hydrolysis of excreta can also eliminate sure dangers by breaking down constituent contaminants and microorganisms (Junior and Guo 2023). The severe restrictive regulations for pathogen levels and preprocessing needs (such as dehydration and processing) add difficulties and expense for biochar production (Zhao et al. 2023).

Wastewater sludge-based biochar could contain poisonous metals, such as lead or cadmium, and poor-quality biochar could potentially become hazardous to the crops health and soil condition (Mamera et al. 2022). These problems indicate that it is very important to employ healthy raw materials and optimize the conditions for pyrolysis in order to ensure safety and effectiveness of biochar application in agriculture (Jiang et al. 2020; Zhang et al. 2022).

The MSW biochar presents some challenges due to its heterogeneous composition, nutrient properties, and particle shape owing to the municipal waste asymmetric structure. The percentage of carbon and nutrient load of the producing biochar may differ due to presence of household litter, vegetation waste, and other naturally produced materials found in municipal waste (Razzak 2024).

The capacity of biochar to absorb water and fertilizer, derived by means of municipal solid waste (MSW), needs less coverage area when compared to the biochar produced using woody biomass or crop residues (Amalina et al. 2022). Also, the presence of polymers or toxic metals in MSW can complicate the production and use of biochar and necessitate large-scale decontamination and screening to ensure the reliability and safety of the final product (Ahmed et al. 2024).

4 Influence of Pyrolysis Parameters on Biochar Properties

Temperature, residence time, and heating rate are the main factors of pyrolysis that largely determine the biochar properties. These parameters but not limited to, could influence the surface area, porosity, and chemical properties of biochar. In order to produce biochar with intended and constantly reliable properties, each of these properties needs to be quantified and optimized for different feedstocks (Mohammadi et al. 2020).

Biochar pore sizes may vary significantly, from 43 to 186 m2/g depending on origin and production conditions (Batista et al. 2018). The higher temperatures (500–800°C) resulted in better surface area (from 50 to 300 m^2/g), and thereby implemented greater available surfaces for adsorption, but at the same time could reduce nitrogen content and its availability for solute uptake (Leng et al. 2021).

4.1 Influential Factors for Porosity and Surface Area

4.1.1 Pyrolysis Temperature and Physical Characteristics of Biochar

Surface area and pore formation nature of biochar is greatly influenced by the temperature at which pyrolysis process takes place. When temperatures are even higher (400,700°C), biomass forms aromatic organic substances, further increasing its overall surface and forming more holes (Zhang et al. 2022). Correspondingly, the topology of micropores is favored at lower temperatures, and the development of mesopores and macropores is favored at harsher temperatures. The biochar surface area at 600 (0) C was 490.8 m2/g, and chemical treatments (including KOH treatment) increased this value to 3263 m^2/g (Ghorbani et al. 2022). The physical characteristics of biochar based on the feedstock composition and pyrolysis conditions, directly impact its applicability for applications such as pollutant adsorption and carbon sequestration (Visser et al. 2024). Lignocelluloses materials especially biomass derived from wood are perfect for creating biochar because of their substantial porosity and total area. While ash-rich raw materials can block pores and decrease the area covered. Feedstocks with an excessive amount of organic carbon encourage the production of small pores. Biochar made from wood usually has a larger micropore volume than crop leftovers (Tomczyk et al. 2020).

4.1.2 Size of the Particles and Production Environment

Over pyrolysis, the pore structure is influenced by the particles size. While larger particles produce macropores that hold water, smaller particles improve fluid and nutrient transfer in soil composites by increasing inter pore connection. In the process of pyrolysis, smaller particles make more contact with temperatures which promote the conversion of micropores into pore sizes. Biochar produced from smaller biomass particles displayed a lower number of micro pores (pores<2 nm) but a higher mesopore region (2–50 nm) (Ippolito et al. 2020). Similarly, mesopore development is affected by oxygen's availability during pyrolysis process which can improve the growth of pores. The biochar produced with oxygen environments showed increased meso-porosity as compared to neutral atmospheres (Muzyka et al. 2023).

4.1.3 pH and Stability

Stability and pH of biochar are major variables which influence the effectiveness of biochar in soil application. Depending on the pyrolysis conditions and feedstock, biochar will have a pH ranging between 7 and 10 (Premalatha et al. 2023). The alkaline properties of biochar facilitate cation exchange capacity (CEC) and nutrient uptake, so it can be used to treat pH-deficient soils. An example is biochar produced using animal dung, which often includes more ash and has a pH exceeding 10, which can significantly lower the alkality of the soil (Liu et al. 2022).

Biochar plays a vital role in pH modification and improvement of soil due to its overall high pH (Amalina et al. 2022; Anand et al. 2023; Anand et al. 2022; Asaad et al. 2024). Biochar acts as a melting agent, is used to neutralize the pH of alkaline soils, and favors plant growth. It has been found that 1% of biochar applied as chocolate shell can increase pH for alkaline soil by 2. 3 U enhanced microorganism growth and nutrient accessibility (Xiao et al. 2018).

4.1.4 Carbon Retention

The biochar could be a powerful way to continue the carbon sequestration process and improve soil conditioning due to its unique carbon storage capability. In the soil, biochar can also aid in bacterial degradation under 100,000 of years long due to carbon turning from plant-like to solid during hydrolysis (Zhang et al. 2023). Enhanced SOC this is partly because biochar can improve stability of soil organic carbon (SOC) by forming nutrients and salts complexes, which avoid the transformation of SOC (Weng et al. 2022).

The porosity of biochar enhances long-term carbon sequestration by stabilizing leftover microbes and promoting the deposition of SOC by absorbing carbon produced by roots (rhizodeposits) (Joseph et al. 2021). Extreme temperatures pyrolysis enhances carbon endurance by maximizing stable carbon amounts and reducing chemically reactive material. Such properties contribute to the establishment of sustainable soil ecosystems and to a decrease in greenhouse gas emissions (Schofield et al. 2019). Quick pyrolysis decreases the amount of biochar produced, maximizes energy, and slow pyrolysis produces more biochar (30–40%).

4.1.5 Field Performance

The effectiveness of biochar in the field varies with the producing uneven outcomes in usage in agriculture due to the types of substrates, combustion temperatures, and soil composition (Yaashikaa et al. 2020). The decreasing microbial populations or accumulation of water in biochar can also have a detrimental effect on environmental health and productivity. According to some previous research, the yield improvements of 10–20 % for cultivars like rice and millet (Marshall et al. 2019). Biochar increased peanut harvests in tropical areas by improving soil composition because of significant alkalinity (Zilberman et al. 2023). The technological and financial obstacles is shown in Table 1 for biomass pyrolysis in India. The absence of uniform guidelines and variations in the composition of biochar continue to be major obstacles to consistent crop yield worldwide (Yang et al. 2018).

5 Recent Technological and Economic Obstacles in India's Biomass Pyrolysis

The quick pyrolysis generates biological oil as a substitute for combustion oil, but it needs advanced equipment and a constant source of feedstock, both of which are scarce in many parts of India. Figure 3 shows the challenges faced by the Indian biomass industry.

5.1 Lack of Standardization in Production Protocols: The Indian Scenario

India's biomass industry is sharing the same destiny that processors, brewers, and winemakers would have encountered, had they not standardized their production process. It's a bit of a knotted problem that includes differences in production innovations, regulatory barriers, and shortfalls in certifying. Various kinds of biomass processing technologies are practiced in India, which ranges from industrial gasifiers to household kilns and may have different environmental foot prints for efficiency and biochar yields (Dotaniya et al. 2024). To illustrate, long-term pyrolysis is commonly employed to produce biochar, yet it is not as efficient as rapid pyrolysis, which is more appropriate to generate bio-oil uses (Negi et al. 2023). This technological variability affects the nutritional value and commercial acceptability of biomass goods and complicates regulation.

Table 1 The technological and financial obstacles unique to biomass pyrolysis in India

Barrier	Description	Impact on India	References
High Production costs	• the economies related to scale are advantageous for big-scale operations (>30,000 TPY) but also need a consistent supply of biomass • human resources and capital expenditure are expensive for small-scale systems (less than 20,000 TPY)	• small-scale businesses pyrolysis facilities across India suffer from a lack of efficiency and expensive labor, e.g., the irregular accessibility of feedstock frequently prevents agricultural biomass systems from scaling	Negi et al. 2023
Scale dependency	• because of expensive capital costs and limited automation, small-scale structures are less practical	• India's small-scale gasification facilities struggle with economy and adaptability	Wiatrowski et al. 2022
Equipment expenses	• the annual expenditure of emissions-control mechanisms and pyrolysis boilers exceeds a million bucks per dry ton of feedstock each hour	• in India, small-scale enterprises are not economically feasible due to excessive prices for machinery and unable to deploy modern plants due to the expensive nature of procurement	Tshikovhi and Motaung, 2023
Energy capture challenges	• the overall energy production of smaller appliances is decreased by inappropriate use of thermal waste	• The inability of many Indian enterprises to maximize the recycling of energy results in decreased total effectiveness. For example, Uttar Pradesh's char industry produces poorer results due to ineffective energy collection devices	Damian et al. 2024

5.2 Regulatory Hurdles and Certification Gaps

Then despite the other technology that has been pursued in instance open burn open combustion air safety permits for small scale or flexile biofuel systems/plants are still a lengthy and expensive process (Varkolu et al. 2025). The missions SAMARTH incorporating biomass co-firing in thermopower plants with renewable purchase exchange mechanisms, this is some of them changing India's legislative stance on biomass usage (Sharma et al. 2024).

However, the limitations of market economy and the different regulation in various places make it difficult to be carried out. On the similar scale, there is a severe limitation in certifying of GHG emissions incurred by production biochar (Yaashikaa et al. 2020). The absence of a definite norm for verification is detrimental in India to

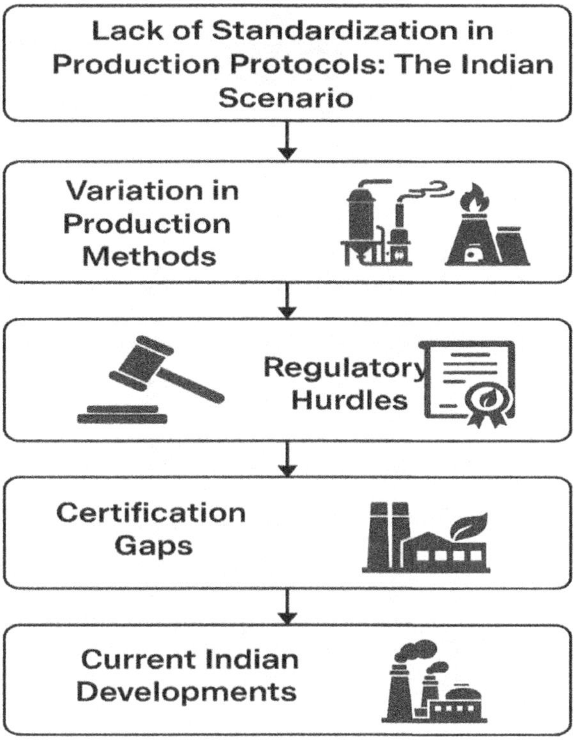

Fig. 3 Challenges faced by the Indian biomass industries

develop a vibrant carbon market for biochar, but there is an increasing trend toward people becoming environment-friendly (Anand et al. 2022). This detracts from the actual environmental advantages of biochar works aside from its feasibility.

6 Soil Related Challenges in Biochar Application

The use of biochar in soils poses a number of concerns relevant to soil interactions and characteristics. Here are a few of the main issues with soil.

6.1 Variability Effectiveness across Soil Types and Climatic Conditions

The value of biochar in improving crop output is contingent upon the type of soil and weather. Location-specific evaluations are essential for the optimal use of biochar in maximum crop production. The biochar increases the porosity of sandy soils,

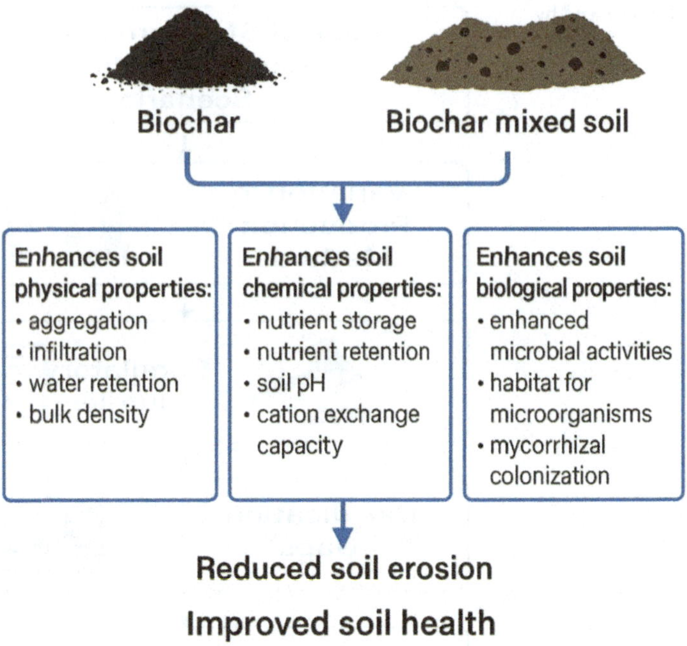

Fig. 4 Multifaceted benefits of biochar on soil properties and erosion control

thus improving nutrient and water accessibility by plants and can also improve its structure and mass density.

Conversely, biochar can also enhance water uptake and ventilation of clay soils by reducing porosity and raising permeation. The adsorbent capacity, nutrient preservation capacity, and porosity of the biochar influence with different soils (Khan et al. 2024; Fig. 4). Climatic factors such as biochar ability to retain humidity may also have different impacts in a desert region than in a highly precipitating environment (Alkharabsheh et al. 2021).

Moreover, relative to rich-soil, BC effect can be more pronounced in degraded or nutrient-poor ones. Optimal application of biochar to soil maximum benefits. For optimal use of the potential in biochar for soil improvement, attention should be paid to the local environmental conditions (Levesque et al. 2020).

6.2 Challenges with Nutrient Dynamics and Compatibility with Fertilizers (e.g., Immobilization, Leaching)

Biochar can preserve and immobilize nutrients, depending upon its nature. When biochar is used as agricultural products, it can lead to reduced access to fertilizers by agricultural plants, without proper management (Kabir et al. 2023). The effect of immobilization on nitrogen is greater since biochar has the potential to alter the microbial mechanisms that help regulate the nitrogen cycle, thus reducing the quantity of nitrogen released to vegetation (Levesque et al. 2020) but it can, conversely, reduce nutrient loss by enhancing the capacity of soils to retain water and additional nutrients. However, the impact differs according to the kind of soil and the properties of the biochar. The combination of biochar with both organic and inorganic fertilizers can improve the availability and retention of nutrients, hence increasing soil fertility. Biochar improves soil fertility and microbial diversity by raising the pH, nitrogen, phosphorus, and potassium content of soil. Additionally, using biochar in conjunction with fertilizers can lessen the need for large amounts of synthetic fertilizers, which could lessen negative environmental effects including greenhouse gas emissions and the leaching of nutrients (Agarwal et al. 2022). Similar, applying salts or biochar or alone, showed that coupling the two enhanced crop production by 179.6%, especially if the biochar formed at 401–500 °C with an excessive C:N ratio (Bai et al. 2022).

7 Environmental and Eco Toxicological Concerns of Biochar

Although, biochar has many uses in soil and ecology, based on its raw material and processing circumstances but it may contain hazardous substances and can change ecosystems, with environmental and ecotoxicological problem. The presence of heavy metals, perfluorochemicals (PFCs), dioxins, ecologically persistent free radicals (EPFRs), polycyclic aromatic hydrocarbons (PAHs), and volatile organic solvents (VOCs) are some other dangerous substances that can cause environmental and ecotoxicological effects (Xiang et al. 2021). The releasing of toxic compounds into the natural environment, such as PAHs and toxic metals from the biochar can also damage biological ecosystems, lower soil quality, and have a detrimental effect on the soil microbiology. The heavy metals derived from fuel or produced through burning like cadmium (Cd), lead (Pb), and arsenic (As), can build up in crop residue and endanger plant absorption. Biochar made from natural feedstocks and by using wastewater sludge may have higher concentrations of heavy metals (Wang et al. 2022 aand b).The feedstock category, decomposition the environment, and soil characteristics are some of the variables that affect the adverse effects on the environment which is given Table 2. Applying biochar can have complicated effects on the quality of groundwater, mostly due to the leaching of pollutants and changes in water characteristics (Rizwan et al. 2023).

Table 2 Potential environmental impacts of harmful substances in biochar

Pollutant	Source	Environmental impact	Mitigation strategy	References
Heavy metals	Feedstock, pyrolysis	Microorganism harmful effects, consumption of plants and pollution of the soil	Proper feedstock selection and ideal atmosphere for pyrolysis	Song et al. 2024
PAHs	Incomplete combustion	Pollution of soil and water, harm to biological life	High temperature and extended duration of residence are ideal pyrolysis conditions	Murtaza et al. 2024
(CO) carbon monoxide	The pyrolysis process produces carbon monoxide because of incomplete combustion	This harmful gas has a negative impact on air quality and is harmful to human health	During pyrolysis, CO emissions can be decreased by ensuring full combustion and a sufficient oxygen supply	Dong et al. 2023
(SOx) Sulfur oxides	When sulfur-rich raw materials are utilized to produce biochar, sulfur oxides are released.	Acid rain, health issues, and environmental deterioration are all caused by these pollutants	SOx emissions can be reduced by using sulfur-free raw materials and desulfurization processes	Jha et al. 2023
Volatile organic compounds (VOCs)	Feedstock, pyrolysis	Pollution in the atmosphere and organism toxicity	Optimized pyrolysis conditions (high temperature, long residence time)	Beljin et al. 2024
EPFRs	High-temperature pyrolysis	Enhanced oxidative stress in microbes	Low pyrolysis conditions are used	Tisserant and Cherubini, 2019
Dioxins	Inappropriate choice of biomass feedstocks	Dangerous to human health as well as having an adverse effect on many environmental media microbes and plants	Choose the right kind of feedstock from biomass by considering factors like the degree of pyrolysis and the amount of toxic substances present	Jha et al. 2023

8 Influence on Soil Microbial Diversity and Ecosystem Functions

The bacteria, archaea, and fungi are among the incredibly diverse microbes living in soil that are essential to the operation of ecological systems and important biological cycles. Microbiological community, climate, soil characteristics, and the distribution of soil are some of the elements that affect soil shifts (Tecon and Or, 2017; Chauhan et al. 2024). For example, the microbial diversity differs greatly among biomes, mostly because of soil moisture and nutrient access (Wang et al. 2022 a, b; Chauhan et al. 2024). The breakdown of organic material, nitrogen cycling, and general soil condition can all be impacted by these changes in microbial populations. The biochar can slow the microbe degradation procedures, increasing the pollutants and decreasing the discharge of nutrients which could affect the general purity of the soil and the functioning of the plants.

9 Socio-Economic and Technical Barriers

There are numerous socioeconomic and technical barriers toward the biochar production and utilization, which cause severe challenges to the biochar business. There are several types of problems. Despite the potential of biochar to improve soils and sequester carbon, the trade is facing serious economic and technical barriers that hinder its wide-scale commercialization.

Socioeconomics problems comprise lack of coherent market demand caused by unsteady tactics and minimal understanding of consumers, expensive operation costs with low profit margin, and limited accessibility to finance to grow business. Moreover, insufficient investment in research and development constrains quality standardization in developing nations, whereas strong verification procedures and unfavorable government-related help prevent commercialization in other territories (Chakma et al. 2025).

Technical challenges the technical or practical aspects of biochar application include operation issues (pollution control and machine servicing) while the variability in feedstock is also a problem, e.g., the inefficacy of pyrolysis due to high moisture tropical crop residue (Koné et al. Many 2025).

9.1 Market and Policy Challenges

Biochar certainly has environmental merit but both market circumstance and policy harmonization makes it far from a worldwide commercial activity. High production costs, irregular value and poor costumer recognition are some of the market challenges (Heinrich et al. 2023). Big companies, for instance, find it difficult to scale up production in response to demand, while smaller businesses struggle with moisture control and feedstock variability.

Asian-Pacific region countries, where agriculture is a traditional practice (Thengane et al. 2021), the adoption of biochar is low due to poor understanding of its long-term benefits and alternative soil amendments. Further, the passive carbon flows which are promoted by biochar as CO_2 sinks (CDR). passivity of the biochar acceptance. Plus, there are additional economic matters which manufacturers have to grapple with thanks to all the certification and permission requirements.

Biochar has been on the trading scene in over 90 % of world from places whose C (carbon), comes with no traditional sequestration claims, yet not having classic (CC) promises, it remains impediment to at very least some very high C economies. To solve these problems and translate sturdy policies into customer demand, it's going to require a coordinated push from everyone involved.

10 Conclusion

Biochar has great potential as a sustainable approach to improve soil health, crop productivity, and mitigate global environmental issues especially climate change. The upcycling of biomass waste is imperative in addressing the complex realities of waste management in the developing world.

The many uses of biochar and its sustainable application so far emphasize the role biochar can play in contributing to a sustainable future on earth as we move into the UN Sustainable Development Goals (SDGs) 2030. Its efficiency results in different soil-plant systems: byproducts of feedstock and primary pyrolysis parameters alter the physicochemical properties. The downside of its widespread adoption is that it's a type of natural variability. Biochar is a long-lasting soil conditioner, and improving the structure of soil has been shown to increase nutrient availability, water-holding capacity and crop growth, regulate soil pH, and promote microbial diversity—key aspects for long-term agricultural productivity and environmental sustainability.

Given its potential for sequestration of carbon and mitigation of greenhouse gas, its value grows for the control of climate change. There are still, however, important information deficits that need to be addressed; these include the longer-term ecological and human health effects and the best rates of use for defined agroecological zones. The feedstock and pyrolysis conditions (applicable to its intended use or location of the site) under which biochar benefits were optimized determined the process chosen. It also combines efficient waste management, digital innovation and sustainability in a holistic way. Future work should put emphasis on the optimization of biochar formulations, the search for synergisms with other soil amendments and that standardized treatments are available allowing easy, safe and large-scale application. Supporting this research and information needs will be critical to unlocking the potential of biochar for agricultural sustainability and climate adaptation.

11 Future Directions for Biochar's Potential Applications in Agriculture and Environment

In both the agricultural and environmental domains, biochar holds great promise as an economical and effective way to enhance soil fertility, soil quality, soil and water treatment, retention of soil carbon, and greenhouse gas control (Nepal et al. 2023). In agricultural applications, it is still crucial to use biochar for crop-soil treatment methods, especially when using it on a wide scale (Afshar and Mofatteh 2024). Biochar's nanoscale reduction in size may have substantial environmental remediation potential due to its enormously increased surface area for the adsorption and chemical reactivity of pollutants in soil and water sources. The air quality protection and deployment difficulties may be mitigated by long-term regulatory frameworks designed for mobile industrial facilities (Liu et al. 2022). Policies that value biochar's greater environmental advantages could make it more profitable in carbon markets (Murtaza et al. 2023). Advances in nanobiotechnology and pyrolysis techniques, such as electromagnetic pyrolysis, may yield better biochar with more uses (Bhandari et al. 2023). Portable units and better small-scale businesses production systems could lower costs and increase availability (Haider et al. 2022). The improved climate effects and fossil fuel offsets could result from increased energy storage from biochar ecosystems (Kumar and Panwar 2024).

Acknowledgments The author AG is highly thankful to the Department of Science and Technology, Govt. of India, New Delhi, for providing financial assistance in the form of Inspire Faculty Fellowship [DST/INSPIRE/04/2021/ 003653].

Author's Contributions Shradha Sharma and Shagun Sharma contributed to the original draft and conceptualization; Harshad Kulkarni providing supervision, review, and editing; and Anand Giri contributed to supervision, review, and final approval.

References

Adak E, Halder S, Koley B, Biswas S, Sengupta S, Kundu S, Sarkar T (2024) An overview of the importance of biochar in sustainable agriculture. J Adv Biol Biotechnol 27(6):924–937

Afshar M, Mofatteh S (2024) Biochar for a sustainable future: environmentally friendly production and diverse applications. Results Eng 23:102433

Agarwal H, Kashyap VH, Mishra A, Bordoloi S, Singh PK, Joshi NC (2022) Biochar-based fertilizers and their applications in plant growth promotion and protection. 3 Biotech 12(6):1–19

Ahmed SF, Mehejabin F, Chowdhury AA, Almomani F, Khan NA, Badruddin IA, Kamangar S (2024) Biochar produced from waste-based feedstocks: mechanisms, affecting factors, economy, utilization, challenges, and prospects. GCB Bioenergy 16(8):1–37

Albalasmeh AA, Quzaih MZ, Gharaibeh MA, Rusan M, Mohawesh OE, Rababah SR, Alqudah A, Alghamdi AG, Naserin A (2024) Significance of pyrolytic temperature, application rate and incubation period of biochar in improving hydro-physical properties of calcareous sandy loam soil. Sci Rep 14(1):1–13

Alkharabsheh HM, Seleiman MF, Battaglia ML, Shami A, Jalal RS, Alhammad BA, Almutairi KF, Al-Saif AM (2021) Biochar and its broad impacts in soil quality and fertility, nutrient leaching and crop productivity: a review. Agronomy 11(5)

Amalina F, Razak ASA, Krishnan S, Sulaiman H, Zularisam AW, Nasrullah M (2022) Biochar production techniques utilizing biomass waste-derived materials and environmental applications—a review. J Hazard Mater Adv 7:100134

Anand A, Gautam S, Ram LC (2023) Feedstock and pyrolysis conditions affect suitability of biochar for various sustainable energy and environmental applications. J Anal Appl Pyrolysis 170:105881

Anand A, Pathak S, Kumar V, Kaushal P (2022) Biochar production from crop residues, its characterization and utilization for electricity generation in India. J Clean Prod 368:133074

AsaadSM,InayatA,Rocha-MenesesL,JamilF,GhenaiC hanablehA(2022)Prospectiveofresponsesurfacemethodologyasanoptimizationtoolforbiomassgasificationprocess.Energies16(1):p.40

Asaad SM, Tawalbeh M, Ali A, Al-Othman A (2024) Applications of bioenergy. In: Renewable energy—volume 2: wave, geothermal, and bioenergy. Elsevier, pp 275–293

Ayaz M, Feizienė D, Tilvikienė V, Akhtar K, Stulpinaitė U, Iqbal R (2021) Biochar role in the sustainability of agriculture and environment. Sustainability 13(3):1330

Babu S, Singh R, Kumar S, Rathore SS, Yadav D, Yadav SK, Yadav V, Ansari MA, Das A, Rajanna GA, Wani OA, Raj R, Yadav DK, Singh VK (2023) Biochar implications in cleaner agricultural production and environmental sustainability. Environ Sci Adv 2(8):1042–1059

Bai SH, Omidvar N, Gallart M, Kämper W, Tahmasbian I, Farrar MB, Singh K, Zhou G, Muqadass B, Xu C-Y, Koech R, Li Y, Nguyen TTN, van Zwieten L (2022) Combined effects of biochar and fertilizer applications on yield: a review and meta-analysis. Sci Total Environ 808:152073

Barbhuiya S, Bhusan Das B, Kanavaris F (2024) Biochar-concrete: a comprehensive review of properties, production and sustainability. Case Stud Constr Mater 20:e02859

Batista EMCC, Shultz J, Matos TTS, Fornari MR, Ferreira TM, Szpoganicz B, de Freitas RA, Mangrich AS (2018) Effect of surface and porosity of biochar on water holding capacity aiming indirectly at preservation of the Amazon biome. Sci Rep 8(1):10677

Bekchanova M, Kuppens T, Cuypers A, Jozefczak M, Malina R (2024) Biochar's effect on the soil carbon cycle: a rapid review and meta-analysis. Biochar 6(1):1–19

Beljin J, Đukanović N, Anojčić J, Simetić T, Apostolović T, Mutić S, Maletić S (2024) Biochar in the remediation of organic pollutants in water: a review of polycyclic aromatic hydrocarbon and pesticide removal. Nano 15(1):26

Bhandari G, Gangola S, Dhasmana A, Rajput V, Gupta S, Malik S, Slama P (2023) Nano-biochar: recent progress, challenges, and opportunities for sustainable environmental remediation. Front Microbiol 14:1–16

Campion L, Bekchanova M, Malina R, Kuppens T (2023) The costs and benefits of biochar production and use: a systematic review. J Clean Prod 408:137138

Chakma S, Hasan M, Rakshit SK, Kozinski J, Kang K (2025) Review of recent advances in the design, synthesis, and modification of biochar for remediation of heavy metal pollution in water. Can J Chem Engineering

Chaudhary P, Singh S, Chaudhary A, Agri U, Bhandari G (2022) Chapter 3 agrochemicals and their effects on soil microbial population. In: Soni R, Suyal D, Goel R (eds) Plant protection: from chemicals to biologicals. De Gruyter, Berlin, Boston, pp 45–58

Chauhan PK, Pathak HK, Dubey G, Sharma H, Upadhyay SK (2024) Impact of bacillus cereus SPB-10 on growth promotion of wheat (Triticum aestivum L.) under arsenic contaminated soil. Curr Microbiol. https://doi.org/10.1007/s00284-024-03673-1

Chauhan PK, Upadhyay SK, Chauhan A, Bhutiani R, Sikarwar RLS, Tiwari MK (2025) A review on simultaneous use of microbes and biochar for removal of toxic heavy metals: recent updates and future outlooks. Physiol Mol Biol Plants 31(3):1–15. https://doi.org/10.1007/s12298-025-01649-4

Chauhan PK, Upadhyay SK, Rajput VD et al (2024) Fostering plant growth performance under drought stress using rhizospheric microbes, their gene editing, and biochar. Environ Geochem Health 46:41. https://doi.org/10.1007/s10653-023-01823-1

Chun Y, Lee SK, Yoo HY, Kim SW (2021) Recent advancements in biochar production according to feedstock classification, pyrolysis conditions, and applications: a review. Bio Resour 16(3):6512–6547

Damian CS, Devarajan Y, Raja T, Jayabal R (2024) A comprehensive review of biomass pyrolysis for hydrogen production in India. Process Saf Environ Prot 190:646–662

Das S, Mohanty S, Sahu G, Rana M, Pilli K (2021) Biochar: a sustainable approach for improving soil health and environment. Soil Erosion—Current Chall Futur Perspect Chang World:1–18

Di Domenico G, Bianchini L, Di Stefano V, Venanzi R, Lo Monaco A, Colantoni A, Picchio R (2024) New frontiers for raw wooden residues, biochar production as a resource for environmental challenges. C 10(2):54

Dong M, He L, Jiang M, Zhu Y, Wang J, Gustave W, Wang S, Deng Y, Zhang X, Wang Z (2023) Biochar for the removal of emerging pollutants from aquatic systems: a review. Int J Environ Res Public Health 20(3):1679

Dotaniya ML, Meena MD, Choudhary RL, Meena MK, Meena VD, Singh H, Lakaria BL, Jat RS, Rai PK, Kumar K, Doutaniya RK, Singh H (2024) Dynamics of major plant nutrients and enzymatic activities in soil influenced by application of biochar and organic waste. PLoS One 19(10):1–13

de Filho MN, C, Melo LCA, Lustosa Filho JF, de Castro Paes É, de Oliveira Dias F, Lino Gomes J, Nick Gomes C (2024) Improved tomato development by biochar soil amendment and foliar application of potassium under different available soil water contents. J Plant Nutr 47(16):2620–2644

Gholamahmadi B, Jeffery S, Gonzalez-Pelayo O, Prats SA, Bastos AC, Keizer JJ, Verheijen FGA (2023) Biochar impacts on runoff and soil erosion by water: a systematic global scale meta-analysis. Sci Total Environ 871:161860

Ghorbani M, Amirahmadi E, Neugschwandtner RW, Konvalina P, Kopecký M, Moudrý J, Perná K, Murindangabo YT (2022) The impact of pyrolysis temperature on biochar properties and its effects on soil hydrological properties. Sustainability 14(22):14722

Ghorbani M, Konvalina P, Walkiewicz A, Neugschwandtner RW, Kopecký M, Zamanian K, Chen WH, Bucur D (2022) Feasibility of biochar derived from sewage sludge to promote sustainable agriculture and mitigate GHG emissions—a review. Int J Environ Res Public Health 19(19):p.12983

Ghosh S, Nandasana M, Webster TJ, Thongmee S (2023) A growaste-generated biochar for the sustainable remediation of refractory pollutants. Front Chem 11:1–19

Glaser B, Lehmann J, Zech W (2002) Ameliorating physical and chemical properties of highly weathered soils in the tropics with charcoal—a review. Biol Fertil Soils 35(4):219–230

Goswami M, Pant G, Mansotra DK, Sharma S, Joshi PC (2021) Biochar: a carbon negative technology for combating climate change. In: Pant D, Kumar Nadda A, Pant KK, Agarwal AK (eds) Advances in carbon capture and utilization, Energy, environment, and sustainability. Springer, Singapore

Gui X, Xu X, Zhang Z, Hu L, Huang W, Zhao L, Cao X (2025) Biochar-amended soil can further sorb atmospheric CO_2 for more carbon sequestration. Commun Earth Environ 6(1):5

Haider FU, Wang X, Zulfiqar U, Farooq M, Hussain S, Mehmood T, Naveed M, Li Y, Liqun C, Saeed Q, Ahmad I, Mustafa A (2022) Biochar application for remediation of organic toxic pollutants in contaminated soils: an update. Ecotoxicol Environ Saf 248:114322

Heinrich T, Park H, Orozco R, Ding Z, Álvarez-López V, Mosquera-Losada MR, Steinbeis L, Hoffmann T (2023) Biochar production from late-harvest grass—challenges and potential for farm-scale implementation. Sustainable Prod Consumption 37:256–267

Ibitoye SE, Loha C, Mahamood RM, Jen T-C, Alam M, Sarkar I, Das P, Akinlabi ET (2024) An overview of biochar production techniques and application in iron and steel industries. Bioresour Bioprocess 11(1):65

Inyang MI, Gao B, Yao Y, Xue Y, Zimmerman A, Mosa A, Pullammanappallil P, Ok YS, Cao X (2016) A review of biochar as a low-cost adsorbent for aqueous heavy metal removal. CritAl Rev Environ Sci Technol 46(4):406–433

Ippolito JA, Cui L, Kammann C, Wrage-Mönnig N, Estavillo JM, Fuertes-Mendizabal T, Cayuela ML, Sigua G, Novak J, Spokas K, Borchard N (2020) Feedstock choice, pyrolysis temperature and type influence biochar characteristics: a comprehensive meta-data analysis review. Biochar 2(4):421–438

Jeffery S, Verheijen FGA, van der Velde M, Bastos AC (2011) A quantitative review of the effects of biochar application to soils on crop productivity using meta-analysis. Agric Ecosyst Environ 144(1):175–187

Jha S, Gaur R, Shahabuddin S, Tyagi I (2023) Biochar as sustainable alternative and green adsorbent for the remediation of noxious pollutants: a comprehensive review. Toxics 11(2):117

Jiang Z, Lian F, Wang Z, Xing B (2020) The role of biochars in sustainable crop production and soil resiliency. J Exp Bot 71(2):520–542

Joseph S, Cowie AL, Van Zwieten L, Bolan N, Budai A, Buss W, Cayuela ML, Graber ER, Ippolito JA, Kuzyakov Y, Luo Y, Ok YS, Palansooriya KN, Shepherd J, Stephens S, Weng Z, Lehmann J (2021) How biochar works, and when it doesn't: a review of mechanisms controlling soil and plant responses to biochar. GCB Bioenergy 13(11):1731–1764

Junior A, Guo M (2023) Efficacy of sewage sludge derived biochar on enhancing soil health and crop productivity in strongly acidic soil. Front Soil Sci 3:1–8

Kabir E, Kim KH, Kwon EE (2023) Biochar as a tool for the improvement of soil and environment. Front Environ Sci 11:1–17

Kalla A, Mayilswamy N, Kandasubramanian B, Mahajan- Tatpate P (2024) Biochar: a sustainable and an eco-friendly material for energy storage applications. Int J Green Energy 21(7):1695–1709

Kalu S, Kulmala L, Zrim J, Peltokangas K, Tammeorg P, Rasa K, Kitzler B, Pihlatie M, Karhu K (2022) Potential of biochar to reduce greenhouse gas emissions and increase nitrogen use efficiency in boreal arable soils in the long-term. Front Environ Sci 10:914766

Kamarudin NS, Dahalan FA, Hasan M, An OS, Parmin NA, Ibrahim N, Hamdzah M, Zain NAM, Muda K, Wikurendra EA (2022) Biochar: a review of its history, characteristics, factors that influence its yield, methods of production, application in wastewater treatment and recent development. Biointerface Res Appl Chem 12(6):7914–7926

Kamyab H, Chelliapan S, Khalili E, Rezania S, Balasubramanian B, Taheri MM, Simancas-Racines D, Rajendran S, Yusuf M (2025) Biochar as a carrier for plant growth-promoting bacteria in phytoremediation of pesticides. J Hazard Mater Adv 18:100673

Kant Bhatia S, Palai AK, Kumar A, Kant Bhatia R, Kumar Patel A, Kumar Thakur V, Yang YH (2021) Trends in renewable energy production employing biomass-based biochar. Bioresour Technol 340:125644

Kapoor RT, Ahmad P, Rafatullah M (2024) Insights into biochar applications: a sustainable strategy toward carbon neutrality and circular economy. ACS Symposium Series, 1479 Vol 2, 1–30

Kaur N, Kieffer C, Ren W, Hui D (2023) How much is soil nitrous oxide emission reduced with biochar application? An evaluation of meta-analyses. GCB Bioenergy 15(1):24–37

Khan S, Irshad S, Mehmood K, Hasnain Z, Nawaz M, Rais A, Gul S, Wahid MA, Hashem A, Abd_Allah EF, Ibrar D (2024a) Biochar production and characteristics, its impacts on soil health, crop production, and yield enhancement: a review. Plants 13(2):166

Khan S, Irshad S, Mehmood K, Hasnain Z, Nawaz M, Rais A, Gul S, Wahid MA, Hashem A, Fathi E, Ibrar D (2024b) Health, crop production, and yield enhancement: a review. Plants 13(166):1–18

Koné S, Galiegue X, Gwenzi W (2025) Techno-economic assessment of biochar systems: state-of-the-art and future research directions. In: Biochar for environmental remediation. Elsevier, pp 447–459

Kumar KV, Panwar NL (2024) Pyrolysis technologies for biochar production in waste management: a review. Clean Energy 8(4):61–78

Leng L, Xiong Q, Yang L, Li H, Zhou Y, Zhang W, Jiang S, Li H, Huang H (2021) An overview on engineering the surface area and porosity of biochar. Sci Total Environ 763:144204

Lévesque V, Oelbermann M, Ziadi N (2020) Biochar in temperate soils: opportunities and challenges. Can J Soil Sci 102(1):1–26

Li L, Long A, Fossum B, Kaiser M (2023) Effects of pyrolysis temperature and feedstock type on biochar characteristics pertinent to soil carbon and soil health: a meta-analysis. Soil Use Manag 39(1):43–52

Li S, Tasnady D (2023) Biochar for soil carbon sequestration: current knowledge, mechanisms, and future perspectives. C 9(3):67

Liu Y, Ma J, Gao J, Chen X, Ouyang X, Weng L, Li H, Chen Y, Li Y (2022) Stability and interaction of biochar and iron mineral nanoparticles: effect of pH, ionic strength, and dissolved organic matter. Biochar 4(1)

Luo J, Lin L, Liu C, Jia C, Chen T, Yang Y, Shen M, Shang H, Zhou S, Huang M, Wang Y, Zhou D, Fan J, Clark JH, Zhang S, Zhu X (2020) Reveal a hidden highly toxic substance in biochar to support its effective elimination strategy. J Hazard Mater 399:123055

Mamera M, van Tol JJ, Aghoghovwia MP (2022) Treatment of faecal sludge and sewage effluent by pinewood biochar to reduce wastewater bacteria and inorganic contaminants leaching. Water Res 221:118775

Marshall J, Muhlack R, Morton BJ, Dunnigan L, Chittleborough D, Kwong CW (2019) Pyrolysis temperature effects on biochar–water interactions and application for improved water holding capacity in vineyard soils. Soil Syst 3(2):1–14

Mohammadi A, Khoshnevisan B, Venkatesh G, Eskandari S (2020) A critical review on advancement and challenges of biochar application in Paddy fields: environmental and life cycle cost analysis. PRO 8(10):1–21

Mohan D, Sarswat A, Ok YS, Pittman CU (2014) Organic and inorganic contaminants removal from water with biochar, a renewable, low cost and sustainable adsorbent—a critical review. Bioresour Technol 160:191–202

Moreno-Riascos S, Ghneim-Herrera T (2020) Impact of biochar use on agricultural production and climate change: a review. Agronomía Colombiana 38(3):367–381

Murtaza G, Ahmed Z, Eldin SM, Ali B, Bawazeer S, Usman M, Iqbal R, Neupane D, Ullah A, Khan A, Hassan MU, Ali I, Tariq A (2023) Biochar-soil-plant interactions: a cross talk for sustainable agriculture under changing climate. Front Environ Sci 11

Murtaza G, Ahmed Z, Valipour M, Ali I, Usman M, Iqbal R, Zulfiqar U, Rizwan M, Mahmood S, Ullah A, Arslan M, Rehman un MH, Ditta A, Tariq A (2024) Recent trends and economic significance of modified/functionalized biochars for remediation of environmental pollutants. Sci Rep 14(1): Nature Publishing Group UK

Mustafa A, Brtnicky M, Hammerschmiedt T, Kucerik J, Kintl A, Chorazy T, Naveed M, Skarpa P, Baltazar T, Malicek O, Holatko J (2022) Food and agricultural wastes-derived biochars in combination with mineral fertilizer as sustainable soil amendments to enhance soil microbiological activity, nutrient cycling and crop production. Front Plant Sci 13:1–11

Muzyka R, Misztal E, Hrabak J, Banks SW, Sajdak M (2023) Various biomass pyrolysis conditions influence the porosity and pore size distribution of biochar. Energy 263:126128

Negi H, Suyal DC, Soni R, Giri K, Goel R (2023) Indian scenario of biomass availability and its bioenergy-conversion potential. Energies 16(15):5805

Nematian M, Keske C, Ng'ombe, J. N. (2021) A techno-economic analysis of biochar production and the bioeconomy for orchard biomass. Waste Manag 135:467–477

Nepal J, Ahmad W, Munsif F, Khan A, Zou Z (2023) Advances and prospects of biochar in improving soil fertility, biochemical quality, and environmental applications. Front Environ Sci 11

Nsamba HK, Hale SE, Cornelissen G, Bachmann RT (2015) Sustainable technologies for small-scale biochar production a review. J Sustain Bioenergy Syst 05(01):10–31

Oni BA, Oziegbe O, Olawole OO (2021) Significance of biochar application to the environment and economy. Ann Agric Sci 64(2):222–236

Osman AI, Lai ZY, Farghali M, Yiin CL, Elgarahy AM, Hammad A, Ihara I, Al-Fatesh AS, Rooney DW, Yap PS (2023) Optimizing biomass pathways to bioenergy and biochar application in electricity generation, biodiesel production, and biohydrogen production. Environ Chem Lett 21(5): Springer International Publishing

Patel MR, Panwar NL (2023) Biochar from agricultural crop residues: environmental, production, and life cycle assessment overview. Resour, Conserv Recycl Adv 19:200173

Pathak HK, Chauhan PK, Seth CS, Dubey G, Upadhyay SK (2024) Mechanistic and future prospects in rhizospheric engineering for agricultural contaminants removal, soil health restoration, and management of climate change stress. Sci Total Environ 927:172116. https://doi.org/10.1016/j.scitotenv.2024.172116

Pathak HK, Seth CS, Chauhan PK, Dubey G, Singh G, Jain D, Upadhyay SK, Dwivedi P, Khoo KS (2024) Recent advancement of nano-biochar for the remediation of heavy metals and emerging contaminants: mechanism, adsorption kinetic model, plant growth and development. Environ Res 255:119136. https://doi.org/10.1016/j.envres.2024.119136

Premalatha RP, Poorna Bindu J, Nivetha E, Malarvizhi P, Manorama K, Parameswari E, Davamani V (2023) A review on biochar's effect on soil properties and crop growth. Front Energy Res 11:1–19

Ramamoorthy P, Mary PCN, Bose KSC, Karthikeyan B, Manivelan K (2022) Effect of biochar on soil health and crop productivity: a review. Agric Rev 45(1):121–126

Razzak SA (2024) Municipal solid and plastic waste derived high-performance biochar production: a comprehensive review. J Anal Appl Pyrolysis 181:106622

Rex P, Mohammed Ismail KR, Meenakshisundaram N, Barmavatu P, Sai Bharadwaj AVSL (2023) Agricultural biomass waste to biochar: a review on biochar applications using machine learning approach and circular economy. Chem Eng 7(3)

Rizwan M, Murtaza G, Zulfiqar F, Moosa A, Iqbal R, Ahmed Z, Irshad S, Khan I, Li T, Chen J, Zhang M, Siddique KHM, Leng L, Li H (2023) Sustainable manufacture and application of biochar to improve soil properties and remediate soil contaminated with organic impurities: a systematic review. Front Environ Sci 11

Rončák P, Németová Z, Vitková J, Danáčová M, Toková L, Aydin E, Valent P, Honek D, Igaz D (2023) Effects of the application of biochar on the soil erosion of plots of sloping agricultural and with silt loam soil. J Hydrol Hydromech 71(4):356–368

Sarfraz R, Hussain A, Sabir A, Ben Fekih I, Ditta A, Xing S (2019) Role of biochar and plant growth promoting rhizobacteria to enhance soil carbon sequestration—a review. Environ Monit Assess 191(4)

Schofield HK, Pettitt TR, Tappin AD, Rollinson GK, Fitzsimons MF (2019) Biochar incorporation increased nitrogen and carbon retention in a waste-derived soil. Sci Total Environ 690:1228–1236

Shahhoseini E, Arefifard M, Karimi M (2023) Biochar for climate change mitigation. In: Materials horizons: from nature to nanomaterials, Part F1442, 123–143

Sharma P (2024) Biochar application for sustainable soil erosion control: a review of current research and future perspectives. Front Environ Sci 12:1–12

Sharma P, Abrol V, Sharma V, Chaddha S, Srinivasa Rao C, Ganie AQ, Ingo Hefft D, El-Sheikh MA, Mansoor S (2021) Effectiveness of biochar and compost on improving soil hydro-physical properties, crop yield and monetary returns in inceptisol subtropics. Saudi J Biol Sci 28(12):7539–7549

Sharma P, Abrol V, Sharma N, Sharma R, Chadha D, Anand S, Khenrab S, Maanik SH, Singh P, Kumari S, Verma D (2024) Policies and strategies for sustainable use of biochar in Indian agriculture. Bio Resour 19(3):6946–6960

Song Q, Kong F, Liu B-F, Song X, Ren H-Y (2024) Biochar-based composites for removing chlorinated organic pollutants: applications, mechanisms, and perspectives. Environ Sci Ecotechnology 21:100420

Sugiarto Y, Sunyoto NMS, Zhu M, Jones I, Zhang D (2021) Effect of biochar addition on microbial community and methane production during anaerobic digestion of food wastes: the role of minerals in biochar. Bioresour Technol 323:124585

Tang H, Hu J, Li B, Liu Y, Tong WK, Yue MR, Wang J, Wang W, Gao M, Liu N, Li J (2025) Degradation-resistant biochar improves soil organic carbon storage: promoting autotrophic metabolism & increasing refractory organic carbon. Bioresour Technol 428:132452

Tecon R, Or D (2017) Biophysical processes supporting the diversity of microbial life in soil. FEMS Microbiol Rev 41(5):599–623

Thengane SK, Kung K, Hunt J, Gilani HR, Lim CJ, Sokhansanj S, Sanchez DL (2021) Market prospects for biochar production and application in California. Biofuels Bioprod Biorefin 15(6):1802–1819

Tisserant A, Cherubini F (2019) Potentials, limitations, co-benefits, and trade-offs of biochar applications to soils for climate change mitigation. Land 8(12):179

Tomczyk A, Sokołowska Z, Boguta P (2020) Biochar physicochemical properties: pyrolysis temperature and feedstock kind effects. Rev Environ Sci Biotechnol 19(1):191–215

Tshikovhi A, Motaung TE (2023) Technologies and innovations for biomass energy production. Sustainability 15(16):12121

Varkolu M, Gundekari S, Omvesh, Palla VCS, Kumar P, Bhattacharjee S, Vinodkumar T (2025) Recent advances in biochar production, characterization, and environmental applications. Catalysts 15(3):243

Visser ED, Seroka NS, Khotseng L (2024) Recent advances in biochar: synthesis techniques, properties, applications, and hydrogen production. PRO 12(6)

Wang J, Li W, Zhao Z, Musoke FSN, Wu X (2022) Ultrasonic activated biochar and its removal of harmful substances in environment. Microorganisms 10(8):1593

Wang H, Zhang R, Zhao Y, Shi H, Liu G (2022) Effect of biochar on rhizosphere soil microbial diversity and metabolism in tobacco-growing soil. Ecologies 3(4):539–556

Wani OA, Akhter F, Kumar SS, Babu S, Kanth RH, Mir SA, Mahdi SS, Malik AR, Bangroo S, Gaafar A-RZ (2023) Mitigating soil erosion through biomass-derived biochar: exploring the influence of feedstock types and pyrolysis temperature. Land 12(12):2111

Wei J, Li Y, Li J, Yan W, Yin Q, Jiang H, Li Z, Jiang J, Li Y, Yao Y, Zhang L (2024) A 2-year pure biochar addition enhances soil carbon sequestration and reduces aggregate stability in understory conditions. Sci Rep 14(1):1–10

Weng Z, Van Zwieten L, Tavakkoli E, Rose MT, Singh BP, Joseph S, Macdonald LM, Kimber S, Morris S, Rose TJ, Archanjo BS, Tang C, Franks AE, Diao H, Schweizer S, Tobin MJ, Klein AR, Vongsvivut J, Chang SLY et al (2022) Microspectroscopic visualization of how biochar lifts the soil organic carbon ceiling. Nat Commun 13(1):5177

Wiatrowski M, Klein BC, Davis RW, Quiroz-Arita C, Tan ECD, Hunt RW, Davis RE (2022) Techno-economic assessment for the production of algal fuels and value-added products: opportunities for high-protein microalgae conversion. Biotechnol Biofuels Bioprod 15(1):8

Xiang L, Liu S, Ye S, Yang H, Song B, Qin F, Shen M, Tan C, Zeng G, Tan X (2021) Potential hazards of biochar: the negative environmental impacts of biochar applications. J Hazard Mater 420:126611

Xiao X, Chen B, Chen Z, Zhu L, Schnoor JL (2018) Insight into multiple and multilevel structures of biochars and their potential environmental applications: a critical review. Environ Sci & Technol 52(9):5027–5047

Yaashikaa PR, Kumar PS, Varjani S, Saravanan A (2020) A critical review on the biochar production techniques, characterization, stability and applications for circular bioeconomy. Biotechnol Rep 28:e00570

Yang S, Jiang Z, Sun X, Ding J, Xu J (2018) Effects of biochar amendment on co_2 emissions from paddy fields under water-saving irrigation. Int J Environ Res Public Health 15(11):2580

Yuan X, Cao Y, Li J, Patel AK, Di Dong C, Jin X, Gu C, Yip ACK, Tsang DCW, Ok YS (2023) Recent advancements and challenges in emerging applications of biochar-based catalysts. Biotechnol Adv 67:108181

Zanutel M, Bielders CL (2023) Contrasted effects of biochar application on interrill erosion depending on age, application rate and soil type. Geoderma Regnal 34:e00706

Zhang C, Ji Y, Li C, Zhang Y, Sun S, Xu Y, Jiang L, Wu C (2023) The application of biochar for co_2 capture: influence of biochar preparation and co_2 capture reactors. Ind Eng Chem Res 62(42):17168–17181

Zhang J, Zhang S, Niu C, Jiang J, Sun H (2022) Positive effects of biochar on the degraded forest soil and tree growth in China: a systematic review. Phyton-Int J Exp Bot 91(8):1601–1616

Zhao L, Sun ZF, Pan XW, Tan JY, Yang SS, Wu JT, Chen C, Yuan Y, Ren NQ (2023) Sewage sludge derived biochar for environmental improvement: advances, challenges, and solutions. Water Res X 18:100167

Zilberman D, Laird D, Rainey C, Song J, Kahn G (2023) Biochar supply-chain and challenges to commercialization. GCB Bioenergy 15(1):7–23

Cutting Edge Innovations: New Approach of Developments in Biochar Technology

Govinda Pal and B. L. N. K. Madhurangi

Abstract Biochar is a promising approach in environmental remediation, CO_2–capture, and soil improvement. This chapter provides a comprehensive overview of the recent progresses in this domain and novel integrations among three main aspects: (i) latest biochar synthesis approaches such as fast pyrolysis and hydrothermal carbonization; (ii) emerging functionalization strategies for transforming biochar, namely mineral-doped biochar, nanobiochar, and composite based of biochar; (iii) novel, interdisciplinary applications including precision agriculture, water purification engineering advances in multiple types of systems. By the integration of three pillars, this intermediate way surmounts the blanks in today's literature due to emerging a single or other technically relevant as well real-world relevant point of views, and broadens their spectre on a complete starting point and further development for biochar usage toward sustainable societies. This chapter also includes environmental and scalability toward advanced biochar systems, including nanoparticle toxicity and their energy content of production. It also touches on the financial feasibility and regulation fallout from widespread implementation. The chapter is unique compared to the available literature in that it "connects" to state-of-the-art room scale developments for the targeted end-use applications. It places biochar as a soil amendment and multi-service, climate adjusted circular economy enabler technology. This is intended to guide future research and policy, as well as industry innovation in the sustainable and responsible development of biochar technologies.

Keywords Biochar innovation · Pollutant remediation · Engineered biochar · Nanomaterials · Pyrolysis technology · Environmental sustainability

G. Pal
Department of Agricultural Engineering, Haldia Institute of Technology, Haldia, West Bengal, India

B. L. N. K. Madhurangi (✉)
Faculty of Graduate Studies, University of Kelaniya, Kelaniya, Sri Lanka
e-mail: nimalikmliyanage@gmail.com

A. Chauhan et al. (eds.), *Biochar As Remediation Tools for Pollutant Removal*,
https://doi.org/10.1007/978-3-032-16746-0_11

1 Introduction

Biochar (BC) is a product of heating biomass in the absence of or with limited air to above 250 °C, a process called charring or pyrolysis (Lehmann and Joseph 2024). The concept of BC originated from the ancient Amazonia's practices of creating ''Terra preta," meaning the black soils of the Indians, which has naturally occurred in nature and also been artificially created by indigenous communities combining low-temperature charcoal with biomass such as compost, manure plant residuals, feces, bones, and many more (Kamarudin et al. 2022).

Figure 1: Spectral processes of the conversion of biomass into biochar (BC). In this case, the process starts with selecting and preparing feedstock (agricultural residues, forestry residues, or organic municipal waste). Using thermochemical conversion (commonly pyrolysis, at 300–700 °C and under a low-oxygen environment), biomass can be converted into bio-oil, solid biochar, and syngas, and the yield and properties can be affected by the pyrolysis techniques (slow, fast, or microwave-assisted). With post-treatment cooling, size reduction, and chemical or physical activation (e.g., acid washing, impregnation), biochar quality is further optimized for soil amendment, carbon capture, and water treatment (Chauhan et al. 2025).

In the modern context, BC stretches its ancient heritage to multifunctional material with uses in environmental remediation, carbon sequestration, waste management, renewable energy generation, and agriculture enhancement.

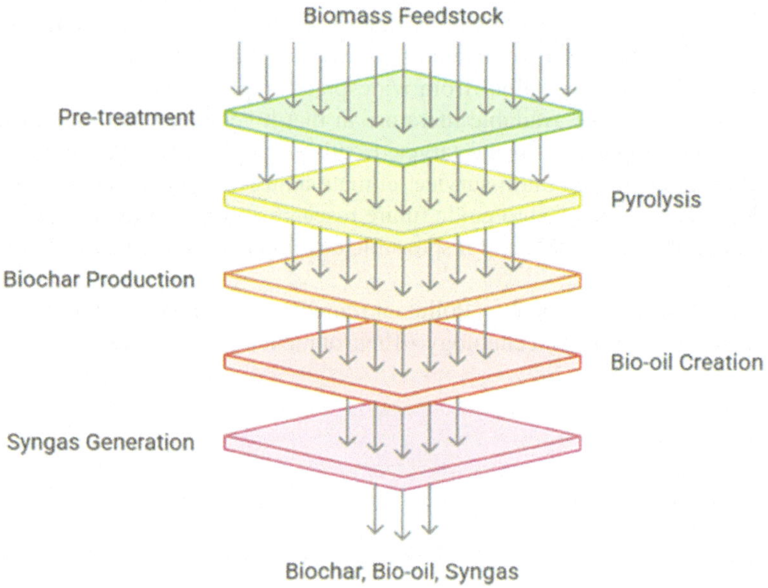

Fig. 1 Biochar production various process

The basic mechanism of BC production is pyrolysis; the thermal decomposition of organic material at high temperature (250–800 °C) in the absence of oxygen. The process converts biomass into main products: heat, syngas, bio-oil (condensable vapors), and solid BC (carbonaceous residue). The yield and characteristics of these products are greatly influenced by pyrolysis conditions, especially temperature, heating rate, residence time, and the type of feedstock material (Aghbashlo et al. 2021; Al-Rumaihi et al. 2022; Safarian 2023).

Recent years have witnessed exponential growth in BC research and development, driven by the urgent need for sustainable solutions to environmental challenges such as climate change, soil degradation, water pollution, and waste management (Pathak et al. 2024; Pal and Patel 2021). Global BC production has expanded from small-scale artisanal methods to sophisticated industrial operations with a market revenue of USD 38.8 million in 2022 and is expected to reach USD 74.7 million by 2030, growing at a compound annual growth rate of 8.5% by 2030 (Grand View Research 2025).

In the twenty-first century, environmental pollution is one of the most significant worldwide problems and pollutants can be heavy metals, POPs, or modern sources of introduction like micro-pollutants which have potential risk to ecosystem as well human health (Ahmad et al. 2022). These remediation options are not cost-effective, as well as produce side wastes and high energy demands (Lim et al. 2016; Sharma et al. 2018; Wanninayake 2021). In this sense, BC is being considered as one of the potential natural remediation agent due to its unique properties and characteristics that consider it a promising environmental solution.

On the other hand, despite a great progress in both studies and practical use of BC, there are still some constraints to be overcome for a full scale application and exploitation of BC as remediation element. These limitations necessitate creative improvements if BC is to be applied as an eco-friendly remedial technology effectively.

2 Recent Advances in BC Production Technologies

The numerous pyrolysis technologies under development present a range of potential routes for BC production, affording scope to tailor product properties and enhancing process efficiency—it can play an important role in achieving sustainable development goals. These novel methods are very unlike the traditional kiln process and present better production rates, lower pollution to the environment, and BC with easily predictable properties for particular remediation purposes.

2.1 Advanced Pyrolysis Techniques

Among the most appealing is the fast pyrolysis at moderate temperature (400–600 °C) and low residence time (minutes instead of hours). In addition, various products arise from the application of this technology to slow pyrolysis, frequently biased by bio-oil production, thus producing BC as a co-product (Liu et al. 2023). Other more recent developments for fast pyrolysis systems have included better control on temperature behavior and improved product consistency associated to enhanced heat transfer in the fluidized bed reactor. One of these is the Empyre demonstration plant in the Netherlands that can handle up to 25,000 ton biomass per year also with higher efficiency than conventional systems (Wang and Wang 2019).

HTC is an unconventional thermal treatment in which hot water (or steam) instead of dry heat is used as the processing medium and working temperatures are between 180 and 260 °C. This so-called "Wet pyrolysis" of HTC is of particular interest for feedstocks with high water contents like sewage sludge, algae, or food waste which avoids the energy intensive pretreatment step \u2013 predrying \u2013 in conventional pyrolysis pathways (Hitzl et al. 2018). The product is based on the typical acronym "Hydro-char" as opposed to dry pyrolysis BC (i.e., higher oxygen, more existing groups in surface, and different types of minerals). These render hydrochar attractive for specific applications, like heavy metal sorption and slow-release fertilizer. HTC scaleup and demonstration have been dramatically accelerated in the last decade or so, with one of the most spectacular cases being TerraNova Ultra plant situated in Germany and processing municipal sewage sludge into hydro-char (20,000 t/a when fully scaled) (Lucian and Fiori 2017).

2.2 Use of Catalysts and Additives for Enhanced BC Properties

Pyrolysis of BC on catalyst support is also rapidly emerging as an effective method for tailor-making intrinsic properties of the BC for targeted applications. The surface activity, porosity, and inorganic content of catalysts may advantageously be enhanced by including catalytic components during a manufacturing step.

Among them, nickel, copper, and iron were found to be very efficient at tailoring BC properties. During pyrolysis, incorporated as catalysts in a variety of processes that determine carbon structure formation and functional group generation (Wang and Wang 2019). For example, iron-catalyzed pyrolysis can produce magnetic recoverability-enhanced BC that is readily reclaimable after use in water treatment applications. Another breakthrough is the introduction of iron oxide nanoparticles into bamboo pyrolysis for the generation of magnetic BC, which exhibited an arsenic-removal efficiency higher than 98% and a recoverability via simply utilizing a magnetic separator (Liu et al. 2023).

Alkali and alkaline earth metals (Na, K, Ca, and Mg) are another essential class of catalysts that affect the yield and nature of BC. These metals facilitate biomass decomposition and the formation of aromatic compounds, leading to increased stability of BC (Han et al. 2021). Recently, it has been demonstrated that calcined BC has efficient phosphate adsorption and therefore is environmentally friendly for WT. Industrialization of the process has partially been developed through collaboration of wastewater plant settings and BC producers in the development of tailored phosphorus sorbing material made of calcium waste, such as egg shells and limestone processing waste (Han et al. 2021).

2.3 Sustainable Feedstock Selection and Modification

BC technology has expanded the spectrum of accepted feedstocks from essentially wood-based biomass to a variety of waste streams. In addition to the usual catalysts, other additives have been researched to incorporate beneficial properties into BC. Silicon compounds, for example, can be added during pyrolysis to obtain a high water retention in the BC. These BC can be used in regions where it is relatively dry. Similarly, nitrogen-based compounds such as amino acids and urea can increase nitrogen content and basicity in BC for applications that need basic surface functionalization, such as CO_2 capture and acid dye adsorption (Dissanayake et al. 2020). Thus, solving waste disposal problems and producing a valuable product in the form of BC. It's not only an environmental imperative, it's a commercial opportunity for the new circular bioeconomy.

The crop residues, like stems, husks, and shells of cereal crops, are abundant and widely spread materials. Prior works have attempted to address the difficulty of processing such residues, including high ash content and non-uniform composition (Dissanayake et al. 2020). For instance, specific pre-treatment of rice husk can include controlled leaching steps to decrease the silica content in the rice husk before heating, whereby the pyrolysis yields a higher level of carbon within the BC while reducing the ash interference. Likewise, being one of the most abundant cereal residues in the world, corn stover has been effectively transformed into high-performance BC with an optimized pyrolysis process, circumventing its inherent compositional disadvantage (Razzak 2024).

Another major feedstock group is animal wastes, and manure, bone, and processing residues, in particular, are of interest for the production of BC. Livestock manure is the source of such important nutrients as phosphorus and nitrogen, which are retained to some extent in the produced BC (Sajjad et al. 2024). New technologies have emerged, such as temperature-stage pyrolysis facilities specifically designed to treat manure for nutrient conservation and optimal successful carbonization. Some similar technologies are now being commercialized in parts of high livestock density, making what was otherwise an environmental waste concern into a valuable planting material (Sheets et al. 2015). Municipal solid waste (MSW) streams, including the organic fraction of household, yard, and food waste, are a growing feedstock source for BC production. The heterogeneous nature of such materials presents considerable challenges, necessitating the innovation of new preprocessing and conversion technologies (Choudhary et al. 2022). Recent developments include automated sorting facilities coupled with specialized pyrolysis units designed to handle moisture content and composition variability. Singapore's Regrow program is a case study of the successful implementation of this approach, with separated organic MSW being converted into BC for urban greening applications and, in the process, reducing landfill demands (Choudhary et al. 2022).

3 Engineered BC for Enhanced Performance

BC has garnered global interest in environmental remediation, soil amendment, and sequestering carbon. However, conventional BC sourced through biomass pyrolysis might not exhibit the required physicochemical traits for certain uses. Advances in material science have opened the door for engineered BC in the hopes of customizing its surface chemistry, porous structure, and functional groups. Tailoring enhanced the BC's performance in removing contaminants, nutrient retention, and catalysis.

3.1 Functionalized BC for Targeted Pollutant Removal

Biochar (BC), as a traditional sorbent, exhibits excellent potential for environmental treatment, but its nonspecific adsorption is commonly observed and prevents it from being focused on the enhancement of targeted pollutants or the specific purpose. Recently, there has been immense focus on the surface modification or functionalization of natural BC to transform it into high-performance application-oriented materials to tackle these constraints.

One of the conventional methods to improve BC properties is chemical activation. This method often requires that the samples be exposed to other activators, such as KOH, H_3PO_4 or $ZnCl_2$ and heating. These two treatments significantly enhanced the surface area and porosity of BC, which had the adsorption capacity that was similar to those of commercial activated carbons (Wang et al. 2023). More recently, another two breakthroughs, the so-called microwave heating-assisted chemical activation was discovered by Ma et al. (2024), and a same to that reportable result but just with one-thousandth of energy consumption, ~ 1800 m^2/g or so for processing time shortened early from 3 h to only 15 min. Acid or base modification is another approach to modify BC properties by the incorporation of specific functional groups to facilitate pollutant adsorption. The acid treatment is beneficial because more oxygen-containing groups (such as carboxyl, hydroxyl, etc.) can be generated in the materials to make them more active toward metal cations and base pollutants. Further, the base treatments help. An interesting variant of that is gradient modified BC, where different parts of a single BC-particle are modified chemically to a different extent. The product is a multi-application adsorbent for simultaneously removing different types of pollutants. These multifunctional BCs have been widely commercialized and applied in wastewater plants for the broadened contaminant adsorption compared with single-function adsorbents used previously (Xiao et al. 2023).

Mineral impregnation is also a successful strategy for the targeted functionalization. In this context, some minerals are incorporated into the BC structure to improve their adsorption selectivity for some water pollutants. For example, iron oxide-impregnated BC was proved to be an efficient sorbent material for arsenic and phosphate, and the manganese oxide-modified BC showed excellent performance to remove phenol (Yang et al. 2023). The advancement of precipitation control tactics has even enabled the homogenous mineral distribution in the BC matrix (rather than only on the surface) to improve the adsorption efficiency and stability of the material. Such high-tech solutions were already implemented in the laboratory and the field, for instance, at the arsenic-removal projects in Bangladesh and some areas of India where iron-modified biochar (BC) is used (Mridha et al. 2024).

Organic molecule grafting is the frontier of BC functionalization, and the organic molecules can be specifically attached to the surface of BC in a covalent way, making the adsorption highly selective. Amine-functionalized BCs exhibit higher adsorption performance toward heavy metals and anion dyes after reactions with ethylenediamine or polyethyleneimine (Yang et al. 2023). Besides that, the insertion of sulfonic acid groups can enhance the adsorption of cationic pollutants for BC. Recently, development has been achieved in the so-called "molecular recognition" BC's where the synthetic receptors are anchored on the surface of BC to allow selective accumulation and removal of pharmaceuticals and EWMs at trace levels (Kang et al. 2024).

Finally, biological functionalization has emerged as a new, original, and interdisciplinary approach. BC was mixed with viable microorganisms or biologically active molecules in this strategy to form biochar biological composites, in which dynamic and versatile properties are acquired. For example, BC colonized fungi have been demonstrated to enhance the elimination of F232 (a typical recalcitrant organic contaminant) through co-adsorption and biodegradation pathways (Yang et al. 2023). In addition, BC inoculated with specific bacterial species has been shown to potentially contribute to the positive effect on the improvement of nitrogen cycling and nutrient supply in agroecosystems (Chauhan et al. 2024). Resultant biologic materials produced in this manner stand in stark contrast to static physicochemical constructions and instead become dynamic remediation devices, able to interface and potentially feed back into complex environmental systems (Chauhan and Upadhyay 2023, 2024).

3.2 *Nano-BC and Its Emerging Applications*

The nano-BC innovation involves BC particles with sizes in the nanoscale range, which is a breakthrough in BC technology that enables new applications and improved performance in others. The nanoscale size imparts special characteristics, such as much greater specific surface area, increased reactivity, and facilitated mobility in environmental matrices.

Manufacturing of nano-BC has improved by leaps and bounds in recent years. Ball milling is still commonly employed among the most straightforward methods, whereby BC particle size is physically reduced by high-energy application. Some recent developments include cryogenic ball milling, where BC is milled at extremely low temperatures to make it more brittle for size reduction with decreased loss of surface functionality (Bangar et al. 2023). An additional important ultrasonic-wave technique is sonication-based methods for dispersing the BC particles in liquid. Recent progress would be scaling up sonication for the continuous process developed to produce the nano-BC with monodispersed size and at an industrial scale (Chauhan et al. 2023).

In addition to the other above-mentioned modifications, the chemical synthesis for nano-BC has been significantly improved. Pickling treatments, such as controlled acid-etching, can selectively remove or stabilize regions of the BC, creating recently designed nanoscale particles with specific traits. Similarly, template-assisted synthesis strategies employ sacrificial templates to fabricate nano-BC with perfectly controlled morphology and pore structures (Xu et al. 2024). Nevertheless, a quite promising emerging approach is based on the direct pyrolysis of nanoscale biomass feedstocks, such as cellulose nanofibers or lignin nanoparticles, to afford nano-BC with outstanding uniformity and performance properties.

The environmental remediation using nano-BCs has expanded significantly due to their higher adsorption capacity and reactivity. Additionally, nano-BC has been highly effective in removing heavy metals in water treatment systems, such as three times higher heavy metal removal for lead, cadmium, and arsenic contaminants compared to conventional BC (Pathak et al. 2024). For hydrophilic organic contaminants like PPCPs, nano-BC possesses accelerated removal kinetics, assures a limited diffusion length, and offers enhanced site accessibility. Commercialization is already conducted in the area of POU water filters, comprising nano-BC cartridges to improve performance with less material required than currently.

Agricultural applications of nano-BC capitalize on its enhanced nutrient retention capacity and mobility in soil. When applied to agricultural soils, nano-BC disperses more uniformly through the soil profile compared to conventional BC, providing more widespread root zone benefits (Shafiq et al. 2023). Recent field trials demonstrate that nano-BC treatments can reduce fertilizer requirements by up to 25% without a loss, and even with an increase in crop yields through enhanced nutrient use efficiency. Nano-BC can also be used to supply micronutrients to plants, with recent studies demonstrating the effective application of zinc and iron provided via nano-BC carriers in otherwise deficient soils (Muhammad et al. 2024).

Biomedical uses are one of the frontier areas for nano-BC synthesis. Current studies illustrate prospective uses in drug delivery systems, where nano-BC's surface functionality and porosity facilitate the controlled release of therapeutic molecules (Garg et al. 2024). Likewise, nano-BC shows promise as a contrast agent in medical imaging and an ingredient in antimicrobial dressings for wound healing. Although mostly at the research stage, these uses point to the immense potential of engineered nano-BC for non-traditional environmental applications.

Despite these encouraging advances, there are still obstacles to the use of nano-BC, specifically regarding perceived environmental and health effects. The same mobility and reactivity that render nano-BC so useful also invite concern about long-term destiny and prospective unexpected effects in environmental systems. To date, studies have approached the issue of "environmentally conscious" nano-BC design, incorporating inherent stability controls or traceability markers to enable monitoring following application (Deb 2024). Likewise, surface passivation methods were created to minimize any possible cytotoxicity issues without compromising the desired functionality for a specific application.

3.3 BC Composites and Hybrid Materials

Among the most promising future trends in BC technology is its combination with other complementary substances that are used to create hybrids and composite structures. These manufactured products combine the BC's inherent strengths with the associated material's specific features, constructing synergistic systems with higher performance profiles than the components themselves.

BC-clay composite has been found most suitable for environmental remediation due to its adsorption capacity and high surface area of the BC combined with structural integrity and cation exchange capacity of clay minerals such as montmorillonite, bentonite, and kaolinite (Gao and Goldfarb 2021). New developments encompass intercalated BC-clay composite in which the clay mineral layers are separated by the particles of the BC, resulting in higher material stability as well as contaminant selectivity. Field-scale implementation confirms the composite outperforms the direct use of clay and that of BC with respect to adsorption capacity, mechanical stability, hydraulic performance, and their use in permeable reactive barriers for the removal of contaminated groundwater (Zhang et al. 2022).

Another very valuable advancement is the BC-polymer composites for water treatment and agriculture. BC have been synthesized by incorporating it into alginate, chitosan, or synthetic resin matrices in this way the yields products with improved mechanical stability, control release and durability (Sim et al. 2021). Recent advances have focused on stimuli-responsive BC-polymer composites that respond to environmental changes in pH, temperature, or the presence of specific pollutants. As an example, researchers from the University of California developed a pH-sensitive BC-hydrogel composite with sequential nutrient releasing basing on soil pH for receiving increasing nutrients throughout growing period (Mansouri et al. 2023).

BC-metal oxides possess the adsorption of BC and reactivity as well as catalytic properties of metal oxides (TiO_2, MnO_2, and Fe_3O_4). The composites have demonstrated excellent activity in AOPs, photodegradation of pollutants, and redox-driven pollutant transformation (Mansouri et al. 2023). An advancement in this category is the formation of core–shell materials, where nanoscale metal oxide film carriers disposed on BC particles are used to obtain maximum reactivity with minimal application of material. In fact, industrialization can be witnessed in advanced wastewater treatment units, where the composites allow adsorption and destruction of persistent organic pollutants like PFAS compounds and drug residues (Baskar et al. 2022).

The former, i.e., BC nanomaterial hybrids are the leading-edge of BC composite technology, combining the sustainable and low-cost feature of BC with the excellent electronic, mechanical and optical properties provided by carbon nanosize-based materials (Verma et al. 2023). Progresses consist of hierarchical materials in which multimicrometer BC surface is directly coated with graphene sheets or nanotubes by catalytic chemical vapor deposition to provide extremely large specific surface and, especially, seamlessly integrated hybrid structures exhibiting electric conductivity. Applications vary from use (references) in supercapacitors and electrochemical sensors to high-performance filtration membranes and electromagnetic interference shielding materials (Mercante 2023).

Living composites may be the most exciting interface in the hybridization of BC and a living life form such as bacterium, fungus, or microalga are integrated into BC to achieve self-replenishing biological entities. These characteristics incorporate the chemical activity and structural integrity of BC with the metabolically active adaptability of the microorganism (Duan et al. 2023).

Examples in the pipeline include cyanobacteria-driven oxygen-generating photosynthetic BC composites that continually biodegrade organic contaminants and fungal-BC composites that continually refill adsorption capacity through biological reconstitution of adsorbed contaminants. While still in their nascent laboratory stage, these living materials have vast potential for applications in next-generation bioremediation as well as resource recovery measures.

4 Application of BC in Pollutant Remediation

The remediation of a wide range of environmental pollutants in both soil and water systems is one of the important topics on which much effort has been directed to using biochar (BC) as a promising adsorbent. Its high efficiency is mainly due to its high specific surface area, natural porosity, and modifiable surface functional groups that have contributed to significant interactions with pollutants. In particular, BC removal efficiencies above 90% have been reached at laboratory scale in the case of heavy metals, polyaromatic hydrocarbons (PAHs), and pesticides (Beljin et al. 2024). To improve its performance, several physico-chemical modifications have been designed to transform BC and attain advanced composites with high adsorptive power, adapted to industrial and agricultural wastewaters (Kapoor and Zdarta 2024; Verma et al. 2024). But the problem then is going from the lab to the field. The intricate nature of environmental native conditions and the economic variability of up-scaled production often contribute to delineating the lab-optimized BC systems as non-directly scalable (Beljin et al. 2024). Furthermore, process-based production aspects of BC such as pyrolysis temperature, raw material used, and preparation methods are widely reported for the retention on pollutants by BC (Murtaza et al. 2023; Verma et al. 2024). A sustainable perspective for these parameters is a must if BC at large scale and one with responsibility in environmental clean up work should be applied in its ecological setting (Kapoor and Zdarta 2024; Murtaza et al. 2023).

In addition to possessing physical–chemical versatility, BC is also recognized as a multifunctional material for the adsorption of organic and inorganic pollutants. Owing to its high specific surface area and adsorption capacity, it can efficiently absorb pollutants such as pesticides, pharmaceuticals, heavy metals in polluted soil and aquatic systems (Zheng et al. 2020; Zhang et al. 2013). So, it is due to ion exchange, electrostatic attraction, and complexation for the removal of heavy metal(s) and hydrogen bonding and hydrophobic interaction for organic pollutant uptake (Islam et al. 2021). In addition, BC enhances soil health due to higher SWHC and low SS which are relevant for improving drought tolerance (Lohar et al. 2024; Lyu et al. 2016). Its alkalinity is also the most critical factor for immobilization of heavy metals through precipitation reactions, and in reducing phytoextraction of heavy metal, affected by increasing soil pH and microbial activity (Kar et al. 2016; Zhang et al. 2013). All of these properties altogether demonstrate the potential prospect of BC as an environmentally friendly multifunctional material for all-purpose environmental remediation.

4.1 Removal of Organic Pollutants (E.G., Pesticides, Pharmaceuticals)

Instead, it is imperative to rapidly eliminate organic pollutants, which have received abundant attention and become problematic because they are resistant to the environment (e.g., pesticides and pharmaceuticals) and traditional sewage treatment technology. HTC has been proven to have high removal efficiency up to 89.8% for pharmaceuticals with optimal removal conditions (220 °C, 360 min), while the degradation of pesticides was accomplished (Miserli et al. 2022). Likewise, similar degradation of a wide range of organic pollutants, such as pharmaceuticals and mycotoxins, at subcritical water treatment has been achieved by fine-tuning the treatment parameters in the non-catalytic regime (Švarc-Gajić et al. 2024).

However, nano-adsorbents offer an innovative approach to this, with their large surface area, tailored properties increasing the adsorption of these complex superflotants (Salles et al. 2023). Moreover, molecular imprinting and ozone oxidation-based novel systems have also been proposed to enhance the selective remediation of organic contaminants in solid and liquid matrices (Xiongyuan et al. 2016). In a review paper, many techniques, such as bioremediation and chemical oxidation, have been reported to treat pesticide pollution (Marican and Durán-Lara 2018).

4.2 Heavy Metal Adsorption and Immobilization

Adsorption and fixation of heavy metals are the two important approaches in environmental remediation, which have used diverse novel techniques to ameliorate pollution. An efficient method is to solidify waste steel slag, rubbish phosphate, retarders into a solidified body immobilizing heavy metal ions by physically package wrapping, adsorption, and the chemical bond in one body, which is significantly reducing its solubility (Wang et al. 2023).

Another approach is modified steel slag mixed with attapulgite for the treatment of subgrade soil, improving its strength and immobilizing heavy metals (Xu et al. 2019). Furthermore, BCs obtained from broiler litter are potentially useful for immobilizing heavy metals (e.g., Cu-II and Pb-II) in water and soil, with application effectiveness depending on the pyrolysis temperature and pH (Uchimiya et al. 2010). Also, biogenic Mn oxide immobilized on zeolite has greater adsorption capacities for heavy metals than ordinary zeolite (Kim et al. 2016). Finally, a method of treating the sludge with clinoptilolite and low crystallinity minerals has been found to sequester and leach heavy metals effectively, indicating that this method may offer economical and environmental advantages (Perera et al. 2024).

4.3 Role of BC in Mitigating Salinity and Improving Soil Health

BC is a relatively new material with important properties that contribute significantly to mitigation of soil salinity and restoration of the health soil (through improving physicochemical properties, plant growth, and microbial activities). Furthermore, application of BC was found to reduce soil EC (the significant parameter for decaling of saline sodic soils) by 13.2% after 90 days, and thus make the soil plant-friendly (Wang et al. 2023). BC also reduces the toxic harmful ions to the plant growth of Na^+ and Cl^-, that is hazardous to the health of the plant by reducing preferential salinity induced stress (Gao et. al. 2024). The physical adsorption and ion exchange ability of BC facilitate the mitigating soil salt-affected through improving the aeration, aggregation of soil particles and water holding (Duan et al. 2021).

BC not only reduces salinity but it also maintains soil fertility by accelerating soil organic matter, nutrient availability, and extracellular enzyme activity (Wang et al. 2024). Its application was found to increase CEC of soil by 17% thus enhanced the uptake as well as retention of nutrient in plants (Wang et al. 2024). In addition, BC can enhance the enrichment of salt-tolerant microbes, reduce soil metabolic pathways, and alleviate the inhibitory effect from salt ions on microbial activities (Wang et al. 2024; Chauhan and Upadhyay 2024). These microbe–microbe interactions are important in maintaining functioning of the soil in salinity.

BC enhances nutrient and water uptake and decreases electrolyte leakage, preserving the photosynthetic system of the plant and resulting in better development

(Gao et al. 2024). In medicinal plants, BC improves the plant growth and biomass by increasing the availability of soil water and nutrients; hence, the chemical fertilizer requirement is decreased (Patani et al. 2023). Nevertheless, the performance of BC application strategies on BC will vary with both baseline soil properties and BC types; additional studies are needed to determine when (doses) such applications should be broadly adopted in different agricultural systems and crops (Wang et al. 2024). These studies are crucial to bridge the gap (knowledge) that impedes BC, as a sustainable option for both soil reclamation and food production.

4.4 BC's Impact on Drought Resilience and Water Retention

In some researches, BC is found to play a key role in improving the drought resistance as well as water retention of crop soils. It further enhances soil porosity, nutrient availability, and soil structure that is also useful for meeting the water needs of crops (Naeem et al. 2024; Li et al. 2021). For example, addition of activated BC increased corn yield and physiological responses under drought stress by increasing protein content and antioxidant enzyme activity (Naeem et al. 2024).

In addition pelletized BC and carboxymethylcellulose have improved significant maize growth parameters, i.e., shoot and root weight under drought as potential to ameliorate water availability (Danish et al. 2024). Meanwhile, the high pore size of BC and its hydrophilic functional groups, including when such BC is produced at pyrolysis temperature, has been linked with improved soil water holding capacity and plant growth under drought stress (Li et al. 2021). BC is an alternative amendment for improving drought resilience in crops in general (Li et al. 2021; Anwar et al. 2024), but even more so in the more drought-prone sandy soil types (Li et al. 2021; Anwar et al. 2024).

5 BC-Based Technologies in Industrial and Agricultural Sectors

Variety of applications to BC technologies across sectors and ag settings. The emphasis needs to be on sustainable life and human-induced climate change. BC has been reported to work as a soil amendment in agriculture promoting soil health, use efficiency on nutrient and development of crops, and growth medium for hydroponic cultivation with less chemical fertilizer dependence (Jagnade et al. 2022; Liu et al. 2023). At industrial level, being BC produced via processes Pyr and gasification is obtained from organic wastes (Köves et al. 2024; Nyambo et al. 2023), the carbonaceous matrix can be used in different applications from thermal insulation to air pollution control. Such waste management model integrates BC into more than pollution abatement and bioenergy recovery of the waste control system by

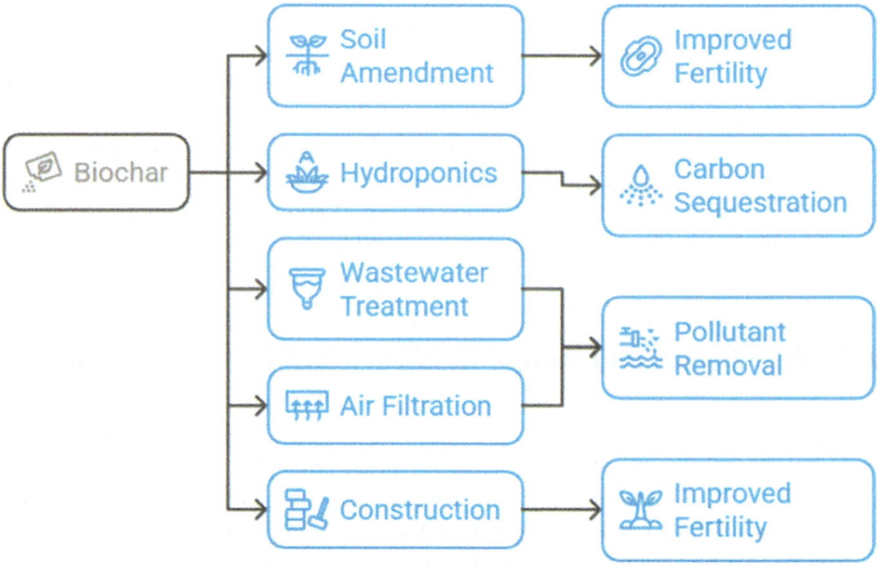

Fig. 2 Biochar application pathways

effectively valorization of the waste, thereby enhancing product resource efficiency (Nyambo et al. 2023). Additionally, these attributes of BC could be appropriate for the purpose of sustainability covered by EU addressing reduction in losses of natural resources and the formation biodegradable products (Gąsior and Tic 2017). In the broadest of strokes, BC would appear to be one solution in improving farming and responsible industry.

Figure 2: Schematic of the biochar value chain in a circular bioeconomy involving the conversion of biomass feedstock (i.e., agricultural or organic waste) to biochar using pyrolysis and its application to sectors such as agriculture, environmental engineering, and construction. In this way, soil health is enhanced, the nutrients are kept, carbon is sequestered, the waste, spent biochar can be re-pyrolyzed or composted, and the recycling plant contributes to waste valorization, resource efficiency, and environmental sustainability.

5.1 Integration of BC in Wastewater Treatment Systems

BC-based applications are attracting more and more attention in agriculture and industry due to the promising effects of BC on soil improvement, organic waste recycling, and climate change abatement. BC possesses remarkable physicochemical characteristics, including high surface area, porosity, and carbon sequestration ability, which may implicate its potential for commercial usage (Liu et al. 2023). For agriculture, the use of BC as a soil amendment is instrumental for maintaining soil, nutrients, and crop productivity and is also a sink of carbon dioxide as a result of the process called "Role of BC in Agriculture to Enhance Crop Productivity: An Overview" in 2023. Furthermore, by turning agricultural wastes into BC, an alternative wastewater management practice is being sought. This would also minimize the need for synthetic fertilizers and the emission of greenhouse gases (Nyambo et al. 2023). Alternatively, in addition to soil application, BC applications to hydroponics and composting can potentially aid nutrient recovery, reduce non-renewable resources, and combat climate-responsive food production (Li et al. 2021).

In commercial operation, BC already can be generated by means of diverse thermochemical processes (e.g., pyrolysis/gasification); both have to be adapted to the raw material/bioreaction conditions and requests on product performance (Köves et al. 2024). Production of BC using organic waste is not only a waste-to-energy concept, but also leads to value process including valuable products for both energy production and soil treatment (Nyambo et al. 2023). Light and strong insulating materials of BC were also considered (BC is an insulation material), which could lead to energy-saving technologies in construction (Gąsior and Tic 2017). Nevertheless, there are several barriers for the spread of BC technology including its relatively high production cost, variety in quality of feedstock, and poor end-user knowledge. The regulation of these bottlenecks is indispensable to realize the potential of BC in sustainable agricultural and industrial practices (Liu et al. 2023; Köves et al. 2024). To further enhance the value of BC as a resource for health and sustainable development of communities, more research and formulation of policies on BC to coordinate production, economy and application of BC in various sectors will be required.

5.2 *Biochar as a Soil Amendment in Precision Agriculture*

BC-derived technologies have emerged as potential global solutions, in both agricultural and industrial sectors, with several environmental and agronomic benefits resulting from their intrinsic physico-chemical properties. In agroecosystems, BC application can improve C and N cycling, increase soil fertility (Liu et al. 2023). This use of agricultural waste to produce BC is a classic example of circular economy in the reduction in landfills, artificial fertilizer use, and release of greenhouse gases (Nyambo et al. 2023). BC also has wide application prospects in the hydroponic cultivation system, composting to promote nutrient circulation and is a key factor for sustainable intensive agricultural development (Li et al. 2021).

The techniques for industrial production of BC vary greatly, and pyrolysis and gasification are the two major precursors in it, each has its own superiority according to type of feedstock knowledge (Köves et al. 2024). The value addition on organic waste stream through conversion into biomass may resolve serious waste management problems, and also produce soil amendment products and energy (Nyambo et al. 2023). BC's insulation and carbon negativity has also initiated research in the application of BC as a construction material, namely for its thermal properties and carbon negativity among ecoconstruction solutions (Gąsior and Tic 2017).

However, while these benefits have been documented and confirmed, the widespread application of BC technologies faces a number of obstacles, including environmental, technological, cost-benefit analysis, and lack of awareness from end-users. While the material continues to show great promise as a sustainable solution to many of our planetary problems, addressing these barriers will require simultaneous work on technology innovation, policy instruments, and markets from a wide range of sectors. It is important to optimize the production processes of BC, enhance the economic feasibility of BC usage, and develop standard application protocols to promote its use in agriculture and industry in subsequent studies.

5.3 Circular Economy and Waste Valorization Through BC

BC production has been recognized as a promising circular economy system for the sustainable valorization of biomass waste and possesses multiple environmental benefits. Demonstrating great feedstock flexibility, those technologies could convert a variety of organic waste streams, e.g., municipal green waste, agricultural residues, and waste-activated sludge, to valuable commodities (Li et al. 2024). Significant long-term carbon sequestration and greenhouse gas emission reduction potential is realized with this conversion to offset landfill disposal and help achieve these BC systems' long-term carbon sink potential. The economic dimension of BC systems is further supported by the possibility to produce a complementary revenue from synergistic use, such as the cost offset for waste management and the production of products for agronomic and ecological siderurgical purposes.

Applications of BC span different fields, and the use of BC as a soil amender seems to be the most developed. BC is also commonly used for agricultural systems to enrich physicochemical soil properties, enhance the soil nutrient cation exchange capacity, and reduce dependence on synthetic fertilizers, classifying it as part of circular economy activities for sustainable intensification of crop production. It has recently been reported that the WS quality organic dyes can be used to treat wastewater because of its high specific surface and porosity, which effectively adsorb many pollutants. Furthermore, BC production could also be integrated with bioenergy systems through cogeneration, in which BC is coupled with existing bioenergy and waste management opportunities for enhancing the renewable energy portfolio. Although there are many potential advantages of using BC technology, process standardization and product quality control represent two major constraints toward wider application, due in large to the inherent variability of feedstock composition and pyrolysis conditions. The industry itself recognizes that overcoming these technical challenges through R&D and setting industry standards will be essential for BC to fulfill its potential as a circular economy solution. In the future, production and quality assurance protocols will require optimization for the different waste streams, and economic sustainability throughout the value chain will be enhanced by the creation of strong markets for BC products (Table 1).

Table 1 Summary of BC innovations and applications

Category	Key innovations/findings	Applications	Challenges
Production technologies	Advanced pyrolysis (fast, hydrothermal) (Bridgwater 2022; Lucian and Fiori 2017), catalytic modification (Han et al. 2021), sustainable feedstock (Dissanayake et al. 2020)	Customizable BC for remediation, energy generation (Grand View Research 2025)	Scalability (Tang et al. 2024), cost efficiency (Dickinson et al. 2015)
Engineered BC	Functionalization (chemical activation (Wang et al. 2023), mineral impregnation (Yang et al. 2023), nano-BC (Pathak et al. 2024), hybrid composites (Gao and Goldfarb 2021)	Targeted pollutant removal (Beljin et al. 2024), biomedical uses (Garg et al. 2024)	Long-term environmental effects (Wang et al. 2022), nanomaterial risks (Deb 2024)
Pollutant remediation	High removal efficiency (> 90%) for heavy metals (Uchimiya et al. 2010), organic pollutants (Miserli et al. 2022), salinity mitigation (Wang et al. 2024)	Water purification (Kapoor and Zdarta 2024), soil health (Gao et al. 2024)	Field-scale variability (Amalina et al. 2023), secondary waste (Lim et al. 2016)
Agricultural use	Drought resilience (Naeem et al. 2024), water retention (Li and Tan 2021), salt stress alleviation (Patani et al. 2023)	Precision agriculture (Liu et al. 2023), circular economy (Singh et al. 2022)	Economic viability (Gong et al. 2024), regional suitability (Zhang et al. 2024)
Industrial integration	Wastewater treatment, construction materials (Gąsior and Tic 2017), bioenergy (Nyambo et al. 2023)	Sustainable waste management (Pavesi et al. 2024)	Standardization (Franco et al. 2024), policy gaps (Cavallin 2025)
Sustainability	Carbon sequestration (Lehmann and Joseph 2024), LCA (Wewer 2023), circular economy (Carvalho et al. 2022)	Climate change mitigation (Gupta and Kapoor 2023)	Regulatory frameworks (Vereš et al. 2014), long-term studies (Masud et al. 2023)

6 Environmental and Economic Sustainability of BC Technologies

BC technologies promise to improve environmental and economic sustainability on a sector-wide scale. While the use of BC in agricultural systems like rice–wheat rotations has shown progress in reducing greenhouse gas emissions and improving soil health, economic factors are still a challenge due to the expenses that come with BC amendments (Gong et al. 2024). Furthermore, BC possesses some positive properties with respect to the organic carbon content and porosity which make it useful in combating environmental pollution as well as climate change (Wangd 2023;

Gupta and Kapoor 2023). Based on life cycle assessments of BC production, this approach is in line with circular economies principles as it was resource efficient and carbon neutral, thus helps us to design an environmentally friendly economic batch (Carvalho et al. 2022). In addition, BC has immense potential in various industrial applications such as agriculture and construction sectors due to its adaptability for suitable solutions toward sustainable development goals and waste disposal (Gąsior and Tic 2017). In summary, while BC technologies offer readily anticipated benefits for environmental and economic performance in products and services, further research, best practice analysis including field studies will help to deliver their cost effectiveness as well as longer-term sustainability.

Figure 3: Circularity of the biochar (BC) economy model, starting with biomass feed stocks, (agricultural residues, forestry waste, organic municipal waste) being converted thermochemically (e.g., pyrolysis) into BC, bio-oil, and syngas; the bio-oil is used for power production and heat in a cogeneration plant and the resulting BC can be employed across the economy (agriculture and food production, environmental remediation for heavy metal, and contaminant removal (soil and water) and materials for the construction industry as composites additives), the used BC and residues can be looped back into feed stocks, building a regenerative loop that can be quantified by the LCA and carbon footprint analysis to ensure GHG reduction, nutrient recycling, and final waste valorization, capture the principles of resource efficiency, carbon neutrality, and system resilience in a circular economy.

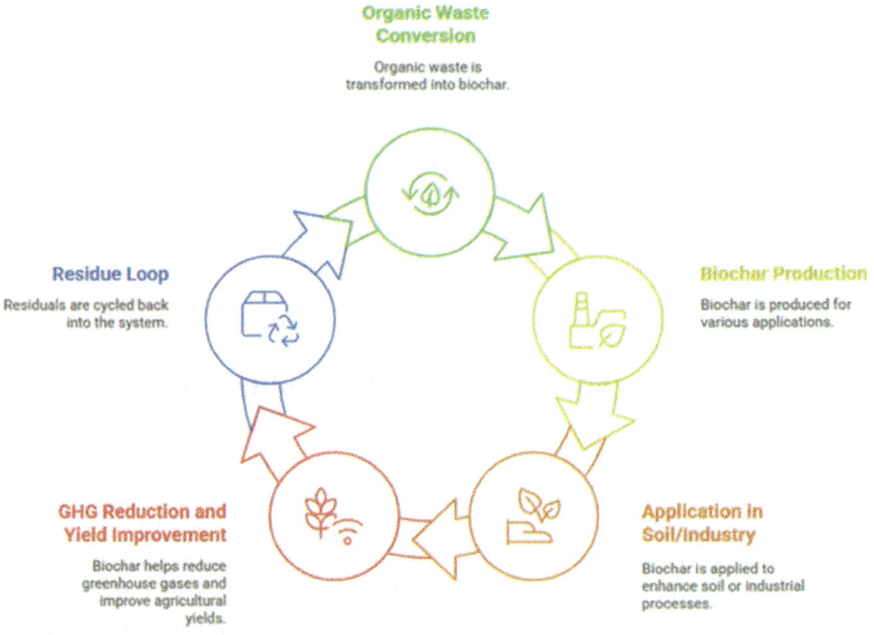

Fig. 3 Biochar circular economy cycle

6.1 Life Cycle Assessment and Carbon Footprint Analysis

Life cycle assessment (LCA) is a comprehensive method for evaluating the environmental impacts of products and services over their entire life cycle, including resource extraction up to end-of-life (Wewer 2023). This is particularly powerful for the carbon footprint analysis (the amount of greenhouse gases emitted) with several processes. For example, one study has recently examined post-operative telemedicine, which found that remote follow up resulted in lower carbon emissions (41.2 kgCO_2/ patient on average) per patient tree for every 6.9 patients the celebrated (Lathan et al. 2024), and was also equivalent to planting a tree for every 6.9 patients reviewed. In addition, a substation's carbon footprint was studied and the results reveal construction as contributing 59.52% of total emissions, therefore, further illustrating the material selection is important as it will also determine the total amount of carbon emission output (Wang et al. 2023). The amounts of raw materials used in biomass power plants were attributed to emissions and consequently also showed the requirements for sustainable supply (Kaewmanee et al. 2023). In addition, semi-transparent PV systems were identified to have smaller carbon footprints than conventional systems in a comparison of photovoltaic (PV) technology options, illustrating that design changes can help reduce emissions (Gonçalves et al. 2021). In summary, LCA is a wide-spread method which in fine may contribute to finding ways of reducing the carbon footprint in various respects.

6.2 Cost-Benefit Analysis of BC Production and Application

BC production and application: A comparative CBA analysis across different contexts BC, derived from a variety of biomass-source materials, improves soil fertility and agricultural performance with net benefits reported between 353,436 and 1.44 million CFA, depending on the feedstock (e.g., rice husk and corn cob) (Fru et al. 2018). For BC application in sub-Saharan Africa, increased cereal grain production translated to positive net present values (NPV) when benefits are assessed over long periods of time, especially in cases where the effect on yield was retained for 30 years (Dickinson et al. 2015). In contrast, in North-Western Europe, NPVs are still negative, showing how their economic viability differs between regions (Dickinson et al. 2015). Moreover, from life cycle assessments (LCA), it has been revealed that although BC production involves environmental littorals, BC application can provide substantially more carbon sequestration advantages compared to production effects (You and Wang 2019; Sparrevik et al. 2014). BC has different economic feasibility in terms of local agricultural settings, production methods, and market factors; these factors must be evaluated to ascertain the availability of implementing BC at a certain location (Bergman et al. 2022).

6.3 Policy and Regulatory Perspectives on BC Utilization

The policy and regulatory environment for BC is complex and geographically variable. In the United States, BC production methods especially those that reduce open burning are regulated under the Clean Air Act, but differing definitions and permitting requirements in various states create an unclear situation for producers (Franco et al. 2024). Globally, there is an increasing awareness of the potential of BC to sequester carbon and improve soil health, highlighting the need for well-defined certification frameworks to guarantee its sustainability and effectiveness (Vereš et al. 2014). In Europe, current regulations and policies on BC focus on the use of charcoal as a soil amendment; indeed, there are no detailed quality specifications for using manure-based BC (Ecochar) that could support its use as fertilizer (Gollenbeek et al. 2018). In addition, the inclusion of BC into agricultural policy and in particular the work of the Common Agricultural Policy) reinforces BC's role in sustainability and achieving environmental goals (Cavallin 2025). In conclusion, well-defined regulation and policy development could help with BC ecosystem expansion in composting, agriculture, and food packaging (Bhattacharya et al. 2024).

7 Challenges and Future Perspectives

BC technology challenges, future perspectives, and its multifunctional applications toward environmental sustainability and energy production. Functionalized BC has recently emerged as a promising new platform to boost catalytic activity and energy storage, two key characteristics for applications such as wastewater treatment and CO_2 capture (Yameen et al. 2024). Nonetheless, food production needs to increase, and greenhouse gas emissions need to decrease by 2050; these two challenges will be addressed through the large-scale manufacturing of BC (Chhetri 2020). It further emphasizes the promising potential of BC in wastewater pollutant removal due to its effectiveness compared to conventional treatment methods (Amalina et al. 2023). In fact, it is not the nanomaterials themselves that are the products, but rather the phenomena that occur when two or more nanomaterials are used together (Dash et al. 2022), as the application of BC as a soil amendment (BC) and pollutant remediation (BC) holds great promise, the unintended consequences of its application, or the BC itself, may involve risks to human health and the environment, as noted, we need to do more research in advance on how to optimize the use of BC as a soil amendment and pollutant remediation tool (Wang et al. 2022). Thus, engineered BC's properties and role in sustainable development must be studied in future directions, while maximizing its benefits and minimizing related risks (Cheng et al. 2023).

7.1 Limitations in Large-Scale BC Implementation

Scaling up BC use for C sequestration has several challenges that limit its large-scale adoption. Economic viability is a major limitation since the pricing of feedstock supply and production, and byproduct pricing play crucial roles in scalability (Tang et al. 2024). Moreover, another aspect that makes BC less effective in different soil types and environmental conditions is the variation of its properties, which depends on feedstock types and pyrolysis processes (Zhang et al. 2024; Amalina et al. 2023). Moreover, there are some possible safety issues regarding the use of BC in agriculture, mainly the potential risks of toxic element leaching and PM pollution (Upadhyay et al. 2024). The uncertainty around BC environmental impacts and efficacy is also driven by a lack of comprehensive long-term studies and standardized protocols for BC application (Masud et al. 2023; Amalina et al. 2023). International Efforts to Undertake a Comprehensive BC Global Atlas: Challenges to Development and Examples of Development Projects. One of the issues of the BC map is also to design what areas globally have a comparative advantage in producing BC.

7.2 Knowledge Gaps and Research Directions in BC Innovation

Therefore, a comprehensive review of BC innovation, indicating several existing critical knowledge gaps and perspectives in the future, should be done across multiple BC applications, including asphalt engineering, soil carbon sequestration in agriculture, and environmental management. Despite this recognized potential, comprehensive reviews call for systematic studies to assess and reveal the potential of BC applicable to asphalt pavements in terms of its properties, production methods, and role as a modifying additive based on life cycle assessments and economics (Zhou 2024). Critical challenges in soil applications include understanding BC soil interactions, the long-term effects on carbon cycling, understanding the mechanisms involved, and modeling results to develop protocols for standardized application (Luo et al. 2023). Furthermore, although BC can help improve soil properties and offer protection against abiotic stresses, to maximize the benefits of BC application in agriculture, especially in regard to the crop yield in a stressed environment, warranting more investigations on the effects of BC on crop yield under abiotic stressed conditions; (Das et al. 2020; Rathinapriya et al. 2025). In summary, filling these gaps is a prerequisite to enabling the wide use of BC technologies in sustainable environmental and agricultural practices.

7.3 Future Trends in BC Technology and Commercialization

Innovations in production and increasing recognition of its environmental benefits underpin the future of BC technology and its commercialization. Novel techniques, such as microwave-assisted pyrolysis (MAP) and gasification, are also being used because of their energy conservation efficiency to produce BC with desirable properties for advancing specific applications in soil amendment and contaminant sequestration (Selvam and Balasubramanian 2023; Wagh et al. 2024). The increasing demand for sustainable biosolids management, especially with increasing regulations on several contaminants like PFAS, is driving the implementation of technologies that produce biochar, as these provide cost-effective methods while reducing greenhouse gas emissions (Abu-Orf et al. 2023). And the global problem of needing to produce twice as much food by 2050 without polluting the atmosphere offers potential for BC as a carbon-sucking resource that could also become widely adopted (Chhetri 2020). Novel production methods, regulatory policies, and market drivers for BC integration in various industries will significantly influence the commercialization of BC (Cheng et al. 2023; Wagh et al. 2024).

7.4 Environmental Risks and Considerations in Emerging BC Technologies

BC is beginning to attract international attention; as a consequence, evaluating the environmental positive and negative impact of BC is vital for sustainable development and policy application. The sequestration of carbon is one of the more widely known environmental services in BC. The use of BC in soil has been reported to stabilize organic carbon for a longer term, influencing global warming, and lower atmospheric GHG contents (Naseem et al. 2024). Also, BC improves the soil fertility and consequently the crop yield by elevating the circulation and decreasing the nutrient leaching from soil profiles (Qadeer et al. 2017). In addition to being highly available and porous, the GO provides efficient adsorption for heavy metals, pesticides, and other pollutants; thus, it can be an attractive material for polluted soils and water bodies (Qadeer et al. 2017).

But BC is not devoid of environmental hazards. And the production of BC, especially through high-temperature pyrolysis, can emit poisonous substrates, such as heavy metals and PAHs, which have the potential to be leached into the surrounding soil and water, resulting in an ecological disaster (Xu 2023; Dong et al. 2025). Apart from production-related effects, BC application may lead to soil health changes as well, mediated indirectly, such as interference with the soil microbiome and potential soil acidification or disproportions in availability of nutrients (Naseem et al. 2024; Dong et al. 2025). BC production could also be water demanding, but particularly in steam-assisted or hydrothermal approaches, which could be challenging where water is scarce (Naseem et al. 2024).

In conclusion, BC is a promising candidate for supporting environmental resilience, such as C sequestration, soil amelioration, and pollution attenuation, although it should be used with full awareness of potential risks. There is an urgent requirement for more studies, environmental monitoring, quality, and safety criteria to fulfill long-term benefits, which might not be worse than the possible risks of BC technologies.

7.5 Environmental Implications of Nanotechnology and Advanced Pyrolysis

Greater attention to the environmental impacts of emerging technologies such as nanotechnology and fast pyrolysis is required now, as they are likely to be used more extensively for both BC production and BC-assisted environmental remediation. These advancements have promising potential for sustainable technological innovation; however, they involve environmental risks with regard to their reasonable application that must be mitigated to ensure such.

One if the major considerations is the toxicity of commonly used nanoparticles in engineered BC, e.g., nanoparticles such as silver nanoparticle (Ag, N, P, S) can be passed on in terms of bio-accumulation causes higher toxic effects and disturbance of the aquatic eco-system. Moreover, they possess nanometric size capable of restraining the bio-barriers and killing cells. However, the literature reports also claims that small size NPs can induce oxidative stress and DNA damage and may have adverse effects on both humans and wildlife species. Moreover, few nanomaterials like CNTs may have negligible acute toxicities but could be potentially hazardous with respect to their environmental longevity and long range ecological effects (Prasad 2023).

Environmental issues related to the energy-intensive pyrolysis methods for the production of advanced BC are also applicable. Despite these methods enhancing the efficiency and tunability of the BC properties, substantial energy input (especially based on high temp/microwave-assisted treatments) is generally involved. If that energy is derived from sources that are not renewable, the climate savings could be eroded very quickly by greenhouse gas emissions. Furthermore, the pyrolysis of contaminated or mixed blends results in the production of dangerous pollutants such as volatile organic compounds and PM, therefore forcing regulated emission approaches (Anwar and Khitab 2019; Yang and Lin 2014).

For this reason technology must advance together with a green alternative. This will necessitate all-encompassing and well-structured regulations, lifecycle assessments, as well as deliberate research efforts to control the adverse effects. Knitting stewardship and technological innovation is a key challenge of implementing nano-technology and advanced pyrolysis in a sustainable BC system.

7.6 Bridging Lab-Scale Innovations of BC and Field-Scale Applications

The transfer of lab-based developments (such as nano-biochar, NBC) into field-scale agronomic applications seems to be a promising horizon for promoting sustainable soil management and agricultural processes. The modified BC named NBBC has improved physical and chemical properties, and it can show better soil matrix interaction potentiality. These attributes render NBC a powerful instrument to face some of the most challenging problems in modern agriculture like nutrient underutilization, soil erosion and climate adaptability.

NFC is a soil biofertilizer whose surface area is known to be significantly more than the soil systems, thus improving water and nutrient constants in the soil system. This physical improve extends the capacity of soil water control and nutrient that can be absorbed by plant. Furthermore, the chemical inertness of nano-BC facilitates in-soil advances and long lasting interactions that favor micro biotic such effects on soil flora and fertility (Kumari and Garg 2024). These properties make NBC a better amendment for agricultural use.

The above-mentioned applications where empirical evidence has been well established are also indicative of the worth of NBC in soil treatment. It can reduce soil salinity and cation exchange capacity, remediate the degraded soil and increase organic matter. It has also been found to induce growth promotion, disease resistance, abiotic stress tolerance, and beneficial effects on the overall crop (Chausali et al. 2021). The application of NBC contributes to carbon sequestration, climate change abatement, and improvement in soil health (Kumari and Garg 2024). Furthermore, due to its excellent capability of nutrient and pollutant adsorption, it is a valuable and environment-friendly option in waste management for agricultural use (Saraugi and Routray 2024).

Although NBC platform technologies offer significant benefits, many obstacles remain in scaling up NBC-based solutions from the controlled settings of the laboratory to real-world industrial environments. The barriers include high production costs, limited access for small-scale farmers, and the absence of long-term field studies. Subsequent work should prioritize these areas, including developing cost-effective production strategies, field validation across a range of agroecological contexts and complete assessment of long-term environmental impacts, to guarantee the sustainable and equitable deployment of NBC technologies.

8 Conclusion

As one of the most exciting developments in the field of BC use, one of the most innovative and environmentally sustainable examples of BC technology has developed dramatically in recent years. Notable innovations encompassed improved pyrolysis approaches (fast and hydrothermal carbonization), catalytic modification, and the development of engineered and nano-BC with enhanced adsorption properties. These advancements have facilitated targeted pollutant removal, increased soil fertility, enhanced water retention, and effective carbon sequestration.

Additionally, innovation has been significant in realizing the full environmental potential of BC. BC has transformed from a simple soil amendment into a multidimensional platform technology to solve challenges with the mother technologies—Material science, nanotechnology and biotechnology in removal, as good bioenhancers in agro and in industrial systems. Creative design concepts such as living composites, BC metal hybrids, and stimuli responsive materials offer pathways to enhance the potentiality of BC for multiple applications.

The future is bright for ECSs to sustainable and scalable BC applications. BC and circular economy driven waste management practices, precision agriculture, and climate smart waste management policy are hopes for resilient global environment. But unlocking it for the masses means overcoming economic, regulatory, and technical challenges. (CB: by the pedestal-based biomass to high-quality, truly sustainable chemical). Accordingly, BC production is an ongoing focus for research and development, and the need to adopt policy interventions and investments that position it as a key cornerstone for sustainable environmental and agronomic policies worldwide is important.

Acknowledgements The authors acknowledge the editor, for providing opportunity. The authors like to thank Haldia Institute of Technology for their support in carrying out this work.

Authors Contribution G.P. conceptualization, writing the article (including picture); methodology, investigation; model design and implementation. GP also contributed in supervision and visualization development source; GP was responsible for significantly working to accuracy and integrity of all portions of the work which involves writing the first draft of the manuscript, substantial contribution toward review/revision. B.L.N.K.M had edited, written, and critically reviewed the manuscript.

References

Abu-Orf M, Blate M, Dursun D (2023) A new biosolids management trend is emerging: biochar producing technologies are rising to the top, wonder why? Proc Water Environ Fed. https://doi.org/10.2175/193864718825159012

Aghbashlo M, Almasi F, Jafari A, Nadian MH, Soltanian S, Lam SS, Tabatabaei M (2021) Describing biomass pyrolysis kinetics using a generic hybrid intelligent model: a critical stage in sustainable waste-oriented biorefineries. Renew Energy 170:81–91. https://doi.org/10.1016/j.renene.2021.01.111

Ahmad F, Saeed Q, Shah SMU, Gondal MA, Mumtaz S (2022) Environmental sustainability: challenges and approaches. In: Natural resources conservation and advances for sustainability. Elsevier, pp 243–270. https://doi.org/10.1016/B978-0-12-822976-7.00019-3

Al-Rumaihi A, Shahbaz M, Mckay G, Mackey H, Al-Ansari T (2022) A review of pyrolysis technologies and feedstock: a blending approach for plastic and biomass towards optimum biochar yield. Renew Sustain Energy Rev 167:112715. https://doi.org/10.1016/j.rser.2022.112715

Amalina F, Abd Razak AS, Zularisam AW, Aziz MAA, Krishnan S, Nasrullah M (2023) Comprehensive assessment of biochar integration in agricultural soil conditioning: advantages, drawbacks, and future prospectives. https://doi.org/10.1016/j.pce.2023.103508

Amalina F, Krishnan S, Zularisam AW, Nasrullah M (2023) Biochar and sustainable environmental development towards adsorptive removal of pollutants: modern advancements and future insight. Chem Eng Res des 173:715–728. https://doi.org/10.1016/j.psep.2023.03.069

Anwar T, Qureshi H, Gull S, Siddiqi EH, Chaudhary T, Ali HM (2024) Enhancing Zea mays growth and drought resilience by synergistic application of Rhizobacteria-Loaded Biochar (RBC) and externally applied Gibberellic Acid (GA). https://doi.org/10.1016/j.eti.2023.103517

Anwar W, Khitab A (2019) Nanotechnology from engineers to toxicologists: risks and remedial measures. Int J Appl Nanotechnol Res (IJANR) 4(2):1–25. https://doi.org/10.4018/IJANR.2019070101

Bangar SP, Singh A, Ashogbon AO, Bobade H (2023) Ball-milling: a sustainable and green approach for starch modification. Int J Biol Macromol 237:124069. https://doi.org/10.1016/J.IJBIOMAC.2023.124069

Baskar AV, Bolan N, Hoang SA, Sooriyakumar P, Kumar M, Singh L, Jasemizad T, Padhye LP, Singh G, Vinu A, Sarkar B, Kirkhąm MB, Rinklebe J, Wang S, Wang H, Balasubramanian R, Siddique KHM (2022) Recovery, regeneration and sustainable management of spent adsorbents from wastewater treatment streams: a review. Sci Total Environ 822:153555. https://doi.org/10.1016/J.SCITOTENV.2022.153555

Beljin J, Đukanović N, Anojčić J, Simetić T, Apostolović T, Mutić S, Maletić S (2024) Biochar in the remediation of organic pollutants in water: a review of polycyclic aromatic hydrocarbon and pesticide removal. Nanomaterials 15(1):26. https://doi.org/10.3390/nano15010026

Bergman R, Sahoo K, Englund K, Mousavi-Avval SH (2022) Lifecycle assessment and techno-economic analysis of biochar pellet production from forest residues and field application. Energies 15(4):1559. https://doi.org/10.3390/en15041559

Bhattacharya T, Khan A, Ghosh T, Kim JT, Rhim J-W (2024) Advances and prospects for biochar utilization in food processing and packaging applications. Sustain Mater Technol. https://doi.org/10.1016/j.susmat.2024.e00831

Biochar enhancing soil resilience: a dual strategy for mitigating heavy metal contamination and drought stress (2024) Lett Appl NanoBioSci 13(2):70. https://doi.org/10.33263/lianbs132.070

Carvalho J, Nascimento L, Soares M, Valério N, Ribeiro A, Faria L, Silva A, Pacheco N, Araujo J, Vilarinho C (2022) Life cycle assessment (LCA) of biochar production from a circular economy perspective. Processes 10(12):2684. https://doi.org/10.3390/pr10122684

Cavallin E (2025) Farming and biochar in the EU and the road to sustainability: drawing connections through the Common Agricultural Policy and organic and carbon farming regulation. Rev Eur Comp Int Environ Law. https://doi.org/10.1111/reel.12592

Chauhan PK, Upadhyay SK (2024) Mixed consortium of salt-tolerant phosphate solubilizing bacteria improves maize (*Zea mays*) plant growth and soil health under saline conditions. Mol Biotechnol 66:489–499. https://doi.org/10.1007/s12033-023-00771-6

Chauhan P, Imam A, Kanaujia PK, Suman SK (2023) Nano-bioremediation: an eco-friendly and effective step towards petroleum hydrocarbon removal from environment. Environ Res 231:116224. https://doi.org/10.1016/J.ENVRES.2023.116224

Chauhan PK, Upadhyay SK, Rajput VD (2024) Fostering plant growth performance under drought stress using rhizospheric microbes, their gene editing, and biochar. Environ Geochem Health 46:41. https://doi.org/10.1007/s10653-023-01823-1

Chauhan PK, Pathak HK, Dubey G, Sharma H, Upadhyay SK (2024) Impact of Bacillus cereus SPB-10 on growth promotion of wheat (*Triticum aestivum L.*) under arsenic contaminated soil. Curr Microbiol. https://doi.org/10.1007/s00284-024-03673-1

Chauhan PK, Upadhyay SK, Chauhan A, Bhutiani R, Sikarwar RLS, Tiwari MK (2025) A review on simultaneous use of microbes and biochar for removal of toxic heavy metals: recent updates and future outlooks. Physiol Mol Biol Plants 31(3):1–15. https://doi.org/10.1007/s12298-025-01649-4

Chauhan PK, Upadhyay SK (2023). Exo-polysaccharide producing bacteria can induce maize plant growth and soil health under saline conditions. Biotechnol Genetic Eng Rev 1–20

Chausali N, Saxena J, Prasad R (2021) Nanobiochar and biochar based nanocomposites: advances and applications. J Agric Food Res 5:100191. https://doi.org/10.1016/J.JAFR.2021.100191

Cheng H, Song Y, Bolan N (2023) Biochar for future and futuristic biochar. Pedosphere 33(5):680–682. https://doi.org/10.1016/j.pedsph.2023.07.023

Chhetri NB (2020) Leveraging current knowledge and exploring future potential of biochar. Int J Environ Sci 24(1):9–12. https://doi.org/10.19080/IJESNR.2020.24.556127

Choudhary K, DeCost B, Chen C, Jain A, Tavazza F, Cohn R, Park CW, Choudhary A, Agrawal A, Billinge SJL, Holm E, Ong SP, Wolverton C (2022) Recent advances and applications of deep learning methods in materials science. NPJ Comput Mater 8(1):1–26. https://doi.org/10.1038/s41524-022-00734-6

Danish S, Hasnain Z, Dawar K, Fahad S, Shah AN, Salmen SH, Ansari MJ (2024) Enhancing maize resilience to drought stress: the synergistic impact of deashed biochar and carboxymethyl cellulose amendment. BMC Plant Biol 24(1):139

Das SK, Das SK, Ghosh GK, Avasthe R (2020) Application of biochar in agriculture and environment, and its safety issues. Biomass Conv Biorefinery 1–11. https://doi.org/10.1007/S13399-020-01013-4

Deb P (2024) Enhancing the nutrient use efficiency through nano-biochar. Nanomater Nano-Biochar Reduc Soil Stress 115–136. https://doi.org/10.1201/9781003503453-6

Dickinson D, Balduccio L, Buysse J, Ronsse F, Van Huylenbroeck G, Prins W (2015) Cost-benefit analysis of using biochar to improve cereals agriculture. Gcb Bioenergy 7(4):850–864. https://doi.org/10.1111/GCBB.12180

Dissanayake PD, You S, Igalavithana AD, Xia Y, Bhatnagar A, Gupta S, Kua HW, Kim S, Kwon JH, Tsang DCW, Ok YS (2020) Biochar-based adsorbents for carbon dioxide capture: a critical review. Renew Sustain Energy Rev 119:109582. https://doi.org/10.1016/J.RSER.2019.109582

Dong M, Jiang M, He L, Zhang Z, Gustave W, Vithanage M, Niazi NK, Chen B, Zhang X, Wang H, He F (2025) Challenges in safe environmental applications of biochar: identifying risks and unintended consequence. Biochar 7(1). https://doi.org/10.1007/s42773-024-00412-4

Duan M, Liu G, Zhou B, Chen X, Wang Q, Zhu H, Li Z (2021) Effects of modified biochar on water and salt distribution and water-stable macro-aggregates in saline-alkaline soil. J Soils Sediments 21:2192–2202

Duan M, Yan R, Wang Q, Zhou B, Zhu H, Liu G, Guo X, Zhang Z (2023) Integrated microbiological and metabolomics analyses to understand the mechanism that allows modified biochar to affect the alkalinity of saline soil and winter wheat growth. Sci Total Environ 866:161330. https://doi.org/10.1016/J.SCITOTENV.2022.161330

Franco CR, Page-Dumroese DS, Pierson D, Miller M, Miles T (2024) Policy and regulations for mobile biochar production in the United States of America. Forests. https://doi.org/10.3390/f15010192

Fru BS, Angwafo TE, Martin TN, Francis NA, Ngome TP (2018) Environmental and socio-economic feasibility of biochar application for cassava production in the bimodal rainforest zone of cameroon. Environ Health Res 2(1):1–9. https://doi.org/10.22161/IJREH.2.1.1

Gao Z, Ding J, Ali B, Nawaz M, Hassan MU, Ali A, Rasheed A, Khan MN, Özdemir FA, Iqbal R, Çığ A, Ercışlı S, Sabagh A (2024) Putting biochar in action: a black gold for efficient mitigation of salinity stress in plants. Review and future directions. ACS Omega. https://doi.org/10.1021/acsomega.3c07921

Gao L, Goldfarb JL (2021) Characterization and adsorption applications of composite biochars of clay minerals and biomass. Environ Sci Pollut Res 28(32):44277–44287. https://doi.org/10.1007/S11356-021-13858-X/METRICS

Garg S, Bhavya Surendran VS, Avanthi A (2024) Recent trends and applications of biochar and nanoparticles from plant biomass, pp 203–216. https://doi.org/10.1007/978-3-031-64601-0_14

Gąsior D, Tic WJ (2017) Application of the biochar-based technologies as the way of realization of the sustainable development strategy. Econ Environ Stud 17(43):597–611. https://doi.org/10.25167/EES.2017.43.9

Gollenbeek L, Ehlert P, de Buisonjé F (2018) Perspectives of ecochar in Europe: uses and regulatory requirements. Wageningen Livestock Research. https://research.wur.nl/en/publications/perspectives-of-ecochar-in-europe-uses-and-regulatory-requirement

Gonçalves GL, da Silva LP, de Araújo Santos PR, Carvalho M, Abrahão R (2021) Comparaçãoambiental entre sistemafotovoltaicoconvencional e semitransparente 1(53):103. https://doi.org/10.18265/1517-0306A2020V1N53P103-111

Gong X, Shi W, Wu J, Qin JJ, Huang W, Feng Y, Sun H, Zheng J, Cheng K, Joseph S, Chen J, Bian R, Li L, Pan G (2024) Converting biochar into biochar-based urea promotes environmental and economic sustainability in rice-wheat rotation system. GCB Bioenergy 17(1). https://doi.org/10.1111/gcbb.70014

Grand View Research (2025) Biochar market size, share & trends analysis report by technology (gasification, pyrolysis), by application (agriculture, others), by region (North America, Asia Pacific), and segment forecasts, 2024–2030

Gu X, Zhang X, Lyu J, Wang T (2016) Steel slag subgrade soil treatment method by heavy metal immobilization and adsorption

Gupta V, Kapoor RT (2023) Biochar technology for sustainable environment. Elsevier BV, pp 99–111. https://doi.org/10.1016/b978-0-443-14013-6.00007-x

Han L, Sun H, Sun K, Yang Y, Fang L, Xing B (2021) Effect of Fe and Al ions on the production of biochar from agricultural biomass: properties, stability and adsorption efficiency of biochar. Renew Sustain Energy Rev 145:111133. https://doi.org/10.1016/J.RSER.2021.111133

Hitzl M, Mendez A, Owsianiak M, Renz M (2018) Making hydrochar suitable for agricultural soil: a thermal treatment to remove organic phytotoxic compounds. J Environ Chem Eng 6(6):7029–7034. https://doi.org/10.1016/j.jece.2018.10.064

Jagnade P, Panwar NL, Gupta T, Agrawal C (2022) Role of biochar in agriculture to enhance crop productivity: an overview. Biointerface Res Appl Chem 13:429

Joshi M, Bhatt D, Srivastava A (2023) Enhanced adsorption efficiency through biochar modification: a comprehensive review. Ind Eng Chem Res 62(35):13748–13761. https://doi.org/10.1021/ACS.IECR.3C02368/ASSET/IMAGES/MEDIUM/IE3C02368_0007.GIF

Kaewmanee J, Pradabphetrat P, Rangpan V (2023) Carbon footprint assessment based on life cycle assessment of biomass power plant. ASEAN J Sci Technol Rep. https://doi.org/10.55164/ajstr.v27i1.250754

Kamarudin NS, Dahalan FA, Hasan M, An OS, Parmin NA, Ibrahim N, Hamdzah M, Zain NAM, Muda K, Wikurendra EA (2022) Biochar: a review of its history, characteristics, factors that influence its yield, methods of production, application in wastewater treatment and recent development. In: Biointerface research in applied chemistry, vol 12, Issue 6. AMG Transcend Association, pp 7914–7926. https://doi.org/10.33263/BRIAC126.79147926

Kang KH, Saifuddin M, Chon K, Bae S, Kim YM (2024) Recent advances in the application of magnetic materials for the management of perfluoroalkyl substances in aqueous phases. Chemosphere 352:141522. https://doi.org/10.1016/J.CHEMOSPHERE.2024.141522

Kapoor RT, Zdarta J (2024) Fabrication of engineered biochar for remediation of toxic contaminants in soil matrices and soil valorization. Chemosphere 142101. https://doi.org/10.1016/j.chemosphere.2024.142101

Kerner P, Struhs E, Mirkouei A, Aho K, Lohse KA, Dungan RS, You Y (2023) Microbial responses to biochar soil amendment and influential factors: a three-level meta-analysis. Environ Sci Technol 57(48):19838–19848. https://doi.org/10.1021/acs.est.3c04201

Kim D-G, Nhung TT, Ko S-O (2016) Enhanced adsorption of heavy metals with biogenic manganese oxide immobilized on zeolite. KSCE J Civ Eng 20(6):2189–2196. https://doi.org/10.1007/S12205-016-0356-1

Köves M, Madár V, Ringer M, Kocsis T (2024) Overview of traditional and contemporary industrial production technologies for biochar along with quality standardization methods. Land 13(9):1388. https://doi.org/10.3390/land13091388
Kumari S, Garg MC (2024) Role of nano-biochar and biochar-based nanocomposites for improving soil health: application and benefits 71–92. https://doi.org/10.1201/9781003503453-4
Lathan R, Walshaw J, Hitchman L, Ravindhran B, Carradice D, Smith GE, Chetter I, Lockwood S, Yiasemido M (2024) O22: postoperative telemedicine is feasible and sustainable; a prospective life-cycle assessment approach to carbon footprint analysis. Br J Surg. https://doi.org/10.1093/bjs/znae046.056
Lehmann J, Joseph S (2024) Biochar for environmental management: science, technology and implementation, 3rd, illustrated edn. Taylor & Francis
Li H, Tan Z (2021) Preparation of high water-retaining biochar and its mechanism of alleviating drought stress in the soil and plant system. Biochar 3(4):1–12. https://doi.org/10.1007/S42773-021-00107-0
Li L, Zhang Y-J, Novak A, Yang Y, Wang J (2021) Role of biochar in improving sandy soil water retention and resilience to drought. Water 13(4):407. https://doi.org/10.3390/W13040407
Li G, Du R, Cao Z, Li C, Xue J, Ma X, Wang S (2024) Research progress in graphene-based adsorbents for wastewater treatment: preparation, adsorption properties and mechanisms for inorganic and organic pollutants C 10(3):78
Li X, Shen X, Jiang W, Xi Y, Li S (2024) Comprehensive review of emerging contaminants: detection technologies, environmental impact, and management strategies. Ecotoxicol Environ Saf 278:116420. https://doi.org/10.1016/j.ecoenv.2024.116420
Lim MW, Von Lau E, Poh PE (2016) A comprehensive guide of remediation technologies for oil-contaminated soil—present works and future directions. Mar Pollut Bull 109(1):14–45. https://doi.org/10.1016/j.marpolbul.2016.04.023
Liu X, Liu C, Pan G, Clarke N (2023) Biochar-based technology in food production, climate change mitigation, and sustainable agricultural soil management: post terra preta era. Springer, pp 93–112. https://doi.org/10.1007/978-981-99-2828-6_5
Liu X, Wang D, Li Z, Ouyang W, Bao Y, Gu C (2023) Efficient separation of iron elements from steel slag based on magnetic separation process. J Market Res 23:2362–2370. https://doi.org/10.1016/j.jmrt.2023.01.186
Lohar S, Kumari P, Sharma A, Godara V, Nagar V, Prajapati M, Singh M (2024) Biochar enhancing soil resilience: a dual strategy for mitigating heavy metal contamination and drought stress. Lett Appl NanobioSci 13:70
Lucian M, Fiori L (2017) Hydrothermal carbonization of waste biomass: process design, modelling, energy efficiency cost analysis. Energies 10(2):211. https://doi.org/10.3390/en10020211
Luo L, Wang J, Lv J, Liu Z, Sun T, Yang Y, Zhu Y-G (2023) Carbon sequestration strategies in soil using biochar: advances, challenges, and opportunities. Environ Sci Technol. https://doi.org/10.1021/acs.est.3c02620
Lyu H, Gong Y, Gurav R, Tang J (2016) Potential application of biochar for bioremediation of contaminated systems. Elsevier, pp 221–246. https://doi.org/10.1016/B978-0-12-803433-0.00009-6
Ma H, Zhang B, Wang S, Liu C, Zhu L, Zhao Z, Li W, Shao Z, Liu X, Dai Y (2024) Enhanced removal of tetracycline by vitamin C-modified cow manure biochar in water. Sci Rep 14(1):1–16. https://doi.org/10.1038/s41598-024-73210-4
Mansouri H, Ait Said H, Noukrati H, Oukarroum A, Ben youcef H, Perreault F (2023) Advances in controlled release fertilizers: cost-effective coating techniques and smart stimuli-responsive hydrogels. Adv Sustain Syst 7(9):2300149. https://doi.org/10.1002/ADSU.202300149
Marican A, Durán-Lara EF (2018) A review on pesticide removal through different processes. Environ Sci Pollut Res 25(3):2051–2064. https://doi.org/10.1007/S11356-017-0796-2
Masud MdAA, Shin WS, Sarker A, Septian A, Das K, Deepo DM, Iqbal MA, Islam ARMT, Malafaia G (2023) A critical review of sustainable application of biochar for green remediation: research

uncertainty and future directions. Sci Total Environ 166813. https://doi.org/10.1016/j.scitotenv.2023.166813

Medyńska-Juraszek A (2016) Biochar as a soil amendment. Soil Sci 67(3):151–157. https://doi.org/10.1515/SSA-2016-0018

Mercante LA, Andre RS, Facure MHM, Correa DS, Mattoso LHC (2023) Recent progress in conductive electrospun materials for flexible electronics: energy, sensing, and electromagnetic shielding applications. Chem Eng J 465:142847. https://doi.org/10.1016/J.CEJ.2023.142847

Miserli K, Nastopoulou A, Konstantinou I (2022) Removal of organic pollutants pharmaceuticals and pesticides) from sewage sludge by hydrothermal carbonization using response surface methodology (RSM). J Chem Technol Biotechnol 97(11):3111–3120. https://doi.org/10.1002/jctb.7178

Mishra S, Sundaram B (2023) Fate, transport, and toxicity of nanoparticles: an emerging pollutant on biotic factors. Chem Eng Res des 174:595–607. https://doi.org/10.1016/j.psep.2023.04.037

Mridha D, Sarkar J, Majumdar A, Sarkar K, Maiti A, Acharya K, Das M, Chen H, Niazi NK, Roychowdhury T (2024) Evaluation of iron-modified biochar on arsenic accumulation by rice: a pathway to assess human health risk from cooked rice. Environ Sci Pollut Res 31(16):23549–23567. https://doi.org/10.1007/S11356-024-32644-Z/METRICS

Muhammad MH, Yasin Ashraf M, Iqra Almas H, Zaib-un-Nisa Ali N, Khaliq B, Ahmad Ansari M, Singh R, Gul S (2024) Synergistic effects of soil and foliar nano-biochar on growth, nitrogen metabolism and mineral uptake in wheat varieties. J King Saud Univ Sci 36(9):103392. https://doi.org/10.1016/J.JKSUS.2024.103392

Murtaza G, Ahmed Z, Eldin SM, Ali I, Usman M, Iqbal R, Rizwan M, Abdel-Hameed UK, Haider AA, Tariq A (2023) Biochar as a green sorbent for remediation of polluted soils and associated toxicity risks: a critical review. Separations 10(3):197. https://doi.org/10.3390/separations10030197

Naeem MB, Jahan S, Rashid A, Shah AA, Raja V, El-Sheikh MA (2024) Improving maize yield and drought tolerance in field conditions through activated biochar application. Dent Sci Rep 14(1). https://doi.org/10.1038/s41598-024-76082-w

Naseem M, Iqbal S, Malik H, Awais M, Jehan S, Jabeen S (2024) Environ Implications Biochar 109–125. https://doi.org/10.2174/97898152380681240100009

Nyambo P, Zhou L, Nciizah AD, Mupambwa HA (2023) Waste conversion into biochar: an option for sustainable valorisation. Elsevier BV, pp 155–170. https://doi.org/10.1016/b978-0-443-15206-1.00018-9

Pal G, Patel T (2021) Heat stress's impact on agricultural worker's health, productivity, and its effective prevention measures: a review and meta-analysis. Int J Agric Syst 9(2):51–79

Patani AA, Prajapati D, Singh S, Enakiev Y, Bozhkov S, Jabborova D, Joshi C (2023) Beneficial effects of biochar application on mitigating the drought and salinity stress implications on plants. Plant Sci Today. https://doi.org/10.14719/pst.2591

Pathak HK, Chauhan PK, Seth CS, Dubey G, Upadhyay SK (2024) Mechanistic and future prospects in rhizospheric engineering for agricultural contaminants removal, soil health restoration, and management of climate change stress. Sci Total Environ 927:172116. https://doi.org/10.1016/j.scitotenv.2024.172116

Pathak HK, Seth CS, Chauhan PK, Dubey G, Singh G, Jain D, Upadhyay SK, Dwivedi P, Khoo KS (2024) Recent advancement of nano-biochar for the remediation of heavy metals and emerging contaminants: mechanism, adsorption kinetic model, plant growth and development. Environ Res 255:119136. https://doi.org/10.1016/j.envres.2024.119136

Perera HCS, Gurunanthanan V, Singh A, Mantilaka MMMGPG, Das G, Arya S (2024) Magnesium oxide (MgO) nanoadsorbents in wastewater treatment: a comprehensive review. J Magnes Alloys

Prasad A (2023) The dual edge of nanotechnology: investigating nanoparticles' environmental risks and benefits. Knowledg Res 2(1):64–72. https://doi.org/10.57067/hgxsr168

Qadeer S, Anjum M, Khalid A, Waqas M, Batool A, Mahmood T (2017) A dialogue on perspectives of biochar applications and its environmental risks. Water Air Soil Pollut 228(8):281. https://doi.org/10.1007/S11270-017-3428-Z

Rathinapriya P, Maharajan T, Jothi R, Prabakaran M, Lee IB, Yi P-H, Jeong ST (2025) Unlocking biochar impacts on abiotic stress dynamics: a systematic review of soil quality and crop improvement. Front Plant Sci 15. https://doi.org/10.3389/fpls.2024.1479925

Razzak SA (2024) Municipal solid and plastic waste derived high-performance biochar production: a comprehensive review. J Anal Appl Pyrol 181:106622. https://doi.org/10.1016/J.JAAP.2024.106622

Role of biochar in agriculture to enhance crop productivity: an overview (2023) Biointerface Res Appl Chem 13(5):429. https://doi.org/10.33263/briac135.429

Role of biochar in polyaromatic hydrocarbons remediation and environment management (2023) Elsevier eBooks, pp 365–385. https://doi.org/10.1016/b978-0-323-91873-2.00004-2

Safarian S (2023) Performance analysis of sustainable technologies for biochar production: a comprehensive review. Energy Rep 9:4574–4593. https://doi.org/10.1016/j.egyr.2023.03.111

Sajjad M, Huang Q, Khan S, Nawab J, Khan MA, Ali A, Ullah R, Kubar AA, Guo G, Yaseen M, Sajjad M (2024) Methods for the removal and recovery of nitrogen and phosphorus nutrients from animal waste: a critical review. Ecol Front 44(1):2–14. https://doi.org/10.1016/J.CHNAES.2023.05.003

Salles TDR, Bruckmann FDS, Rhoden CRB (2023) Nanoadsorbents in the removal of organic pollutants. Elsevier BV, pp 429–442. https://doi.org/10.1016/b978-0-443-18456-7.00019-5

Saraugi SS, & Routray, W. (2024). Advances in sustainable production and applications of nano-biochar. Sci Total Environ 176883. https://doi.org/10.1016/j.scitotenv.2024.176883

Saxena P, Miglani R, Shah D, Rajput VD, Minkina T (2024) Role of nanoparticles and nano-biochar in reducing salinity stress for improving soil health and sustainable crop production. Nanomater Nano-Biochar Reducing Soil Stress 137–159. https://doi.org/10.1201/9781003503453-7

Selvam M, Balasubramanian P (2023) Emerging trends and research frontiers of biochar derived through microwave assisted pyrolysis: a scientometric review. Bioresour Technol Rep 24:101601. https://doi.org/10.1016/j.biteb.2023.101601

Shafiq F, Anwar S, Firdaus-e-Bareen Zhang L, Ashraf M (2023) Nano-biochar: properties and prospects for sustainable agriculture. Land Degrad Dev 34(9):2445–2463. https://doi.org/10.1002/LDR.4620

Sharma S, Tiwari S, Hasan A, Saxena V, Pandey LM (2018) Recent advances in conventional and contemporary methods for remediation of heavy metal-contaminated soils. 3 Biotech 8(4). https://doi.org/10.1007/s13205-018-1237-8

Sheets JP, Yang L, Ge X, Wang Z, Li Y (2015) Beyond land application: emerging technologies for the treatment and reuse of anaerobically digested agricultural and food waste. Waste Manage 44:94–115. https://doi.org/10.1016/J.WASMAN.2015.07.037

Shiralian M (2016) Biochar production and uses. Res Rev J Ecol Environ Sci 4(1):40–42. https://www.rroij.com/open-access/biochar-production-and-uses-.pdf

Sim DHH, Tan IAW, Lim LLP, Hameed BH (2021) Encapsulated biochar-based sustained release fertilizer for precision agriculture: a review. J Clean Prod 303:127018. https://doi.org/10.1016/J.JCLEPRO.2021.127018

Singh J, Soni SK, Ranjan R (2023) Biochar for the improvement of crop production, pp 297–317. https://doi.org/10.1007/978-981-99-4090-5_13

Sparrevik M, Lindhjem H, Andria V, MagerholmFet A, Cornelissen G, Cornelissen G, Cornelissen G (2014) Environmental and socioeconomic impacts of utilizing waste for biochar in rural areas in Indonesia—a systems perspective. Environ Sci Technol 48(9):4664–4671. https://doi.org/10.1021/ES405190Q

Švarc-Gajić J, Brezo-Borjan T, Jakšić S, Despotović V, Finčur N, Bognár S, Jovanović D, Šojić Merkulov D (2024) Removal of organic pollutants by subcritical water under moderate conditions. https://doi.org/10.20944/preprints202406.0004.v1

Tan H, Ong PY, Klemeš JJ, Bong CPC, Li C, Gao Y, Lee CT (2021) Mitigation of soil salinity using biochar derived from lignocellulosic biomass. Chem Eng Trans 83:235–240. https://doi.org/10.3303/CET2183040

Tang Y, Li Y, Cockerill TT (2024) Environmental and economic spatial analysis system for biochar production—case studies in the East of England and the East Midlands. Biomass Bioenerg. https://doi.org/10.1016/j.biombioe.2024.107187

Uchimiya M, Lima IM, Klasson KT, Chang S, Wartelle LH, Rodgers J (2010) Immobilization of heavy metal ions (CuII, CdII, NiII, and PbII) by broiler litter-derived biochars in water and soil. J Agric Food Chem 58(9):5538–5544. https://doi.org/10.1021/JF9044217

Upadhyay VK, Choudhary KK, Agrawal SB (2024) Use of biochar as a sustainable agronomic tool, its limitations and impact on environment: a review. Discov Agric 2(1). https://doi.org/10.1007/s44279-024-00033-2

Vereš J, Kolonicny J, Ochodek T (2014) Biochar status under international law and regulatory issues for the practical application. Chem Eng Trans 37:799–804

Verma R, Dutta S, Kumar A, Dabodiya TS, Kumar N, Karuppasamy KSK, Sangmesh B, Jaiswal A, Jaiswal KK (2023) Biochar-based nanocomposite materials: types, characteristics, physical activation, and diverse application scenarios. Adv Sci Technol Innov Part F741:3–18. https://doi.org/10.1007/978-3-031-28873-9_1

Verma A, Sharma G, Kumar A, Dhiman P, Stadler FJ (2024) Recent advancements in biochar and its composite for the remediation of hazardous pollutants. Curr Anal Chem 20. https://doi.org/10.2174/0115734110286724240318051113

Wagh P, Nawani N, Vaidya V (2024) Biochar: exploring futuristic trends in advanced biochar research, pp 1–10. https://doi.org/10.58532/v3bfbt10p1ch1

Wang J, Wang S (2019) Preparation, modification and environmental application of biochar: a review. J Clean Prod 227:1002–1022. https://doi.org/10.1016/J.JCLEPRO.2019.04.282

Wang B, Lan J, Bo C, Gong B, Ou J (2023) Adsorption of heavy metal onto biomass-derived activated carbon: review. RSC Adv 13(7):4275–4302. https://doi.org/10.1039/D2RA07911A

Wang K, Yang X, Zhu L, Li J, He W (2023) Study on life-cycle carbon footprint assessment of substation 12804, 128040T. https://doi.org/10.1117/12.2688314

Wang Y, Li R, Qiao J (2023) Solidification of heavy metals in municipal solid waste incineration washed fly ash by asphalt mixture. Chemosphere 343:140281

Wang X, Ding J, Han L, Tan J, Ge X, Nan Q (2024) Biochar addition reduces salinity in salt-affected soils with no impact on soil pH: a meta-analysis. Geoderma. https://doi.org/10.1016/j.geoderma.2024.116845

Wang Y-J, Wu P, Bolan N, Wang H (2022) The potential agronomic and environmental applications of biochar: prospects and challenges. Encycl Soils Environ. https://doi.org/10.1016/b978-0-12-822974-3.00027-6

Wangd D (2023) Sustainability assessment of biochar applications. In: Current developments in biotechnology and bioengineering: biochar towards sustainable environment, pp 415–441

Wanninayake DM (2021) Comparison of currently available PFAS remediation technologies in water: a review. J Environ Manage 283:111977. https://doi.org/10.1016/J.JENVMAN.2021.111977

Wewer G (2023) Life cycle assessment. Elsevier eBooks. https://doi.org/10.1016/b978-0-12-824315-2.00735-1

Xiao W, Sun R, Hu S, Meng C, Xie B, Yi M, Wu Y (2023) Recent advances and future perspective on lignocellulose-based materials as adsorbents in diverse water treatment applications. Int J Biol Macromol 253:126984. https://doi.org/10.1016/J.IJBIOMAC.2023.126984

Xiongyuan S, Tingting S, Youbin S, Huarong T, Kang W, Huimin X, Huiqing W (2016) Organic pollutant removal system and organic pollutant removal method

Xu C, Qi J, Yang W, Chen Y, Yang C, He Y, Lin A (2019) Immobilization of heavy metals in vegetable-growing soils using nano zero-valent iron modified attapulgite clay. Sci Total Environ 686:476–483

Xu J, Xie Y, Yao Q, Lv L, Chu H (2024) Advances in sustainable nano-biochar: precursors, synthesis methods and applications. Nanoscale 16(32):15009–15032. https://doi.org/10.1039/D4NR01694G

Xu M (2023) Research progress of environmental risk substances in the preparation of biochar. Highlights Sci Eng Technol 40:245–250. https://doi.org/10.54097/hset.v40i.6660

Yameen MZ, Naqvi SR, Juchelková D, Khan MNA (2024) Harnessing the power of functionalized biochar: progress, challenges, and future perspectives in energy, water treatment, and environmental sustainability. Biochar 6:1–80. https://doi.org/10.1007/s42773-024-00316-3

Yang K, Lin D (2014) Environmental risks of engineered nanomaterials. J Zhejiang Univ Sci 15(8):547–551. https://doi.org/10.1631/JZUS.A1400219

Yang X, Wan Y, Zheng Y, He F, Yu Z, Huang J, Wang H, Ok YS, Jiang Y, Gao B (2019) Surface functional groups of carbon-based adsorbents and their roles in the removal of heavy metals from aqueous solutions: a critical review. Chem Eng J 366:608–621. https://doi.org/10.1016/j.cej.2019.02.119

Yang J, Sun X, Li Q, Cheng L, Wang R, An L, Chen W, Wei L, An Q-D (2023) Fabrication of multifunction-integrated adsorbent based on zwitterion functionalized cellulose with high adsorption performance for anionic/cationic dyes and heavy metal ions. https://doi.org/10.2139/SSRN.4616233

You S, Wang X (2019) On the carbon abatement potential and economic viability of biochar production systems: cost-benefit and life cycle assessment. Elsevier, pp 385–408. https://doi.org/10.1016/B978-0-12-811729-3.00020-0

Zhang X, Wang H, He L, Lu K, Sarmah AK, Li J, Bolan N, Pei J, Huang H (2013) Using biochar for remediation of soils contaminated with heavy metals and organic pollutants. Environ Sci Pollut Res 20(12):8472–8483. https://doi.org/10.1007/S11356-013-1659-0

Zhang Y, Cao B, Yin H, Meng L, Jin W, Wang F, Xu J, Al-Tabbaa A (2022) Application of zeolites in permeable reactive barriers (PRBs) for in-situ groundwater remediation: a critical review. Chemosphere 308:136290. https://doi.org/10.1016/J.CHEMOSPHERE.2022.136290

Zhang X, Zhou Q, Wang L, Wan B, Yang Q (2024) How biochar addition affects denitrification and the microbial electron transport system (ETSA): a meta-analysis based on a global scale. Agriculture 14(12):2320. https://doi.org/10.3390/agriculture14122320

Zheng H, Zhang C, Zhang C, Liu B, Liu G, Zhao M, Zhao M, Xu G, Luo X, Li F, Xing B (2020) Biochar for water and soil remediation: production, characterization, and application. Springer, Singapore, pp 153–196. https://doi.org/10.1007/978-981-13-9447-8_11

Zhou L (2024) A review of biomass-derived biochar and its potential in asphalt pavement engineering. https://doi.org/10.2478/msp-2024-0022

Effects of Biochar on Biochemical Cycles

Souravi Bardhan, Dipak Kr Chanda, and Samuele Barrili

Abstract Biochar is a carbon-rich substance produced during pyrolysis of various types of biomass, which has become a multifunctional product with far-reaching implications on soil health, nutrient cycles, and environmental sustainability worldwide. In addition to its classical use as a soil amendment, biochar is actively involved in the regulation of key biogeochemical cycles (carbon, nitrogen, phosphorus, sulfur, and micronutrients) and has been demonstrated to influence ecosystem processes. Organic carbon in soil and biochar supports long-term carbon sequestration as opposed to release the greenhouse gases, especially such as the co/the methane. Its high surface area, porosity, and alkalinity promote nutrient retention, alter soil pH, and increase availability for micronutrients like nitrogen and phosphorus. Moreover, biochar interacts with sulfur compounds and micronutrients by decreasing toxicity and alleviating deficiency through adsorption and buffer effects. It is also important in causing soil microbial communities and enzyme activity to accelerate nutrient transformation processes, such as nitrification, denitrification, and phosphorus solubilization. Nonetheless, the scale and nature of these advantages are highly contextualized depending on the type of feedstock, state of pyrolysis, and environmental uses. This chapter reviews recent developments found in current research, restates, and maps possible courses of future improvement of biochar-based interactions in ecosystem management. Finally, biochar provides a bright opportunity to improve the biogeochemical cycle and add the sustainable agriculture, climate stability, and global environmental recovery.

S. Bardhan (✉)
Department of Environmental Science, Netaji Nagar College for Women, University of Calcutta, Kolkata, West Bengal, India
e-mail: bardhan666srv@gmail.com

Department of Physics, Jadavpur University, Kolkata, West Bengal, India

D. K. Chanda
Somal Refractory, Ludhiana, Punjab, India

Matrexia, Hoogly, India

S. Barrili
Waste Management, Marketing Strategist, M4W-Marketing for Waste, Rome, Italy

A. Chauhan et al. (eds.), *Biochar As Remediation Tools for Pollutant Removal*,
https://doi.org/10.1007/978-3-032-16746-0_12

Keywords Biochar application · Biogeochemical cycles · Carbon sequestration · Nitrogen cycling · Microbial activity · Nutrient retention

1 Introduction

Comprehensive literature showed the central position of biochar in climate change reduction, returning soil health, and ensuring nutrient balance (Nair et al. 2017; Lehmann et al. 2021). Hornung et al. (2024) characterized biochar as thermally stable carbonaceous material derived following pyrolysis of organic-based biomass in the absence of oxygen. Due to its specific physicochemical properties, biochar has evolved as multi-functional avenues in soil fertility improvement (Ding et al. 2016), waste treatment (Xiang et al. 2020), carbon sequestration (Bolan et al. 2022), and in the environmental remediation of contaminants (Hu et al. 2020) significantly beyond serving as a soil amendment or carbon sink, biochar experts fundamental and complex influences on orbicular biogeochemical cycle.

Biogeochemical cycle of carbon, nitrogen, phosphorous, and sulfur as well as other key elements control fluxes and movements of materials in ecosystems (Berhe et al. 2018) (Fig. 1). These cycles control soil and water quality, affecting greenhouse gases fluxes, and define the fate of environmental pollutants. Nevertheless, the human interventions such as intensive agriculture, deforestation, and fossil fuel combustions have disturbed this cycle, resulting in climatic and ecologically damaging consequences (Singh 2024).

That involves leaching of nutrients and encourages algal blooms, climate change, soil erosion (Zaman et al. 2025). As a complex additive, biochar interacts with these cycles potentially reducing or inhibiting effects based on the feedstock, production methods, and site-specific situation. Biochar's capacity to influence biochemical cycles is derived from several fundamental attributes. Its porosity and large surface area enable it to take up materials and provide habitats for microbes (Wong and Ogbonnaya 2021; Pathak et al. 2024a). It possesses alkaline pH and mineral ash, which make soil acidic and contribute necessary nutrients for plants into the soil (Bedassa 2020). Munera-Echeverri et al. (2018) and Pathak et al. (2024b) found that the cation-exchange capacity (CEC) of biochar can hold nutrients and control ion availability, while its stable carbon structure is conducive to long-term C fixation. These properties collectively have a strong impact on the cycling of major and trace elements via physical, chemical, and biological processes.

Biochar adds carbon to the carbon cycle as a sink and mediator of SOM. It reduces CO_2 and CH_4 emissions by stimulating the stabilization of unstable organic matter and leading to anaerobic conditions (Li et al. 2021), but also promotes microbial activity and increases the input of plant-root-driven carbon into soils (Pan et al. 2021).

The addition of biochar causes the utilization ratio of nitrogen to increase by fixing ammonium and nitrate, which leads to a decrease in leaching, volatilization, and nitrous oxide release (Liu et al. 2017). It is also beneficial for phosphorus by

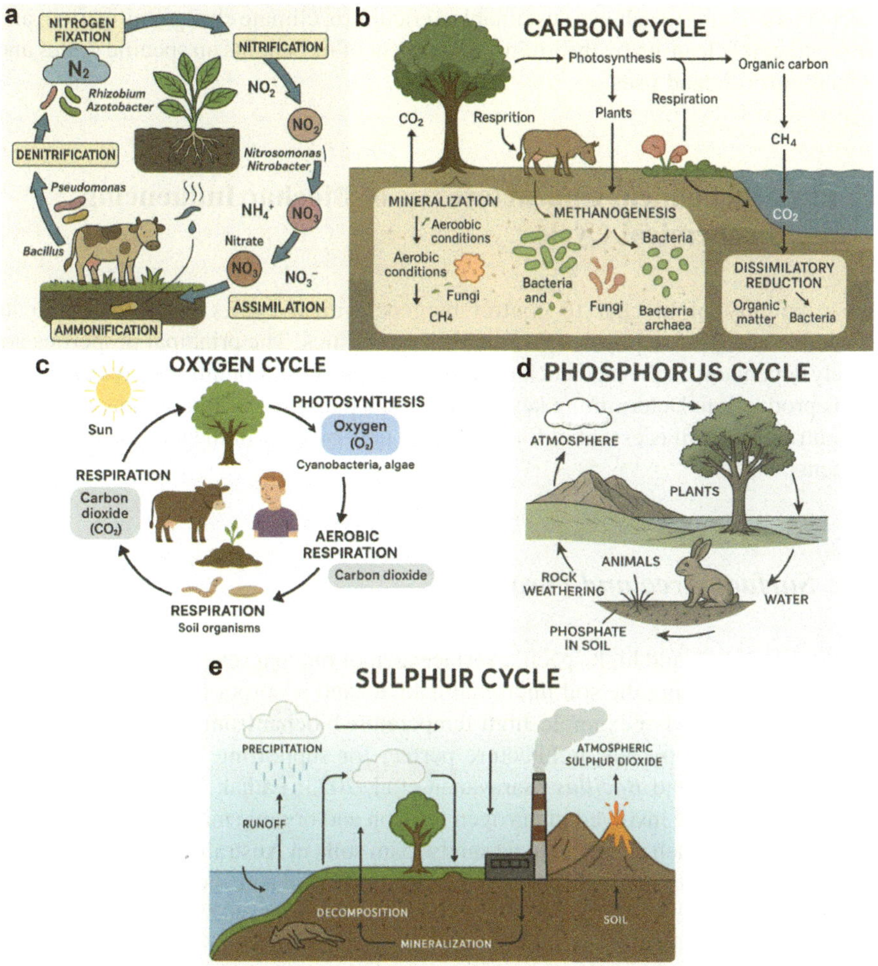

Fig. 1 Various biogeochemical cycles in the environment **a** Nitrogen; **b** Carbon; **c** Oxygen; **d** Phosphorus, and **e** Sulfur cycle

decreasing fixation and stimulating phosphate solubilizing microbes, and in sulfur cycle it reduces sulfate losses along with promoting microbial transformations (Leng et al. 2022). For micronutrient; Fe, Mn, Zn, and Cu, biochar bufferage were diagnosed even deficiency and toxicity of these elements at different soil conditions (Younas et al. 2024).

The overall objective of this chapter is to systematically review the various effects of biochar on main biochemical cycles. It combines existing and future research to identify how biochar influences elemental cycling in terrestrial systems. This chapter emphasizes identifying combined beneficial effects and potential side effects of biochar application. The chapter therefore adds to the overarching aim of using biochar as a means to achieve sustainable agriculture, climate change mitigation, and environmental clean-up by informing assessment of its benefits on specific scales and in dual-purpose land uses.

2 Physicochemical Characteristics of Biochar Influencing the Biochemical Cycles

The capability of biochar to control biogeochemical cycles is attributed to its adjustable and diverse physicochemical characteristics. The principal properties are largely dependent on the feedstock's biomass and pyrolysis conditions which are used for its production. Determining key biochar traits that drive its interaction with biota and nutrients in soil is essential for predicting how biochar influences the cycling of elements.

2.1 *Surface Area and Porosity*

The porous nature and high specific surface area of biochar (even 500 m^2/g) are key factors in determining the soil microbial habitats and adsorption of nutrients in soil (Zhang et al. 2025). For example, high-temperature biochar from hardwood develops a micro- and mesoporous architecture perfect for supporting beneficial microbes like *Azospirillum* and *Bacillus* (Saravanan et al. 2023; Pathak et al. 2024a). These microorganisms are involved in nitrogen fixation and organic matter decomposition. In field tests, one such experiment on sandy loam soils in Australia demonstrated that the addition of 10 t/ha of high-porosity biochar enhanced soil water holding capacity by 15% and nitrate leaching by 25% (Alkharabsheh et al. 2021). The pores act as a physical refuge for microbes from predation and water stress, which indirectly enhances nutrient mineralization and stabilization.

2.2 *pH and Alkalinity*

Biochar generally has natural alkalinity, typically in the range of 7–11 pH, making it a significant candidate for acid soil amendment (Dai et al. 2017). Mosharrof et al. (2021) reported that soil pH from the rice husk biochar of 500 °C increased tropical acid soil pH in a range of 4.8–6.5, which ameliorated phosphorus availability and reduced toxicity solubility of aluminium Cs. A higher pH value also promotes the development of beneficial microbial populations such as Nitrosomonas, involved in nitrification. On acidic forest soils, liming with biochar has been found to enhance microbial biomass carbon and phosphorus cycling-related enzymatic activities (Zhang et al. 2019; Chauhan et al. 2024a, 2024b).

2.3 Cation Exchange Capacity (CEC)

The CEC of biochar increases over time as it oxidizes in the soil matrix and forms additional functional groups like carboxyls (Munera-Echeverri et al. 2018). High-CEC biochars such as those derived from manure can retain nutrients like ammonium and potassium, hence reducing nutrient loss by leaching (Alkharabsheh et al. 2021). For example, Kätterer et al. (2022) reported that in a Kenyan maize field trial, biochar enhanced potassium retention by 30% without reducing fertilizer application at the cost of yield. In temperate soils, CEC optimization preserves the availability of nutrients in terms of seasons such that microbes and plants have consistent access to the needed ions.

2.4 Functional Groups and Surface Chemistry

Functional groups such as hydroxyl (-OH), carboxyl (-COOH), and phenolic (-PhOH) on the biochar surface catalyze metal ion complexation and organic nutrient complexity (Li et al. 2023; Pathak et al. 2024a). The groups are more prevalent in low-temperature biochars and become unstable with high pyrolysis. For example, carboxyl group availability enhances ammonium and transition metal adsorption, and phenolic groups enhance microbial adhesion. This chemistry facilitates interaction with microbial exoenzymes like urease and phosphatase, which play a crucial role in cycling nitrogen and phosphorus (Ghosh et al. 2024). Biochar with a higher concentration of surface functional groups has been found to facilitate fungal colonization and enzymatic hydrolysis of organic matter.

2.5 Mineral Content and Nutrient Enrichment

Mineral ash content in biochar also varies with feedstock-poultry litter biochars, for instance, are particularly rich in phosphorus (as high as 2% by weight), along with potassium, calcium, and minor nutrients. Biochars high in minerals are applied as slow-release fertilizers (Freitas et al. 2020). For instance, poultry litter biochar applied to P-deficient soils in Bangladesh enhanced P uptake by 60% and biomass of the crop by 40% (Al Mamun et al. 2022). Conversely, calcium and magnesium from ash also contribute to pH buffering and soil particle flocculation, which improves soil structure and microbial habitat (Guo et al. 2020). Such nutrient-dense biochars are increasingly being engineered into degraded soils or organic farming systems.

2.6 *Stability and Recalcitrance*

Carbon in biochar is largely in stable aromatic forms. Such stability locks carbon for centuries, which is highly important in carbon sequestration. For example, Amazonian Terra Preta soils preserve ancient charcoal-derived biochars that have structure and fertility (Lorenz et al. 2018). In a Canadian field trial, biochar addition at 20 t/ha reduced soil CO_2 efflux by 18% during three years, showing suppression of microbial respiration through recalcitrant carbon (Allaire et al. 2015). Additionally, stability reduces priming effects whereby microbes degrade native organic matter, thus conserving soil organic carbon and controlling carbon cycling.

2.7 *Electrical Conductivity (EC)*

Electrical conductivity (EC) of biochar is a measure of soluble ion content and is strongly correlated with feedstock origin. For instance, seaweed biochar may have EC > 10 dS/m as a result of high salt content, which is not desirable for seed germination (Chanthini et al. 2024). Wood biochars usually have lower EC values (< 1 dS/m), which is safer for large-scale application (Crespo-Barreiro et al. 2023). EC regulates microbial community functions by modifying osmotic potential and nutrient gradients. Tomato plant experiment in a greenhouse experiment indicated that moderate EC biochar promoted plant growth and fruiting by raising nitrogen and phosphorus uptake without causing salt stress (Ud Din et al. 2023). Modulation of EC is necessary when biochar is applied to saline or semi-arid soils.

Together, these characteristics are dynamic; they synergistically interact and evolve with time, weathering, and biologically, as they continue to form intricate, heterogeneous microenvironments. Understanding and modifying these characteristics, biochar may be used strategically to control soil processes, enhance nutrient use efficiency, and produce sustainable agricultural and environmental outcomes.

3 Impacts of Biochar on Biochemical Cycles

Biochar has multi-directional impacts on various biochemical cycles apart from the carbon cycle, such as the nitrogen, phosphorus, sulfur, and micronutrient cycles. These are impacts arising from its physicochemical nature, microbial activity, and impact on soil properties. The function of these impacts must be appreciated to maximize the utilization of biochar in agriculture and remediation (Table 1).

Table 1 Various feedstocks used in biochar production and consecutive impact on the biochemical cycles and on the environment

Feedstock type	Materials	Primary nutrient elements	Biochemical cycles involved	Impact on environment	References
Agricultural residues	Rice husk, wheat straw, corn stover	Carbon (C), nitrogen (N), phosphorus (P), potassium (K)	Carbon, nitrogen, phosphorus, potassium cycles	Rich in lignocellulosic biomass, these residues influence the carbon cycle by contributing to long-term carbon sequestration. Their carbon-to-nitrogen (C:N) ratios affect nitrogen immobilization or mineralization, and biochar from these sources often enhances soil fertility by improving nutrient retention, cation exchange capacity (CEC), and microbial biomass	Chamoli et al. (2025)
Forestry residues	Sawdust, bark, wood chips	Carbon (C), calcium (ca), magnesium (Mg)	Carbon cycle, partially calcium cycle (soil buffering)	Typically high in lignin and low in ash content, these feedstocks produce biochar with a highly stable aromatic carbon structure. They improve soil buffering capacity through the addition of base cations like Ca^{2+} and Mg^{2+}, and provide habitat for soil microbial consortia essential to nutrient cycling	Chamoli et al. (2025)
Animal manure	Cow dung, poultry litter, pig manure	Nitrogen (N), phosphorus (P), Sulfur (S), potassium (K)	Nitrogen, phosphorus, Sulfur, potassium cycles	Rich in nutrients and microbial content, manure-derived biochar can significantly influence nitrogen and phosphorus cycling. When properly pyrolyzed, it immobilizes ammonia (NH_4^+), reduces leaching, and enhances microbial nitrification and denitrification processes, thus regulating N_2O emissions	Kacprzak et al. (2023)

(continued)

Table 1 (continued)

Feedstock type	Materials	Primary nutrient elements	Biochemical cycles involved	Impact on environment	References
Sewage sludge	Treated municipal biosolids	Nitrogen (N), phosphorus (P), trace metals (Zn, cu, Pb)	Nitrogen, phosphorus, trace element cycles	Sewage sludge is nutrient-dense but also contains heavy metals. Pyrolysis reduces pathogenicity and stabilizes heavy metals, minimizing leaching. The resulting biochar improves nutrient cycling and organic matter stabilization but must be applied cautiously to prevent potential metal toxicity in soils	Gopinath et al. (2021)
Industrial biomass waste	Paper sludge, sugarcane bagasse	Carbon (C), trace minerals	Carbon cycle	Industrial biomass waste is generally consistent in high carbon content and yields biochar suitable for carbon sequestration. When sourced from clean production, it can enhance organic carbon pools and microbial substrate availability in degraded soils. Regulatory monitoring is needed due to potential chemical contamination	Fawzy et al. (2021)
Aquatic biomass	Algae, water hyacinth	Nitrogen (N), phosphorus (P), potassium (K), calcium (ca)	Nitrogen, phosphorus, potassium, carbon cycles	Due to rapid growth and high nutrient accumulation, aquatic biomass contributes significantly to nutrient cycling. Biochar produced can remove excess nutrients from eutrophic systems and reduce nitrate and phosphate runoff through sorption and enhanced microbial assimilation	Vasseghian et al. (2024)

(continued)

Table 1 (continued)

Feedstock type	Materials	Primary nutrient elements	Biochemical cycles involved	Impact on environment	References
Urban organic waste	Food scraps, garden trimmings	Carbon (C), nitrogen (N), potassium (K), micronutrients	Carbon, nitrogen, micronutrient cycles	Highly heterogeneous in composition, this feedstock can produce nutrient-rich biochar that enhances soil organic matter, supports microbial activity, and reduces methane emissions when diverted from landfills. Proper sorting and contamination control are essential	Razzak (2024)
Energy crops	Miscanthus, switchgrass, bamboo	Carbon (C), potassium (K), silicon (Si)	Carbon, potassium, silicon cycles	These fast-growing crops provide consistent biomass and produce biochar with high surface area and porosity. They aid in carbon sequestration, potassium recycling, and soil water retention, while also supporting silicon cycling, which is crucial for some plant defense mechanisms	Muradov et al. (2012)
Agro-industrial waste	Coffee husks, coconut shells, palm kernel shell	Carbon (C), Phenolics, potassium (K)	Carbon, potassium, aromatic compound transformation cycles	This waste stream is valuable for producing high-quality biochar rich in recalcitrant carbon. It contributes to the stabilization of soil organic matter, adsorbs organic pollutants due to its high aromaticity, and enhances microbial resilience in stressed soils	Khedulkar et al. (2023)
Municipal solid waste (organic fraction)	Vegetable waste, fruit peels	Carbon (C), nitrogen (N), phosphorus (P), heavy metals	Carbon, nitrogen, phosphorus, trace elements cycles	When properly segregated and processed, this feedstock offers a pathway for nutrient recycling and organic waste management. Biochar reduces GHG emissions from landfills, enhances composting processes, and improves soil amendment quality. However, contamination risk (plastic, metals) must be managed	Razzak (2024)

3.1 Impact on the Carbon Cycle

Biochar is the pioneering factor in the carbon (C) cycle as it affects carbon storage, emission, and pathways of carbon transformation in terrestrial ecosystems. Greater content of carbon and high aromaticity confer on it a superior stability in soil with long-term sequestration capacity for carbon along with radical change in turnover patterns of soil organic carbon (SOC) (Kan et al. 2022).

3.1.1 Carbon Sequestration Potential

One of the most significant uses of biochar is to sequester atmospheric carbon in a long-term stable form for centuries to millennia (Lorenz and Lal 2014). Biochar is made up of highly resistant aromatic molecules that are condensed, and the mean residence times vary from hundreds of years to thousands of years depending on environmental conditions of biochar. The stability is an extremely potent climate change mitigation tool. For instance, temperate and boreal soils which have both been amended with biochar have registered up to 40% overall decadal soil carbon gain (Weng et al. 2017).

3.1.2 Impact on Soil Organic Matter (SOM)

The incorporation of biochar into the soil increases SOM by engaging with indigenous organic matter and influencing microbial metabolism. It can also be a 'carbon sponge' that absorbs dissolved organic carbon (DOC), hence preventing its leaching and promoting its microbial mineralization to humic substances (Sun et al. 2021). Sometimes biochar causes a 'positive priming effect,' enhancing microbial activity and SOM decomposition (Rasul et al. 2022). However, most frequently, it has a 'negative priming effect' whereby the decomposition rate slows down through microbial preference for labile carbon in biochar, leading to net soil C stock accumulations (Whitman et al. 2015).

3.1.3 Reducing Greenhouse Gases

Biochar lowers greenhouse gas emissions, such as CO_2, CH_4, and N_2O, substantially. In anaerobic rice soils, biochar is an electron shuttle that improves the reduction of Fe (III) at the cost of methane (CH_4)-producing methanogenesis (Liu et al. 2022). Field trials have established that methane emission can be minimized by up to 70% when rice paddies are treated with biochar (Dong et al. 2013). Surface properties of biochar also promotes adsorption of nitrate and ammonium, minimizing denitrification-linked N_2O emissions (Chang et al. 2023).

In temperate agro-ecosystems, biochar has been found to decrease CO_2 efflux, partly because it is recalcitrant and partly because it can modulate microbial respiration. As such, it is a valuable amendment for both soil health and reduction of agricultural carbon balances.

3.1.4 Microbial Carbon Use Efficiency (CUE)

Biochar also influences microbial carbon use efficiency, a critical factor in determining whether or not carbon is respired or sequestered. Biochar has the capacity to provide habitat/niches and an influence on redox potential. Biochar conditions microbial communities to immobilize more carbon into biomass production than respiration (Palansooriya et al. 2019). The process of remodulation enhances microbial necromass accumulation, persistent compounds of soil organic matter, thereby promoting greater long-term carbon sequestration in soils.

3.1.5 Root Exudates Interactions and Rhizosphere Dynamics

Biochar impacts carbon cycling in the rhizosphere through interactions with root exudates. The low-molecular-weight organic compounds emitted by plant roots are adsorbed on biochar surfaces, shielding them from premature microbial degradation and allowing them to be released slowly (Pei et al. 2020; Upadhyay et al. 2022). Not only is microbial colonization favored but also symbiotic associations, such as between mycorrhizal fungi and plant roots (Warnock et al. 2007; Chauhan and Upadhyay 2024). Thus, rhizospheres of biochar-fertilized soils have improved belowground carbon fixation, nutrient acquisition, and carbon transfer (Fig. 2).

3.1.6 Implications at the Field Level

At the practical level, field experiments have demonstrated that biochar can add 10–40% to soil carbon stores over a few years, depending on application rate, biochar quality, and soil texture (Lorenz and Lal 2014). For instance, in a German long-term experiment, application of 20 t/ha wood biochar provided a net increase of 4.5 t C/ha of carbon after five years, in addition to increased soil aggregation and water retention (Sohi et al. 2009). Such gains make biochar a key element of carbon-bright agriculture management.

Thus, biochar is not only a permanent carbon sink but also fundamentally restructures carbon cycle processes by interacting with microbial communities, plant systems, and soil organic matter. Application of biochar provides a multi-modal approach to soil health improvement, agricultural productivity increase, and climate change mitigation through the control of terrestrial carbon fluxes.

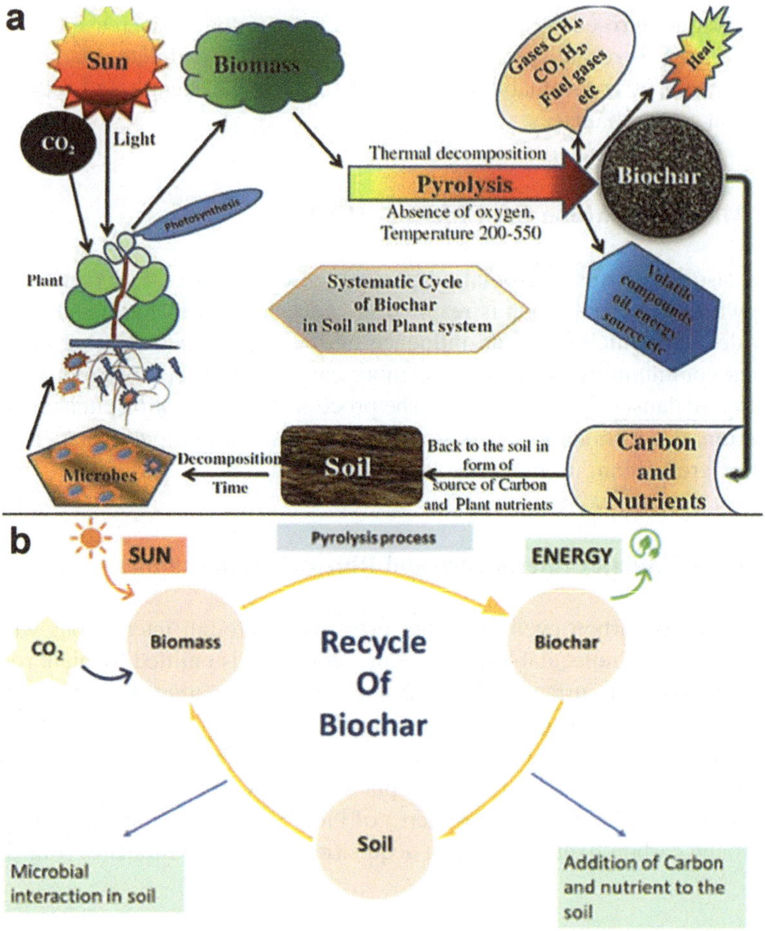

Fig. 2 **a** Biochar in carbon cycle, influence on plants and animals in the carbon cycle; **b** recycling of biochar in the environment during carbon cycle. Reprinted from Jatav et al. 2020. Open Access

3.2 Impact and Control of Biochar on the Nitrogen Cycle

Nitrogen is a highly useful element in plant produce and ecosystem processes but highly mobile and reactive in the environment, resulting in huge losses and pollution as leaching of nitrate, volatilization of ammonia, and emission of nitrous oxide (Ghaly and Ramakrishnan 2015). Biochar has been found to have potential for being a useful amendment to control nitrogen transformations and promote nitrogen use efficiency in soils.

3.2.1 Improvement of Nitrogen Retention

Biochar improves the retention of nitrogen in soils significantly by adsorbing nitrate (NO_3^-) and ammonium (NH_4^+) ions onto the surface area of its porous matrix and functional groups. Biochar possesses high cation exchange capacity (CEC) and surface area that allow it to capture mobile forms of nitrogen that are less likely to leach (Graber et al. 2017). It is a very significant mechanism through which sandy or low-organic-matter soils are likely to lose nitrogen.

3.2.2 Effects on Nitrification and Denitrification

The biological processes of nitrogen transformation, such as nitrification (oxidation of ammonium to nitrate) and denitrification (reduction of nitrate to nitrogen gases), are affected by biochar. Studies by Zainudin et al. (2020) and Parasar and Agarwala (2025) noted that the biochar changes the community composition of nitrifiers (Nitrosomonas, Nitrobacter) and denitrifiers (Pseudomonas, Bacillus), resulting in faster or slower processes based on its chemical characteristics as well as soil properties.

It saves N_2O emission to a large extent. Biochar is known to serve as an electron shuttle that encourages full denitrification to harmless N_2 in contrast with atmospheric N_2O, a strong greenhouse gas (Zhang et al. 2021). Furthermore, soil redox status is modified by biochar, inhibiting nitrification under anaerobic conditions and thus further reducing N_2O production.

3.2.3 Nitrogen Mineralization and Immobilization Influence

Decomposition of organic matter returns inorganic nitrogen to the system through mineralization, a process regulated by microbial activity and substrate quality. Rates of mineralization are affected by biochar by modifying microbial enzyme production and availability of nitrogen (Li et al. 2020). Depending on its C/N ratio, biochar can stimulate or suppress mineralization.

High C/N ratio biochar can lead to short-term immobilization of nitrogen, in which microbes consume available nitrogen to degrade carbonaceous biochar, thereby decreasing short-term nitrogen availability to crops (Fatima et al. 2021). But in most cases, it is a temporary process and results in long-term nitrogen immobilization of soils. Low C/N ratio or nitrogen-rich feedstock-enriched biochars, however, can release available nitrogen into the soil (de Oliveira Paiva et al. 2024).

3.2.4 Synergistic Effects with Organic Amendments

Biochar, along with organic manure and composts, has synergistic effects to boost nitrogen cycling. Biochar is a physical and chemical buffer that enhances the retardation of nitrogen losses by volatilization, particularly for ammonium fertilizer (Zainudin et al. 2020). It also forms microhabitats for nitrogen-active microbial populations that favor improved microbial nitrogen transformations.

For instance, simultaneous use of poultry litter and biochar has been found to minimize ammonia volatilization and maximize nitrogen retention in the soil (Agyarko-Mintah et al. 2017). The couple maximizes nitrogen uptake by crops and minimizes environmental pollution possibilities.

3.2.5 Improvement of Biological Nitrogen Fixation

Biochar also stimulates symbiotic nitrogen fixation, especially under legume-cropping conditions. Its porous nature and ability to retain nutrients facilitate root elongation and root-nodule colonization by rhizobia bacteria (Xiu et al. 2021). Furthermore, through the promotion of soil pH and reduction of toxic metals (Chamoli et al. 2025), biochar creates a conducive environment for nitrogen-fixing organisms (Fig. 3).

Soil and field trials of groundnut, soybean, and other legumes indicated enhanced nodule growth and nitrogen fixation content upon the application of biochar. Such observations not only enhance agricultural production but also render the addition of nitrogen more sustainable in agroecosystems.

3.2.6 Environmental Implications

The environmental consequence of the impact of biochar on the nitrogen cycle is extensive. The avoidance of nitrogen leaching and gas loss through the use of biochar inhibits global warming and maintains water quality. Greater retention of nitrogen also translates into reduced application of fertilizers, with subsequent savings in resources and farm expense reduction.

3.3 Effect and Control of Biochar over the Oxygen Cycle

Oxygen, a critical component of aerobic respiration and many biochemical transformations, is a key player in soils' biogeochemical processes and world element cycling. Even though oxygen is not a nutrient technically, it has a great effect on redox potential, microbial metabolism, decomposition of organic matter, and carbon, nitrogen, sulfur, and phosphorus cycling (Patrick Jr. and DeLaune 1977). The use of

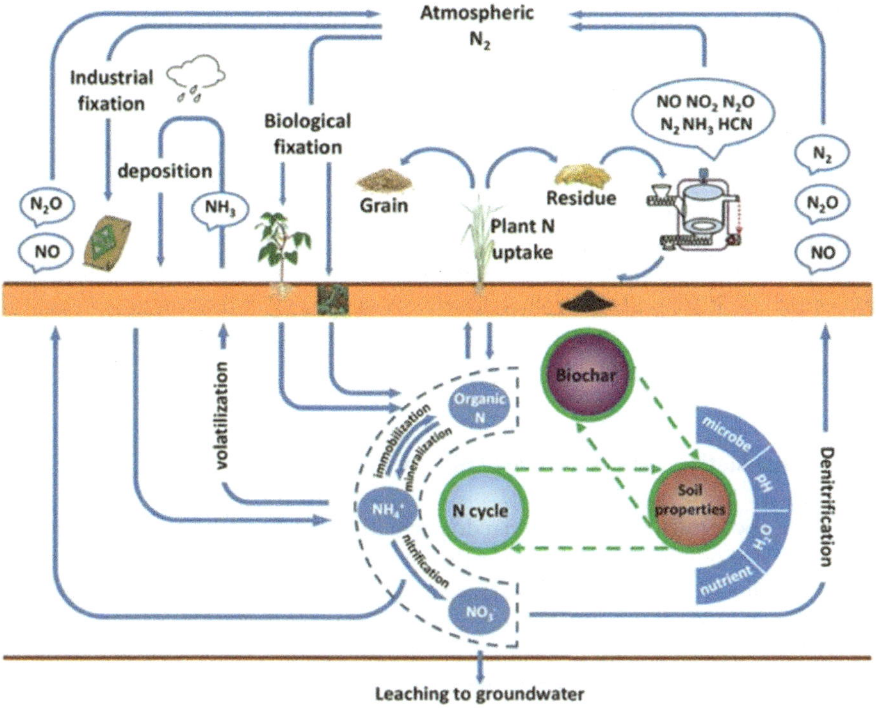

Fig. 3 Influence of biochar in nitrogen cycle, especially in soil ecosystem. Reproduced with permission from Liu et al. 2018. Springer Nature. License number: 6052220913378

biochar may indirectly influence soil system oxygen dynamics by modifying aerobic and anaerobic conditions as well as oxidative-reductive reactions.

3.3.1 Soil Aeration and Biochar

The biochar's physical porosity and structure can increase aeration of the soil substantially, particularly for compacted and waterlogged soils. Through enhanced porosity and reduced bulk density, biochar creates porous structures that are organized and permit oxygen movement through the soil profile. Enhanced oxygen diffusion leads to enhanced aerobic microbial respiration, plant root respiration, and oxidation of reduced compounds (Laanbroek 1990) such as ammonium, ferrous iron, and sulfides.

For instance, oxygen penetration depth was enhanced by 20–40% in clay soils treated with biochar, expanding the aerobic zone and enhancing aerobic decomposition reactions (Li et al. 2022). This adjustment has a cascading effect on other oxygen-dependent biogeochemical processes.

3.3.2 Effect on Microbial Respiration and Community Structure

Oxygen availability is a key regulator of microbial community structure and function. Biochar-amendment soil favors aerobic heterotroph that enhances organic matter mineralization and nutrient cycling (Dai et al. 2021), whereas low oxygen soils foster anaerobic conditions that slow decomposition and increase methane and nitrous oxide emissions (Chunmei et al. 2018).

Biochar exhibits redox buffering capacity, indirectly regulating oxygen dynamics in the rhizosphere. By adsorbing oxygen-inhibiting compounds such as phenolics and ammonia, it supports microbial activity even in oxygen-limiting microenvironments (Ro et al. 2015). Enhanced oxygen fluxes in the rhizosphere further strengthen symbiotic interactions with plant roots.

3.3.3 Interaction with Redox-Sensitive Nutrient Cycles

Many biogeochemical processes are redox-sensitive and therefore directly influenced by the levels of oxygen. Biochar indirectly regulates most such processes by adjusting the oxygen levels in the soil. For instance, nitrification, the oxidation of ferrous iron to ferric iron, and sulfide oxidation to sulfate are oxygen-dependent processes which are stimulated in soils with biochar (Liu et al. 2025).

Additionally, biochar alleviates anoxia stress in plant roots and rhizosphere microbes and stabilizes oxidative rhizosphere status (Mushtaq et al. 2024). The effect supports optimal functioning of aerobic enzymatic activities such as phosphatase and urease activities.

3.3.4 Greenhouse Gas Emission Mitigation

One of the most severe environmental problems with regard to low oxygen contents is the emission of very reactive greenhouse gases such as methane (CH_4) and nitrous oxide (N_2O). Through the promotion of aerobic conditions, biochar can suppress methanogenesis and incomplete denitrification that are typical in anaerobic conditions (Zhang et al. 2023).

A few experiments in rice paddy and flooded soil field trials have reported a 30–70% decrease in methane emission after adding biochar (Dong et al. 2013; Xiao et al. 2019). The decreases are a function of enhanced oxygen penetration and the stimulation of methanotrophic bacteria to oxidize methane prior to escaping to the atmosphere.

3.3.5 Long-Term Soil Health and Oxidative Stability

The long-term impact of biochar on oxygen dynamics contributes to improved soil oxidative stability. Resistance and stability of biochar toward microorganisms render it a long-term aeration amendment, especially when used in combination with conservation tillage and cover cropping (Thapa and Mowrer 2024). This synergistic effect improves long-term soil health, plant productivity, and nutrient cycling.

In addition, through the mitigation of hypoxic stress and the enhancement of redox balance, biochar facilitates a sustainable agroecosystem, particularly against climate waterlogging or extreme rainfall events.

3.4 *Biochar Role and Contribution to Phosphorus Cycle*

Phosphorus (P) is an essential macronutrient for plant growth and energy transfer, cellular constituent, nucleic acid biosynthesis, and membrane function (Malhotra et al. 2018). Despite its essentiality to plant life, phosphorus in soils is usually of limited availability because of its reactivity with soil minerals and possible immobilization in insoluble compounds. Biochar is a potentially valuable soil amendment that has the potential to increase phosphorus availability, reduce P loss, and shift the phosphorus cycle by changing several chemical, physical, and biological processes (Fig. 4).

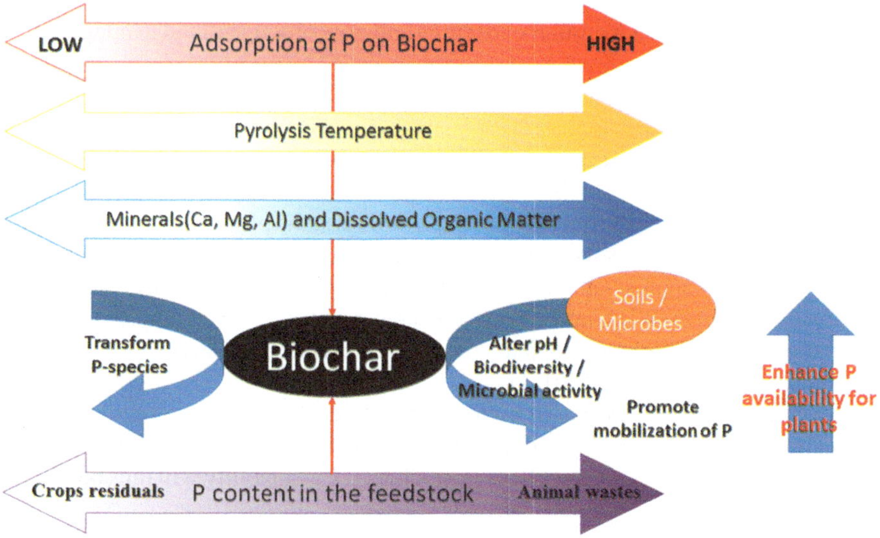

Fig. 4 Influence of biochar on environmental phosphorus cycle. Reproduced from Pan et al. 2021. Open Access

3.4.1 Biochar as Phosphorus Source

Phosphorus-rich biochars from materials like poultry litter, bone meal, or manure have the potential to be a direct plant-available phosphorus source. Pyrolysis can convert inorganic and organic P to stable but accessible forms of phosphates bound to calcium (Uchimiya et al. 2015). These are slow-release fertilizers that release P to crops for extended time frames. For instance, poultry litter biochar has been proven to contain extremely high levels of P and have excellent agronomic value in P-deficient soils.

3.4.2 Sorption and Desorption Dynamics

The surface functional groups and porosity of biochar enable adsorption of phosphate ion from the soil solution and prevention of leaching and enhanced retention of phosphorus. Ligand exchange, electrostatic attraction, or pH- and mineral-dependent precipitation reactions adsorb phosphorus. The buffering effect provides equilibrium of required concentrations of phosphorus in the root zone (Räty et al. 2010).

On the other hand, in polluted phosphorus water bodies or soils, biochar can act as an adsorbent which sequesters P and prevents surrounding water bodies from becoming eutrophic (Novais et al. 2018). Modified biochars (e.g., Mg-, Fe-, or La-treated) have higher P adsorption capacities and are consequently also applicable in water treatment.

3.4.3 Soil pH and Availability of Phosphorus Impact

The alkalinity in the majority of the biochars will raise the pH of soil and has a highly significant impact on phosphorus solubility and speciation. Under acid soils, biochar application will suppress aluminum and iron ions with extremely high affinity for phosphate and thereby make P more available to plants. For example, hardwood biochar application for acid tropical soil was reported to increase crop P uptake by more than 25% (Ch'ng et al. 2014).

Nevertheless, pH shifts caused by biochar in high-pH soils may have the capacity to precipitate phosphorus as calcium phosphates and limit its availability in the short term (Buss et al. 2018). General effect of biochar on P availability will thus be biochar and soil property dependent.

3.4.4 Microbial and Enzymatic Interactions

Biochar influences the phosphorus cycle and microbial community structure and function by enhancing more microbial activity. Biochar-enriched soil encourages microbes like arbuscular mycorrhizal fungi (AMF) (Ralebitso-Senior and Orr 2016; Upahyay and Chauhan 2022) and phosphate-solubilizing bacteria (PSB) (Chauhan and Upadhyay 2023; Aziz et al. 2024) to become enzymatically more active (i.e., phosphatases) during organic P mineralization.

For instance, there has been evidence of enhanced populations of PSB such as *Pseudomonas* and *Bacillus* in soils where biochar was added, especially where this has been combined with the application of organic fertilizer. These microorganisms release organic acids and enzymes which are used in immobilized phosphorus solubilization, resulting in enhanced plant uptake and availability (Ajeng et al. 2023).

3.4.5 Mitigation of P Loss and Environmental Hazards

Phosphorus loss through leachate and runoff is the most potentially hazardous to aquatic ecosystems. The capacity of biochar to retain P within the soil profile limits nutrient movement, especially in rain conditions. This role is of particular relevance in agroecosystems where fertilizer surplus could potentially lead to pollution by nutrients.

Other than that, biochar can also be used in buffer strips or wetlands to trap P before it flows into sensitive waters (Hoffmann et al. 2009). Other than water quality mitigation, such applications entail recycling of nutrients back into farm production.

3.4.6 Implication for Sustainable Agriculture

By regulating phosphorus cycling, biochar promotes more efficient use of nutrients. Biochar optimizes fertilizer performance, minimizes reliance on finite phosphate rock deposits, and maintains integrated soil fertility management. Production of biochar from existing biomass also boosts the circular economy and climate-resilient agriculture concept.

Long-term research on production of tailored biochar, co-fertilization of inorganic and organic fertilizers, and long-term plot experiments would be necessary to realize its highest potentiality in phosphorus cycle and pollution avoidance (Ahmed et al. 2024).

Overall, biochar releases and regulates phosphorus in soils and alters its availability, mobility, and conversion through a range of multidimensional chemical, physical, and biological processes. Strategic utilization of biochar has huge potential as an effective tool for promoting efficiency and sustainability of phosphorus in agroecosystems.

3.5 Impact of Biochar on the Sulfur Cycle

Sulfur (S) is a critical macronutrient for plant development and microbial respiration and plays a critical role in the most crucial biological reactions: redox regulation, enzyme performance, and amino acid biology (methionine and cysteine) (Narayan et al. 2023). Needed in smaller quantities compared to nitrogen or phosphorus, sulfur deficiency is also increasingly common in agricultural crops with reduced atmospheric deposition and enhanced cropping intensity. Biochar use has been shown to potentially control the sulfur cycle of the soil microbial ecosystem through several physicochemical and biological mechanisms and hence improve soil fertility and environmental stability.

3.5.1 Biochar as a Sulfur Source and Sink

Biochars with significant amounts of sulfur from manure-, sewage sludge-, or specific plant residues may directly act as a sulfur input. Organic sulfur, by pyrolysis, is transformed into more stable inorganic forms of sulfur like sulfates, which become readily available to plants. Manure biochar, for example, has been reported to improve crops' sulfur concentration and yield in sulfur-poor soils (Leng et al. 2022).

Alternatively, biochar can also act as a sulfur sink via sorption of sulfur species such as sulfate and hydrogen sulfide (Shang et al. 2013). Surface functional groups, metal oxides, and porosity improve sulfur compound sorption and sequestration, particularly in contaminated soils or wastewater treatment processes. The dual role enables the ability of biochar to supply and sequester sulfur based on the environmental setting.

3.5.2 Prevention of Sulfur Sequestration and Leaching

Sulfur leaching in the form of sulfate can lead to enormous losses of nutrients and water pollution. Incorporation of biochar has been found to reduce leaching of sulfur by increasing the water-holding capacity of the soil and establishing extensive interactions with the sulfur compounds (Taheri et al. 2024). In coarse-textured or sandy soils where there is high mobility of sulfur, the porous structure of biochar retains sulfates, keeping them within the rhizosphere for very long periods.

A field trial demonstrated that woodchip biochar decreased sulfate leaching by more than 40% from control treatments, particularly where rainfall is heavy. This role is beneficial to improve sulfur utilization efficiency and off-site environmental contamination prevention.

3.5.3 Soil pH and Sulfur Speciation Effects

Biochar's pH-buffering effect also impacts sulfur speciation and availability. In acidic soils, pH is increased by biochar, affecting the conversion of sulfur from its oxidized (sulfate) to reduced (sulfide) forms and vice versa (Leng et al. 2022). Under alkaline conditions, sulfate is optimal for plant uptake.

However, under certain redox and pH conditions, biochar also enhances the transformation of sulfur to less bioavailable forms and therefore must be accounted for by selection and use depending on soil type. For example, in flooded or wetland soil, biochar will affect sulfide production and reduction of sulfate and therefore alter sulfur bioavailability as well as microbial community (Wang et al. 2021).

3.5.4 Biochar–Microbe Interactions in Sulfur Cycling

Soil microbes are of significant influence in processes of sulfur transformation, such as oxidation of sulfur (by *Thiobacillus*spp.) and reduction of sulfate (by *Desulfovibrio*spp.) (Gupta and Germida 2021). Application of biochar tends to enhance microbial biomass and the activities of enzymes, including those of sulfur cycling.

Studies have shown that biochar enhances the density of sulfur-oxidizing bacteria in aerobic soils to enhance the oxidation of elemental sulfur to recoverable sulfate by plants (Amin and Mihoub 2021). In anaerobic soils like rice paddies, it has been shown to change microbial community structure to inhibit or stimulate sulfate-reducing bacteria, with redox conditions and organic matter content dominating.

Besides, biochar application was also found to induce arylsulfatase activity, an important enzyme in the organic sulfur mineralization process, and hence enhance sulfur availability to the plant (Khadem et al. 2019).

3.5.5 Function in Mitigation of Sulfur-Related Pollution

Anaerobic poisonous gas hydrogen sulfide (H_2S) is poisonous to the environment and human species. Its high surface area and toxicity make it highly suitable for the adsorption and immobilization of poisonous H_2S in polluted soils and biogas systems.

Biochar-added composting activities have demonstrated reduced H_2S and other sulfur-rich gas emissions, enhancing waste management system environmental performance (DeLuca et al. 2024). In addition, engineered biochars metal oxide surface-modified (e.g., Fe, Zn) have exhibited more promise in the adsorption of volatile sulfur compounds from wastewater and landfills.

3.5.6 Implications for Sustainable Soil and Water Management

Biochar plays a multifaceted role in the sulfur cycle, supporting its use as a sustainable soil amendment for fertilizer efficiency, pollution control, and water quality regulation. By enhancing sulfur storage and optimizing its release, biochar can reduce fertilizer dependency, boost crop productivity, and minimize environmental sulfur losses. Yuan et al. (2024) highlighted its synergistic effects with sulfur fertilizers, microbial inoculants, and other organic amendments. However, understanding biochar's role in sulfur dynamics under diverse environmental and management conditions requires mechanistic studies and long-term field trials. Overall, biochar functions as both a regulator and mediator of the sulfur cycle, controlling sulfur supply, microbial conversion, and loss pathways, with precision applications offering great potential for sustainable agriculture and natural resources conservation.

4 Interactions of Micronutrients

Micronutrients like iron-Fe, manganese-Mn, zinc-Zn, copper-Cu, boron-B, molybdenum-Mo, and chlorine-Cl are all part of plant metabolic processes but in extremely small quantities. Toxin deficiency or overdoses can have gigantic effects on crop yield and ecosystem homeostasis. Biochar, as a result of its multi-generic physicochemical property and interaction with soil components, has a central role in regulating the availability, mobility, and bioaccessibility of micronutrients within the soil system.

4.1 Biochar as Micronutrient Carrier and Source

Pyrolysis temperature and feedstock utilized in the production of biochar influence micronutrient composition. Biochars obtained from nutrient-rich feedstocks such as animal manure or crop residues have high levels of Fe, Zn, and Cu. Such biochars are slow-release micronutrient carriers that provide micronutrients to plants for a long duration of time (Lu et al. 2013). Pyrolysis temperature also controls nutrient retention lower to moderate temperatures will retain greater percentages of micronutrients in bioavailable forms compared to high-temperature biochars with the ability to sequester them into stable carbon pools.

4.2 Impacts on Redox Conditions and Soil pH

Biochar can potentially increase soil pH as it is alkaline in nature, and it has a direct influence on the speciation and solubility of micronutrients. For instance, Fe and Mn become immobile in alkaline soil through precipitation of insoluble hydroxides. Or, molybdenum is more accessible with rising pH (Riaz et al. 2020). Therefore, application of biochar needs to be suited to satisfy the strongly specific micronutrient demand of target crops and soil regime. In anaerobic or water-logging situation, biochar can change redox dynamics, and this will have the potential to change oxidation-reduction equilibrium of micronutrient forms, primarily Fe and Mn (Zhang and Furman 2021).

4.3 Immobilization and Adsorption of Micronutrients

Biochar's high surface area, pore size, and functional group density offer a site for immobilization and adsorption of both desirable and undesirable metals. The sorption is capable in contaminated soil where micronutrient elements like Cu or Zn at toxic concentration levels are present (Xu et al. 2022). For instance, the use of biochar in contaminated soils has been shown to decrease the bioavailability of the metals; therefore, plant health and phyto-toxicity are safeguarded (Mohamed et al. 2017). The same mechanism of adsorption can, nonetheless, limit the availability of nutrients for plant growth if not properly balanced.

4.4 Interaction with Soil Organic Matter and Microorganisms

Biochar affects soil organic matter dynamics, and this affects chelation and mobility of micronutrients directly. It affects microbial activity and community structure, which is stimulatory or inhibitory to micronutrient biogeochemical cycling. For example, biochar has been shown to stimulate siderophore bacteria that enhance iron solubilization and consequently enhance plant uptake (Rafique et al. 2017; Singh et al. 2022) (Fig. 5). Additionally, mycorrhizal fungi involved in micronutrient uptake can be enhanced with enhanced habitat quality by means of biochar.

4.5 Agricultural Implications

Under acid tropical soils, biochar application was observed to remedy Zn and B deficiency and produce an extremely high increment in maize and soybean crops (Choudhary et al. 2021). Under calcareous soils, whereby the micronutrients are low in availability due to their high pH levels, the efficacy of biochar is a function of form and functionalization. The further boost of biochar amendment by nutrient supplementation or surface modification (e.g., by chelating agent or mineral precipitate) is a new frontier for micronutrient optimization to the ultimate level (Younas et al. 2024).

Co-fertilization of $ZnSO_4$ with biochar, for instance, was found to be superior in Zn uptake promotion in rice cultivation to sole application (Raza et al. 2021). Fe-loaded biochar has also been shown to be effective for Fe-deficient soil reclamation (Ullah et al. 2023).

In addition to nutrient management, the role of biochar in trace element interactions provides ecological remediation and food safety benefits. Through control of the migration of beneficial and toxic trace elements, biochar minimizes threats to

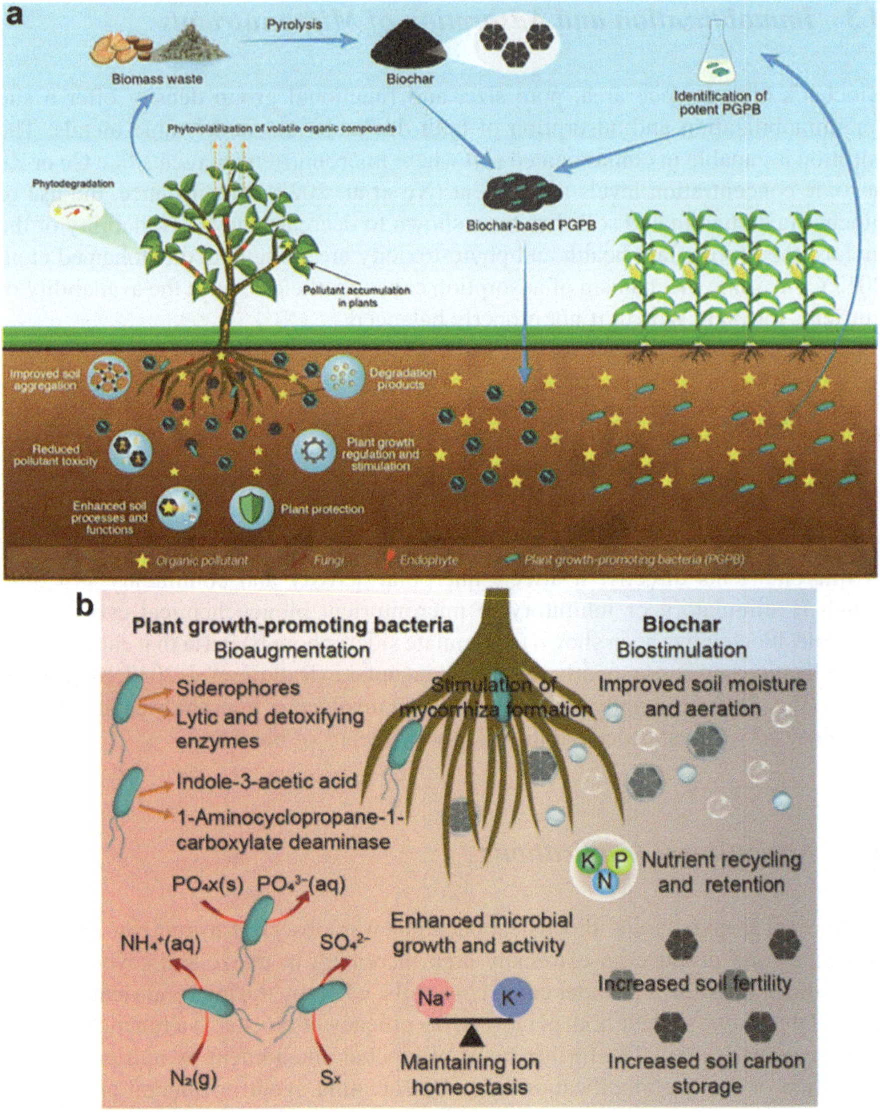

Fig. 5 **a** Diagram illustrating bioaugmentation using plant-growth-promoting bacteria (PGPB) which improves soil quality by fixing nitrogen, solubilizing nutrients (phosphate and potassium), producing allelochemicals (e.g., siderophores) and growth-regulating compounds, and enhancing mycorrhizal symbiosis. Biochar supports microbial activity, acts as a slow-release nutrient source, improves soil structure, and regulates ion uptake, and they optimize phytoremediation by boosting plant-microbe interactions and soil health (Xiang et al. 2022)

trace element-polluted water bodies and the food chain. However, enhanced micronutrient availability improves plant health, disease resistance, and nutrient quality, and facilitates sustainable agriculture.

Overall, the role of biochar in micronutrient processes is complex and location-dependent. When properly selected and applied, biochar is a great asset as a method for mitigating imbalances, improving nutrient use efficiency, and reducing the toxicity of heavy metals, thereby improving resilient agroecosystems.

5 Microbial Mediation of Biogeochemical Cycles

Microbes are the 'stealthy engineers' of biogeochemical cycling, facilitating conversion, mobilization, and immobilization of the elements on which ecosystem functioning hinges. Biochar amendment exerts a tremendous influence on such microbial communities and their metabolism and therefore controls rates and directions of biogeochemical cycling. Effects of biochar on microbial mediation of carbon, nitrogen, phosphorus, sulfur, and micronutrient cycling are addressed in this chapter with focus on current advances and emerging trends.

5.1 Biochar as Microbial Niche Modifier and Habitat

The biochar's heterogeneous surface chemistry and micropore structure create perfect microclimates for microbial colonization. The biochar's micropores and mesopores offer protection against toxic chemicals, environmental stress, and predation and offer room for a wide range of microbial consortia to settle. The microhabitats protect microbes from desiccation, toxic chemicals, and heat and effectively maximize microbial functional capacity and survival in stressful soils.

For instance, particles of biochar are the habitat of archaea, fungi, and bacteria at the core of nutrient cycling. As discussed earlier, various experiments suggest that the addition of biochar activates microbial biomass and community with discriminatory preference toward beneficial groups such as sulfur-oxidizing microbes (*Thiobacillus*), phosphorus-solubilizing bacteria (*Pseudomonas, Bacillus*), and nitrifying bacteria (*Nitrosomonas, Nitrobacter*). Spatial heterogeneity created by biochar triggers niche differentiation conducive to coexistence and synergism for promoting turnover of nutrients (Duan et al. 2023).

5.2 Impact on Microbial Metabolic Processes and Enzyme Activity

Biochar incorporation affects enzymatic decomposition in soil and processes of conversion of organic matter and nutrients. Specific enzymes, including dehydrogenases (total microbial activity), urease (mineralization of nitrogen), phosphatases (mineralization of phosphorus), and arylsulfatases (mineralization of sulfur), would likely be stimulated with the addition of biochar (Chettri et al. 2021).

The process is linked with direct and indirect effects. Biochar can adsorb inhibitory chemicals that restrain enzyme activity, chelate necessary cofactors, or alter soil pH to optimal levels for enzymatic activity. For instance, enhanced phosphatase activity following biochar addition is linked to enhanced phosphorus availability to plants, an important benefit in phosphorus-limited soils.

5.3 Modulation of Microbial Community Structure and Functional Genes

Besides activity and biomass, biochar also affects gene expression patterns and microbial community structure. High-throughput sequencing research revealed that biochar enhances relative gene abundance of nitrogen-fixation-related genes (*nirH*), nitrification-related genes (a*moA*), denitrification-related genes (*nirK, nosZ*), and phosphorus-solubilization-related genes (Pan et al. 2021), as illustrated in Fig. 6.

Through preferential enrichment of active microbial communities, biochar influences processes such as nitrification and denitrification and therefore changes nitrogen dynamics in the greenhouse and soil and greenhouse gas releases. For example, it has been reported to suppress numbers of denitrifiers emitting nitrous oxide (N_2O) and thereby limit climate change effects (Zhang et al. 2021).

5.4 Biochar and Microbial Interactions in Carbon Sequestration

Microbial decomposition of organic products is directly accountable for soil sequestration of carbon. Microbial decomposition rates are regulated by biochar through adsorption of the labile fractions of the organic compounds that either sequester carbon from microbial mineralization or act as a substrate for microbial growth (Ling et al. 2022). Either process generates a dynamic balance in soil turnover of carbon.

Apart from that, biochar can induce microbial taxa for long-term stabilized carbon sequestration and humus. Soil organic matter is stabilized through exudation and sorption of microbial necromass on the surface of the biochar.

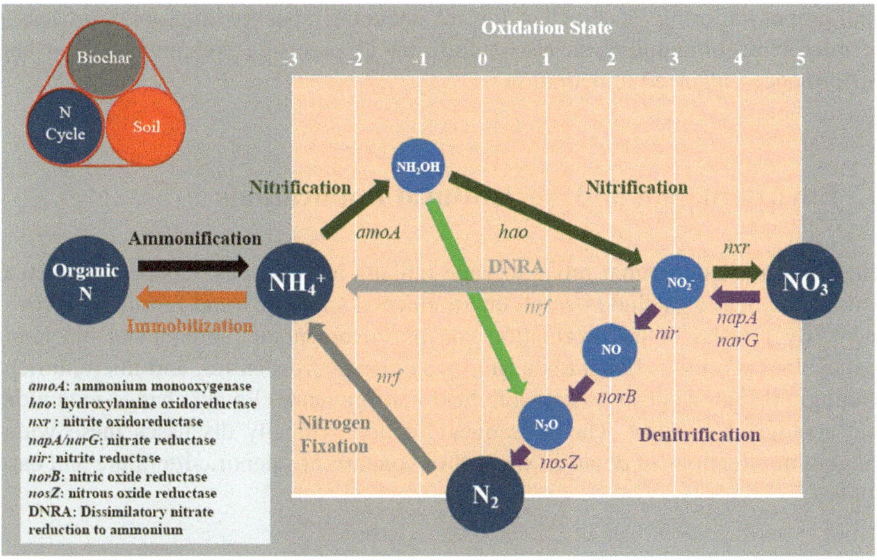

Fig. 6 Illustration of the enzymes associated with nitrogen cycle and color-coded arrows represent the different nitrogen transformation processes (Pan et al. 2021)

5.5 *Synergies Between Biochars and Microbes on Degradation of Pollutants*

Biochar enhances the bioremediation potential of microorganisms through immobilization of toxic contaminants and provision of an appropriate microenvironment for the degrading microbes. For example, biochar additions trigger the activation of hydrocarbon-degrading, pesticide-degrading (Yang et al. 2024), and heavy metal-degrading (Koh and Khor 2023) microbe populations.

Biochar can further facilitate electron transfer reactions by which microbial reduction of contaminants is enabled as a redox conductive matrix. This facilitates greater degradation rates and greater numbers of remedial contaminants through the electrochemical-microbial interface.

5.6 *Challenges and Research Frontiers*

Disagreements in prediction occur under conditions of heterogeneity in microbial assemblages, biochar, and soils in spite of recent advances realized through biochar-microbe interactions. Future research will try to integrate multi-omics approaches (metagenomics, metatranscriptomics, metabolomics) for reporting mechanistic details on biochar effects on microbial processes (Satya et al. 2024). Designing and proper tailoring of biochar toward inducing specific microbial processes or microbial communities is also significant to precision soil management and performance-enhanced remediation.

6 Environmental and Agricultural Implications

The application of biochar refers to wide-ranging implications outside of universal soil amendment to include overall environmental suitability and farming improvement. By process of transformation and improvement of fundamental biochemical cycling processes like carbon, nitrogen, phosphorus, sulfur, and micronutrient cycling, biochar supports sustainable land management, climate change reduction, and ecological stability. The subsequent section critically discusses these wide-ranging implications in detail in support of concurrent scientific literature and case studies.

6.1 Improvement of Soil Fertility and Crop Yield

Specific physicochemical properties of biochar effectively improve the fertility of the soil by enhancing nutrient storage, availability, and efficiency of nutrient uptake. Biochar promotes adsorption of major nutrients more efficiently due to high surface area, porosity, and functionality, along with lower leaching loss overcoming a humongous problem in traditional agriculture.

6.1.1 Nitrogen and Phosphorus Efficiency

In tropical soils themselves frequently highly weathered and low in P, biochar additions have been found to enhance P availability through complexation and pH adjustment. In one study in Brazilian Oxisols, rice crops in soil to which biochar had been added took up to 25% more phosphorus than the controls, leading to highly significantly higher grain yield (Morais et al. 2023). This was due to the capacity of the biochar to suppress the common phosphorus fixation property of aluminum and iron oxides in acid soils.

In the same way, biochar regulates the nitrogen cycle by preventing volatilization and leaching and induction of nitrification to promote nitrogen availability. Lower rates of synthetic N fertilizers were kept by biochar under testing with a temperate wheat cropping system with yield retained and 30% leaching of nitrate reduced (Haider et al. 2017). This mix is affordable to farmers and also reduces pollution in the environment.

6.1.2 Micronutrient Availability

In addition to macronutrients, biochar affects micronutrient cycling including zinc (Zn), copper (Cu), and iron (Fe). In the calcareous soils that are generally low in micronutrients, poultry litter-biochar amendments have been found to enhance Zn and Fe availability and thus enhance micronutrient concentration of wheat grain (Masud et al. 2020). Such improvements contribute to addressing 'hidden hunger,' which is a worldwide issue where micronutrient deficiency exists in spite of caloric adequacy.

6.1.3 Soil Physical Properties and Water Retention

Biochar enhances soil porosity, texture, and aggregation. Addition of biochar to the southwestern United States' sandy soils, for instance, added 15–20% water-holding capacity, resulting in increased drought tolerance of maize (Ndubo 2023). Water storage in this case is invaluable in arid and semi-arid areas, where water scarcity diminishes farm yield.

On acidic soils, the liming effect of biochar increases pH to promote root growth and microbial activity. For example, acid soils of sub-Saharan Africa with wood-derived biochar additions had more than 30% more maize yield as a result of increased nutrient availability, among others, and reduced aluminum toxicity (Mashamaite et al. 2024).

6.2 Reducing Greenhouse Gases and Climate Change

Biochar is crucial for climate change mitigation as it changes soil biochemical processes emitting greenhouse gases (GHGs) such as carbon dioxide (CO_2), nitrous oxide (N_2O), and methane (CH_4).

6.2.1 Carbon Sequestration

One of the most well-documented advantages of biochar is that it sequesters carbon in long-lasting, recalcitrant form (Ghosh et al. 2025). In contrast with labile soil organic carbon to be decomposed within a very short time frame, carbon sequestered in biochar lasts for decades to centuries. In temperate grassland long-term experiments, for instance, additions of biochar have added 10–25% extra pools of soil organic carbon, which will deliver humongous feedback into drawdown of CO_2 in the atmosphere (Vetter et al. 2022).

6.2.2 Nitrous Oxide Emissions

Nitrous oxide, formed mainly through microbial nitrification and denitrification, also shares a similar global warming potential of around 300 times CO_2. The application of biochar inhibits N_2O emission by altering soil aeration, raising the pH, and altering microbial community structure. Application of biochar to Chinese rice paddies lowered N_2O emission by as much as 50% due to greater oxygen diffusion while limiting denitrifiers under improved redox conditions (Li et al. 2021).

6.2.3 Methane Emissions

CH_4 is highly potent greenhouse gas predominantly emitted from wetlands and rice flooded paddies. Biochar has been shown to stimulate methane oxidation by enhancing activity of methano-trophic bacteria. A field trial in the Mekong Delta demonstrated that biochar application reduced methane emission nearly 40% thereby promoting climate resilient crop production (Farmer et al. 2023). Seeking these GHG mitigation advantages makes biochar a very useful component of climate-smart agriculture, which is incompatible with the world's carbon-neutrality plans.

6.3 Improved Soil and Water Quality

Biochar's effects on microbially mediated processes and nutrient dynamics radiate outward to soil and water quality advantages critical to ecosystem and human health.

6.3.1 Hydrological Benefits

Enhanced soil structure by biochar enhances the permeability of water and diminishes surface flow, lowering sediment and nutrient loss to the water body. This is beneficial for flood hazard management and groundwater preservation, particularly in highly agricultural landscapes.

6.3.2 Nutrient Runoff and Eutrophication

Excess nitrate and phosphorus in runoff from agriculture cause eutrophication of water bodies with the formation of toxic algal blooms and dead zones. Biochar restricts the mobility of nutrients by storing and adsorbing nutrients. A field experiment in the Chesapeake Bay watershed demonstrated that 35% nitrate leaching reduction was accomplished by biochar amendments, thus preventing pollution by nutrients (Lee et al. 2016).

6.3.3 Heavy Metal and Organic Contaminant Immobilization

The availability of high surface area and charge functional groups in biochar promotes heavy metal adsorption of such elements as lead (Pb), cadmium (Cd), and arsenic (As) and organic pollutants such as pesticides and polycyclic aromatic hydrocarbons (PAHs). Biochar application decreased the bioavailable concentration of Pb by 60% in an Australia mining-affected contaminated soil, restricting plant uptake and food chain entry (Ghani et al. 2024).

6.3.4 Improved Microbial Well-being of the Soil

Healthy soil microbial communities are critical for nutrient cycling and pollutant degradation. Microbes have habitat and energy source provided by biochar, stimulating enzymatic activity and community diversity. Biostimulation improves natural attenuation of pollutants and soil resilience.

6.4 Encouraging Circular Economy and Sustainable Agriculture

Biochar aligns with the principles of circular bioeconomy and sustainable agriculture because it upcycles waste biomass into valuable soil conditioners or amendments, thereby closes nutrient loops.

6.4.1 Waste Valorization

Forest and agricultural residues, wood products, and waste that would otherwise decompose and produce GHGs or contaminate ecosystems are being converted to biochar by pyrolysis. Rice husk in Southeast Asia (Mohammadi et al. 2017) and sugarcane bagasse in Brazil (Zafeer et al. 2024), for instance, were pyrolyzed into biochar, providing locally sourced soil amendments that substitute for imported fertilizers.

6.4.2 Reducing Use of Chemical Fertilizer

With improved nutrient utilization efficiency, biochar minimizes the application of synthetic fertilizers and hence the cost of production and greenhouse gas emissions associated with the production of fertilizers. This is especially crucial in agricultural-based economies that are resource-scarce.

6.4.3 Improving Food and Nutritional Security

Biochar enhances soil health by enhanced nutrient cycling and fertility, enhancing crop yields and micronutrient quality. This has a direct impact on nutritional quality and food security, counteracting malnutrition and population issues.

6.4.4 Integrative Strategies

Blending biochar with microbial inoculants, compost, and precision fertilizer offers the most profound agronomic and ecological benefits. For example Jalal et al. (2022) demonstrated that co-application of biochar with nitrogen-fixing bacteria significantly enhances legume productivity while simultaneously improving soil nitrogen sustainability, underscoring the synergistic potential of integrated biochar-based strategies.

6.5 Challenges and Future Directions

The application of biochar is nevertheless constrained by several challenges:

- *Feedstock and production variability: Differences* in feedstock type, pyrolysis temperature, and processing conditions yield biochar with diverse characteristics, complicating standardization and recommendations.
- *Context-dependent efficacy:* Biochar performance varies with soil types, climate, crop species, and management practices requiring site-specific validation.
- *Economic and Logistical Limitations*: high costs of production and transport, especially in remote or resource-limited regions, may hinder widespread adaptation.

Future research should therefore prioritize:

- Long-term field trials: To assess the longevity and cumulative effects of biochar spanning several crop cycles.
- Mechanistic investigations: Directed to biochar-microbe-plant interaction for understanding underlying biochemical pathways.
- Life cycle assessments: For full scope to ensure environmental sustainability and economic viability.
- Customized biochars: Manipulating biochar to target particular nutrient cycles or pollutant recovery.

7 Conclusion

The various interactions of biochars with biogeochemical cycling underlie its potential in the context of sustainable agriculture, environmental stewardship, and climate adaptation. Due to its high cation exchange capacity (CEC), porosity, surface area, and stable carbon content, biochar acts as carriers, buffers, and mediators of soil biochemical processes. It locks away carbon in the long term, decreases CO2 and greenhouse gases emissions, increases nitrogen retention with minimal leaching, volatilization and microbial losses, enhances phosphorus availability through changing soil chemistry, and manages sulfur as well as oxygen dynamics to sustain plants and microbes. Among the most important impacts of biochar is its ability to modify microbiological communities and exercise desirable microbial function. These changes result in large conversions of nutrient pools, pollutant bioavailability and ecosystem recovery. Soil physical and chemical properties significantly change with addition of biochar leading to increased fertility, reduced leaching of contaminants, and enhanced crop production.

Impact of biochar on ecology and biota is huge. Biochar not only helps in soil fertility and quality improvement, but also provides a solution to world issues such as soil erosion, water pollution, and hunger. Efficient and targeted utilization of biochar, where the origin feedstock, pyrolysis conditions, and local soil conditions are considered, can make it a powerful tool for circular economy and regenerative land-use management. However, large-scale field experiments, routine use, and cooperation among stakeholders are still required to realize the full potential of biochar. New production and use of biochar, such as product-specific formulations, biochar-added composite products, or synergy with organic or microbial additives, are more likely to enhance the application. Therefore, biochar may be an agent to drive and synchronize biochemical cycles toward ecological health and sustainable agriculture. The inclusion of land-use planning could be the make or break for progress in responding to climate change, biodiversity loss, and food system resilience.

References

Agyarko-Mintah E, Cowie A, Van Zwieten L, Singh BP, Smillie R, Harden S, Fornasier F (2017) Biochar lowers ammonia emission and improves nitrogen retention in poultry litter composting. Waste Manag 61:129–137

Ahmed N, Deng L, Wang C, Shah ZUH, Deng L, Li Y, Tu P (2024) Advancements in biochar modification for enhanced phosphorus utilization in agriculture. Land 13(5):644

Ajeng AA, Abdullah R, Ling TC (2023) Biochar-bacillus consortium for a sustainable agriculture: physicochemical and soil stability analyses. Biochar 5(1):17

Al Mamun S, Saha S, Ferdush J, Tusher TR, Islam MS (2022) Organic amendments for crop production, phosphorus bioavailability and heavy metal immobilisation: a review. Crop Pasture Sci

Alkharabsheh HM, Seleiman MF, Battaglia ML, Shami A, Jalal RS, Alhammad BA, Al-Saif AM (2021) Biochar and its broad impacts in soil quality and fertility, nutrient leaching and crop productivity: a review. Agronomy 11(5):993

Allaire SE, Baril B, Vanasse A, Lange SF, MacKay J, Smith DL (2015) Carbon dynamics in a biochar-amended loamy soil under switchgrass. Can J Soil Sci 95(1):1–13

Amin AEEAZ, Mihoub A (2021) Effect of sulfur-enriched biochar in combination with sulfur-oxidizing bacterium (Thiobacillus spp.) on release and distribution of phosphorus in high calcareous p-fixing soils. J Soil Sci Plant Nutr 21(3):2041–2047

Aziz S, Bibi S, Hasan MM, Biswas P, Ali MI, Bilal M, Maitra S (2024) A review on influence of biochar amendment on soil processes and environmental remediation. Biotechnol Genet Eng Rev 40(4):3270–3304

Bedassa M (2020) Soil acid management using biochar: review. Int J Agric Sci Food Technol 6:211–217

Berhe AA, Barnes RT, Six J, Marín-Spiotta E (2018) Role of soil erosion in biogeochemical cycling of essential elements: carbon, nitrogen, and phosphorus. Annu Rev Earth Planet Sci 46(1):521–548

Bolan N, Hoang SA, Beiyuan J, Gupta S, Hou D, Karakoti A, Van Zwieten L (2022) Multifunctional applications of biochar beyond carbon storage. Int Mater Rev 67(2):150–200

Buss W, Assavavittayanon K, Shepherd JG, Heal KV, Sohi S (2018) Biochar phosphorus release is limited by high pH and excess calcium. J Environ Qual 47(5):1298–1303

Chamoli A, Karn SK, Kumari M, Sivaramasamy E (2025) Biochar mediated fixation of nitrogen compounds (ammonia and nitrite) in soil: a review. Biodegradation 36(2):22

Chang JH, Sivasubramanian P, Dong CD, Kumar M (2023) Study on adsorption of ammonium and nitrate in wastewater by modified biochar. Bioresour Technol Rep 21:101346

Chanthini KMP, Pavithra GS, Murugan P, Malarvizhi P, Deva-Andrews A, Ramasubramanian R, Prockow J (2024) Management of excessive soil H+ ion induced toxicities by application of organic seaweed amendment enhances photosynthesis and resource use efficiencies in rice (Oryza sativa). Environ Res 247:118179

Chauhan PK, Upadhyay SK (2023) Exo-polysaccharide producing bacteria can induce maize plant growth and soil health under saline conditions. Biotechnol Genet Eng Rev:1–20

Chauhan PK, Upadhyay SK (2024) Mixed consortium of salt-tolerant phosphate solubilizing bacteria improves maize (Zea mays) plant growth and soil health under saline conditions. Mol Biotechnol 66:489–499

Chauhan PK, Upadhyay SK, Rajput VD et al (2024a) Fostering plant growth performance under drought stress using rhizospheric microbes, their gene editing, and biochar. Environ Geochem Health 46:41

Chettri D, Sharma B, Verma AK, Verma AK (2021) Significance of microbial enzyme activities in agriculture. Microbiol Act Soil Plant Health Manag:351–373

Ch'ng HY, Ahmed OH, Majid NMA (2014) Biochar and compost influence the phosphorus availability, nutrients uptake, and growth of maize (Zea mays L.) in tropical acid soil. Pak J Agric Sci 51(4)

Choudhary TK, Khan KS, Hussain Q, Ashfaq M (2021) Nutrient availability to maize crop (Zea mays L.) in biochar amended alkaline subtropical soil. J Soil Sci Plant Nutr 21(2):1293–1306

Chuahan PK, Pathak HK, Dubey G, Sharma H, Upadhyay SK (2024b) Impact of Bacillus cereus SPB-10 on growth promotion of wheat (Triticum aestivum L.) under arsenic contaminated soil, current microbiology. https://doi.org/10.1007/s00284-024-03673-1

Chunmei X, Liping C, Song C, Guang C, Xiufu Z, Danying W (2018) Effects of soil microbes on methane emissions from paddy fields under varying soil oxygen conditions. Agron J 110(5):1738–1747

Crespo-Barreiro A, Gómez N, González-Arias J, Ortiz-Liébana N, González-Andrés F, Cara-Jiménez J (2023) Scaling-up of the production of biochar from olive tree pruning for agricultural use: evaluation of biochar characteristics and phytotoxicity. Agriculture 13(5):1064

Dai Z, Xiong X, Zhu H, Xu H, Leng P, Li J, Xu J (2021) Association of biochar properties with changes in soil bacterial, fungal and fauna communities and nutrient cycling processes. Biochar 3:239–254
Dai Z, Zhang X, Tang C, Muhammad N, Wu J, Brookes PC, Xu J (2017) Potential role of biochars in decreasing soil acidification-a critical review. Sci Total Environ 581:601–611
DeLuca TH, Gundale M, MacKenzie MD, Gao S, Jones D (2024) Biochar effects on soil nutrient transformations. Taylor & Francis, In Biochar for environmental management
Ding Y, Liu Y, Liu S, Li Z, Tan X, Huang X, Zheng B (2016) Biochar to improve soil fertility. A review. Agron Sustain Dev 36:1–18
Dong D, Yang M, Wang C, Wang H, Li Y, Luo J, Wu W (2013) Responses of methane emissions and rice yield to applications of biochar and straw in a paddy field. J Soils Sediments 13:1450–1460
Duan Y, Awasthi MK, Yang J, Tian Y, Li H, Cao S, Ravindran B (2023) Bacterial community dynamics and co-occurrence network patterns during different stages of biochar-driven composting. Bioresour Technol 384:129358
Farmer J, Ghosh S, Rajakaruna P, Holmatov B (2023) A review of greenhouse gas emissions from selected wetlands in the Vietnamese Mekong Delta. https://hdl.handle.net/10568/139412
Fatima S, Riaz M, Al-Wabel MI, Arif MS, Yasmeen T, Hussain Q, Arif M (2021) Higher biochar rate strongly reduced decomposition of soil organic matter to enhance C and N sequestration in nutrient-poor alkaline calcareous soil. J Soils Sediments 21:148–162
Fawzy S, Osman AI, Yang H, Doran J, Rooney DW (2021) Industrial biochar systems for atmospheric carbon removal: a review. Environ Chem Lett 19:3023–3055
Freitas AM, Nair VD, Harris WG (2020) Biochar as influenced by feedstock variability: implications and opportunities for phosphorus management. Front Sustain Food Syst 4:510982
Ghaly AE, Ramakrishnan VV (2015) Nitrogen sources and cycling in the ecosystem and its role in air, water and soil pollution: a critical review. J Pollut Eff Control 3(2):1–26
Ghani J, Nawab J, Khan S, Khan MA, Ahmad I, Ali HM, Dinelli E (2024) Organic amendments minimize the migration of potentially toxic elements in soil–plant system in degraded agricultural lands. Biomass Convers Biorefinery 14(5):6547–6565
Ghosh D, Page-Dumroese DS, Han HS, Anderson N (2025) Role of biochar made from low-value woody forest residues in ecological sustainability and carbon neutrality. Soil Sci Soc Am J 89(1):e20793
Ghosh A, Sah D, Chakraborty M, Rai JPN (2024) Mechanism and application of bacterial exopolysaccharides: an advanced approach for sustainable heavy metal abolition from soil. Carbohydr Res 544:109247
Gopinath A, Divyapriya G, Srivastava V, Laiju AR, Nidheesh PV, Kumar MS (2021) Conversion of sewage sludge into biochar: a potential resource in water and wastewater treatment. Environ Res 194:110656
Graber ER, Singh B, Hanley K, Lehmann J (2017) Determination of cation exchange capacity in biochar. Biochar: Guid Anal Methods 1:74–107
Guo M, Song W, Tian J (2020) Biochar-facilitated soil remediation: mechanisms and efficacy variations. Front Environ Sci 8:521512
Gupta VVSR, Germida JJ (2021) Microbial transformations of sulfur in soil. In: Principles and applications of soil microbiology. Elsevier, pp 489–522
Haider G, Steffens D, Moser G, Müller C, Kammann CI (2017) Biochar reduced nitrate leaching and improved soil moisture content without yield improvements in a four-year field study. Agric Ecosyst Environ 237:80–94
Hoffmann CC, Kjaergaard C, Uusi-Kämppä J, Hansen HCB, Kronvang B (2009) Phosphorus retention in riparian buffers: review of their efficiency. J Environ Qual 38(5):1942–1955
Hornung A, Stenzel F, Grunwald J (2024) Biochar—just a black matter is not enough. Biomass Convers Biorefinery 14(5):5889–5900
Hu B, Ai Y, Jin J, Hayat T, Alsaedi A, Zhuang L, Wang X (2020) Efficient elimination of organic and inorganic pollutants by biochar and biochar-based materials. Biochar 2:47–64

Jalal A, Filho MCMT, da Silva EC, da Silva Oliveira CE, Freitas LA, do Nascimento, V (2022) Plant growth-promoting bacteria and nitrogen fixing bacteria: sustainability of non-legume crops. In: Nitrogen fixing bacteria: sustainable growth of non-legumes. Singapore, Springer Nature Singapore, pp 233–275

Jatav HS, Singh SK, Jatav SS, Rajput VD, Parihar M, Mahawer SK, Singhal RK (2020) Importance of biochar in agriculture and its consequence. Appl Biochar Environ Saf:109–123

Kacprzak M, Malińska K, Grosser A, Sobik-Szołtysek J, Wystalska K, Dróżdż D, Meers E (2023) Cycles of carbon, nitrogen and phosphorus in poultry manure management technologies–environmental aspects. Crit Rev Environ Sci Technol 53(8):914–938

Kan ZR, Liu WX, Liu WS, Lal R, Dang YP, Zhao X, Zhang HL (2022) Mechanisms of soil organic carbon stability and its response to no-till: a global synthesis and perspective. Glob Chang Biol 28(3):693–710

Kätterer T, Roobroeck D, Kimutai G, Karltun E, Nyberg G, Sundberg C, de Nowina KR (2022) Maize grain yield responses to realistic biochar application rates on smallholder farms in Kenya. Agron Sustain Dev 42(4):63

Khadem A, Besharati H, Khalaj MA (2019) Biochar application changed arylsulfatase activity, kinetic and thermodynamic aspects. Eur J Soil Biol 95:103134

Khedulkar AP, Pandit B, Doong RA (2023) Agricultural waste to real worth biochar as a sustainable material for supercapacitor. Sci Total Environ 869:161441

Koh LM, Khor SM (2023) Biodegradation process: basics, factors affecting, and industrial applications. In: Handbook of biodegradable materials. Springer International Publishing, Cham, pp 19–56

Laanbroek HJ (1990) Bacterial cycling of minerals that affect plant growth in waterlogged soils: a review. Aquat Bot 38(1):109–125

Lee S, Yeo IY, Sadeghi AM, McCarty GW, Hively WD, Lang MW (2016) Impacts of watershed characteristics and crop rotations on winter cover crop nitrate-nitrogen uptake capacity within agricultural watersheds in the Chesapeake Bay region. PloS one 11(6):e0157637

Lehmann J, Cowie A, Masiello CA, Kammann C, Woolf D, Amonette JE, Whitman T (2021) Biochar in climate change mitigation. Nat Geosci 14(12):883–892

Leng L, Liu R, Xu S, Mohamed BA, Yang Z, Hu Y, Li H (2022) An overview of sulfur-functional groups in biochar from pyrolysis of biomass. J Environ Chem Eng 10(2):107185

Li H, Meng J, Liu Z, Lan Y, Yang X, Huang Y, Chen W (2021) Effects of biochar on N_2O emission in denitrification pathway from paddy soil: a drying incubation study. Sci Total Environ 787:147591

Li S, Tasnady D (2023) Biochar for soil carbon sequestration: current knowledge, mechanisms, and future perspectives. C 9(3):67

Li S, Wang S, Fan M, Wu Y, Shangguan Z (2020) Interactions between biochar and nitrogen impact soil carbon mineralization and the microbial community. Soil Tillage Res 196:104437

Li M, Wang J, Shen H, He Z, Zhong H, Sun W, Tang Y (2023) Removal of benzohydroxamic acid-metal complexes pollution from beneficiation wastewater by metal-biochar/peroxymonosulfate system: behaviors investigation and mechanism exploration. Chem Eng J 461:142008

Li Y, Xiong X, Zhang C, Liu A (2022) Sustainable restoration of anoxic freshwater using environmentally-compatible oxygen-carrying biochar: performance and mechanisms. Water Res 214:118204

Ling L, Luo Y, Jiang B, Lv J, Meng C, Liao Y, Xu J (2022) Biochar induces mineralization of soil recalcitrant components by activation of biochar responsive bacteria groups. Soil Biol Biochem 172:108778

Liu Z, He T, Cao T, Yang T, Meng J, Chen W (2017) Effects of biochar application on nitrogen leaching, ammonia volatilization and nitrogen use efficiency in two distinct soils. J Soil Sci Plant Nutr 17(2):515–528

Liu Q, Li Y, Liu S, Gao W, Shen J, Zhang G, Wu J (2022) Anaerobic primed CO_2 and CH_4 in paddy soil are driven by Fe reduction and stimulated by biochar. Sci Total Environ 808:151911

Liu P, Wen S, Zhu S, Hu X, Wang Y (2025) Microbial degradation of soil organic pollutants: mechanisms, challenges, and advances in Forest ecosystem management. Processes 13(3):916

Liu Q, Zhang Y, Liu B, Amonette JE, Lin Z, Liu G, Xie Z (2018) How does biochar influence soil N cycle? A meta-analysis. Plant Soil 426:211–225

Lorenz K, Lal R (2014) Biochar application to soil for climate change mitigation by soil organic carbon sequestration. J Plant Nutr Soil Sci 177(5):651–670

Lorenz K, Lal R, Lorenz K, Lal R (2018) Biochar. Carbon sequestration in agricultural ecosystems pp 301–355

Lu J, Liu S, Chen W, Meng J (2023) Study on the mechanism of biochar affecting the effectiveness of phosphate solubilizing bacteria. World J Microbiol Biotechnol 39(3):87

Lu H, Zhang W, Wang S, Zhuang L, Yang Y, Qiu R (2013) Characterization of sewage sludge-derived biochars from different feedstocks and pyrolysis temperatures. J Anal Appl Pyrolysis 102:137–143

Malhotra H, Vandana, Sharma S, Pandey R (2018) Phosphorus nutrition: plant growth in response to deficiency and excess. Plant Nutr Abiotic Stress Toler:171–190

Mashamaite CV, Motsi H, Manyevere A, Poswa SB (2024) Assessing the potential of biochar as a viable alternative to synthetic fertilizers in sub-Saharan Africa smallholder farming: a review. Agronomy 14(6):1215

Masud MM, Abdulaha-Al Baquy M, Akhter S, Sen R, Barman A, Khatun MR (2020) Liming effects of poultry litter derived biochar on soil acidity amelioration and maize growth. Ecotoxicol Environ Saf 202:110865

Mohamed BA, Ellis N, Kim CS, Bi X (2017) The role of tailored biochar in increasing plant growth, and reducing bioavailability, phytotoxicity, and uptake of heavy metals in contaminated soil. Environ Pollut 230:329–338

Mohammadi A, Cowie AL, Mai TLA, Brandao M, de la Rosa RA, Kristiansen P, Joseph S (2017) Climate-change and health effects of using rice husk for biochar-compost: comparing three pyrolysis systems. J Clean Prod 162:260–272

Morais EGD, Jindo K, Silva CA (2023) Biochar-based phosphate fertilizers: synthesis, properties, kinetics of P release and recommendation for crops grown in Oxisols. Agronomy 13(2):326

Mosharrof M, Uddin MK, Sulaiman MF, Mia S, Shamsuzzaman SM, Haque ANA (2021) Combined application of rice husk biochar and lime increases phosphorus availability and maize yield in an acidic soil. Agriculture 11(8):793

Munera-Echeverri JL, Martinsen V, Strand LT, Zivanovic V, Cornelissen G, Mulder J (2018) Cation exchange capacity of biochar: an urgent method modification. Sci Total Environ 642:190–197

Muradov N, Fidalgo B, Gujar AC, Garceau N, T-Raissi, A. (2012) Production and characterization of Lemna minor bio-char and its catalytic application for biogas reforming. Biomass Bioenergy 42:123–131

Mushtaq T, Bano A, Ullah A (2024) Effects of Rhizospheric microbes, growth regulators, and biochar in modulating antioxidant machinery of plants under stress. J Plant Growth Regul:1–22

Nair VD, Nair PR, Dari B, Freitas AM, Chatterjee N, Pinheiro FM (2017) Biochar in the agroecosystem–climate-change–sustainability nexus. Front Plant Sci 8:2051

Narayan OP, Kumar P, Yadav B, Dua M, Johri AK (2023) Sulfur nutrition and its role in plant growth and development. Plant Signaling Behav 18(1):2030082

Ndubo OT (2023) Effect of compost amendments on soil physicochemical properties, plant growth and efficient use of water, nitrogen and phosphorus in maize (zea mays l). (Doctoral dissertation, Botswana University of Agriculture & Natural Resources)

Novais SV, Zenero MDO, Barreto MSC, Montes CR, Cerri CEP (2018) Phosphorus removal from eutrophic water using modified biochar. Sci Total Environ 633:825–835

de Oliveira Paiva I, de Morais EG, Jindo K, Silva CA (2024) Biochar N content, pools and aromaticity as affected by feedstock and pyrolysis temperature. Waste Biomass Valorization 15(6):3599–3619

Palansooriya KN, Wong JTF, Hashimoto Y, Huang L, Rinklebe J, Chang SX et al (2019) Response of microbial communities to biochar-amended soils: a critical review. Biochar 1:3–22

Pan SY, Dong CD, Su JF, Wang PY, Chen CW, Chang JS et al (2021) The role of biochar in regulating the carbon, phosphorus, and nitrogen cycles exemplified by soil systems. Sustainability 13(10):5612

Parasar BJ, Agarwala N (2025) Unravelling the role of biochar-microbe-soil tripartite interaction in regulating soil carbon and nitrogen budget: a panacea to soil sustainability. Biochar 7(1):37

Parvage MM, Herbert R (2023) Sequential removal of nitrate and sulfate in woodchip and hematite–coated biochar bioreactor. Environ Sci: Water Res Technol 9(2):489–499

Pathak HK, Chauhan PK, Seth CS, Dubey G, Upadhyay SK (2024a) Mechanistic and future prospects in rhizospheric engineering for agricultural contaminants removal, soil health restoration, and management of climate change stress. Sci Total Environ 927:172116. https://doi.org/10.1016/j.scitotenv.2024.172116

Pathak HK, Seth CS, Chauhan PK, Dubey G, Singh G, Jain D, Upadhyay SK, Dwivedi P, Khoo KS (2024b) Recent advancement of nano-biochar for the remediation of heavy metals and emerging contaminants: mechanism, adsorption kinetic model, plant growth and development. Environ Res 255:119136. https://doi.org/10.1016/j.envres.2024.119136

Patrick WH Jr, DeLaune RD (1977) Chemical and biological redox systems affecting nutrient availability in the coastal wetlands. Geosci Man 18(13):137

Pei J, Li J, Fang C, Zhao J, Nie M, Wu J (2020) Different responses of root exudates to biochar application under elevated CO_2. Agric Ecosyst Environ 301:107061

Rafique M, Sultan T, Ortas I, Chaudhary HJ (2017) Enhancement of maize plant growth with inoculation of phosphate-solubilizing bacteria and biochar amendment in soil. Soil Sci Plant Nutr 63(5):460–469

Ralebitso-Senior TK, Orr CH (2016) Biochar application: essential soil microbial ecology. Elsevier

Rasul M, Cho J, Shin HS, Hur J (2022) Biochar-induced priming effects in soil via modifying the status of soil organic matter and microflora: a review. Sci Total Environ 805:150304

Räty M, Uusi-Kämppä J, Yli-Halla M, Rasa K, Pietola L (2010) Phosphorus and nitrogen cycles in the vegetation of differently managed buffer zones. Nutr Cycl Agroecosyst 86:121–132

Raza HMA, Bashir MA, Rehim A, Jan M, Raza QUA, Berlyn GP (2021) Potassium and zinc co-fertilization provide new insights to improve maize (Zea mays L.) physiology and productivity. Pak J Bot 53(6):2059–2065

Razzak SA (2024) Municipal solid and plastic waste derived high-performance biochar production: a comprehensive review. J Anal Appl Pyrolysis 106622

Riaz MU, Ayub MA, Khalid H, ul Haq MA, Rasul A, ur Rehman MZ, Ali S (2020) Fate of micronutrients in alkaline soils. Resour Use Effic Agric:577–613

Ro KS, Lima IM, Reddy GB, Jackson MA, Gao B (2015) Removing gaseous NH_3 using biochar as an adsorbent. Agriculture 5(4):991–1002

Saravanan A, Swaminaathan P, Kumar PS, Yaashikaa PR, Kamalesh R, Rangasamy G (2023) A comprehensive review on immobilized microbes-biochar and their environmental remediation: mechanism, challenges and future perspectives. Environ Res 236:116723

Satya S, Sharma S, Choudhary G, Kaushik G (2024) Advances in environmental microbiology: a multi-omic perspective. In: Microbial omics in environment and health. Singapore, Springer Nature Singapore, pp 175–204

Shang G, Shen G, Liu L, Chen Q, Xu Z (2013) Kinetics and mechanisms of hydrogen sulfide adsorption by biochars. Bioresour Technol 133:495–499

Singh V (2024) Biogeochemical cycles. In: Textbook of environment and ecology. Singapore, Springer Nature Singapore, pp 75–94

Singh P, Chauhan PK, Upadhyay SK, Singh RK, Dwivedi P, Wang J, Jain D, Jiang M (2022) Mechanistic insights and potential use of siderophores producing microbes in rhizosphere for mitigation of stress in plants grown in degraded land. Front Microbiol 13:898979

Sohi S, Lopez-Capel E, Krull E, Bol R (2009) Biochar, climate change and soil: a review to guide future research. CSIRO Land Water Sci Rep 5(09):17–31

Song B, Tang J, Zhen M, Liu X (2019) Influence of graphene oxide and biochar on anaerobic degradation of petroleum hydrocarbons. J Biosci Bioeng 128(1):72–79

Sun Y, Xiong X, He M, Xu Z, Hou D, Zhang W et al (2021) Roles of biochar-derived dissolved organic matter in soil amendment and environmental remediation: a critical review. Chem Eng J 424:130387

Taheri MAR, Astaraei AR, Lakzian A, Emami H (2024) The role of biochar and sulfur-modified biochar on soil water content, biochemical properties and millet crop under saline-sodic and calcareous soil. Plant Soil 499(1):221–236

Thapa B, Mowrer J (2024) Soil carbon and aggregate stability are positively related and increased under combined soil amendment, tillage, and cover cropping practices. Soil Sci Soc Am J 88(3):730–744

Thies JE, Rillig MC (2012) Characteristics of biochar: biological properties. In: Biochar for environmental management. Routledge, pp 117–138

Uchimiya M, Hiradate S, Antal MJ Jr (2015) Dissolved phosphorus speciation of flash carbonization, slow pyrolysis, and fast pyrolysis biochars. ACS Sustain Chem Eng 3(7):1642–1649

Ud Din MM, Khan MI, Azam M, Ali MH, Qadri R, Naveed M, Nasir A (2023) Effect of biochar and compost addition on mitigating salinity stress and improving fruit quality of tomato. Agronomy 13(9):2197

Ullah S, Ul Haq T, Shehzad MA, Ahmad KS, Imran M, Matloob A et al (2023) Fe and Zn-loaded cotton-sticks biochar and organic amendments improves chickpea production by enhancing ionic uptake and chemical properties of sandy soils. Cogent Food Agric 9(2):2265110

Upadhyay SK, Chauhan PK (2022) Optimization of eco-friendly amendments as sustainable asset for salt-tolerant plant growth-promoting bacteria mediated maize (Zea Mays L.) plant growth, Na uptake reduction and saline soil restoration. Environ Res 211:113081. https://doi.org/10.1016/j.envres.2022.113081

Upadhyay SK, Srivastava AK, Rajput VD, Chauhan PK, Bhojiya AA, Jain D, Chaubey G, Dwivedi P, Sharma B, Minkina T (2022) Root exudates: mechanistic insight of plant growth promoting Rhizobacteria for sustainable crop production. Front Microbiol 13:916488. https://doi.org/10.3389/fmicb.2022.916488

Vasseghian Y, Nadagouda MM, Aminabhavi TM (2024) Biochar-enhanced bioremediation of eutrophic waters impacted by algal blooms. J Environ Manag 367:122044

Vetter SH, Abdalla M, Kuhnert M, Smith P (2022) Soil carbon sequestration and biochar, pp 194–243. https://doi.org/10.1039/9781839165245-00194

Wang Y, Zhang Y, Ok YS, Jiang T, Liu P, Shu R, Zhong H (2021) Biochar-impacted sulfur cycling affects methylmercury phytoavailability in soils under different redox conditions. J Hazard Mater 407:124397

Warnock DD, Lehmann J, Kuyper TW, Rillig MC (2007) Mycorrhizal responses to biochar in soil–concepts and mechanisms. Plant Soil 300:9–20

Weng Z, Van Zwieten L, Singh BP, Tavakkoli E, Joseph S, Macdonald LM, Cowie A (2017) Biochar built soil carbon over a decade by stabilizing rhizodeposits. Nat Clim Chang 7(5):371–376

Whitman T, Singh BP, Zimmerman AR (2015) Priming effects in biochar-amended soils: implications of biochar-soil organic matter interactions for carbon storage. Biochar Environ Manag:455–487

Wong JW, Ogbonnaya UO (2021) Biochar porosity: a nature-based dependent parameter to deliver microorganisms to soils for land restoration. Environ Sci Pollut Res 28:46894–46909

Xiang L, Harindintwali JD, Wang F, Redmile-Gordon M, Chang SX, Fu Y, Xing B (2022) Integrating biochar, bacteria, and plants for sustainable remediation of soils contaminated with organic pollutants. Environ Sci Technol 56(23):16546–16566

Xiang W, Zhang X, Chen J, Zou W, He F, Hu X et al (2020) Biochar technology in wastewater treatment: a critical review. Chemosphere 252:126539

Xiao YN, Sun X, Ding J, Jiang Z, Xu J (2019) Biochar improved rice yield and mitigated CH_4 and N_2O emissions from paddy field under controlled irrigation in the Taihu Lake region of China. Atmos Environ 200:69–77

Xiu L, Zhang W, Wu D, Sun Y, Zhang H, Gu W, Chen W (2021) Biochar can improve biological nitrogen fixation by altering the root growth strategy of soybean in Albic soil. Sci Total Environ 773:144564

Xu Q, Xu Q, Zhu H, Li H, Yin W, Feng K, Wang X (2022) Does biochar application in heavy metal-contaminated soils affect soil micronutrient dynamics? Chemosphere 290:133349

Yang Y, Wang X, Wang Y, Saleem M, Mu Y, Zheng Y, Zhang Q (2024) Pesticide contamination remediation by biochar-immobilized microorganisms: a review. Int J Environ Sci Technol 21(2):2225–2238

Younas Z, Rahman U, Ikram M, Raja NI (2024) Biochar and micronutrients availability: problem and future prospects. In: Biochar-assisted remediation of contaminated soils under changing climate. Elsevier, pp 101–127

Yuan J, Chen H, Chen G, Pokharel P, Chang SX, Wang Y, Wang Y (2024) Long-term biochar application influences phosphsorus and associated iron and sulfur transformations in the rhizosphere. Carbon Res 3(1):25

Zafeer MK, Menezes RA, Venkatachalam H, Bhat KS (2024) Sugarcane bagasse-based biochar and its potential applications: a review. Emerg Mater 7(1):133–161

Zainudin MH, Mustapha NA, Maeda T, Ramli N, Sakai K, Hassan M (2020) Biochar enhanced the nitrifying and denitrifying bacterial communities during the composting of poultry manure and rice straw. Waste Manag 106:240–249

Zaman W, Ayaz A, Puppe D (2025) Biogeochemical cycles in plant–soil systems: significance for agriculture, interconnections, and anthropogenic disruptions. Biology 14(4):433

Zhang K, Deng Y, Liu Z, Feng Y, Hu C, Wang Z (2023) Biochar facilitated direct interspecies electron transfer in anaerobic digestion to alleviate antibiotics inhibition and enhance methanogenesis: a review. Int J Environ Res Public Health 20(3):2296

Zhang Z, Furman A (2021) Soil redox dynamics under dynamic hydrologic regimes-A review. Sci Total Environ 763:143026

Zhang J, Liu C, Ling J, Zhou W, Wang Y, Cheng H, Dong J (2025) Revealing the potential of biochar for heavy metal polluted seagrass remediation from microbial perspective. Ecotoxicol Environ Saf 292:117991

Zhang L, Xiang Y, Jing Y, Zhang R (2019) Biochar amendment effects on the activities of soil carbon, nitrogen, and phosphorus hydrolytic enzymes: a meta-analysis. Environ Sci Pollut Res 26:22990–23001

Zhang Y, Zhang Z, Chen Y (2021) Biochar mitigates N2O emission of microbial denitrification through modulating carbon metabolism and allocation of reducing power. Environ Sci Technol 55(12):8068–8078

Future Prospects and Emerging Trends of Biochar

Somvir Singh, Arti Thakur, Vishav Kiran, Kanika Dulta, Samson Karugu Kamau, Soumya Ghosh, Sakshi Tomar, and Gopal Dubey

Abstract Control of the environment through biochar as a sustainable agricultural product and its control over climatic change is increasingly making it an acceptable compound. Not only can it lock up heavy metals, break down natural contaminants, but it could also provide an answer to the newcomers such as microplastic pollution due to its porous structure and chemical resistance. Biochar is significant in the agricultural land in the promotion of the quality of the soil by enhancing the soil texture, trapping of desirable nutrients, and the acquisition of the salutary macrobiotic effect. The properties are not only making the soil fertile, but also allowing crops to resist the climatic stresses like salinity, droughts and plant diseases and hence biochar creates a death grip in the struggle to foster condition-resistant agriculture. The best benefit of biochar is that it can potentially bind carbon and is therefore a good candidate of a negative emission technology that can be employed to offset emission of greenhouse gases around the globe. Other than environmental application, the

S. Singh (✉)
Department of Biosciences, University Institute of Biotechnology, Chandigarh University, Gharuan, Mohali 140413, India
e-mail: samrajput0002@gmail.com

A. Thakur
Department of Botany, Shoolini Institute of Life Sciences and Business Management, Solan, India

V. Kiran
LR Institute of Pharmacy Jabil Kyar, Solan, India

K. Dulta
Centre of Advanced Innovation Technologies, VSB-Technical University of Ostrava, Ostrava-Poruba, Czech Republic

S. K. Kamau
Department of Applied Life Sciences, Gyeongsang National University, Jinju, Korea

S. Ghosh
Natural and Medical Sciences Research Centre, University of Nizwa, Nizwa, Oman

S. Tomar
Department of Environmental Science, AKS University, Satna, Madhya Pradesh, India

G. Dubey
Department of Environmental Science, VBS Purvanchal University, Uttar Pradesh, Jaunpur, India

A. Chauhan et al. (eds.), *Biochar As Remediation Tools for Pollutant Removal*,
https://doi.org/10.1007/978-3-032-16746-0_13

new studies have now switched to engineered biochar that is also finding application in the biomedical field. Researchers indicate that it could be used to deliver drugs, heal wound infections and act as an antimicrobial due to its potential applications; the possibilities of using the compound extend beyond the traditional agricultural or environmental use. In addition to direct applications, biochar plays a major role in the global economy in the transformation of waste products like crop residues, sewage sludge, and products to useful resources. This process does not only contribute to a reduction of environmental strain but also leads to the achievement of a number of the United Nations Sustainable Development Goals (SDGs). Specifically, biochar fits into SDG 7 on access to affordable and clean energy as well as SDG 12 on responsible consumption and production and SDG 13 on climate action, but also has a wider scope as part of sustainable development. The future prospects of biochar will be determined to a large extent by developments regarding the material being modified, cost effective and large-scale production and comprehensive life cycle evaluation. These measures play a vital role when it comes to closing the divide between laboratory tests and practice, so that the innovative applications of the biochar could be expanded and potentially it would be integrated into the industrial systems.

Keywords Biochar technology · Emerging applications · Environmental remediation · Carbon sequestration · Climate-smart agriculture · Circular economy

1 Introduction

The twenty first century faces various challenges related to global food, water, and energy demands while mitigating climate change and reversing environmental degradation. This degradation manifested through the loss of soil organic matters, soil contamination, overuse of synthetic fertilizers, soil erosion, acidification, salinization, and erosion of genetic diversity are threats to crop productivity and food security (Kopittke et al. 2019). As the global population surpasses 8 billion, ensuring sustaining agricultural production in the face of limited land resources and escalating ecological stress presents a promising challenge (Ortiz-Bobea et al. 2021). Since over 98.8% of human nutrition depends on topsoil and soils provide multiple ecosystem services their degradation directly undermines global food security and ecological stability (Kopittke et al. 2019; Chauhan et al. 2025a, b). Innovative, sustainable, and circular approaches are therefore essential.

Among the most promising solutions, biochar has emerged as a multifunctional, carbon rich material capable of addressing interconnected challenges in agriculture, environment, energy, and health. Produced mainly by pyrolysis of organic residues such as crop waste, forestry biomass, and sewage biochar transforms waste streams into stable, porous and highly functional, materials (Pathak et al. 2024a). Conventional pyrolysis (350–700 °C) remains the most widely applied industrial method, but newer processes such as hydrothermal carbonization, microwave assisted pyrolysis and solar driven systems are improving energy efficiency and reducing fossil fuel dependence (Pathak et al. 2024a). These technological advancements are not only enhancing biochar properties but also positioning it as a scalable solution within the circular economy.

The emerging role of biochar in environmental remediation is central to its growing significance. High sorption capacity enables the immobilization of toxic metals, pesticides, dyes, pharmaceuticals, and organic pollutants in soils and waters, thereby reducing ecological risks (Chauhan et al. 2025a, b). Recent developments in engineered biochar through acid alkali activation, surface oxidation, or microbial functionalization further enhance its adsorption capacity and catalytic reactivity, making it a frontier material in sustainable remediation technologies (Pathak et al. 2024a).

Biocompatibility, tenable surface characteristics and leaky structure of biochar are increasingly being utilized in medical and pharmaceutical studies. Such characteristics cause it to be suitable in the creation of drug delivery tools, antimicrobial scaffolds to support wound remedy, detoxification treatments, and tissue engineering remedies. The latest studies also point to its possible use in treatment of cancer and in fighting antimicrobial resistance, which means that biochar research takes a new revolutionary turn (Min et al. 2025).

The biochar has already proven effective in the past in the field of agriculture whereby it can be applied to condition the soil that increases the capability of soil to hold nutrients, efficiency in the use of water and microbial diversity. The use of biochar, which is capable of overcoming abiotic stressors such as drought, salinity, and heavy metal toxicity, and has synergistic effects with plant growth promoting rhizobacteria (PGPR) and organic amends, is a key to climatic smarts agriculture (Pathak et al. 2024b; Chauhan et al. 2025a, b). It can therefore be considered as a resilience-building program of world food security and a soil amendment.

Biochar is one of the most plausible negative emission solutions as much as amelioration of climate change is concerned. Biochar provides long-term carbon sequestration via stabilisation of organic carbon to recalcitrant forms that are stored in soils over centuries to millennia (Pathak et al. 2024b). It is estimated that such a large adoption has the power to offset hundred of gigatons of CO_2 equivalent emissions per annum, which is consistent with the global reduction of the climate changes, as well as offers agronomic advantages (Weng and Cowie 2025).

The opportunities of biochar are more than within the frame of the circular economy and waste management paradigm. The conversion of agricultural residues, organic municipal wastes and sewage sludge into biochar combats the two thorns in the flesh, waste disposal, and resources recovery. The products obtained as a result of the enhancement of soil to carbon neutral fuel help directly toward SDGs 7, 12, and 13 (Pathak et al. 2024a). This combination shows in a single holistic form.

Realizing such benefits at scale requires supportive policy and regulatory frameworks. Standard for quality, application safety, and environmental performance must be harmonized, while carbon credits markets and financial incentives can accelerate adoption. Governments and international agencies are beginning to incorporate biochar into carbon reduction strategies, signalling growing policy recognition of its role in sustainable development (Salma et al. 2024). The expansion of biochar application in sustainable development is also notable. Beyond soils and remediation bicohar is now being incorporated into green construction materials, energy storage systems, livestock nutrition and urban infrastructure, showcasing its adaptability and wide ranging societal benefits (Afshar and Mofatteh 2024). Yet, challenges remain. The scalability, economic feasibility, and long-term impacts of large-scale biochar deployment still demand rigorous research. Further directions must emphasize interdisciplinary approaches, advanced feedstock utilization, and international collaboration to ensure that biochar fulfils its transformative potential.

In view of the emerging global significance of biochar, the present chapter aims (a) to evaluate its role in environmental remediation and sustainable resource management, (b) to examine its applications in medicine, pharmaceuticals, and agriculture within a circular economy framework, and (c) to analyze emerging trends of biochar in climate mitigation and its contribution to global sustainability goals.

2 Emerging Role of Biochar in Environmental Remediation

2.1 Heavy Metal and Organic Contamination Detoxification in Soil

Biochar has emerged as a transformative, affordable, and eco-sustainable material for restoring soils contaminated with heavy metals and organic pollutants. Its exceptionally high surface area, porous structure, and diverse oxygen containing functional groups such as aldehyde, ketone, phenol, carboxyl, hydroxyl, and aromatic structures serve as active binding sites that drive pollutant immobilization and detoxification (Pathak et al. 2024a). Lead (Pb) is immobilized through surface complexation and interaction with $-COO^-$, $-O^-$, and C–π bonds (Sun et al. 2021), mercury (Hg) forms stable complexes with sulfoxidethiol, and carboxyl groups (Liu et al. 2016), nickel (Ni) binds with C–O, C=O, hydroxyl, and carboxyl functionalities (Zhu et al. 2021), while zinc (Zn) associates with alcohol, hydroxyl, carbonyl, and phenolic hydroxyl groups (Soria et al. 2020). Biochar too enhances soil organic carbon, which decreases mercury bioavailability and methylation (Hu et al. 2023). Phenol –OH and C=O groups oxidize As^{3+} to less toxic As^{5+} (Wen et al. 2021) and Mn^-, Fe^-, and Ce^- modified biochar further decrease arsenic mobility through adsorption, co-precipitation, and complexation (Gong et al. 2022). Together, these mechanisms cause adsorption, precipitation, reduction, complexation and several of these mechanisms are significant in the reduction of the toxicity and mobility of the heavy metals.

Beyond metals, biochar exerts a decisive influence on organic pollutant removal by functioning both as an adsorbent and a catalyst. Free radicals generated on its surface react with oxygen to from hydroxyl radicals (·OH), eliminating the need for additional oxidants (Ruan et al. 2019), while phenolic hydroxyl groups act as electron shuttles that intensify the production of OH and sulphate radicals (SO_4^-) in the presence of an oxidant (Fang et al. 2015). Moreover, biochar enhances photocatalytic degradation by serving as an electron acceptor and providing structural support to phtocatalysts (Javanmard et al. 2024). Its porous matrix and carbon content also stimulate microbial activity and biodiversity, accelerating biodegradation processes; low temperature biochars, rich in nutrient, encourage microbial proliferation, whereas high temperature biochars with greater sorption capacity, effectively immobilize contaminants (Nehela et al. 2021). At the same time, biochar improves soil fertility by increasing organic carbon, nutrient levels, $_{p}$H and cation exchange capacity, thereby restructuring microbial communities and enhancing in situ biodegradation (Khan et al. 2024).

Ultimately, the efficiency of these multifunctional pathways depends on biochars origin, structural characteristics, oxygen containing functional groups redox potential, and environmental context (Bano et al. 2025). Altogether biochar integrates adsorption, redox transformation, catalysis, precipitation and microbial stimulation into a unified system, positioning it as a versatile and highly influential tool for heavy metal immobilization, organic pollutant degradation, soil fertility restoration, and long-term environmental sustainability.

2.2 *Mitigation of Climate Change and Carbon Capture*

Biochar is increasingly recognized as one of the most powerful nature-based strategies to combat climate change. Unlike short-term mitigation technologies, biochar provides a durable pathway for carbon capture by converting biomass into a highly stable carbon from through pyrolysis under limited oxygen conditions. This process transforms easily degradable organic matter into aromatic recalcitrant carbon structures that remain in soils for hundreds to thousands of years, thereby locking atmospheric CO_2 into terrestrial systems. At the same time, biochar enhances soil fertility and reduces greenhouse gas (GHG) emissions, establishing itself as a multifunctional solution for climate resilient agriculture (Shoudho et al. 2024).

The mechanism of biochar-mediated carbon capture is both direct and indirect. Directly, biochar sequesters carbon in its condensed, aromatic lattice that resists microbial breakdown. Indirectly, it modifies soil environments in ways that suppress GHG release (Li and Tasnady 2023). By improving porosity, aeration and water retention, biochar reduces anaerobic conditions that promote methane (CH_4) generation. Its high cation exchange capacity facilitates efficient nutrient cycling, thereby lowering the demand for synthetic fertilizers whose production and application are carbon intensive.

Evidence from global assessments underscores the transformative potential of biochar. Biochar shows a global climate mitigation potential of 2.6–10.3 Pg CO_2eq yr^{-1} through durable carbon sequestration, reduced soil GHG emissions, fossil fuel displacement, and residue utilization (Weng and Cowie 2025). Similarly Shoudho et al. (2024) studied that biochar offers a transformative pathway for climate mitigation, capturing 1–35 $GtCO_2$ this century.

Carbon sequestration in the soils lessens green house gas emissions and increases fertility, water retention and yields. The twofold positive effect of biochar as an emission-reduction and productivity enhancer further qualifies biochar as an essential instrument in striking a balance between food and environmental concerns.

New laws and market procedures highlight the increased role of biochar. Methods of validating the carbon sequestration have been developed by the European Biochar Certificate (EBC) and U.S. Biochar Initiative, and these operations can receive carbon credits as a part of voluntary market (Salma et al. 2024). This change offers economic benefits to farmers and industries to incorporate biochar in climate smart operation.

Datta et al. (2025) showed that India's 156 Mt of surplus crop waste may be converted into 34.5 Tg of biochar, providing significant climatic and agricultural advantage. This shift could cut 405 Tg CO_2eq emissions annually, sequester 7.5 Tg carbon, and enhance soil fertility and water retention. With Punjab's rice residues offering the greatest potential, replacing residue burning with biochar systems position India to combat air pollution, rejuvenate soils, and emerge as a global leader in sustainable climate action.

2.3 Commercial and Industrial Uses

The building industry, which inevitably has a major detrimental impact on the environment worldwide, uses cement and concrete extensively. In addition to particulate matter and potentially hazardous substances that affect human health, the production of cement and concrete is responsible for approximately 8% of all anthropogenic CO_2 emissions, making it the most intractable environmental issue (Miller et al. 2018). In newly constructed structures, concrete is responsible for at least 70% of CO_2 emissions and this percentage is around 40%. Transportation of raw materials and production of aggregates come next (Habert et al. 2020). Adding 1% (v/v) of biochar to cement significantly reduced its electrical resistivity (42%) while maintaining high mechanical properties (Zhang et al. 2022).

2.4 Packaging Composites Based on Biochar

A higher biochar content led to a higher composite density because the particles of biochar had a higher mass and packing density inside the polymer matrix. However, the mechanical properties did not steadily get better as the biochar content rose. Instead, the amount of biochar used and the polymer matrix had distinct impacts on these properties. Although biochar was able to increase the material's density, its impacts on mechanical behaviour were complex and multifaceted. The carbon reinforcement frequently reduced the plastic properties of the composites, but the addition of 10% polyethylene glycol (PEG) helped restore them (Alves et al. 2024).

2.5 *Energy Generation and Feedstock for Bio-refineries*

Harvesting the biomass's multiple energy sources is doable with the suitable technology. Utilising the correct conversion technologies is vital to extracting as much energy and useful chemicals as possible from biomass. Ensuring that energy conversion technologies are both economically viable and environmentally sustainable—with minimal or zero emissions—is crucial. Using biochar as a waste management approach can cleanse broad regions, particularly agricultural fields, of undesired chemicals including pollution and heavy metals (HMs). According to Mishra et al. (2023), it is also touted as a natural wastewater treatment filter that eliminates medicines, pesticides, and antibiotics.

3 Current Trends of Biochar in Medical and Pharmaceutical Industry

3.1 *Biochar as a Carrier for a Drug Delivery System*

Biochar is a transporter, a catalyst and an absorbent. Despite having a wide surface area, a diversity of functional molecules, easy tailor ability and cost-effective qualities that make it a promising carrier, there is not much study on using biochar as a molecular delivery vehicle. Sashidhar et al. conducted a comprehensive evaluation of the use of biochar as a carrier in soil remediation in 2020. They underlined that because biochar naturally contains a variety of functional groups, it has clear advantages over conventional carrier materials. Additionally, as a carrier, biochar has shown a greater buffering capacity, which is clearly advantageous for the regulated release of chemicals. Thus, when employed as nanocarriers, biochar may have enormous promise for medication delivery (Zhuo et al. 2023).

3.2 *Use in Wound Healing and Tissue Regeneration*

A hydrogel based on cow dung biochar demonstrated promise in the production and release of active iodine in situ, in addition to its photodynamic therapy and photothermal therapeutic capabilities. This significantly hastened the full-thickness skin infection with MRSA. When exposed to visible light, the BC in the hydrogel helped produce reactive oxygen species (ROS) and when exposed to near-infrared light, it provided photo-thermal characteristics (Borjihan et al. 2023).

3.3 *Applications in Biomedicine (Toxin Adsorption, Biofilters)*

In engineered minimal impact development systems (such as bio retention systems), biochar has become a common filter medium for controlling hydrology and purifying urban rainfall. Biochar-amended biofiltration systems demonstrate high efficacy in removing diverse pollutants from urban runoff. Reported performance metrics include microbial contaminant reductions ranging from 0.78 to 4.23 log_{10} units, heavy metal removal efficiencies between 27 and 100%, organic compound elimination rates spanning 54–100%, total nitrogen reductions of 32–61%, and total phosphorus removal efficiencies ranging from 45 to 94%.The reasons underlying the variances in biofiltration performance are the chemistry of the storm water, the amount of dissolved organic matter, and the properties of the biochar. Filtration/straining is the principal way for eliminating bacteria, whereas sorption, ion exchange, and biotransformation are the primary strategies for eliminating chemical pollutants (Biswal et al. 2022).

4 Emerging Applications of Biochar in Agriculture

4.1 *Biochar as Soil Amendment*

Biochar has emerged as a powerful, nature-based solution for enhancing plant resilience against environmental, biotic, and abiotic stresses. Beyond its role in carbon sequestration, biochar functions as a soil enhancer that transforms the rhizosphere by improving microbial activity, aeration, cation exchange capacity, and buffering soil acidity (Omokaro et al. 2025). These modifications not only increase soil fertility and structure but also prepare plants to deal with contaminants and changing environmental condition (Mansoor et al. 2021). Its influence, however, is context-dependent determined by biochar type, application rate, soil quality, and local climate.

The mechanism of biochar at work to improve plant performance has various variants such as improving nutritional content, promoting nutrient cycling, promoting microbial diversity and triggering antioxidant defences that are essential during stress (Ighalo et al. 2024). It substantially improves the water conductance under the drought conditions, and it improves the efficiency of water usage (Ali et al. 2022; Ghazouani et al. 2023). On the physiological level, it reduces oxidative stress through reduction in malondialdehyde, electrolyte leakage, hydrogen peroxide and antioxidant enzyme activity to stabilize plant metabolism (Rizwan et al. 2018).

Biochar in saline soils regulates ionic equilibrium by enhancing the adsorption of K^+ ions and reducing the accumulation of Na^+ that is accompanied by further antioxidative enzymes biosynthesis and production that ultimately enhances plant growth (Wu et al. 2025). It increases the pH waterlogged soils, enhances nutrient uptake and high yield potential (Jahromi et al. 2020). Simultaneously, biochar eliminates greenhouse gas emission due to a decrease in the release of N_2O and CH_4, as well as the increase in the long-term carbon storage of the soil (Pineiro et al. 2025).

Biochar is also effective in protection against biotic stress. Biochar made with wheat straw prevents tomato bacterial wilt by promoting positive rhizospheric microbes (Tian et al. 2021) and maize stalk biochar by mitigating Phytophthora blight in pepper by boosting bio-control fungi (Wang et al. 2020). *Bacillus subtilis* and Trichoderma harziamm mixed with rice husk biochar enhanced disease resistance, which was related to the activation of defense-related genes, including PR-1b and JERF39 (Arshad et al. 2021). In the same way, it reduces the Fusarium solani in apple trees and enhances growth of *Radtx pseudostellariae* through inhibition of pathogen and enhancement of metabolites (Wang et al. 2020).

Taken together, these results indicate that biochar is much more than soil amendment; it is a driver of climate resilient, sustainable agriculture. Biochar offers viable solution to the food system protection and at the same time realizes the world climate objectives through the prevention of drought, salinity, floods, and pathogens in crops.

4.2 Impact on Soil Physical Properties

The benefit of application of the biochar in sandy and coarse-textured soils is already proven, especially related to enhancing the physical aspects of soils to increase water content and reduce bulk density (Das et al. 2021; Joseph et al. 2021). Biochar is however highly dependent on the feedstock type and soil properties to determine its efficiency as a soil amendment. It is specifically worth mentioning that the use of biochar treatment has been found to enhance the soil aggregates (Yan et al. 2022). As an example, the long-term use of straw and biochar made of straw improved the 0–40 cm layer soil aggregate ions in a six-year rice–wheat rotation (Bai et al. 2019).Similarly, in a rapeseed–potato cropping system, the addition of wheat straw biochar (pH 10.35) at a rate of 2–40 t ha^{-1} for one year improved aggregate stability and increased microaggregate content, while simultaneously enhancing the yields of both rapeseed and sweet potato (Liu et al. 2014). Furthermore, biochar has been shown to improve soil water retention by increasing pore size and aggregate stability (Kang et al. 2022).

4.3 Impact on Soil Chemical Properties

Biochar interacts with soil chemistry mainly through dissolution processes (Joseph et al. 2021). Its capacity to retain carbon and nitrogen not only enhances soil fertility over time but too contributes to reducing both short and semipermanent environmental degradation, thereby mitigating adverse impacts on human and animal health (Verheijen et al. 2019).As a result, biochar increases soil nutrient retention by reducing vaporization and leaching of nutrients and agricultural pollutants into surface and groundwater and improving soil absorption and retention of these substances. These variables, however, are highly dependent on the source and features of the applied biochar (Weber and Quicker 2018), as well as soil parameters, plant reactions to the environment, interactions between the biochar and the soil, and the biochar's capacity to absorb nutrients (Ahmad-Bhat et al. 2022).In the first 30 days after application, wood biochar had the highest water retention and the least amount of NO_3 leaching (Jones et al. 2012), but rice husk biochar resulted in considerable N leaching (Widowati et al. 2014).The use of biochar improved the soil's fertility (Ali et al. 2022) and extractable nutrients (Laird et al. 2010a; b). The pH and electrical conductivity of the soil were significantly reduced by the addition of biochar with tiny and medium particle sizes (Zeeshan et al. 2020).

4.4 Impact on Soil Biological Properties

The biochar was viewed as enhancing the main properties of soil owing to its structure and high alkalinity level. It also enhances porosity and water retention of the soil which ensured the growth of microorganism. It is the case even with contaminated with metals (loid)s (Gorovtsov et al. 2020) and microplastics (Palansooriya et al. 2022) soils. Even though it is not clear how biochar affects soil microbes directly, one should remember that it has a synergistic effect that becomes greater. Biochar also affects the physical, chemical, and natural characteristics of soil especially the pH, aggregation, and water retention that are commonly perceived to be major initiators of microbial growth, soil respiration and microbial biomass carbon as well as the total microbial diversity (Diatta et al. 2020). Biochar has also been linked with enhanced crop overall toughness and prevented reliable soil based illnesses by enhancing natural health of soil (Tan et al. 2022). Further, due to its role in sequestration of carbon in the soil and enhancement of soil fertility, biochar has been demonstrated to enhance the proliferation of mycorrhizal fungi (Warnock et al. 2007).

5 Emerging Trends of Biochar in Climate Change Mitigation

5.1 Biochar Potential in Reduction of Emission of Greenhouse Gases

Applying biochar as a soil supplement has been shown in recent research to help lower releases greenhouse gas from soil towards environment (Xu et al. 2021). Application of 2–5% wood chip biochar in soil shows dramatically lower greenhouse gas emissions in rice and beet crops (Lai et al. 2013). Biochar contributes to the reduction of greenhouse gas emissions from agricultural soils through two simple mechanisms: improving soil moisture retention and enhancing plant nitrogen uptake. These processes jointly lead to significant decreases in nitric oxide (N_2O) emissions, with reductions of 20–30% under waterless conditions and 40–50% under muddy conditions (Saarnio et al. 2013). In maize cultivation systems using mulching and drip irrigation, screaky application rates of biochar (30 t ha^{-1}) have too been reported to reduce methane (CH_4) emissions by up to 134% (Yang et al. 2020).

5.2 *Biochar in Reducing Water Pollution*

Water pollution management particularly remediation of tainted wastewater from various sources, is one of the newer uses for biochar. Engineered biochars with increased porosity, limited surface area, profuse operative groups, and greater adsorption capacity are often further operative than clean biochars. Such improvements can be achieved through pre-treatment of feedstocks or post treatment strategies aimed at contaminant removal (Xiang et al. 2020). For instance, coconut shell-derived biochar has been shown to significantly reduce Pb concentrations in lead contaminated water by facilitating Pb complexation within its surface and porous structure (Hao et al. 2021). Similarly, KOH activated highly porous biochar has demonstrated effective removal of chromium and naphthalene from wastewater. In another study, biochar derived from hydrothermal carbonization of sugarcane bagasse exhibited severe adsorption and removal of sulfamethoxazole from wastewater, mainly through charge assisted hydrogen bonding and π–π interactions with graphitized carbon (Prasannamedha et al. 2021).

5.3 *Biochar and Emerging-Hazardous Wastes Such as Organic Pollutants and Microplastics*

One of the emerging applications of biochar is in the treatment of extremely dangerous wastes, including organic pollutants and microplastics, which have become an outstanding global environmental concern (Kumar et al. 2023). The remediation of microplastic residues is especially grievous for improving soil and water quality, as the considerable use of plastics in both urban and agricultural systems has led to widespread contamination of mundane and subaqueous environments (Ye et al. 2020). Biochar derived from corn straw and hardwood has been reported to efficaciously remove and immobilize microplastic particles (10–22 μm) through entrapment within its honeycomb-like porous structures, achieving removal efficiencies >95% when applied as a filtration medium (Wang et al. 2020). In addition, catalytic approaches, such as the utilization of attractable biochar to trigger oxidation processes, have shown promise for separating and degrading microplastics in aquatic systems (Ye et al. 2020).

6 Future Trends of Biochar in Circular Economy and Waste Management

6.1 Biochar Production from Municipal and Agricultural Waste

Open dumping of municipal solid waste in the developing countries is a highly used waste disposal method leading to pollution caused by the gaseous emissions and leachate discharge of the landfills. Biomass may be transformed into biochar, a carbonaceous product that could be used to reduce pollution and recover the resources found in the municipal solid waste (Gunarathne et al. 2019).

6.2 Application of Biochar in Supporting Microbial Biofilms for Bioremediation

As has been shown, biochar-based microbial consortia may be used to stimulate the biodegradation of petroleum hydrocarbons, chlorinated compounds and polycyclic aromatic hydrocarbons (PAHs) (Zhang et al. 2010; Chen et al. 2015). Also, the biochar adsorption of toxins and the reduction of their bioavailability benefits bioremediation processes and mitigates unfavourable environmental factors to the activity of microbes. Biochar-biofilm systems can be designed to target remediation of the soil and aquatic environment through the attachment of target microbial strains to biochar. It is a green and cheap alternative to the conventional remediation plans.

7 Policy and Regulatory Framework for Biochar Production and Its Uses

7.1 Global Regulations Related to the Production and Application of Biochar

The IPCC (Intergovernmental Panel on Climate Change of the United Nations) has acknowledged biochar as a negative emissions technology (IPCC 2019). Australia and other nations have accepted biochar into the Emissions Reduction Fund (ERF), issuing credits for projects showing carbon abatement. The European Biochar Certificate (EBC) proposes quality and environmental safety standards for biochar utilization in the EU in a bid to ensure sustainable production practice assurance. In paddy fields, the "Japan Biochar Association" is already utilizing biochar to enhance soil fertility and mitigate methane emissions. In spite of this, international harmonisation of standards, lifecycle assessment procedures and regulatory policy is lacking at present. As biochar moves from a research issue to a commercial opportunity, and global standards will be required to realize its full benefits while maintaining environmental integrity.

7.2 *Sustainability Certifications and Environmental Laws*

Feedstock acquisition, emissions from pyrolysis, and the prospective influence of biochar on water and soil systems are the major topics of environmental regulations. Emission standards for the production of biochar have been formulated by the Environmental Protection Agency (EPA) under the Clean Air Act, targeting particulate matter and volatile organic compounds (VOCs) emitted during pyrolysis. Carbon sequestration outcomes must be reported in detail as the Biochar Carbon Offset Method in Australia is governed by the Emissions Reduction Fund (ERF). Other voluntary certification schemes also have been developed to verify the quality and sustainability of the product. The most popular standard used is the European Biochar Certificate (EBC). It ensures the environment safety and carbon content of the manufactured biochar and production parameters and feedstock origin. The International Biochar Initiative (IBI) also offers a system of certification on the basis of rigorous testing procedures considering parameters such as pH and ash content and nutrients content and potential contaminations.

7.3 *Financial Viability and Costs*

Financial viability of the production systems of the biochar is the greatest barrier (Cowie et al. 2015; Shackley et al. 2015). Transport cost and cost of feedstock most affect financial viability.

7.4 *Feedstock Acquisition*

This is the opportunity cost of the diversion of a marketable feedstock to conversion to the pyrolitic oil or the production cost in the case of a feedstock being manufactured for the purpose of pyrolysis (Kung et al. 2013). For example, the procurement cost of corn stover is \$82 to \$100 per tonne, its market price (Thompson and Tyner 2014).

Diverting a brewery's waste grains to bioenergy or animal production has a very high opportunity cost of $91 (Shackley et al. 2015). There may be a negative price for the feedstock, however, if it is a waste stream that would otherwise be costly to treat (e.g. municipal solid waste).

The purchase of feedstock is one of the most sensitive parameters in the cost–benefit analysis of biochar, according to many studies (Shackley et al. 2015). Shifting away from marketable feedstocks such as brewery grain and maize stover and toward waste stream feedstocks such as garden debris and forest thinning could be one of the secrets of a viable pyrolysis unit. The possible benefits of biochar in forestry, when feedstock acquisition could be costly.

7.5 Feedstock Transportation

The transportation cost of feedstock is linear and is a function of the distance between the origin of the feedstock and the pyrolysis location. Accordingly, later cost–benefit analyses have not deemed it a sensitive parameter (Campbell et al. 2018). The transport of biomass is costly, although economies of scale can be beneficial. Assuming that the feedstock and the pyrolysis operation are, on average, 9.22 miles apart, the cost of transportation may add 11% to the cost of production (Kung et al. 2013). As such, the production of biochar will "likely be a much localised market in its initial technological dissemination phase, so the location of feedstock supply will be an important consideration" (Dumortier et al. 2020). The nearness of the feedstock might lose importance as mobile pyrolyzers prove to be useful. Alternatively, a firm operating in a related industry (e.g., wood and paper products) can opt to have a pyrolysis unit installed on their grounds, which would obviate transportation costs (Kim et al. 2015). In the meantime, only feedstocks near their location can be entertained for pyrolysis activities.

7.6 Standardization of Biochar

A sustainable utilisation approach that fully addresses spatial heterogeneity and encompasses field-to-regional dimensions within the pertinent socio-economic framework is necessary due to the variability of biochar and the variety in soil properties across spatial and temporal scales. In this system the model considers the greenhouse gas (GHG) emissions generated during the production process the greenhouse gas and the crops production from the production site. Feedstock quality and resources competitiveness also play a role. Certification of the biochar offers an operational methodology that is the center of application of the sustainability objectives. Design of the certification system guides the system performance and its application on a stand-alone mode or as a subsystem to other methodologies like contamination control measures or zero-waste initiative or life cycle assessment (LCA) (Cobut et al. 2013; Verheijen et al. 2015).

7.7 *Certification Scheme*

Organizations such as the International Biochar Initiative (IBI), the European Biochar Certificate (EBC), and the Biochar Quality Mandate (BQM) establish standards and guidelines for biochar quality and production practices have created guidelines that control the certification and selling of biochar (EBC 2022; Liu et al. 2023). Additionally, biochar is a component of the national regulatory systems for composts, soil additives, and fertilisers in Germany, Austria, Switzerland, and Italy. This subject comprises technological feasibility, growth potential and cost in soils. It involves long-term carbon sequestration along with reliable systems for verification and ongoing monitoring, and the benefits of applying biochar to various carbon repositories. According to Meyer et al. (2017) and Fawzy et al. (2022), the standards for biochar contain information on the feedstock attributes used, the distance travelled during production and transportation, and criteria on the quality of the biochar. They also include controls on the levels of organic pollutants and heavy metals in the environment.

7.8 *The International Biochar Initiative*

IBI Biochar Standards are standards that detail and direct the test procedures during the production of the biochar. The standards also purport to provide commercial businesses and companies with a specific approach on the determinations of the aforementioned specific properties and the properties of the biochar whenever the use of the soil amendment application involves the use of the biochar. The use of the IBI Biochar Certification Program is the objective of the IBI Biochar Standards. It is obtained from the feedstocks with the help of the various thermochemical conversion technologies and the latter provides the resulting biochar with various properties. The raw or the refined feedstock and the mix of the feedstock and the diluents or the feedstock alone makes up the feedstock. The diluents of 10% or higher material of the feedstock all must be so specified explicitly and the feedstock be of 2% or less weight impurities (Novotny et al. 2023).

Biosolids, rice hulls, straws and other such products are all considered unprocessed feedstocks. On the other hand, different types of manure and related materials make up processed feedstock. According to IBI standards, biochar cannot be described as any pyrolytic product with a molar H/Corg ratio >0.7 or $<10\%$ (w %) of organic carbon (Corg) (IBI 2015). The molar H/Corg ratio is the proportion of hydrogen atoms to carbon atoms in a material's organic composition. One employs this ratio to understand the chemical composition of substances. Generally speaking, a greater

H/Corg ratio implies that there is more hydrogen than carbon, which could signify that the chemical is less aromatic or has not suffered as much thermal alteration. The IBI Biochar Standards specify that there are three test categories for biochar: category A covers the core functional qualities that all biochar must have, category B covers the assessment of toxicants, and category C covers advanced evaluation and soil enhancing properties.

7.9 The European Biochar Certificate

While the European Biochar Certificate (EBC) serves as a voluntary industry standard in Europe and Switzerland, its compliance is required for the agricultural application of biochar (EBC 2022). There are six types of biochar, according to the EBC: "Feed," "Agro Organic," "Urban," "Consumer Materials," and "Basic Materials." To receive the certificate, the biosolids feedstock, production method, biochar characteristics, and application procedure must all meet EBC requirements.

7.10 Biochar Quality Mandate

BQM outlines the water-holding and cation-exchange capacities of the biochar as well as the Maximum Permissible Limits (MPLs) for toxicants in order to establish a two-tiered (standard and high grade) quality grading system with specific requirements for each. Based on feedstock sustainability assessments, this framework includes life cycle assessment methodology for reducing greenhouse gas emissions, chain of custody, and proof of sustainable and legal management. Drawing on existing UK and EU regulations, it incorporates detailed emission standards for the manufacture of the biochar to protect human health and maintain the integrity of the ecosystem. It also includes regulations for using biochar (Meyer et al. 2017). Proposals to revise and broaden the mandate to include novel biochar uses and advances in the EU have been made, even though the BQM has yet to have any commercially certified products (Shackley et al. 2014).

8 Emerging Applications of Biochar in Sustainable Development

8.1 Mitigation of Greenhouse Gas Emissions

Over the past few decades, the amount of greenhouse gases (GHGs) generated by global development activities has increased considerably. Per current estimates of global warming, the average world temperature will increase by 1.5 °C by 2035 and 2.5 °C by 2100. The intensified greenhouse effect amplifies various atmospheric climate changes, leading to altered rainfall patterns, rising sea levels, accelerated glacier melting, increased occurrences of floods and droughts, habitat fragmentation, loss of biodiversity, and disruptions in microbial processes (due to a variety of diseases), all of which have long-term effects on human health. Using charcoal (BC) as a wastewater treatment technique and soil supplement can reduce the various greenhouse gas emissions created during the pollution removal process. As a result, the future sections will reveal the amount to which BC administration lowers GHG via distinct treatment processes and related mechanisms.

8.2 GHG Mitigation During Soil Amendment

Biochar can be a sustainable material that can fight climate change in numerous ways. The extensive application of BC to farm areas will assist minimize a portion of the created GHGs, which is desirable given farming methods contribute to certain climatic changes. Consequently, various researches have attempted to apply biochar in farm situations, with success in lowering greenhouse gas emissions. In addition, BC has a strong ability to retain water, absorb nutrients, store carbon, release nutrients, treat wastewater and minimize infection, according to Smith et al. (2019). Upon application to soil, it has boosted the stable storage of carbon after degrading organic materials. According to Windeatt et al. (2014), 300 tonnes of BC created each year would help sequester 0.5 billion tonnes of carbon from CO_2 emissions. This represents around 1.5% of CO_2 emissions every year. According to Liu et al. (2015), carbon storage as BC might lead to an annual CO_2 decrease of 0.3 billion tonnes.

8.3 GHG Mitigation During Wastewater Treatment

The primary causes of greenhouse gas emissions are fossil fuel combustion, deforestation, and agricultural discharges. However, septic tanks, drainage, swamps, wastewaters, and wastewater treatment plants also emit CH_4, N_2 and CO (Parravicini et al. 2016). Erroneous wastewater management will lead to a multitude of point and non-point sources of greenhouse gas emissions. Wastewater treatment methods are therefore essential. The primary wastewater treatment plant, WWTPs, emit a large quantity of greenhouse gases throughout their operations. Sludge dewatering, sludge disposal and WWTP operational units (chemical and biological) are the main contributors to greenhouse gas emissions. In addition, the emissions comprised CO_2, CH_4 and N_2O. But the IPCC maintains that the emissions arise from fugacious biogenic sources, omitting CO_2 emissions from the WWTP's GHG. However, anthropogenic CO_2 contributes more to the GHG effect than other minor GHGs due to its high atmospheric concentration (Pahunang et al. 2021).

8.4 SDG as Bioenergy Production

A number of approaches have utilized BC to a great extent in order to produce energy from biomass. BC's biological and thermochemical conversion played a part in the production of bioenergy via various techniques. The thermochemical route provides biomass with heat and temperature, and the final product aids in the development of energy storage systems like super capacitors and electrodes for bio-electrochemical systems, among numerous others (Lee et al. 2018). Conversely, the biological route synthesizes biofuels like ethanol, biodiesel, bio-oil, and bio-hydrogen through the utilization of microbes for the degradation of biomass via fermentation and anaerobic processes (Osman et al. 2023). These technologies synthesize low-cost biochar with a range of promising features, like porosity and SSA, that have opened the gate for subsequent experimental studies and application in the bioenergy generation industries (Cha et al. 2016). Today, the quality of BC has become increasingly necessary, and several research have modified its properties utilising activation techniques (Lee et al. 2017). SDG 7, which seeks clean, affordable energy for environmental goals, is connected with this use of BC. For a number of environmental applications, BC provides a sustainable bioenergy generating alternative. The ensuing subsections have elaborated on BC and its variations for creating sustainable energy in respect to particular sorts of applications.

SDG: Water reclamation (Enhance clean water and sanitation)—The vast volume of wastewater produced has led to the construction of several physicochemical and biological treatment systems over the last few decades (Kamali and Khodaparast 2015). Though advance oxidation process (AOP), ultrasonic irradiation, and others are more prevalent physicochemical treatments, anaerobic digestion and the activated sludge process are two commonly applied biological treatment strategies (Kamali et al. 2021). Interestingly, adsorption has garnered the greatest attention among these techniques for the removal of both organic and inorganic waste water pollutants (Catalkaya and Kargi 2008). Nevertheless, the use of adsorption in the studies was mainly to utilise carbonaceous materials like activated carbon (Wong et al. 2018). Many experimental studies proved that activated carbon is an economical and environmentally friendly method to treat both organic and inorganic wastewater pollutants (Cecen and Aktas 2011). It has become increasingly popular in this arena to give top priority and investigate carbonaceous materials for innovative environmental remediation strategies. Due to this, several studies have recommended and advocated BC as a sustainable solution to combat marine pollution. Comprehensive experimental studies affirmed that BC performs well as an adsorbent to eliminate inorganic pollutants, for example, heavy metals (Shan et al. 2020). Conventional and organic pollutants encompass dyes, insecticides, and various other toxins from different aqueous solutions (Nirmaladevi and Palanisamy 2020).

9 Future Prospects and Research Direction

Biochar has become an all-purpose carbonaceous substance with far-reaching environmental sustainability, agricultural resilience, climate change, and the circular economy implications. Biochar, produced with the help of various feedstocks by using thermochemical methods like pyrolysis and hydrothermal carbonization, represents an opportunity to convert municipal, agricultural, and industrial wastes into valuable products in a sustainable manner. High surface area, porosity, and a wide range of functional groups present in biochar are those properties that allow it to become an effective soil conditioner, pollutant adsorbent, microbial habitat, and permanent carbon sink.

Biochar has shown remarkable properties of immobilising heavy metals, degrading organic pollutants and filtering microplastics in terrestrial and aquatic environments in environmental remediation. Further functionalization and engineering of biochars can augment these properties, providing special use in the treatment of wastewater, stormwater and detoxification of hazardous waste. Biochar has been shown to increase soil fertility, soil structure, promote healthy microbial activity, and confer plant tolerance to abiotic stressors, including salinity, drought and flooding in agricultural practice. All these functions in combination lead to increased productivity and climatic resilient agricultural systems.

Under climate change, biochar is becoming more known as a negative emission technology to the IPCC. Its carbon recalcitrance forms guarantee long-term sequestration and its use in soils and wastewater systems helps mitigate greenhouse gases like N_2O, CH_4 and CO_2. Combined with the contribution to bioenergy production and incorporation into biorefineries, biochar will be an essential gateway between mitigating carbon and creating renewable energy. At the policy level, models like the European Biochar Certificate (EBC), International Biochar Initiative (IBI), and Biochar Quality Mandate (BQM) are at the forefront towards environmental safety, quality and standardization of products. Nevertheless, there are still issues related to the economic viability, feedstock procurement, logistics of transportation, and mass implementation. The barriers need to be addressed with cost effective strategies of production, use of waste derived feedstock, decentralized pyrolysis, and incorporation as part of a localized circular economy.

In prospective, the future of biochar study and use is in several promising avenues; synthesis of nanostructure biochar of adsorption and catalytic activity; bicohar combination with plant growth promoting microbes of soil and crop health; biochar based composite design to be used in construction, packaging and biomedical use; and carbon credits schemes development to encourage wide use. Also vital will be inter-disciplinary coordination's to evaluate the long-term ecological effects, measure the benefits of carbon sequestration and harmonize the international standards of sustainability.

Acknowledgements The authors acknowledge Chandigarh University, Gharuan, Mohali, Punjab and the Department of Biosciences for providing assistance for this work.

Authors' Contributions SS, AT and VK conceptualized the study. KD and SKK handled data. SS, AT, VK, SKK and SG wrote the original draft. SS, AT, ST and RLSS reviewed and edited. All the authors reviewed, commented and edited the paper for submission.

References

Afshar M, Mofatteh S (2024) Biochar for a sustainable future: environmentally friendly production and diverse applications. Results Eng 23:102433. https://doi.org/10.1016/j.rineng.2024.102433

Ahmad-Bhat S, Kuriqi A, Dar MUD, Bhat O, Sammen S, Towfiqul-Islam AR et al (2022) Application of biochar for improving physical, chemical, and hydrological soil properties: a systematic review. Sustainability 14:11104. https://doi.org/10.3390/su141711104

Ali I, Yuan P, Ullah S, Iqbal A, Zhao Q, Liang H et al (2022) Biochar amendment and nitrogen fertilizer contribute to the changes in soil properties and microbial communities in a paddy field. Front Microbiol 13:1154

Alves Z, Brites P, Ferreira NM, Figueiredo G, Otero-Irurueta G, Gonçalves I, Mendo S, Ferreira P, Nunes C (2024) Thermoplastic starch-based films loaded with biochar-ZnO particles for active food packaging. J Food Eng 361:111741

Arshad M, Rizwan M, Ali S, Zia-ur-Rehman M (2021) Responses of antioxidant enzymes to heavy metal stress in plants irrigated with contaminated water. Chemosphere 263:128276

Bai N, Zhang H, Li S, Zheng X, Zhang J, Zhang H et al (2019) Long-term effects of straw and straw-derived biochar on soil aggregation and fungal community in a rice–wheat rotation system. PeerJ 6:e6171. https://doi.org/10.7717/peerj.6171

Bano A, Aziz MK, Prasad B, Ravi R, Shah MP, Lins PVDS, Meili L, Prasad KS (2025) The multifaceted power of biochar: a review on its role in pollution control, sustainable agriculture, and circular economy. Environ Chem Ecotoxicol 7:286–304

Biswal BK, Vijayaraghavan K, Tsen-Tieng DL, Balasubramanian R (2022) Biochar-based bio retention systems for removal of chemical and microbial pollutants from storm water: a critical review. J Hazard Mater 422:126886

Borjihan Q, Meng S, Bai H, Chen T, Hu X, Xiao D, Shi L, Dong A (2023) Active iodine regulated in cow dung biochar-based hydrogel combined with PDT/PTT for MRSA infected wound therapy. Mater des 231:112051

Campbell RM, Anderson NM, Daugaard DE, Naughton HT (2018) Financial viability of biofuel and biochar production from forest biomass in the face of market price volatility and uncertainty. Appl Energy 230:330–343

Catalkaya EC, Kargi F (2008) Advanced oxidation treatment of pulp mill effluent for TOC and toxicity removals. J Environ Manag 87:396–404

Cecen F, Aktas O (2011) Activated carbon for water and wastewater treatment: integration of adsorption and biological treatment. Wiley, Hoboken

Cha JS, Park SH, Jung S-C, Ryu C, Jeon J-K, Shin M-C, Park Y-K (2016) Production and utilization of biochar: a review. J Ind Eng Chem 40:1–15

Chauhan PK, Upadhyay SK, Chauhan A, Bhutiani R, Sikarwar RLS, Tiwari MK (2025a) A review on simultaneous use of microbes and biochar for removal of toxic heavy metals: recent updates and future outlooks. Physiol Mol Biol Plants 14:1–18. https://doi.org/10.1007/s12298-025-01649-4

Chauhan PK, Upadhyay SK, Chauhan A, Bhutiani R, Sikarwar RLS, Tiwari MK (2025b) A review on simultaneous use of microbes and biochar for removal of toxic heavy metals: recent updates and future outlooks. Physiol Mol Biol Plants 31(3):1–15. https://doi.org/10.1007/s12298-025-01649-4

Chen Z, Zhao F-J, Lombi E, McGrath SP (2015) Assessing the bioavailability of heavy metals in contaminated soils using chemical extraction and plant uptake studies. Environ Pollut 200: 117–125

Cobut A, Beauregard R, Blanchet P (2013) Using life cycle thinking to analyse environmental labeling: the case of appearance wood products. Int J Life Cycle Assess 18:722–742

Cowie A, Woolf D, Gaunt J, Brandao M, Rosa RA, Cowie A (2015) Biochar, carbon accounting and climate change. In: Lehmann J, Joseph S (eds) Biochar for environmental management: science, technology and implementation. Taylor and Francis, London, pp 763–794

Das S, Mohanty S, Sahu G, Rana M, Pilli K (2021) Biochar: a sustainable approach for improving soil health and environment. In: Vieira A, Rodrigues SC (eds) Soil erosion current challenges and future perspectives in a changing world. IntechOpen, London

Datta A, Dutta S, Sharma S, Rahman MH (2025) Spatial variation of biochar production potential from surplus crop residues in India. Curr Res Environ Sustain 9:100279

Diatta AA, Fike JH, Battaglia ML, Galbraith JM, Baig MB (2020) Effects of biochar on soil fertility and crop productivity in arid regions: a review. Arab J Geosci 13:595. https://doi.org/10.1007/s12517-020-05586-2

Dumortier J, Dokoohaki H, Elobeid A, Hayes DJ, Laird D, Miguez FE (2020) Global land-use and carbon emission implications from biochar application to cropland in the United States. J Clean Product 258:120684

EBC (2022) Guidelines for a sustainable production of biochar. Version 10.1

Fang G, Liu C, Gao J, Dionysiou DD, Zhou D (2015) Manipulation of persistent free radicals in biochar to activate persulfate for contaminant degradation. Environ Sci Technol 49:5645–5653

Fawzy S, Osman AI, Mehta N, Moran D, Al-Muhtaseb AH, Rooney DW (2022) Atmospheric carbon removal via industrial biochar systems: a techno-economic-environmental study. J Clean Product 371:133660. https://doi.org/10.1016/j.jclepro.2022.133660

Ghazouani H, Ibrahimi K, Amami R, Helaoui S, Boughattas I, Kanzari S, Milham P, Ansar S, Sher F (2023) Integrative effect of activated biochar to reduce water stress impact and enhance antioxidant capacity in crops. Sci Total Environ 905:166950. https://doi.org/10.1016/j.scitotenv.2023.166950

Gong Y, Wang Y, Lin N, Wang R, Wang M, Zhang X (2022) Iron-based materials for simultaneous removal of heavy metal (loid)s and emerging organic contaminants from the aquatic environment: recent advances and perspectives. Environ Pollut 299:118871. https://doi.org/10.1016/j.envpol.2022.118871

Gorovtsov AV, Minkina TM, Mandzhieva SS, Perelomov LV, Soja G, Zamulina IV et al (2020) The mechanisms of biochar interactions with microorganisms in soil. Environ Geochem Health 42:2495–2518. https://doi.org/10.1007/s10653019-00412-5

Gunarathne V, Ashiq A, Ramanayaka S, Wijekoon P, Vithanage M (2019) Biochar from municipal solid waste for resource recovery and pollution remediation. Environ Chem Lett 17:1225–1235

Habert G, Miller SA, John VM, Provis J, Favier A, Horvath A, Scrivener K (2020) Environmental impacts and decarbonization strategies in the cement and concrete industries. Nat Rev Earth Environ 1:93. https://doi.org/10.1038/s43017-020-0093-3

Hao N, Cao J, Ye J, Zhang C, Li C, Bate B (2021) Content and morphology of lead remediated by activated carbon and biochar: a spectral induced polarization study. J Hazard Mater 411:124605. https://doi.org/10.1016/j.jhazmat.2020.124605

Hu H, Gao Y, Tan W, Xi B (2023) Effects of dissolved organic matter on mercury speciation in rice rhizosphere amended with sulfur-rich biochar. Soil Environ Health 1(2):100022. https://doi.org/10.1016/j.seh.2023.100022
IBI – Index of Biotic Integrity (2015) Development and application of multimetric biotic indices for freshwater ecosystem health assessment. Inst Water Res
Ighalo JO, Ohoro CR, Ojukwu VE, Oniye M, Shaikh WA, Biswas JK, Seth CS, Mohan GBM, Chandran SA, Rangabhashiyam S (2024) Biochar for ameliorating soil fertility and microbial diversity: from production to action of the black gold. iScience 28(1):111524
IPCC (2019) Climate change and land: IPCC special report
Jahromi H, et al (2020) Effects of heavy metal contamination on plant physiological traits and soil microbial activity. J Hazard Mater 389:121873
Javanmard A, Daud WMABW, Patah MFA, Zuki FM, Verdugo AS (2024) Harnessing the potential of biochar-based catalysts for sustainable adsorptive and photocatalytic applications: a comprehensive review. Process Saf Environ Protect 189:387–413. https://doi.org/10.1016/j.psep.2024.05.118
Jones DL, Rousk J, Edwards-Jones G, DeLuca TH, Murphy DV (2012) Biochar-mediated changes in soil quality and plant growth in a three year field trial. Soil Biol Biochem 45:113–124. https://doi.org/10.1016/j.soilbio.2011.10.012
Joseph S, Cowie AL, Van Zwieten L, Bolan N, Budai A, Buss W et al (2021) How biochar works, and when it doesn't: a review of mechanisms controlling soil and plant responses to biochar. GCB Bioenergy 13:1731–1764. https://doi.org/10.1111/gcbb.12885
Kamali M, Khodaparast Z (2015) Review on recent developments on pulp and paper mill wastewater treatment. Ecotoxicol Environ Saf 114:326–342
Kamali M, Appels L, Kwon EE, Aminabhavi TM, Dewil R (2021) Biochar in water and wastewater treatment: a sustainability assessment. Chem Eng J 420:129946
Kang MW, Yibeltal M, Kim YH, Oh SJ, Lee JC, Kwon EE et al (2022) Enhancement of soil physical properties and soil water retention with biochar-based soil amendments. Sci Total Environ 836:155746. https://doi.org/10.1016/j.scitotenv.2022.155746
Khan S, Irshad S, Mehmood K, Hasnain Z, Nawaz M, Rais A, Gul S, Wahid MA, Hashem A, Abd-Allah EF, Ibrar D (2024) Biochar production and characteristics, its impacts on soil health, crop production, and yield enhancement: a review. Plants 13(2):166. https://doi.org/10.3390/plants13020166
Kim D, Anderson NM, Chung W (2015) Financial performance of a mobile pyrolysis system used to produce biochar from sawmill residues. For Prod J 65(5–6):189–197
Kopittke PM, Menzies NW, Wang P, McKenna BA, Lombi E (2019) Soil and the intensification of agriculture for global food security. Environ Int 132:105078. https://doi.org/10.1016/j.envint.2019.105078
Kumar R, Verma A, Rakib RJ, Gupta PK, Sharma P, Garg A et al (2023) Adsorptive behavior of micro(nano)plastics through biochar: co-existence, consequences, and challenges in contaminated ecosystems. Sci Total Environ 856:159097. https://doi.org/10.1016/j.scitotenv.2022.159097
Kung C-C, McCarl BA, Cao X (2013) Economics of pyrolysis-based energy production and biochar utilization: a case study in Taiwan. Energy Policy 60:317–323
Lai W-Y, Lai C-M, Ke G-R, Chung R-S, Chen C-T, Cheng C-H et al (2013) The effects of woodchip biochar application on crop yield, carbon sequestration and greenhouse gas emissions from soils planted with rice or leaf beet. J Taiwan Inst Chem Eng 44:1039–1044. https://doi.org/10.1016/j.jtice.2013.06.028
Laird D, Fleming P, Wang B, Horton R, Karlen D (2010a) Biochar impact on nutrient leaching from a Midwestern agricultural soil. Geoderma 158:436–442. https://doi.org/10.1016/j.geoderma.2010.05.012
Laird DA, Fleming P, Davis DD, Horton R, Wang B, Karlen DL (2010b) Impact of biochar amendments on the quality of a typical Midwestern agricultural soil. Geo Derma 158:443–449
Lee J, Kim K-H, Kwon EE (2017) Biochar as a catalyst. Renew Sustain Energy Rev 77:70–79

Lee HW, Kim YM, Kim S, Ryu C, Park SH, Park YK (2018) Review of the use of activated biochar for energy and environmental applications. Carbon Lett 26:1–10

Li S, Tasnady D (2023) Biochar for soil carbon sequestration: current knowledge, mechanisms, and future perspectives. C 9(3):67

Liu Z, Chen X, Jing Y, Li Q, Zhang J, Huang Q (2014) Effects of biochar amendment on rapeseed and sweet potato yields and water stable aggregate in upland red soil. CATENA 123:45–51. https://doi.org/10.1016/j.catena.2014.07.005

Liu W-J, Zhang X-Q, Wu S (2015) Soil heavy metal contamination and ecological risk assessment in industrial regions: a meta-analysis. Ecol Ind 56:99–108

Liu P, Ptacek CJ, Blowes DW, Landis RC (2016) Mechanisms of mercury removal by biochars produced from different feedstocks determined using X-ray absorption spectroscopy. J Hazard Mater 308:233–242. https://doi.org/10.1016/j.jhazmat.2016.01.007

Liu QK, Meki H, Zheng Y, Yuan M, Shao X, Luo X, Li Z, Jiang F, Li B, Xing S (2023) Biochar application in remediating salt-affected soil to achieve carbon neutrality and abate climate change. Biochar 5:45. https://doi.org/10.1007/s42773-023-00244-8

Mansoor S, Kour N, Manhas S, Zahid S, Wani OA, Sharma V, Ahmad P (2021) Biochar as a tool for effective management of drought and heavy metal toxicity. Chemosphere 271:129458. https://doi.org/10.1016/j.chemosphere.2020.129458

Meyer S, Genesio L, Vogel I, Schmidt HP, Soja G, Someus E, Shackley S, Verheijen FGA, Glaser B (2017) Biochar standardization and legislation harmonization. J Environ Eng Landsc Manag 25:175–191

Miller SA, John VM, Pacca SA, Horvath A (2018) Carbon dioxide reduction potential in the global cement industry by 2050. Cem Concr Res 114:115–124. https://doi.org/10.1016/j.cemconres.2017.08.026

Min KH, Kim KH, Seo JH, Pack SP (2025) Biochar utilization in antimicrobial, anticancer, and biosensing applications: a review. Biomolecules 15(6):760. https://doi.org/10.3390/biom15060760

Mishra RK, Kumar DJ, Narula A, Chistie SM, Naik SU (2023) Production and beneficial impact of biochar for environmental application: a review on types of feedstocks, chemical compositions, operating parameters, techno-economic study, and life cycle assessment. Fuel 343:127968

Nehela Y, Mazrou YSA, Alshaal T, Rady AMS, El-Sherif AMA, Omara AED, El-Monem AM, Hafez EM (2021) The integrated amendment of sodic-saline soils using biochar and plant growth-promoting rhizobacteria enhances maize (*Zea mays* L.) resilience to water salinity. Plants 10(9):1960

Nirmaladevi S, Palanisamy N (2020) A comparative study of the removal of cationic and anionic dyes from aqueous solutions using biochar as an adsorbent. Desalin Water Treat 175:282–292

Novotný M, Marković M, Raček J, Šipka M, Chorazy T, Tošić I, Hlavínek P (2023) The use of biochar made from biomass and biosolids as a substrate for green infrastructure: a review. Sustain Chem Pharm 32:100999. https://doi.org/10.1016/j.scp.2023.100999

Omokaro GO, Kornev KP, Nafula ZS, Chikukula AA, Osayogie OG, Efeni OS (2025) Biochar for sustainable soil management: enhancing soil fertility, plant growth and climate resilience. Farm Syst 3(4):100167

Ortiz-Bobea A, Ault TR, Carrillo CM, Chambers RG, Lobell DB (2021) Anthropogenic climate change has slowed global agricultural productivity growth. Nat Clim Chang 11:306–312. https://doi.org/10.1038/s41558-021-01000-1

Osman AI, Lai ZY, Farghali M, Yiin CL, Elgarahy AM, Hammad A, Ihara I, Al-Fatesh AS, Rooney DW, Yap PS (2023) Optimizing biomass pathways to bioenergy and biochar application in electricity generation

Pahunang RR, Buonerba A, Senatore V, Oliva G, Ouda M, Zarra T, Munoz R, Puig S, Ballesteros FC, Li CW et al (2021) Advances in technological control of greenhouse gas emissions from wastewater in the context of circular economy. Sci Total Environ 792:148479

Palansooriya KN, Li J, Dissanayake PD, Tsang DCW, Ok YS (2022) Biochar application to contaminated soils: a critical review on heavy metal immobilization and soil health improvement. Environ Pollut 293:118519

Parravicini V, Svardal K, Krampe J (2016) Greenhouse gas emissions from wastewater treatment plants. Energy Proc 97:246–253

Pathak HK, Chauhan PK, Seth CS, Dubey G, Upadhyay SK (2024a) Mechanistic and future prospects in rhizospheric engineering for agricultural contaminants removal, soil health restoration, and management of climate change stress. Sci Total Environ 927:172116. https://doi.org/10.1016/j.scitotenv.2024.172116

Pathak HK, Seth CS, Chauhan PK, Dubey G, Singh G, Jain D, Upadhyay SK, Dwivedi P, Khoo KS (2024b) Recent advancement of nano-biochar for the remediation of heavy metals and emerging contaminants: mechanism, adsorption kinetic model, plant growth and development. Environ Res 255:119136. https://doi.org/10.1016/j.envres.2024.119136

Pineiro R, Smith J, Lee A (2025) Assessment of trace metal bioavailability in agricultural soils under varying land-use systems. Sci Total Environ 877:162845

Prasannamedha G, Kumar PS, Mehala R, Sharumitha TJ, Surendhar D (2021) Enhanced adsorptive removal of sulfamethoxazole from water using biochar derived from hydrothermal carbonization of sugarcane bagasse. J Hazard Mater 407:124825. https://doi.org/10.1016/j.jhazmat.2020.124825

Rizwan M, Ali S, Adrees M, Arshad M, Hussain A (2018) Mechanisms of heavy metal toxicity and sequestration in plants: a review. Environ Sci Pollut Res 25(32):32126–32143

Ruan X, Sun Y, Du W, Tang Y, Liu Q, Zhang Z, Doherty W, Ray L, Qian G, Tsang CW (2019) Formation, characteristics, and applications of environmentally persistent free radicals in biochars: a review. Bioresour Technol 281:457–468. https://doi.org/10.1016/j.biortech.2019.02.105

Saarnio S, Heimonen K, Kettunen R (2013) Biochar addition indirectly affects N_2O emissionsvia-soilmoisture and plant Nuptake. Soil Biol Biochem 58:99–106

Salma A, Fryda L, Djelal H (2024) Biochar: a key player in carbon credits and climate mitigation. Resources 13(2):31. https://doi.org/10.3390/resources13020031

Shackley S, Clare A, Joseph S, McCarl BA, Schmidt HP (2015) Economic evaluation of biochar systems: current evidence and challenges. In: Lehmann J, Joseph S (eds) Biochar for environmental management: science, technology and implementation. Taylor and Francis, London, pp 813–851

Shackley S, Esteinou RI, Hopkins D, Hammond J (2014) Biochar quality mandate (BQM) version 1.0. British Biochar Foundation, Edinburgh, p 55

Shan R, Shi Y, Gu J, Wang Y, Yuan H (2020) Single and competitive adsorption affinity of heavy metals toward peanut shell-derived biochar and its mechanisms in aqueous systems. Chin J Chem Eng 28:1375–1383

Shoudho KN, Khan TH, Ara UR, Khan MR, Shawon ZBZ, Hoque ME (2024) Biochar in global carbon cycle: towards sustainable development goals. Curr Res Green Sustain Chem 8:100409. https://doi.org/10.1016/j.crgsc.2024.100409

Smith P, Adams J, Beerling DJ, Beringer T, Calvin KV, Fuss S, Griscom B, Hagemann N, Kammann C, Kraxner F et al (2019) Land-management options for greenhouse gas removal and their impacts on ecosystem services and the sustainable development goals. Annu Rev Environ Resour 44:255–286

Soria RI, Rolfe SA, Betancourth MP, Thornton SF (2020) The relationship between properties of plant-based biochars and sorption of Cd(II), Pb(II) and Zn(II) in soil model systems. Heliyon 6:05388. https://doi.org/10.1016/j.heliyon.2020.e05388

Sun L, Ataka M, Han M, Han Y, Gan D, Xu T, Guo Y, Zhu B (2021) Root exudation as a major competitive fine-root functional trait of 18 coexisting species in a subtropical forest. New Phytol 229:259–271. https://doi.org/10.1111/nph.16865

Tan S, Narayanan M, Thu-Huong DT, Iito N, Unpaprom Y, Pugazhendhi A et al (2022) A perspective on the interaction between biochar and soil microbes: a way to regain soil eminence. Environ Res 214:113832

Thompson JL, Tyner WE (2014) Corn stover for bioenergy production: cost estimates and farmer supply response. Biomass Bioenergy 62:166–173

Tian S, Chen Y, Liu Q, Wang F (2021) Heavy metal pollution and its impact on soil microbial biomass and enzymatic activity. Environ Pollut 272:115998

Verheijen FG, Bastos AC, Schmidt HP, Brandao M, Jeffery S (2015) Biochar sustainability and certification. Biochar for environmental management. Routledge, London, pp 795–812

Verheijen FGA, Zhuravel A, Silva FC, Amaro A, Ben-Hur M, Keizer JJ (2019) The influence of biochar particle size and concentration on bulk density and maximum water holding capacity of sandy vs sandy loam soil in a column experiment. Geoderma 347:194–202. https://doi.org/10.1016/j.geoderma.2019.03.044

Wang Z, Sedighi M, Lea-Langton A (2020) Filtration of microplastic spheres by biochar: removal efficiency and immobilisation mechanisms. Water Res 184:116165. https://doi.org/10.1016/j.watres.2020.116165

Warnock DD, Lehmann J, Kuyper TW, Rillig MC (2007) Mycorrhizal responses to biochar in soil-concepts and mechanisms. Plant Soil 300:9–20. https://doi.org/10.1007/s11104-007-93915

Weber K, Quicker P (2018) Properties of biochar. Fuel 217:240–261. https://doi.org/10.1016/j.fuel.2017.12.054

Wen E, Yang X, Chen H, Shaheen SM, Sarkar B, Xu S, Song H, Liang Y, Rinklebe J, Hou D, Li Y, Wu F, Pohorely M, Wong JWC, Wang H (2021) Iron-modified biochar and water management regime-induced changes in plant growth, enzyme activities, and phytoavailability of arsenic, cadmium and lead in a paddy soil. J Hazard Mater 407:124344. https://doi.org/10.1016/j.jhazmat.2020.124344

Weng ZH, Cowie AL (2025) Estimates vary but credible evidence points to gigaton-scale climate change mitigation potential of biochar. Commun Earth Environ 6:259. https://doi.org/10.1038/s43247-025-02228-x

Widowati W, Asnah A, Utomo WH (2014) The use of biochar to reduce nitrogen and potassium leaching from soil cultivated with maize. J Degrade Min Land Manage 2(1):211

Windeatt JH, Ross AB, Williams PT, Forster PM, Nahil MA, Singh S (2014) Characteristics of biochars from crop residues: potential for carbon sequestration and soil amendment. J Environ Manage 146:189–197

Wong S, Ngadi N, Inuwa IM, Hassan O (2018) Recent advances in applications of activated carbon from bio-waste for wastewater treatment: a short review. J Clean Prod 175:361–375

Wu Y, Wang Y, Yang W, Zhang J, Wu Y, Li J, Bing H (2025) Multi-element dataset of soil profiles across climatic zones in China's mountains. Earth Syst Sci Data 17:4779–4797. https://doi.org/10.5194/essd-17-4779-2025

Xiang W, Zhang X, Chen J, Zou W, He F, Hu X et al (2020) Biochar technology in wastewater treatment: a critical review. Chemosphere 252:126539. https://doi.org/10.1016/j.chemosphere.2020.126539

Xu H, Cai A, Wu D, Liang G, Xiao J, Xu M et al (2021) Effects of biochar application on crop productivity, soil carbon sequestration, and global warming potential controlled by biochar C:N ratio and soil pH: a global meta-analysis. Soil Tillage Res 213:105125. https://doi.org/10.1016/j.still.2021.105125

Yan M, Li X, Liu Y, Li Y, He L, Zhang J (2022) Biochar enhanced soil aggregation and C-related enzyme activity in post-mining land on the Loess Plateau, China. Land Degrad Dev 33:1054–1061. https://doi.org/10.1002/ldr.4230

Yang W, Feng G, Miles D, Gao L, Jia Y, Li C et al (2020) Impact of biochar on greenhouse gas emissions and soil carbon sequestration in corn grown under drip irrigation with mulching. Sci Total Environ 729:138752. https://doi.org/10.1016/j.scitotenv.2020.138752

Ye S, Cheng M, Zeng G, Tan X, Wu H, Liang J et al (2020) Insights into catalytic removal and separation of attached metals from natural-aged microplastics by magnetic biochar activating oxidation process. Water Res 179:115876. https://doi.org/10.1016/j.watres.2020.115876

Zeeshan M, Ahmad W, Hussain F, Ahamd W, Numan M, Shah M et al (2020) Phytostabilization of the heavy metals in the soil with biochar applications, the impact on chlorophyll, carotene, soil fertility and tomato crop yield. J Clean Prod 255:120318. https://doi.org/10.1016/j.jclepro.2020.120318

Zhang H, et al (2010) Heavy metal accumulation and tolerance in plants: mechanisms and implications for phytoremediation. J Environ Sci 22(1):89–98

Zhang Y, He M, Wang L, Yan J, Ma B, Zhu X, Ok YS, Mechtcherine V, Tsang DC (2022) Biochar as construction materials for achieving carbon neutrality. Biochar 4(1):59

Zhu J, Gao W, Ge L, Zhao W, Zhang G, Niu Y (2021) Immobilization properties and adsorption mechanism of nickel(II) in soil by biochar combined with humic acid-wood vinegar. Ecotoxicol Environ Saf 215:112159. https://doi.org/10.1016/j.ecoenv.2021.112159

Zhuo Q, Liang Y, Hu Y, Shi M, Zhao C, Zhang S (2023) Applications of biochar in medical and related environmental fields: current status and future perspectives. Carbon Res 2(1):32

Negative Impact of Biochar

Brij Bhushan, Arunima Nayak, Richa Bhatt, Anshika Chauhan, Garima Bhatnagar, Priyanka Negi, Sonali Goyal, Priya Chaudhary, and Sami Abou Fayssal

Abstract Biochar has demonstrated unique characteristics like high surface area, rich porous network, favorable surface functionality, as well as material stability; making it amenable to various applications like in wastewater remediation, soil enrichment, agricultural production, waste management, enhancements in biofuels production, control of greenhouse gas emissions (GHG), and many more. Also, production of biochar from various agro-based waste matrices has opened the pathway for sustainability, circular economy, as well as resource and waste management. But there have been well-defined and scientifically identified potential negative features of biochar during its interaction with environment that has caused adverse impacts. During its interaction with soil or with water or with the atmosphere, when biochar is applied as a soil amendment or as adsorbent for wastewater remediation or for capture of GHG, there has been observed leaching of noxious pollutants like heavy metal ions, polycyclic aromatic hydrocarbons, dioxins, toxic gaseous emissions like N_2O, CO_2, etc. Release of such leachates has been observed to cause a significant change in the soil/water/air characteristics, adversely affecting the crops, agricultural produce, water, as well as the air quality. The ultimate repercussion has been observed on the health of the soil/water microbial community, aquatic community, plants, animals, and humans. Biochar application has showed toxic effects to the germination and growth of plants. Details of the harmful components of biochar because of non-judicious selection of precursor during fabrication as well as its adverse characteristics due to its nano-size is not only the basis of this chapter, but also a close discussion on the exact negative impact of biochar on the soil, water, and air dynamics is reviewed. Finally, measures adopted till date to reduce the toxic effects of biochar, viz., during fabrication, as well as during application have been elaborated. The objective of the overview presented is the wider dissemination to adopt future research objectives for either carrying out a production-toxicity inter-relationship

B. Bhushan · A. Nayak (✉) · R. Bhatt · A. Chauhan · G. Bhatnagar · P. Negi · S. Goyal · P. Chaudhary
Department of Chemistry, Graphic Era Deemed to Be University, Dehradun, India
e-mail: arunima.nayak12@gmail.com

S. A. Fayssal
Department of Agronomy, Faculty of Agronomy, University of Forestry, Sofia, Bulgaria

A. Chauhan et al. (eds.), *Biochar As Remediation Tools for Pollutant Removal*,
https://doi.org/10.1007/978-3-032-16746-0_14

assessment or structure–toxicity inter-relationship assessment or elucidation of the mechanistic aspect of the biochar interactions with the biota.

Keywords Biochar · Negative impact · Interaction with soil · Interaction with water · Interaction with air

1 Introduction

Biochar has been successfully applied in avenues ranging from soil amendment leading to enhanced agricultural output (Joseph and Taylor 2014; Yadav et al. 2023; Ulusal et al. 2021), to capture water-based pollutants leading to enhanced water quality (Akhil et al. 2021; He et al. 2022), to waste management and to the capture of greenhouse gas emissions leading to enhanced air quality. A review by Adhikari et al. has exemplified the role of biochar in enhancing the water-holding capacity and for reducing the hydrophobicity of soil. A roadmap has been provided for optimizing biochar for application as a soil amendment (Adhikari et al. 2022). Wang et al. (2022a; b), in his work, have specified that biochar has the potential to alter the physical and chemical characteristics of soil in a positive way, for example, decreasing soil density, increasing aeration, reducing soil dryness and cracking; all these are linked to higher agricultural productivity.

As per Gupta and Gupta, biochar has demonstrated tremendous potential in the remediation of various pollutants from wastewater. Also, biochar has been cited as a low-cost adsorbent that can be fabricated from widely available feedstock and can be easily manipulated via surface modifications for bringing about enhanced performance (Gupta and Gupta 2022; Pathak et al. 2024b).

The success in achieving the desired outputs in environmental remediation, waste management, and in soil conditioning has enabled scientists and researchers to aggressively work on biochar for enhancing its application in the field of soil, water, and air pollution containment (Godlewska et al. 2021). The features that have favored the success of biochar are its high surface area with a rich microporous structure (Wang and Wang 2019). Its high carbon content to the tune of 50–90% has made it an environmentally benign material, and its application in diverse fields has made it an interesting topic of its further research development. It's easy fabrication from diverse bio-based materials as well as from various waste matrices has not only solved the problem of solid waste management but also has opened the pathway for sustainability, circular economy, as well as resource management (Afshar and Mofatteh 2024).

The quality or the physio-chemical characteristics of the biochar is dependent on the quality of the bio-based precursors or the feedstocks and on the pyrolysis conditions used during its fabrication. A discussion on the factors influencing the biochar properties and on its effect on soil conditioning is elaborated in a review by Tan et al. (2017). The nature of the feedstock, as well as the pyrolysis temperature along with residence time, has been identified as crucial in determining core properties of the biochar, like microporosity, surface area, and elemental composition (Sampatrao 2021; Antonangelo et al. 2019; Tomczyk et al. 2020). For example, high pyrolysis temperatures have resulted in low carbon content and have induced high surface area in the fabricated biochar; at the cost of lower yields. Similarly, among the feedstocks, woody biomass has reported higher carbon along with better microstructures as well as higher yields in the biochar. Optimization of conditions has resulted in better prospects in biochar designing as evidenced from various studies. But studies on the effect of biochar under real environmental condition are scarce.

Earlier studies (Buss et al. 2022; Hale et al. 2012) and, more recently, published reviews have reported the detrimental aspects of biochar to the environment (Brtnicky et al. 2021; Godlewska et al. 2021). Such reports have not only given evidence of the presence of toxic organic and inorganic constituents in the biochar but also have explicitly elaborated on the adverse consequences of the effect of such biomass utilization and their interaction with various environmental media. This has roused the concerns of environmentalists toward the safety aspects of biochar, with the impetus for further research investigation. Thus, with the aim to not only make a more focused awareness on various aspects leading to the cause of toxicity of biochar, its potential interaction, and impact on the environmental medium, the chapter has elaborated on the preventive and mitigation strategies that need to be adopted by researchers in the future. A diagrammatic representation is thus projected in Fig. 1 showing an overview of the positive and negative aspects of biochar.

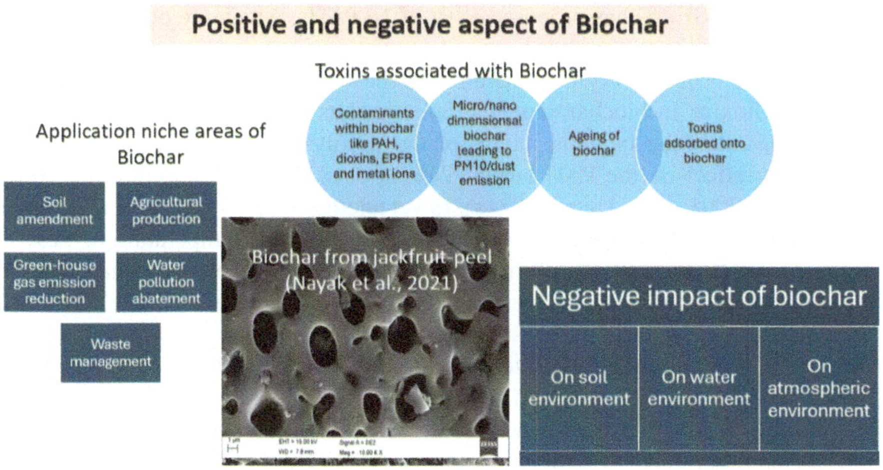

Fig. 1 Graphical representation of the positive and negative aspects of biochar (Nayak et al., 2021)

2 Qualitative Aspect of Biochar Leading to Adverse Effects and Its Causative Analysis

The elemental constituents in biochar have been studied in detail in various studies. There have been reports of the presence of toxins in the fabricated biochar, and in certain instances, the concentration of toxins was reported to be above permissible limits. For example, Hale et al. (2012) fabricated 59 biochar samples from 23 different precursors via slow as well as fast pyrolysis at varying temperatures of 250–900 °C. The idea was to test the presence, concentration of toxins in the samples, and determine the correlation of pyrolysis temperature as well as feedstock precursor on the toxins present. Results revealed the presence of both polyaromatic hydrocarbons (PAH) and dioxins in the samples, and their concentration was found to have a high correlation not only on the precursor type but also on the pyrolysis conditions of temperature and residence time. Among the studied precursors, corn stover recorded the highest PAH concentrations. Fast pyrolysis conditions of enhanced temperature as well as short residence time showed significantly higher PAH levels in the biochar when compared under similar experimental conditions for biochar samples fabricated from slow pyrolysis. Dioxin concentration was reported in all samples, but the highest concentration of 12 pg/g was reported in biochar fabricated at 300 °C.

PAH, as has been reported in biochar samples as per various literary data, is a class of toxin known as polycyclic aromatic hydrocarbons, which are organic compounds having a minimum of two condensed aromatic rings. The United States Environment Pollution Act (USEPA) has proposed a list of 16 PAHs, which are potentially toxic and are of concern to the environment.

There have been literary documentations showing evidence of the presence of PAHs and dioxins in biochar samples. Such reports as tabulated in Table 1 have demonstrated that the inherent nature of the pyrolysis fabrication process has a contributory effect on the PAH levels (Wornat et al. 1987; Hajaligol et al. 2001; Garcia-Perez 2008; Pathak et al. 2024a). In yet another study by Kloss et al., similar observations were demonstrated stating that high pyrolysis temperature of 525 °C irrespective of the precursor selected (straw, spruce, and poplar) resulted in high PAH constituents with naphthalene being the key component in the three fabricated biochar (Kloss et al. 2012). The study reported that irrespective of the precursor selected for study, high pyrolysis temperature has induced incomplete combustion of the carbon-rich precursor leading to a high probability of formation of PAHs. A study by Lyu et al. selected a single precursor, viz., sawdust, and fabricated several biochar samples via pyrolysis at temperature ranging from 250 to 700 °C. PAH and dioxins were present in all samples but the highest PAH and dioxin levels (0.86 and 6.1×10^{-7} mg/kg) were recorded for biochar fabricated at 400 °C. On the other hand, the lowest PAH levels as recorded at 250 °C pyrolysis temperature was 0.19 mg/kg and the same at 700 °C was 70×10^{-9} mg/Kg, respectively (Liu et al. 2018).

A study by Buss et al. (2022) showed that a proper pyrolysis design unit is responsible for the fabrication of a biochar with minimum PAH levels. The study did not favor the feedstock quality or the pyrolytic temperature/residence time/inert gas flow

Table 1 PAH content in biochar from different feedstock under different pyrolytic conditions

Toxin type	Concentration in biochar (mg/Kg)	Biochar feedstock	Pyrolysis temperature	References
PAH				
Phenanthrene	2263×10^{-3}	Sewage sludge	500 °C/N_2	Konczak et al. (2019)
Phenanthrene	1730×10^{-3}	Sewage sludge	600 °C/N_2	
Phenanthrene	1449×10^{-3}	Sewage sludge	700 °C/N_2	
Naphthalene	1290×10^{-3}	Sewage sludge + willow	500 °C/N_2	
Naphthalene	1343×10^{-3}	Sewage sludge + willow	600 °C/N_2	
Naphthalene	1517×10^{-3}	Sewage sludge + willow	700 °C/N_2	
Naphthalene	1482×10^{-3}	Sewage sludge	500 °C/CO_2	
Naphthalene	1125×10^{-3}	Sewage sludge	600 °C/CO_2	
Phenanthrene	715×10^{-3}	Sewage sludge	700 °C/CO_2	
Naphthalene	797×10^{-3}	Sewage sludge + willow	500 °C/CO_2	
Naphthalene	818×10^{-3}	Sewage sludge + willow	600 °C/CO_2	
Naphthalene	938×10^{-3}	Sewage sludge + willow	700 °C/CO_2	
Phenanthrene	1609	Corn stover	350 °C	Hale et al. (2012)
Naphthalene	1959	Corn stover	450 °C	
Naphthalene	1770	Corn stover	550 °C	
Phenanthrene	106	Pine wood	500 °C	
Naphthalene	111	Pine wood	700 °C	
Naphthalene	73	Pine wood	900 °C	
Naphthalene	9556	Wood	450 °C	Quilliam et al. (2013)
Naphthalene	64,650	Rice husk	450 °C	
Pyrene	15,660	Poplar wood	1200 °C	Visioli et al. (2016)
Phenanthrene	53,420	Softwood	550 °C	Buss et al. (2022)

rate conditions in influencing the PAH levels in the fabricated biochar. Thus, the 73 biochar samples fabricated from 14 different feedstocks showed an average PAH level of 27 mg/Kg (ranging from 1.2 to 232 mg/Kg). Among the 73 biochar samples, 10 were identified to have the highest non-naphthalene PAH levels (naphthalene being identified as having the least toxicity among the 16 USEPA identified PAH). While the total PAH level of these 10 biochar samples was 85 mg/Kg, the non-naphthalene level was reported as 58 mg/Kg. The study further reported that the high PAH levels in such 10 biochar samples could not be correlated to a particular feedstock type nor to any pyrolysis condition. The final recommendation was on the focus toward designing the pyrolysis set-up and frequent monitoring.

Leaching behavior of the PAHs from biochar applied to soil samples was studied by various researchers. In this study, Chen et al. fabricated sewage sludge-based biochar via pyrolysis at temperatures ranging from 300 to 700 °C. Highest PAH leaching of 11.5 μg/L was recorded, which was 15% of the PAH level in biochar. It was reported that for the biochar fabricated at 700 °C (Chen et al. 2019a; b).

Besides the organic-based toxins, the presence of inorganic toxins, like heavy metals, has been reported in various studies. Results have been partially represented and tabulated in Table 2. It is a clear indication that feedstock and its chemical constituents have a direct relation to the biochar quality. Various studies have given documentary evidence of the presence of heavy metals in biochar samples. Oleszczuk et al. used commercial biochar not only to estimate the toxin types and levels but also to estimate the toxicity of the biochar on plants, bacteria, alga, protozoa, etc. The test results revealed that besides the presence of 16 different types of PAHs (total sum ranged between 1124 and 28,339 $\times 10^{-3}$ mg/kg), heavy metals detected were Cd, Cu, Ni, Pb, Zn, and Cr. Highest levels in mg/kg in the reported biochar were Zn: 102, Cr: 18, Pb: 22.3, Ni: 9.9, Cu: 2.2, Cd: 0.87 in biochar derived from *Miscanthus* (elephant grass). Other biochar samples derived from coconut shell and from wicker too showed high Zn levels of 37.2 and 97.9 mg/kg. Pb levels in coconut shell biochar and in wicker-based biochar were reported as 23.7 and 20.6 mg/kg, respectively (Oleszczuk et al. 2013). In a study by Devi and Saroha, biochar samples fabricated from sewage sludge via pyrolysis showed the presence of Cu, Pb, and Zn in biochar at high concentration levels. For example, the study reported that when the temperature was increased, Cu, Pb, and Zn content in the fabricated biochar increased to 61, 73, and 65%, respectively, at a pyrolysis temperature of 700 °C (Devi and Saroha 2014).

Besides, there have been reports of the presence of a distinct class of highly reactive toxin identified as environmentally persistent free radicals (EPFR) in biochar samples (Zhang et al. 2023). Such free radicals are organic-based toxins and are characterized by a minimum of a lone electron and hence are highly reactive species. But, unlike other free radicals, EPFRs have high stability, existing in the environment for months. Thus, EPFRs have a high binding tendency to biochar and play a significant role in the production of reactive oxygen species (ROS), leading to various oxidative degradation reactions. Such toxins have showed potential toxic effect to agriculture practices and to the living biota. Ruan et al. (2019) have showed in their studies that both precursor type, as well as pyrolysis temperature conditions, impact the formation of EPFR. Studies have also showed that lignin was the least degraded component under lower pyrolysis temperatures. Being structurally strong and dense, lignin has required high energy and complex process for its decomposition; thereby leading to its least contribution in the formation of EPFR. Lei et al. (2019) demonstrated that pyrolysis of softwood lignin (Fir) showed higher EPFR concentrations in comparison to non-woody (bamboo) and hardwood-based lignin (Eucalyptus). Pyrolysis temperature as well as time showed significant effect on EPFR concentrations in the pyrolyzed biochar. For example, pine tree biochar fabricated via pyrolysis to 500 °C showed high EPFR concentrations as compared to the same pyrolyzed at

Table 2 Heavy metal content in biochar from different feedstock under different pyrolytic conditions

Toxin type	Concentration in biochar (mg/Kg)	Biochar feedstock	Pyrolysis temperature	References
Heavy metal				
Zn, Cu	Zn = 129.24, Cu = 122.89	Pig manure	700 °C	Meng et al. (2017)
Zn, Cu, Pb, Fe	Zn = 2103.6, Cu = 690.8, Pb = 438.3, Fe = 192.8	Sewage sludge	400 °C	Lu et al. (2013)
Cu, Cd	Cu = 21.30, Cd = 6.31	*Pennisetum sinese* plant	NA	Cui et al. (2021)
Cr, Cu, Pb, Zn	Cr = 225; Cu = 248; Pb = 204; Zn = 2355	Sewage sludge	700 °C	
Cd, Cr, Zn, Cu	Cd = 1.10; Cr = 30.83; Zn = 1457; Cu = 1335	Swine manure	800 °C	
Zn, Cu, Pb, Ni	Zn = 332.79; Cu = 146.97; Pb = 52.99; Ni = 20.81	Sewage sludge	700 °C	Devi and Saroha (2014)
Zn, Pb	Zn = 21.60; Pb = 32.90	Wicker plant	650 °C	Oleszczuk et al. (2013)
Zn, Cu, Pb, Ni, Cr	Zn = 102.00; Cu = 2.22; Pb = 22.30; Ni = 9.95; Cr = 18.00	Miscanthus plant	650 °C	Oleszczuk et al. (2013)
Zn, Pb, Fe, Mn	Zn = 0.03; Pb = 0.03; Fe = 4.21; Mn = 0.03	Restaurant food waste	650 °C	Oleszczuk et al. (2013)
Zn, Cu, Mn	Zn = 41.46; Cu = 33.84; Mn = 41.47	Coconut shell	420 °C	Castilla-Caballero et al. (2020)
Cr, Ni, Pb, Cd, Cu, Zn	Cr = 23; Ni = 16; Cd = 0.17; Pb = 35; Cu = 11; Zn = 9	Plant biomass	300 °C	Kujawska (2023)
	Cr = 28; Ni = 21; Cd = 0.15; Pb = 1.7; Cu = 3; Zn = 36	Municipal waste	650 °C	
	Cr = 5; Ni = 4.6; Cd = 0.12; Pb = 1.7; Cu = 4; Zn = 18	Compost	650 °C	
	Cr = 69; Ni = 33; Cd = 0.7; Pb = 33; Cu = 40; Zn = 138	Coal refuse	600 °C	

lower temperatures (Yang et al. 2017). In yet another study, biochar fabricated at low pyrolysis temperatures (200, 300 °C) but time of pyrolysis if elongated (viz., 12 h) showed high EPFR concentrations. Similarly, biochar fabricated at higher pyrolysis ($\geq$500 °C) at low time of 1 h showed high EPFR concentrations (Fang et al. 2015). In fact, the presence of transition metal ions like Fe^{3+} in biochar precursor has also promoted the formation of EPFR (Fang et al. 2014).

During the interaction with the soil or with aqueous environment, leaching of the different contaminants (PAH, EPFRs, metal ions, etc.) present in biochar has been reported in various studies. The presence of such contaminants in the environmental media has been the matter of concern because of the detrimental impact of such toxins to the soil, water quality, as well as to the microbial community. For example, in a study reported by Rinklebe et al., a rice hull-based biochar fabricated at 500 °C pyrolysis temperature when applied to soil at the rate of 5% (w/w) showed the availability of various metal ions like Sb, Sn, and Tl. This was in contrast to a similar soil sample without biochar addition. The authors commented on the effect of redox potential and pH of the soil in initiating the mobility of the metal ions. The redox potential and pH of biochar added soil was reported to be $E_H = -12$ to $+$ 333 mV and 4.9–8.1, respectively. The untreated soil showed redox potential and pH values of $E_H = -30$ to $+218$ mV and 5.9–8.6, respectively. Both redox potential and pH of the biochar added soil were altered and which facilitated the release, solubility, mobility, and phyto-availability of toxins to the plants, to groundwater, to flooded soils, or to constructed wetlands (Rinklebe et al. 2020). Similar case of metal leaching from biochar was reported by von Gunten et al. (2017). Aged biochar has also been reported to bring about leaching of toxins to the environment. Aging has brought about changes in physical and chemical make-up of the biochar leading to leaching of toxins (either inherently present or the adsorbed contaminants) via various mechanisms like oxidization, fragmentation, etc. A study by Xie et al. (2022) showed that the aged corn stover biochar (aged for 7 years in farmland) had a reduced C content by 34% and the O content increased by 8%. The aged biochar showed higher oxygen-containing surface functionalities. Besides, the surface area, porosity increased in the aged biochar. The result was enhanced metal binding capacity in the aged biochar. But in comparison with that of the fresh biochar, the leaching of metal ions (Cd^{2+} and Pb^{2}) from aged biochar was more (39% and 28% respectively). The fresh biochar showed a leaching of Cd^{2+} and Pb^{2+} to the tune of 24% and 21%, respectively. Results thus indicated the reduced stability of aged biochar to capture metal ions from environment.

3 Impact of Biochar on Soil Characteristics and on Soil Biota

Biochar has been widely applied onto soil for bringing about significantly improved soil characteristics like nutrient availability, fertility, water retention capacity, carbon sequestration capability, as well as providing better environment for growth of microorganisms. All these have led to better seed germination and crop growth and/or productivity. But various studies have cited an adverse impact of toxins in biochar on the soil environment. There have been reports about the adverse impact of the application of biochar that was fabricated at high pyrolysis temperatures on soil. The results, as described by Wang et al. (2019a; b), showed that the soil quality had enhanced alkalinity (Wang et al. 2019a). Such pH imbalances have resulted in decreased nutrient availability to the growth of plants (Wang et al. 2019b; Zhang et al. 2020a, b). The addition of biochar has caused decreased nutrient supply but also has caused restricted nutrient absorption leading to reduced plant growth (El-Naggar et al. 2019a, b). Biochar applicability has also resulted in reduced soil tensile strength (Li et al. 2019). Soil tensile strength is a parameter that is used to assess the ability of soil to get ruptured in the presence of external forces. As per a study conducted by Yang and Lu (2021), effect of biochar from five different precursors (viz., straw of rice, wheat and maize, rice husk, and bamboo) was studied on paddy growing soil under laboratory conditions. Results revealed that irrespective of the precursors used for fabrication, addition of biochar decreased the soil tensile strength. Further, the study revealed that the rice straw-derived biochar showed the highest decrease in the soil tensile strength (viz. 63.6%). Studies have also indicated that the bioavailability of certain toxic elements to the soil has been made possible via biochar additions. For example, a study showed the enhancement in the bioavailability of arsenic in sandy loam soil by 101% by the addition of rice straw-derived biochar; which is a leading environmental risk (El-Naggar et al. 2020).

Besides the adverse impact on soil quality, application of toxin-laden biochar to soil has impacted the growth indices of crops grown on such soil. On a more elaborate scale in a study, the fabricated biochar samples showed total PAH levels of 638–12,347 $\times 10^{-3}$ mg/kg. The PAH concentration when measured in 8 vegetable crops that were grown in the biochar added soil showed a maximum level of 565 $\times 10^{-3}$ mg/kg. The health risk assessment tests conducted on the vegetable crops with benzo[*a*]pyrene toxic equivalency quotient (TEQ) and with the incremental lifetime cancer risk (ILCR) showed values that were higher than the maximum contaminant level. These results demonstrated that application of a PAH-laden biochar sample as crop enhancer may not only cause contamination in plants, leading to adverse growth effects, but also may result in life-threatening effects on humans and animals. Another study by Visholi et al. showed that the copper and zinc ions in biochar caused decreased seed germination and root elongation indices in crops like cucumber, watercress, and sorghum. Results revealed that in all crops studied, a dosage application of >5% by weight of Cu laden biochar caused hindered germination and root elongation, whereas a similar negative impact on germination and root elongation was brought about by a higher dosage application (>10% by weight) of Zn laden biochar. The presence of EPFR in biochar sample has also adversely affected the plant growth and the soil microbes, as evident from various studies conducted (Ruan et al. 2019). For example, in a study, Liao et al. fabricated biochar from rice straw via heating at 500 °C. The biochar application showed enhanced phytotoxicity to corn, wheat, and rice seeds; viz., retarded root and stem growth, as well as inhibited germination and damaged plasma membrane, were observed during all seed germination tests (Liao et al. 2014). In a study, as reported by Odinga et al., EPFR from biochar, when applied to soil, may adversely affect the activities of enzymes that are involved in the recycling of nutrients. EPFRs have shown catalytic activity in black carbon during the formation of activated species. Such species have shown to stimulate the biotransformation of pollutants in the soil samples via microbial activity. But during such reactions, EPFRs have demonstrated altered microbial structure leading to induced soil toxicity (Odinga et al. 2020).

Besides the commonly studied toxins like PAH, metal ions, dioxins, and EPFRs on biochar and their impact on the soil biota, there have been reports in which the contributory effect of nanobiochar toward inhibited growth on crops was demonstrated. Nanobiochar having dimensions in the nano-range has exhibited high surface area and has brought about greater growth inhibition (Huang et al. 2020). In a study by Zhang et al., nanobiochar samples were fabricated from two different precursors and using pyrolysis temperatures in the range of 300–700 °C. A total of six biochar samples were fabricated. The feedstock selected was rice straw and wood sawdust. Seedlings of rice, tomato, and reed were selected for testing of the phytotoxicity. Results revealed that when lignin-based plants were pyrolyzed at high temperatures, the nanobiochar thus developed showed inhibited germination of rice seedlings. The reason cited for acute phytotoxicity was due to the phenolic compounds that were formed during pyrolysis, and which were found to be deposited on the surface of nanobiochar (Zhang et al. 2020a; b).

Besides, there has been evidence of the leaching of toxins from biochar, which showed a damaging influence on the soil microbial community. While PAH laden biochar showed mutagenic effect on salmonella (Anjum et al. 2014), EPFRs in biochar showed a significant change in the microbial structure (Balakrishna et al. 2009). Presence of heavy metals like Cd, furans, phenols, etc., in biochar sample showed reduced growth of bacterial and actinomycetes populations at high concentrations (Du et al. 2023).

4 Impact of Biochar on Aquatic Quality and on Aquatic Microbes

Biochar has shown tremendous potential as a cost-effective and sustainable adsorbent when applied to remediate water-based pollutants like heavy metal ions, dyes, pharmaceuticals, etc. (Gupta and Gupta 2022). But certain risks have been demonstrated in various studies during the application of biochar in water treatment. Such risks have been demonstrated especially by biochar samples showing the presence of inorganic mineral elements. For example, two biochar samples were fabricated from two different precursors, viz., rice husk-based, and dairy manure-based. Elemental analysis revealed the presence of mineral content like Ca, K, Mg, Fe, Al, and Na in dairy manure-based biochar sample. Total phosphorus (P) content in the dairy manure-based biochar was 3.23% in comparison with the same in the rice husk-based biochar (Xu et al. 2013). Leaching of such elements tends to cause eutrophication by increasing phosphorus levels in aquatic bodies leading to algal blooms. Another study showed the leaching of ammonium ions from a biochar sample accounting for approximately 4.9% of total ammonium content (Chen et al. 2017). Leaching of phosphorus was observed from two different biochar samples. The biochar was fabricated from different feedstocks like poultry manure and sugarcane straw. Phosphorus leaching was found to be enhanced to 20% in pure water, whereas more than 90% of phosphorus was desorbed in the presence of alkaline water (presence of HCO_3^-) (Novais et al. 2018). Thus, leaching of N, P, ammonium, phosphate, etc., has resulted in incidences of common problems in aquatic bodies like high algal blooms and has depleted oxygen levels in aquatic bodies (Deng et al. 2019).

The biochar has high binding affinity toward various environmental toxins. The exhausted biochar is itself a toxin to the aquatic community. Also, the migration/transport of such biochar samples along with the adsorbed pollutant moieties has resulted in their presence in various water bodies like rivers, lakes, groundwater, and surface water (Chen et al. 2019a; b). Another important aspect is the higher mobility demonstrated by nanobiochar samples. Their high surface reactivity has allowed for not only the capture of various water pollutants but has also caused their enhanced mobility (Liu et al. 2018).

Ecotoxicity tests conducted have demonstrated varied levels of toxicity to different aquatic biota. Tests by Oleszczuk et al. (2013) showed that the commercial biochar samples showed high PAH and heavy metal contents. Ecotoxicity tests were conducted on plants, bacteria, algae, protozoa, and crustaceans. The species selected for plants, bacteria, algae, protozoa, and crustaceans were *L. sativum, V. fischeri, S. capricornutum,* T. *thermophila*, and *Daphnia magna.* Among all the studied species, *D. magna* showed high sensitivity to the biochar. Another important finding was that the level of PAH in the biochar showed direct proportionality to the mortality rate in *D. magna.* In yet another study, EPFR in biochar was found to be highly toxic to aquatic plants via release of hydroxyl free radicals into the aquatic medium. The study showed that the hydroxyl free radicals released into the aquatic medium from EPFR damaged the cell wall and organs of aquatic plants (Odinga et al. 2020).

The study by Odinga et al. revealed that EPFRs from biochar samples have interacted with water to form stable reactive oxidation species (ROS), which in turn have demonstrated long-term toxic effects to the water quality as well as to aquatic species. Nanobiochar samples showed highest level of toxicity to aquatic algal species (Liu et al. 2018). The slow release of pollutants (like dyes, metal ions, pharmaceuticals) adsorbed onto the biochar/nanobiochar into the water has resulted in toxicity to the marine organisms (Freixa et al. 2018).

5 Impact of Biochar on Atmospheric Quality

Biochar has played a significant role in mitigating climate change via bringing about an effective decrease in greenhouse gas emissions (CH_4, N_2O) as well as via control of carbon dioxide emissions to the atmosphere (Lehmann et al. 2021). Such net zero atmospheric emissions have been made possible by the application of biochar to soil. One of the important aspects of biochar is its ability to sequestrate carbon in soil. Because of its long-term stability, biochar has served as a carbon storage solution and further prevents the carbon emission into the environment (Li and Tasnady 2023). The features that have shown favorable aspects of biochar for GHG emission control from soil to air are its high surface area, high cation exchange capacity leading to high water holding capacity, high nutrient binding capacity, and high carbon concentration in soil.

But such positive aspects during biochar application have been observed from short-time laboratory-based studies on soil samples (Lehmann et al. 2021). Studies have indicated the long-term stability of biochar in the environment (Zhao et al. 2020). Thus, aged biochar has shown some negative environmental implications. The potential of such aged biochar for atmospheric mitigation or control of GHG emission was observed to be greatly reduced (Liu et al. 2018). But the number of studies certifying the positive impact of biochar in enhanced GHG emission reduction is scarce. The ability of a biochar sample when used after 7 years showed a 48% decrease of N_2O emissions in acidic soils. N_2O emissions fell to 22% when the same aged biochar was used in alkaline soils. Comparisons were made to the control soil (Zhang et al. 2021).

EPFRs present in biochar are potentially toxic atmospheric pollutants. When inhaled, they have been demonstrated to cause respiratory illness, and in some acute cases have shown cardiovascular problems in humans (Odinga et al. 2020).

It is a scientifically established fact that biochar amendment to soil has reduced atmospheric nitrous oxide emissions (Cayuela et al. 2014). It was demonstrated that a slow pyrolyzed biochar fabricated from wheat straw increased the N_2O emissions by 291% in a 45-day incubation period. The biochar was applied at 4% dosage rate to a rice paddy soil. The study further concluded about the direct link of biochar porosity on the soil pH and in turn on the abundance/growth of ammonium oxidizing bacteria (AOB). The growth of AOB was further linked to the denitrification rate and to the increase in N_2O to atmosphere (Lin et al. 2017). Ribas et al. (2019), proved that the porous sites within biochar can create anoxic conditions under favorable biological conditions for complete denitrification, favoring more uptake of N_2O to form CH_4. Such cases have been observed during the application of biochar to soil in high-temperature regions or during the summer season. Studies have also indicated that the mineral constituents in biomass precursor leading to high ash in biochar sample may cause N_2O emissions during their application in soil (Luz et al. 2013). El-Nagger et al., in his study, showed that the quality of biochar in mitigating CO_2 emissions during soil application was dependent on the soil quality as well as on the biochar feedstock. This can be more specifically explained as follows: three different biochar samples were applied to two infertile soil samples. The application dosage rate was 30 tones/hectare, and the soils selected for the study were sandy loam soil and sandy soil. The incubation period was 90 days at 25 °C. Results revealed that the soil quality had a strong dependence on CO_2 soil emissions. For example, the biochar irrespective of their precursor type when applied to sandy loam soil caused significantly higher (twofold–threefold) CO_2 emissions as compared to the results from sand soil (El-Naggar et al. 2018). A similar finding was demonstrated by Wang et al. (2016) in which it was pointed out that biochar when applied to sandy infertile soil caused increased microbial interactions and increased CO_2 emissions. A detailed negative impact of biochar applicability to soil and to atmosphere pollution has been detailed in various reviews (El-Naggar et al. 2019a, b).

Besides the GHG emissions, biochar application has also resulted in increased particulate matter concentrations (PM_{10}) in the atmosphere, leading to air pollution and serious cytotoxic effects on living biota. Such observation has been backed by various experimental data and by various scientific studies (Li et al. 2018; El-Naggar et al. 2019b; Gelardi et al. 2019; Sigmund et al. 2017; Lyu et al. 2018). Studies have indicated the low density, variable particle dimension, and very high surface reactivity in biochar. During their varied applications, their high mobility and high reactivity have resulted in their fast dispersion into the atmosphere, ultimately leading to increased PM_{10} levels. Also, aging or exposure to atmospheric conditions has resulted in breaking down of biochar to various coarse-sized particles. Their low density and low particle dimensions favor their fast dispersion. Also, their high surface reactivity has caused effective and fast binding to various atmospheric pollutants and, ultimately, via fast dispersions, has exhibited binding to various biological substrates leading to acute cytotoxicity to human and animal health (Gelardi et al. 2019). There have been reports stating that deposition of biochar particulate matter on the inner membrane of the lung has caused damage to lungs with incidences of chronic bronchitis (Yang et al. 2017).

6 Discussion on Risk Minimization Via Biochar Usage

Despite the risks and negativity associated toward the abiotic and biotic factors in soil, water, and air, biochar is a valued, low-cost material having tremendous potential to clean water, soil, and atmosphere. But the detailed scientific evidence of the risks associated with biochar on the health of the environment and on the living biota requires closer introspection on the measures taken to minimize such risks. Extensive research on the risk assessments and its outcomes will help in not only eliminating risks, but also in advancing and expanding the application range and market potential of biochar.

Extensive investigation has revealed that one of the leading causes for toxicity of biochar during its interaction with the environment is the inappropriate selection of the precursor. A study by Quilliam et al. (2013) has rightly proved the direct consequence of precursor on the chemical constituents of biochar. Thus, feedstock is the first step to combat or mitigate the risks associated with biochar usage. The study by Quilliam et al. has amply verified that feedstock with plant-based origin has low probability of harmful constituents like PAH, metal ions, dioxins, etc. Pyrolysis condition is another factor on which the quality of biochar is dependent. Slow pyrolysis involving slow heating rates, heating at low temperatures (300–600 °C) and long residence times has shown lower ecological risks to biochar as compared to fast pyrolysis. The biochar thus fabricated has greater benefits for use as soil amendment (El-Naggar et al. 2019b). Increased residence time as well as higher inert gas flow rate can increase PAH levels in biochar during high temperature pyrolysis (Ko et al. 2018; Madej et al. 2016). Hale et al. (2012) have claimed that low temperature pyrolysis can bring about enhanced PAH levels in biochar samples irrespective of precursor selected for fabrication of biochar. An important factor for consideration is the use of appropriate drying of biomass precursor prior to pyrolysis, as presence of moisture may cause incomplete combustion leading to the formation of PM_{10} (Dunnigan et al. 2018). Studies have indicated that physical property of the biochar can help offset the formation of PM_{10}. Thus, formation of pellets via incorporation of lignin and $Ca(OH)_2$ as binders may increase the mechanical strength of the biochar to resist abrasion and formation of micro/nanobiochar or formation of dust (Reza et al. 2012; Hu et al. 2015).

During the application of biochar to soil and specially to minimize the dust formation or generation of PM_{10}, studies have recommended the biochar to be applied to high moisture content or to liquid slurry prior to application to soil (Silva et al. 2015). Sub-surface banding technique has the potential to minimize dispersion of biochar in wind or in water medium; thereby minimizing biochar loss.

During application in water, appropriate and judicious choice of precursor is essential to minimize the elemental content like N, P, etc. Also, such judicious selection will help to reduce the ash content levels in the biochar (Xu et al. 2013). To minimize biochar dispersion and to minimize the adverse effect of aging of biochar, studies have recommended the use of larger-sized biochar (Liu et al. 2018). Finally, to minimize the toxicity due to ROS and EPFR, studies have recommended the selection of hard wood lignin (Ye et al. 2020).

Despite the precautionary measures that have been undertaken during biochar production and its usage, another important aspect is to measure the risks imposed by biochars under complex real-time environments by using the life cycle assessment tests (LCA). Also, there is widespread information available in the literature focusing on the laboratory-scale production and application of biochar. But reports on large-scale data regarding either the production or application of biochar and their potential environmental/economic impact are distinctly missing.

For example, in a study by Mandal et al., they reported that the pyrolysis temperature had a direct impact on the reduction of GHG emission. Thus, biochar fabricated from rapeseed straw via pyrolysis at 400 and 600 °C had the potential to reduce GHG emissions by 73% and 83%, respectively (Mandal et al. 2016). In yet another study, poultry manure biochar when applied as fertilizer demonstrated high nutrient supply to the plant resulting in high cucumber yield (Solaiman et al. 2020). But the same author conducted a meta-analysis and the results revealed that biochar application demonstrated a significant increase in the soil CO_2 emission by 28%. More recently, the effect of biochar was monitored via parameters like crop yield, GHG and soil nitrogen cycling under lab and field experimental conditions. Results, as revealed by Huang et al., showed significant differences in the ability of the biochar to improve crop yield, reduce GHG emissions, etc. The major factors that showed direct effect on the biochar capability were the different experimental conditions, different feedstocks used, etc. (Huang et al. 2023).

Various LCAs have been conducted showing the benefits of biochar production (Li et al. 2024), yet the results of such studies have been found to be varying across different selection criteria. Thus, a universal design criterion could not be possible for application to a majority of the biochar systems.

7 Conclusion and Future Research Directions

The positive aspect of biochar is well known, for which there is widespread application in diverse areas pertaining to soil amendment, water clean-up, controlling atmospheric GHG, climate change, etc. There is a boom in research activities pertaining to biochar fabrication and testing under diverse conditions for their effective application. Numerous toxins are present in biochar which have an adverse effect during its interaction with soil, water, and air; thereby limiting the large-scale applicability of biochar. Although various studies have been conducted that have shown minimization of risks caused by biochar to the ecology, various strategies need to be undertaken in the future research as bulletined below.

Majority of studies on biochar application have been devoted to lab-scale and pilot-scale, while field studies are scarce. To have a foolproof knowledge system, future studies should be coherence of laboratory-scale and field-scale studies. There should be correlation between factors like feedstock selection, pyrolytic conditions on the soil characteristics, and on the growth conditions of plants. Simultaneous conduction of phytotoxicity is a necessity.

Future studies need to focus on long-term field-scale studies. Current knowledge is limited on how the biochar can have a long-term effect on the health of soil, its microbial community, and on the growth of plants.

While field-scale studies are relevant, future studies should address the biochar effects toward physical, chemical, and biological interactions with the medium in which it is applied. The knowledge regarding the source of toxic effect of biochar needs to be addressed. For example, whether the toxicity of biochar is arising from its size effects, from its internal chemical constituents, or from the external toxins adsorbed onto the biochar surface. Such mechanistic interpretations will help in advancing the state of knowledge about any toxic effect of biochar.

The behavior of the biochar with changing environmental patterns needs to be studied to gain insight on its stability and future application.

References

Adhikari S, Timms W, Parvez Mahmud MA (2022) Optimising water holding capacity and hydrophobicity of biochar for soil amendment: a review. Sci Total Environ 851:158043

Afshar M, Mofatteh S (2024) Biochar for a sustainable future: environmentally friendly production and diverse applications. Results Eng 23:102433

Akhil D, Lakshmi D, Kartik A, Viet DNV (2021) Production, characterization, activation and environmental applications of engineered biochar: a review. Environ Chem Lett 14:1–37. https://doi.org/10.1007/s10311-020-01167-7

Anjum R, Krakat N, Reza MT, Klocke M (2014) Assessment of mutagenic potential of pyrolysis biochar by Ames Salmonella/mammalian-microsomal mutagenicity test. Ecotoxicol Environ Saf 107:306–312. https://doi.org/10.1016/j.ecoenv.2014.06.018

Antonangelo JA, Zhang H, Sun X, Kumar A (2019) Physicochemical properties and morphology of biochars as affected by feedstock sources and pyrolysis temperatures. Biochar 15:1–12. https://doi.org/10.1007/s42773-019-00028-z

Balakrishna S, Lomnicki S, McAvey KM, Cole RB, Dellinger B, Cormier SA (2009) Environmentally persistent free radicals amplify ultrafine particle mediated cellular oxidative stress and cytotoxicity. Part Fibre Toxicol 6:11. https://doi.org/10.1186/1743-8977-6-11

Brtnicky M, Datta R, Holatko J, Bielska L, Gusiatin ZM, Kucerik J, Hammerschmiedt T, Danish S, Radziemska M, Mravcova L, Fahad S, Kintl A, Sudoma M, Ahmed N, Pecina V (2021) A critical review of the possible adverse effects of biochar in the soil environment. Sci Total Environ 796:148756

Buss W, Hilber I, Graham MC, Mašek O (2022) Composition of PAHs in biochar and implications for biochar production. ACS Sustain Chem Eng 10(20):6755–6765. https://doi.org/10.1021/acssuschemeng.2c00952

Castilla-Caballero D, Barraza-Burgos J, Gunasekaran S, Roa-Espinosa A, Colina-Márquez J, Machuca-Martínez F, Hernandez-Ramírez A, Vázquez-Rodríguez S (2020) Experimental data on the production and characterization of biochars derived from coconut-shell wastes obtained from the Colombian pacific coast at low temperature pyrolysis. Data Brief 28:104855

Cayuela ML, van Zwieten L, Singh BP, Jeffery S, Roig A, Sánchez-Monedero MA (2014) Biochar's role in mitigating soil nitrous oxide emissions: a review and meta-analysis. Agricult Ecosyst Environ 191:5–16

Chen L, Chen XL, Zhou CH, Yang HM, Ji SF, Tong DS, Zhong ZK, Yu WH, Chu MQ (2017) Environmental-friendly montmorillonite-biochar composites: facile production and tunable adsorption-release of ammonium and phosphate. J Clean Product 156:648–659

Chen M, Tao X, Wang D, Xu Z, Xu X, Hu X, Xu N, Cao X (2019a) Facilitated transport of cadmium by biochar-Fe_3O_4 nanocomposites in water-saturated natural soils. Sci Total Environ 684:265–275

Chen X, Yang L, Myneni SCB, Deng Y (2019b) Leaching of polycyclic aromatic hydrocarbons (PAHs) from sewage sludge-derived biochar. Chem Eng J 373:840–845

Cui H, Li D, Liu X, Fan Y, Zhang X, Zhang S, Zhou J, Fang G, Zhou J (2021) Dry-wet and freeze-thaw aging activate endogenous copper and cadmium in biochar. J Clean Product 288:125605

Deng R, Huang D, Wan J, Xue W, Lei L, Wen X, Liu X, Chen S, Yang Y, Li Z, Li B (2019) Chloro-phosphate impregnated biochar prepared by co-precipitation for the lead, cadmium and copper synergic scavenging from aqueous solution. Bioresour Technol 293:122102

Devi P, Saroha AK (2014) Risk analysis of pyrolyzed biochar made from paper mill effluent treatment plant sludge for bioavailability and eco-toxicity of heavy metals. Bioresour Technol 162:308–315

Du L, Ahmad S, Liu L, Wang L, Tang J (2023) A review of antibiotics and antibiotic resistance genes (ARGs) adsorption by biochar and modified biochar in water. Sci Total Environ 858:159815. https://doi.org/10.1016/j.scitotenv.2022.159815

Dunnigan L, Morton BJ, Hall PA, Kwong CW (2018) Production of biochar and bioenergy from rice husk: influence of feedstock drying on particulate matter and the associated polycyclic aromatic hydrocarbon emissions. Atmosp Environ 190:218–225

El-Naggar A, Lee SS, Awad YM, Yang X, Ryu C, Rizwan M, Rinklebe J, Tsang DCW, Ok YS (2018) Influence of soil properties and feedstocks on biochar potential for carbon mineralization and improvement of infertile soils. Geoderma 332:100–108

El-Naggar A, El-Naggar AH, Shaheen SM, Sarkar B, Chang SX, Tsang DCW, Rinklebe J, Ok YS (2019a) Biochar composition-dependent impacts on soil nutrient release, carbon mineralization, and potential environmental risk: a review. J Environ Manag 241:458–467. https://doi.org/10.1016/j.jenvman.2019.04.011

El-Naggar A, Lee SS, Rinklebe J, Farooq M, Song H, Sarmah AK, Zimmerman AR, Ahmad M, Shaheen SM, Ok YS (2019b) Biochar application to low fertility soils: a review of current status, and future prospects. Geoderma 337:536–554. https://doi.org/10.1016/j.geoderma.2018.09.034

El-Naggar A, Lee M-H, Hur J, Lee YH, Igalavithana AD, Shaheen SM, Ryu C, Rinklebe J, Tsang DCW, Ok YS (2020) Biochar-induced metal immobilization and soil biogeochemical process: an integrated mechanistic approach. Sci Total Environ 698:134112. https://doi.org/10.1016/j.scitotenv.2019.134112

Fang G, Gao J, Liu C, Dionysiou DD, Wang Y, Zhou D (2014) Key role of persistent free REl-Naggaradicals in hydrogen peroxide activation by biochar: implications to organic contaminant degradation. Environ Sci Technol 48:1902–1910

Fang G, Liu C, Gao J, Dionysiou DD, Zhou D (2015) Manipulation of persistent free radicals in biochar to activate persulfate for contaminant degradation. Environ Sci Technol 49:5645–5653

Freixa A, Acuña V, Sanchís J, Farré M, Barceló D, Sabater S (2018) Ecotoxicological effects of carbon-based nanomaterials in aquatic organisms. Sci Total Environ 619–620:328–337

Garcia-Perez M (2008) The formation of polyaromatic hydrocarbons and dioxins during pyrolysis: a review of the literature with descriptions of biomass composition. Fast Pyrolysis Technol Thermochem React 14:2542

Gelardi DL, Li C, Parikh SJ (2019) An emerging environmental concern: biochar-induced dust emissions and their potentially toxic properties. Sci Total Environ 678:813–820

Godlewska P, Ok YS, Oleszczuk P (2021) The dark side of black gold: ecotoxicological aspects of biochar and biochar-amended soils. J Hazrd Mat 403:123833

Gupta P, Gupta N (2022) Potential role of biochar in water treatment. Desalin Water Treat 251:79–104

Hajaligol M, Waymack B, Kellogg D (2001) Low temperature formation of aromatic hydrocarbon from pyrolysis of cellulosic materials. Fuel 80(12):1799–1807

Hale SE, Lehmann J, Rutherford D, Zimmerman AR, Bachmann RT, Shitumbanuma V, O'Toole A, Sundqvist KL, Arp HPH, Cornelissen G (2012) Quantifying the total and bioavailable polycyclic aromatic hydrocarbons and dioxins in biochars. Environ Sci Technol 46:2830–2838. https://doi.org/10.1021/es203984k

He M, Xu Z, Hou D et al (2022) Waste-derived biochar for water pollution control and sustainable development. Nat Rev Earth Environ 3:444–460. https://doi.org/10.1038/s43017-022-00306-8
Hu Q, Shao J, Yang H, Yao D, Wang X, Chen H (2015) Effects of binders on the properties of bio-char pellets. Appl Energy 157:508–516
Huang X, Zhang C, Zhu S, Chen D, Ho SH (2020) Effects of biochar on microalgal growth: Difference between dissolved and undissolved fractions. ACS Sustain Chem Eng 8:9156–9164. https://doi.org/10.1021/acssuschemeng.0c01694
Huang Y, Tao B, Lal R, Lorenz K, Jacinthe P-A, Shrestha RK, Bai X, Singh MP, Lindsey LE, Ren W (2023) A global synthesis of biochar's sustainability in climate-smart agriculture: evidence from field and laboratory experiments. Renew Sustain Energy Rev 172:113042
Joseph S, Taylor P (2014) The production and application of biochar in soils. In: Advances in biorefineries. Woodhead Publishing Limited, pp 525–555
Kloss S, Zehetner F, Dellantonio A, Hamid R, Ottner F, Liedtke V, Schwanninger M, Gerzabek MH, Soja G (2012) Characterization of slow pyrolysis biochars: effects of feedstocks and pyrolysis temperature on biochar properties. J Environ Qual 41:990–1000. https://doi.org/10.2134/jeq 2011.0070
Ko JH, Wang J, Xu Q (2018) Impact of pyrolysis conditions on polycyclic aromatic hydrocarbons (PAHs) formation in particulate matter (PM) during sewage sludge pyrolysis. Chemosphere 208:108–116
Konczak M, Gao Y, Oleszczuk P (2019) Carbon dioxide as a carrier gas and biomass addition decrease the total and bioavailable polycyclic aromatic hydrocarbons in biochar produced from sewage sludge. Chemosphere 228:26–34. https://doi.org/10.1016/j.chemosphere.2019.04.029
Kujawska J (2023) Content of heavy metals in various biochar and assessment of environmental risk. J Ecol Eng 24(8):287–295
Lehmann J, Cowie A, Masiello CA, Kammann C, Woolf D, Amonette JE et al (2021) Biochar in climate change mitigation. Nat Geosci 14:883–892. https://doi.org/10.1038/s41561-021-008 52-8
Lei M, Wu S, Liang J, Liu C (2019) Comprehensive understanding the chemical structure evolution and crucial intermediate radical in situ observation in enzymatic hydrolysis/mild acidolysis lignin pyrolysis. J Anal Appl Pyrol 138:249–260
Li S, Tasnady D (2023) Biochar for soil carbon sequestration: current knowledge, mechanisms, and future perspectives. Carbon Res 9(3):67. https://doi.org/10.3390/c9030067
Li C, Bair DA, Parikh SJ (2018) Estimating potential dust emissions from biochar amended soils under simulated tillage. Sci Total Environ 625:1093–1101
Li H-D, Tang C-S, Cheng Q, Li S-J, Gong X-P, Shi B (2019) Tensile strength of clayey soil and the strain analysis based on image processing techniques. Eng Geol 253:137–148. https://doi.org/10.1016/j.enggeo.2019.03.008
Li J, Sun W, Lichtfouse E, Maurer C, Liu H (2024) Life cycle assessment of biochar for sustainable agricultural application: a review. Sci Total Environ 951:175448
Liao S, Pan B, Li H, Zhang D, Xing B (2014) Detecting free radicals in biochars and determining their ability to inhibit the germination and growth of corn, wheat and rice seedlings. Environ Sci Technol 48:8581–8587. https://doi.org/10.1021/es501483t
Lin Y, Ding W, Liu D, He T, Yoo G, Yuan J, Chen Z, Fan J (2017) Wheat straw derived biochar amendment stimulated N_2O emissions from rice paddy soils by regulating the amoA genes of ammonia-oxidizing bacteria. Soil Biol Biochem 113:89–98
Liu G, Zheng H, Jiang Z, Zhao J, Wang Z, Pan B, Xing B (2018) Formation and physicochemical characteristics of nano biochar: Insight into chemical and colloidal stability. Environ Sci Technol 52:10369–10379
Lu H, Zhang W, Wang S, Zhuang L, Yang Y, Qiu R (2013) Characterization of sewage sludge-derived biochars from different feedstocks and pyrolysis temperatures. J Anal Appl Pyrol 102:137–143
Luz CM, Angel S-MM, Asuncion R, Kelly H, Akio E, Johannes L (2013) Biochar and denitrification in soils: when, how much and why does biochar reduce N_2O emissions. Sci Rep 3:1732

Lyu Y, Guo H, Cheng T, Li X (2018) Particle size distributions of oxidative potential of lung-deposited particles: assessing contributions from quinones and water-soluble metals. Environ Sci Technol 52:6592–6600

Madej J, Hilber I, Bucheli TD, Oleszczuk P (2016) Biochars with low polycyclic aromatic hydrocarbon concentrations achievable by pyrolysis under high carrier gas flows irrespective of oxygen content or feedstock. J Anal Appl Pyrol 122:365–369

Mandal S, Sarkar B, Bolan N, Novak J, Ok YS, Zwieten LV et al (2016) Designing advanced biochar products for maximizing greenhouse gas mitigation potential. CRC Crit Rev Environ Control 46:1367–1401. https://doi.org/10.1080/10643389.2016.1239975

Meng J, Wang L, Zhong L, Liu X, Brookes PC, Xu J, Chen H (2017) Contrasting effects of composting and pyrolysis on bioavailability and speciation of Cu and Zn in pig manure. Chemosphere 180:93–99

Nayak A, Bhushan B, Gupta V, Kotnala S (2021) Fabrication of microwave assisted biogenic magnetite-biochar nanocomposite: a green adsorbent from jackfruit peel for removal and recovery of nutrients in water sample. J Ind Eng Chem 100(25):134–148

Novais SV, Zenero MDO, Barreto MSC, Montes CR, Cerri CEP (2018) Phosphorus removal from eutrophic water using modified biochar. Sci Total Environ 633:825–835

Odinga ES, Waigi MG, Gudda FO, Wang J, Yang B, Hu X, Li S, Gao Y (2020) Occurrence, formation, environmental fate and risks of environmentally persistent free radicals in biochars. Environ Int 134:105172

Oleszczuk P, Josko I, Kusmierz M (2013) Biochar properties regarding to contaminants content and ecotoxicological assessment. J Hazard Mater 260:375–382

Pathak HK, Seth CS, Chauhan PK, Dubey G, Singh G, Jain D, Upadhyay SK, Dwivedi P, Khoo KS (2024a) Recent advancement of nano-biochar for the remediation of heavy metals and emerging contaminants: mechanism, adsorption kinetic model, plant growth and development. Environ Res 255:119136

Pathak HK, Chauhan PK, Seth CS, Dubey G, Upadhyay SK (2024b) Mechanistic and future prospects in rhizospheric engineering for agricultural contaminants removal, soil health restoration, and management of climate change stress. Sci Total Environ 927:172116. https://doi.org/10.1016/j.scitotenv.2024.172116

Quilliam RS, Rangecroft S, Emmett BA, DeLuca TH, Jones DL (2013) Is biochar a source or sink for polycyclic aromatic hydrocarbon (PAH) compounds in agricultural soils. GCB Bioenergy 5:96–103

Reza MT, Lynam JG, Vasquez VR, Coronella CJ (2012) Pelletization of biochar from hydrothermally carbonized wood. Environ Prog Sustain Energy 31:225–234

Ribas A, Mattana S, Llurba R, Debouk H, Sebastià MT, Domene X (2019) Biochar application and summer temperatures reduce N_2O and enhance CH_4 emissions in a Mediterranean agroecosystem: role of biologically induced anoxic microsites. Sci Total Environ 685:1075–1086

Rinklebe J, Shaheen SM, El-Naggar A, Wang H, Laing GD, Alessi DS, Ok YS (2020) Redox-induced mobilization of Ag, Sb, Sn, and Tl in the dissolved, colloidal and solid phase of a biochar-treated and un-treated mining soil. Environ Int 140:105754

Ruan X, Sun Y, Du W, Tang Y, Liu Q, Zhang Z, Doherty W, Frost RL, Qian G, Tsang DC (2019) Formation, characteristics, and applications of environmentally persistent free radicals in biochars: a review. Bioresour Technol 281:457–468. https://doi.org/10.1016/j.biortech.2019.02.097

Sampatrao G (2021) Review on biomass feedstocks, pyrolysis mechanism and physicochemical properties of biochar: state-of-the-art framework to speed up vision of circular bioeconomy. J Clean Prod 297:126645. https://doi.org/10.1016/j.jclepro.2021.126645

Sigmund G, Huber D, Bucheli TD, Baumann M, Borth N, Guebitz GM, Hofmann T (2017) Cytotoxicity of biochar: a workplace safety concern. Environ Sci Technol Lett 4:362–366

Silva FC, Borrego C, Keizer JJ, Amorim JH, Verheijen FGA (2015) Effects of moisture content on wind erosion thresholds of biochar. Atmosp Environ 123:121–128

Solaiman ZM, Shafi MI, Beamont E, Anawar HM (2020) Poultry litter biochar increases mycorrhizal colonisation, soil fertility and cucumber yield in a fertigation system on sandy soil. Agriculture 10:480. https://doi.org/10.3390/agriculture10100480

Tan Z, Lin CSK, Ji X, Rainey TJ (2017) Returning biochar to fields: a review. Appl Soil Ecol 116:1–11

Tomczyk A, Sokołowska Z, Boguta P (2020) Biochar physicochemical properties: pyrolysis temperature and feedstock kind effects. Rev Environ Sci Bio/technol 19:191–215. https://doi.org/10.1007/s11157-020-09523-3

Ulusal A, Varol EA, Bruckman VJ, Uzun BB (2021) Opportunity for sustainable biomass valorization to produce biochar for improving soil characteristics. Biomass Convers Biorefinery 11:1041–1051. https://doi.org/10.1007/s13399-020-00923-7

Visioli G, Conti FD, Menta C, Bandiera M, Malcevschi A, Jones DL, Vamerali T (2016) Assessing biochar ecotoxicology for soil amendment by root phytotoxicity bioassays. Environ Monit Assess 188:166. https://doi.org/10.1007/s10661-016-5165-3

von Gunten K, Alam MS, Hubmann M, Ok YS, Konhauser KO, Alessi DS (2017) Modified sequential extraction for biochar and petroleum coke: metal release potential and its environmental implications. Bioresour Technol 236:106–110

Wang J, Wang S (2019) Preparation, modification and environmental application of biochar: a review. J Clean Prod 54:1–74. https://doi.org/10.1016/j.jclepro.2019.04.282

Wang J, Xiong Z, Kuzyakov Y (2016) Biochar stability in soil: meta-analysis of decomposition and priming effects. GCB Bioenergy 8:512–523

Wang J, Odinga ES, Zhang W, Zhou X, Yang B, Waigi MG, Gao Y (2019a) Polyaromatic hydrocarbons in biochars and human health risks of food crops grown in biochar-amended soils: a synthesis study. Environ Int 130:104899. https://doi.org/10.1016/j.envint.2019.05.063

Wang X, Chi Q, Liu X, Wang Y (2019b) Influence of pyrolysis temperature on characteristics and environmental risk of heavy metals in pyrolyzed biochar made from hydrothermally treated sewage sludge. Chemosphere 216:698–706. https://doi.org/10.1016/j.chemosphere.2018.10.187

Wang H, Shao D, Ji B, Gu W, Yao M (2022a) Biochar effects on soil properties, water movement and irrigation water use efficiency of cultivated land in Qinghai-Tibet Plateau. Sci Total Environ 829:154520

Wang Y, Gu X, Huang Y, Ding Z, Chen Y, Hu X (2022b) Insight into biomass feedstock on formation of biochar-bound environmentally persistent free radicals under different pyrolysis temperatures. RSC Adv 12:19318–19326

Wornat MJ, Sarofim AF, Longwell JP (1987) Changes in the degree of substitution of polycyclic aromatic-compounds from pyrolysis of a high-volatile bituminous coal. Energy Fuels 1(5):431–437

Xie Y, Guolin-Zhou G, Huang X, Cao X, Ye A, Deng Y, Zhang J, Lin C, Zhang R (2022) Study on the physicochemical properties changes of field aging biochar and its effects on the immobilization mechanism for Cd^{2+} and Pb^{2+}. Ecotoxicol Environ Saf 230:113107

Xu X, Cao X, Zhao L (2013) Comparison of rice husk- and dairy manure-derived biochars for simultaneously removing heavy metals from aqueous solutions: role of mineral components in biochars. Chemosphere 92:955–961. https://doi.org/10.1016/j.chemosphere.2013.03.009

Yadav SPS, Bhandari S, Bhatta D, Poudel A, Bhattarai S, Yadav P, Ghimire N, Paudel P, Paudel P, Shrestha J, Oli B (2023) Biochar application: a sustainable approach to improve soil health. J Agricult Food Res 11:100498

Yang CD, Lu SG (2021) Effects of five different biochars on aggregation, water retention and mechanical properties of paddy soil: a field experiment of three-season crops. Soil Tillage Res 205:104798. https://doi.org/10.1016/j.still.2020.104798

Yang X, Li Q, Tang Z, Zhang W, Yu G, Shen Q, Zhao FJ (2017) Heavy metal concentrations and arsenic speciation in animal manure composts in China. Waste Manag 64:333–339

Ye S, Cheng M, Zeng G, Tan X, Wu H, Liang J, Shen M, Song B, Liu J, Yang H, Zhang Y (2020) Insights into catalytic removal and separation of attached metals from natural-aged microplastics by magnetic biochar activating oxidation process. Water Res 179:115876

Zhang K, Wang Y, Mao J, Chen B (2020a) Effects of biochar nanoparticles on seed germination and seedling growth. Environ Pollut 256:113409. https://doi.org/10.1016/j.envpol.2019.113409

Zhang M, Song G, Gelardi DL, Huang L, Khan E, Mašek O, Parikh SJ, Ok YS (2020b) Evaluating biochar and its modifications for the removal of ammonium, nitrate, and phosphate in water. Water Res 186:116303. https://doi.org/10.1016/j.watres.2020.116303

Zhang Q, Zhang X, Duan P, Jiang X, Shen H, Yan X et al (2021) The effect of long-term biochar amendment on N_2O emissions: experiments with ^{15}N–^{18}O isotopes combined with specific inhibition approaches. Sci Total Environ 769:144533. https://doi.org/10.1016/j.scitotenv.2020.144533

Zhang R, Zhang R, Zimmerman AR, Wang H, Gao B (2023) Applications, impacts, and management of biochar persistent free radicals: a review. Environ Pollut 327:121543

Zhao B, O'Connor D, Shen Z, Tsang DCW, Rinklebe J, Hou D (2020) Sulfur modified biochar as a soil amendment to stabilize mercury pollution: an accelerated simulation of long-term aging effects. Environ Pollut 264:114687. https://doi.org/10.1016/j.envpol.2020.114687

Limitations and Gaps in Current Research for Implications of Biochar for Environmental Sustainability

Souravi Bardhan, Rakesh Kumar, and Dipak Kr Chanda

Abstract Biochar, a pyrolysis product of biomass transformation processes, has gained increasing attention as a low-cost and eco-friendly method for treating pollutants and amending soils. The high porosity, large surface area, and oxygen-functional groups in biochars facilitate their utilization for adsorption of causal pathogens and pollutants. Nevertheless, due to its great adsorption capacity, a number of drawbacks and gaps in knowledge do regulate its extensive use in the field of environmental remediation. The list of challenges mentioned here is critical, as it is based on the heterogeneity of biochar physicochemically, interaction of the pollutants, reusability and regeneration, long-term stability, environmental and regulatory concerns, and others. Selection of suitable feedstock for biochar production is the premier challenge, and then it mostly depends upon pyrolysis conditions and post-treatment processing. Furthermore, the majority of studies are centered on single-pollutant systems, while natural contamination is always a multifaceted mixture, i.e., co-existing pollutants, with diverse chemical characteristics. The long-term biochar stability and degradation properties in the environment are still a challenge, even post-adsorption of any pollutants. Undoubtedly, in scaling up the use of this potential material, reusability and regeneration would be one of the major concerns. Regulatory and environmental issues complicate the application of biochar as well. Overcoming these limitations will be essential in realizing the determined potential of biochar as a green remediation technology for pollutant removal. Here in this chapter, we will discuss the limitations and the knowledge gap in biochar research and development.

S. Bardhan (✉)
Department of Environmental Science, Netaji Nagar College for Women, University of Calcutta, Kolkata, India
e-mail: bardhan666srv@gmail.com

Department of Physics, Jadavpur University, Kolkata, India

R. Kumar
Biosystems Engineering Department, Auburn University, Auburn, USA

D. K. Chanda
Somal Refractory, Ludhiana, Punjab, India

Matrexia, Hoogly, India

A. Chauhan et al. (eds.), *Biochar As Remediation Tools for Pollutant Removal*,
https://doi.org/10.1007/978-3-032-16746-0_15

Keywords Biochar · Feedstocks · Pyrolysis · Stability · Pollutants · Co-contaminant interactions

1 Introduction

Biochar is a dense carbonaceous substrate material developed from biomass via thermochemical conversion under restricted oxygen levels, often by pyrolysis, gasification, or hydrothermal carbonization. The physicochemically inert porous solid material contains a high percentage of carbon, and the physicochemical properties are highly variable depending on feedstock type and production conditions (Brar et al. 2024; Pathak et al. 2024a). Initially, biochar gained significance organically due to its role as a soil fertility factor, as it was conceived from the ancient practice of "Terra Preta" in the Amazon Basin. However, in the past few decades, biochar has evolved from a soil conditioner to a more advanced product with numerous applications in the environment, particularly in pollution management and resource recovery (Fairhead et al. 2012). Biochar can also immobilize toxic heavy metals, biodegrade recalcitrant organic pollutants, and even host beneficial microbial communities for bioremediation (Cheng et al. 2021; Chauhan et al. 2025). Furthermore, biochar thus produced from waste biomass fits the circular economy rationale of converting agricultural, forestry, and municipal waste into higher-value products. The understanding that all of these physical properties can be tailored based on the choice of feedstocks and processing conditions has placed biochar's potential uses at the forefront of global research interest (Kochanek et al. 2022).

Indiscriminate industrialization and urbanization have resulted in a sudden rise in the heavy metal, organic chemical, and drug loading of nutrient pollution of ecology over the last few decades (Saxena 2025). Some of the old-fashioned remedies did work, but they were costly, energy intensive, and caused more pollution in a second layer. The multifunctional use of biochar is a cost-effective reciprocal, eco-benign, and multi-purpose material (Kuppusamy et al. 2016; Kumar et al. 2023a, b). The open porous structure and reactive surface of biochar enable it to adsorb a variety of pollutants. It also has the potential to immobilize toxic heavy metals, biodegrade the stubborn recalcitrant organic pollutants, and even harbor other microbial communities that can be used in bioremediation (Cheng et al. 2021). In addition, biochar production from waste biomass fits with the circular economy concept of converting agriculture, forestry, and municipal waste into higher-value commodities. Additionally, its sequestration form when in water and soil matrices enables it to withhold contaminants over extremely extended periods [23], an advantage in the case that large-scale clean-up efforts with very wide coverage are undertaken. Nevertheless, biochar benefits double in terms of environmental management because, in reducing pollution, it also sequesters carbon, thus aiding the mitigation of climate change (Tisserantand Cherubini 2019).

This biochar attribute qualifies it as a potential tool for the achievement of several Sustainable Development Goals (SDGs), most notably clean water, sustainable cities, climate action, and sustainable consumption and production (Shoudho et al. 2024).

Despite much potential and increasing applications for biochar, research in this area still faces limitations and gaps in knowledge. Much of the literature today emphasizes laboratory investigations without much real-life evidence of actual performance, long-term stability, durability, and environmental implications (Mahmood et al. 2024; Upadhyay et al. 2024). Since the performance of biochar is dependent on varying types of feedstocks, the process of production, or the modification process (Vijayaraghavan 2019), it is not simple to define usual practices or guidelines to apply. Although effective for a broad spectrum of contaminants, ranging from dyes and heavy metals to pharmaceutical residues and new pollutants, mechanistic understanding is still incomplete (Roy et al. 2021; Bardhan et al. 2021; Mondal et al. 2022). Among the required research areas are competitive adsorption, regeneration efficiency, interaction with co-pollutants, and the environmental fate of adsorbed pollutants (Fig. 1).

In this chapter, we emphasize such shortcomings and try to present an overview of the current gaps in biochar research, mainly its environmental applications. This chapter addresses the inconsistencies and lack of evidence in the findings from recent

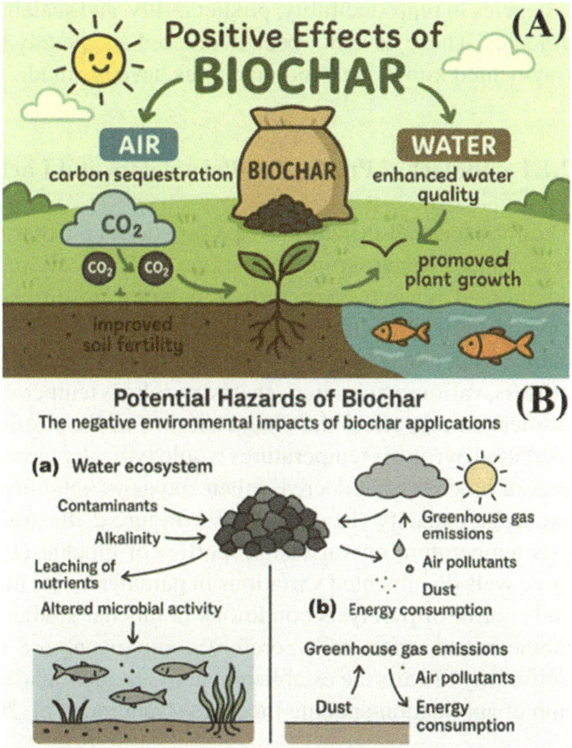

Fig. 1 **A** Diagram illustrating positive and negative aspects of Biochar; **B** Schematic diagram showing various potential negative impacts of biochar on **a** water ecosystem and **b** in the atmosphere

studies and provides future research recommendations. The key concerns in the current research gaps include variation in biochar properties, inconsistent efficiency in removing pollutants, obstacles to widespread application, environmental issues, regulatory vacuum, and ways to overcome such challenges (Table 1).

2 Various Limitations in Current Research on Biochar

2.1 Physicochemical Heterogeneity of Biochar

Variability in different properties of biochar that occur based on feedstock differences and pyrolysis conditions is one of the major challenges. The nature of biomass, anything ranging from cellulose to lignin, and pyrolysis conditions—temperature and residence time, among others—determine the properties of biochar such as surface area, pore structure, and functional groups (Ippolito et al. 2020; Pathak et al. 2024b). Such variation makes it challenging to standardize biochar for commercial use. The physicochemical heterogeneity of biochar is attributed to several factors, including feedstock variation, pyrolysis, and post-treatment modifications (Ahmed et al. 2024). While these factors offer opportunities for tailoring, they create inconsistencies in reproducibility, predictability, and scalability under biochar-based technologies. These have not been addressed adequately by recent research, and hence, fragmented knowledge and outcomes have resulted.

2.1.1 Impact of Production Parameters and Lack of Standardization

The chemical and structural characteristics of biochar, including porosity and surface area, functional group composition, ash content, and carbon stability, are largely determined by pyrolysis, a thermochemical conversion process conducted in an oxygen-limited environment, which produces biochar by decomposing organic matter under elevated temperatures (Tomczyk et al. 2020). Under the pyrolysis process, various parameters, that is, pyrolysis temperature, heating rate, and pyrolysis residence time, affect the characteristics of the resulting biochar. This implies that increased pyrolysis temperatures would typically increase the aromaticity and surface area of biochars, but decrease their oxygen-containing functional groups and cation exchange capacity (Li et al. 2019). Figure 2 illustrates the influence of the pyrolysis temperature on various properties of biochar (Beusch 2021). The majority of these well-documented variations in parameters are not standardized in the reporting and control of pyrolysis conditions in biochar studies. Various studies use different methods and units, and therefore comparisons are typically incorrect. This non-uniformity thwarts the establishment of all-encompassing regulations for the production of application-specific biochars (Genuino et al. 2018). Furthermore, the studies

Table 1 Current applications of biochar in pollutant removal and associated limitations

Application domain	Pollutants targeted	Mechanisms involved	Current advancements	limitations/research gaps	Ref.
Adsorption of heavy metals	Pb^{2+}, Cd^{2+}, As^{3+}, Hg^{2+}, Cu^{2+}, Zn^{2+}	Electrostatic attraction, ion exchange, surface complexation, precipitation	High-temperature biochar; Fe/S/N-doped biochars; enhanced surface area	Poor performance in multi-contaminant systems; challenges in scalability and field applications	Rao et al. 2024
Organic pollutant degradation	Pesticides, PAHs, pharmaceuticals, synthetic dyes	π–π interactions, hydrophobic partitioning, redox reactions, catalytic degradation	Catalytic biochars (e.g., Fe-biochar for persulfate/ H_2O_2 activation); engineered surface functionalities	Limited field studies; insufficient data on long-term fate and byproducts	Dorner et al. 2022
Industrial wastewater remediation	Mixed industrial effluents (organics + heavy metals)	Adsorption, filtration, and integration in constructed wetlands	Biochar filters in textile, leather, and pharmaceutical industries; low-cost alternative to activated carbon	Regeneration and reuse of spent biochar; lack of integrated multi-pollutant treatment strategies	Liu et al. 2024
Removal of emerging pollutants	Antibiotics, EDCs, microplastics, personal care products	Sorption via π–π stacking, hydrogen bonding, surface modifications	Nanoparticle-doped biochars; customized pyrolysis conditions for improved adsorption	Limited environmental concentration data; long-term performance and degradation behavior underexplored	Kumar et al. 2023a; Liu et al. 2024
Recovery of nutrients	Ammonium (NH_4^+), Phosphate (PO_4^{3-})	Electrostatic attraction, ligand exchange	Biochar-fertilizer composites; nutrient-enriched biochar for circular agriculture	Bioavailability of nutrients after adsorption; impact on soil microbiota; desorption control mechanisms not well understood	Xu et al. 2023

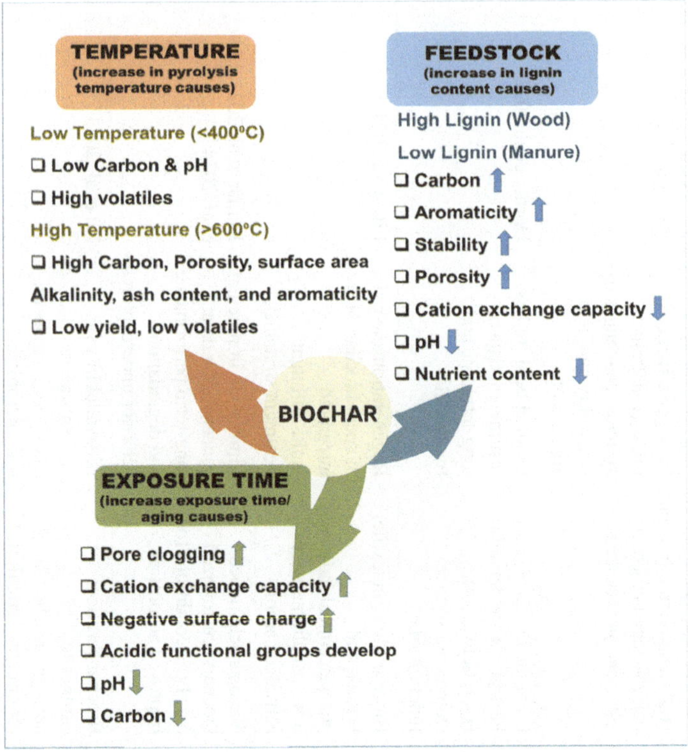

Fig. 2 Illustration of the impact of various parameters on biochar, direction of the arrow indicating increase/decrease of the property

on pyrolysis parameter optimization to determine targeted physicochemical properties (Liu et al. 2018) are also ongoing studies with no best practice established to date.

2.1.2 Biochar Property Variations Due to Feedstock Differences

Another significant source of deviation arises when different feedstocks are utilized. Biomass feedstocks like agricultural residues, forest and wind wastes, manure, and sewage sludge all yield biochars with extremely divergent structural and chemical compositions (Pariyar et al. 2020). Even differences in moisture levels, lignocellulosic content, and levels of contamination may create differences within the same category of feedstock through pyrolysis.

Figure 2 illustrates the influence of lignin content of the feedstock on various properties of biochar (Beusch 2021). These differences also complicate the prediction of biochar behavior in environmental matrices and the relationship between structure and function (Kenney et al. 2013). Apart from that, impurities of the feedstock such as heavy metals, silica, or salt etc. may be collected in biochar which affect its environment friendly nature and adsorption (Murtaza et al. 2023). Nevertheless, these impurities are often ignored or underreported in laboratory studies and it has been argued that the application of biochar may not be reproducible or environmentally sustainable in the long term.

2.1.3 Limitations in Characterization and Analytical Techniques

The uncertified and inconsistent nature of biochar worsens its intrinsic heterogeneity. In most works, the only data provided is limited to basic characteristics such as pH pzc//pH i/e, BET surface area, or elemental composition of samples; media and many other important parameters (pore size distribution, content of functional groups, zeta potential) are often not included (Tan et al. 2021; Leng et al. 2021). This kind of unjustified exclusion necessarily limits the potential of coherent comparisons across studies or predicting how biochar will perform in different environments (Amin et al. 2016). Some of these more sophisticated analytical techniques, such as x-ray photoelectron spectroscopy (XPS), nuclear magnetic resonance (NMR), and synchrotron, can give an in-depth account of biochar structure and functionality (Bolan et al. 2022).

Unfortunately, the expense and unavailability of these advanced analytical methods have rendered them not widely applied, particularly in a low-resource scenario. Therefore, several studies on biochar superficially describe their materials, thereby curtailing one's knowledge of the relationship between certain key structural attributes and environmental performance.

2.1.4 Adsorption Efficiency and Functional Reliability Variability

Physicochemical heterogeneity also carries over to variability in biochar activity performance in pollutant adsorption and degradation. Biochars made under various conditions or using various feedstocks tend to exhibit varying adsorption capacity for a given pollutant, making it extremely unreliable for predictive modeling and difficult to scale laboratory results to field-scale uses (Castiglioni et al. 2021). Additionally, the characteristics of biochar alter after application due to aging and interactions with the environment, leading to changes over time. Yet, long-term experiments that monitor the evolution of biochar properties under real environmental conditions are extremely limited (Leng et al. 2019a, b), and thus, significant uncertainty remains about its durability and reliability for actual application of biochar.

2.2 Feedstock Selection and Optimization Challenges

Feedstock selection is one of the major steps of biochar production that reasonably influences its chemical and physical characteristics, environmental performance, and usability (Mishra et al. 2023). However, its deficiency renders current feedstock selection research challenging and sometimes incomplete for optimum outcome (Mishra et al. 2023), due to variations in raw materials, lack of universal selection criteria, and insufficient consideration of the long-term impacts of different biomass sources (Ippolito et al. 2020). All these are disadvantages in the manufacture of standardization and high-performance biochar to recover resources and remediate the environment. Addressing these problems is critical to the successful integration of biochar-based systems into global sustainability and circular economy directives by valorizing agricultural residues and industrial waste biomass into high-value-added, versatile products that can enhance soil fertility, store carbon, and mitigate pollution.

2.2.1 Feedstock Discrepancies and Inconsistencies

Organic feedstocks are very varied, and this is because biochar can be made out of agricultural wastes such as straw, husk, and stems, forest wastes such as sawdust and bark, cow dung, sewage sludge, and food waste. This demonstrates how diversity strongly affects biochar performance, in particular when considering variability in char C content, ash, volatile matter, and nutrients or contaminants concentration of the final product (loosely following Ippolito et al. 2020), occasionally conflicting with each other. such as wood biomass (plant-based feedstocks), result in biochar with a high carbon stability rating and surface area.

In contrast, biochar sourced from animal materials, such as manure or sewage sludge, has a higher nutrient content and functional groups; these could also include heavy metals like copper and nickel, or persistent organic pollutants like chlorinated pesticides, dioxins, and polycyclic aromatic hydrocarbons, which constitute additional contaminants (Yuan et al. 2016). At present, a standard approach of basing feedstock selection on the intended use of the biochar produced currently does not exist (Masek et al. 2016). The bulk of the research seems to have assumed an ad hoc method when selecting feedstock(s), without capitalizing on existing correlations between feedstock properties and products biochar property (Manya 2012; Ippolito et al. 2020; Mishra et al. 2023). For instance, Ippolito et al. (2020) elucidated the correlation between feedstocks and various biochar physicochemical properties, as well as their influence on the final product. The various types of feedstocks, along with their advantages and drawbacks as reported by Ippolito et al. (2020) and Zhao et al. (2019), have been illustrated in Table 2. Thus, this inconsistency is one reason for the varied outcomes between studies, complicating the development of predictive models, such as Life Cycle Assessment (LCA) tools, sorption and kinetic models, and soil–biochar interaction models for environmental applications. This complexity hinders calibration, limits model transferability, necessitates long-term data, and makes it challenging for a single model to accurately predict all performance outcomes simultaneously.

2.2.2 Co-pyrolysis and Blending

Co-pyrolysis is the combined thermal breakdown of more than one feedstock, which enables the chemical interactions to form one synergistic hybrid biochar with altered properties, whereas blending is the physical blending of individually pyrolyzed biochars without chemical interaction, which has the advantage of ease of use but no synergistic advantage of co-pyrolysis (Kwon et al. 2022). Recent literature suggests that co-pyrolysis or blending of feedstocks could be an avenue to customize biochar properties for some applications, like maintaining an acceptable balance between porosity and density of functional groups (Al-Rumaihi et al. 2022). Table 2 illustrates how different feedstocks interact during pyrolysis. Nevertheless, this is an emerging field with a limited understanding of how the interaction between feedstock during thermal degradation will affect the overall physicochemical structure of the final product (Ahmed and Hameed 2020). Additionally, the synergistic or antagonistic behavior of mixed biomass feedstock to environmental safety, stability, and adsorption of biochar is not well understood. Additional studies are necessary to determine the best combinations of feedstock biomass and pyrolysis conditions that will provide the best efficiency for pollution removal with reduced ecological risk (Esso et al. 2022).

2.2.3 Availability and Sustainability of Feedstocks

Another area that has remained under-addressed is the sustainability and regional availability of biomass resources. Most applications of biochar utilize currently available feedstocks that are regionally prevalent in supply but undoubtedly unscalable and unsustainable outside of those regions (Ahmed et al. 2024). The geographical setting of the feedstocks' source makes it difficult to generalize outcomes of research done on the feedstocks and thus makes it difficult when standardization is used to quantify biochars made from regional biomass material (Fridahl et al. 2020). For example, South and Southeast Asia use more of rice husk as rice farming is common in these areas and readily available as a major biomass material, but the resultant biochar is less reactive and permeable with high silica and ash content (Mohammadi et al. 2017). This limits its efficiency in adsorption of heavy metals or microorganism stimulation in soils. On the other hand, certain areas of Europe or North America mostly use hardwood biochars. The soil of these places can be amended with more nutrient-rich and higher cation-exchange capacity content biochar, and rice husk biochar, on the contrary, would rather cause it to transfer undesirable alkalinity to already neutral or alkaline soils (DeLuca et al. 2024), reducing its agronomic potential. Moreover, energy, agricultural, and biochar production competition for biomass can restrict mass deployment (Lee et al. 2020). In addition, few studies have fully

Table 2 Comparative table on various types of feedstocks used in biochar preparation, their interaction/behavior during pyrolysis process, benefits, and limitations

Feedstock type	Sources	Interaction behavior during co-pyrolysis	Advantages	Limitations
Agricultural residues	Wheat straw, rice husk, corn stover	Can release volatiles that facilitate surface functionalization of lignin-rich feedstocks; moderate synergy	• Abundant and seasonal availability • Moderate nutrient levels • Improves water retention and microbial activity in soil	• High silica or ash in some types • Varies by crop and season • Pre-treatment may be required to remove impurities
Woody biomass	Hardwood, softwood, forestry waste	Provides a rigid carbon framework; interacts well with nutrient-rich or ash-rich feedstocks, improving stability and structure	• High lignin content, which enhances stability and aromatic carbon • High surface area and porosity • Long-term carbon sequestration	• Low nutrient content (N, P, K) • Limited agronomic value in nutrient-poor soils • Slow degradation in soil
Animal manures	Poultry litter, cattle dung, pig manure	Acts as a mineral catalyst; promotes nitrogen doping and enhances porosity in lignocellulosic co-feedstocks	• Rich in essential nutrients (N, P, K) • Boosts soil fertility and cation exchange capacity • Stimulates soil microbial life	• Low carbon stability • Possible odor and pathogen concerns • Potential for heavy metal contamination
Sewage sludge	Municipal biosolids	Catalyzes carbonization of organic feedstocks; may introduce stability or contamination depending on composition	• Resource-efficient waste reuse • High in phosphorus and micronutrients • Suitable for soil remediation	• High in heavy metals and pollutants • Regulatory restrictions • Requires post-treatment and careful application
Green waste	Grass clippings, leaves, garden waste	Can improve hydrogen and volatile production; minor synergy with high-carbon feedstocks	• Easily available in urban areas • Encourages organic recycling • Supports humus and microbial biomass formation	• High moisture content • Variable composition • Lower biochar yield than woody materials

(continued)

Table 2 (continued)

Feedstock type	Sources	Interaction behavior during co-pyrolysis	Advantages	Limitations
Industrial biomass waste	Paper mill sludge, sugarcane bagasse	Can enhance activation or porosity when mixed with wood or agricultural residues; needs quality screening	• Large-scale potential • Converts waste into valuable input • Enhances soil structure	• May contain industrial residues • Highly inconsistent quality • Monitoring needed for environmental safety
Aquatic biomass	Algae, seaweed	May aid mineral dispersion and surface charge development; limited synergy with high-lignin feedstocks	• Renewable and fast-growing • High in minerals (K, Ca, Mg) • Potential for saline or coastal soil amendment	• Low in lignin that makes them less stable carbon • High water content makes processing energy-intensive • Limited field testing

evaluated the environmental trade-offs of sourcing and shipping feedstock, particularly in the case of processing waste materials that may introduce contaminants or necessitate energy-intensive treatments (Campion et al. 2023; Teja et al. 2024). Local and regionally available, affordable feedstock can be sustainable since long-distance transport of the biomass feedstocks will increase carbon emissions, energy use, and logistics costs, thus having a major negative impact on the overall sustainability of biochar production.

2.3 Pollutant-Specific Interactions and Competitive Adsorption

Most studies have been conducted in scenarios involving single pollutants under controlled conditions. But in reality, the contamination would consist of complex mixtures of various pollutants, unlike laboratory settings, as pollutants rarely exist in isolation in nature; rather, they exist as mixtures with chemicals of different properties, concentrations, and interactions. Such combinations influence the effectiveness of biochar, and it is very difficult to replicate exact natural conditions while analyzing biochar performance at lab scale. Thus, what has been observed in laboratory-controlled situations is not necessarily repeatable in nature, the reasons being insufficient knowledge of pollutant-specific interactions involved and further complications arising from competitive adsorption in the multi-contaminant systems (Barquilha and Braga 2021). These shortcomings in existing research thus hinder the credible prediction of biochar in real-field applications. Hence, to fill this research gap, there exists an acute necessity for field-scale investigations that assess biochar performance under practical, multi-contaminant scenarios to guarantee its practical effectiveness and dependability in various environmental applications.

2.3.1 Incomplete Understanding of Pollutant-Specific Interactions

The advancements in the study of biochar application have been hampered by a lack of research on the mechanisms through which different pollutants interact with functional groups and unique structural features of biochar, such as surface area, pore size, and the heterogeneity of surface chemistry (Gęca et al. 2023). Most studies focus on pollutant removal efficiency under simplified and ideal laboratory conditions. They rarely investigate specific interaction pathways such as ion exchange, surface complexation, electrostatic attraction, π–π interactions, or hydrogen bonding. For example, in the case of heavy metals like Cd^{2+} and Pb^{2+}, adsorption has typically been explained through cation exchange or interactions with surface functional groups such as carboxylic and hydroxyl groups (Xie et al. 2022). Nevertheless, the spatial accessibility and distribution of these functional groups on different biochar surfaces remain poorly understood, making generalization or optimization of their performance challenging.

Conversely, organic contaminants such as antibiotics, dyes, and pesticides are often thought to engage with biochar through hydrophobic interactions, π–π stacking, and pore filling (Qiu et al. 2022; Pathak et al. 2024b). For instance, the adsorption of the redolent antibiotic ciprofloxacin may involve π–π interactions with the graphitic regions of biochar (Peng et al. 2016). Likewise, hydrogen bonding can be essential in the elimination of nontropical organics like particular herbicides or drugs. Unfortunately, such mechanisms are often described in broad-minded terms without linking them to particular molecular properties of the pollutants. What makes the problem even harder is that biochar properties are extremely volatile and depend on the feedstock and conditions of pyrolysis with a direct impact on the surface functionality, porosity, and mineral content. Such variability does not allow making a globalized generalization of the models of pollutant- biochar interaction (Geca et al. 2023).

Moreover, the research was also hindered because it is still extremely difficult industrially to distinguish inorganic pollutants (e.g., heavy metals) from organic pollutants (e.g., drugs, dyes, and pesticides) during the mechanistic studies. This is particularly worrying for the new contaminants like PFAS and microplastics, which bring unparalleled testing requirements (Zhao et al. 2021). PFAS molecules are chemically inert, hydrophobic, and resistant to degradation, making them difficult to adsorb through standard mechanisms (Dickman and Aga 2022; Pathak et al. 2024b). Similarly, microplastics vary in size, shape, and surface chemistry, thus necessitating physically based models that incorporate physical entrapment and weak intermolecular interactions (Vieira et al. 2021). Given these limitations, there is an urgent need to develop pollutant-specific adsorption models that integrate the diverse chemistries of both new and conventional pollutants. Such models would need to synthesize sophisticated molecular descriptors, biochar surface characteristics, and real-world environmental variables to enhance predictive accuracy and enable more targeted biochar design.

2.3.2 Limited Research on Competitive Adsorption in Multi-contaminant Systems

In any polluted water, the substances never exist alone. As discussed earlier, they are there in complex mixtures containing heavy metals, nutrients (such as nitrates and phosphates), organic pollutants (such as pesticides or dyes), and emerging contaminants such as medicines and microplastics. The performance of biochar used under these circumstances will also be greatly influenced by competitive adsorption involving these multiple contaminants competing for the same active sites on the biochar surface (Lee and Shin 2021). However, in most laboratory works, biochar performance has been tested under conditions containing only single solutes. This reduce the natural chemical complexity and overestimate the adsorption capacity of biochar (Leng et al. 2019a). In actual field multi-contaminant conditions, strong adsorbates can displace weaker adsorbates, thus unexpectedly reducing adsorption capacity. For example, phosphate and nitrate ions compete for adsorption with heavy metals at some functional groups, while dissolved organic matter can either completely block the sites for adsorption or alter the biochar surface charge (Zhang et al. 2020). Various interaction and adsorption mechanisms of different pollutants with biochar (Ambaye et al. 2021) has been illustrated in Fig. 3. The failure to systematically study the competitive and synergistic interactions among several pollutants remains an important limitation to date.Even the modeling activities aimed at predicting the behavior of biochar in mixed-contaminant matrices are still in their infancy (Tong et al. 2019). Classical models such as Freundlich and Langmuir isotherms are derived for single contaminant systems and are incomplete when applied to multi-solute conditions (Amrutha et al. 2023). The conventional models cannot explain site-specific interactions (selective adsorption of certain pollutants to certain functional groups) or dynamic adsorption properties, both of which play key roles under actual conditionsFor developing and application of the next generation multi-solute adsorption model, which could be used to accurately predict biochar performances under this scenario, it is a great urgency. Such models will have to account for competitive and synergistic effects, pollutant specific affinities as well as the timedependent biochar surface chemistry. Those are the near relevance to scale up biochar applications in real environmental system and to design biochar products for a specific target contaminant mixture.

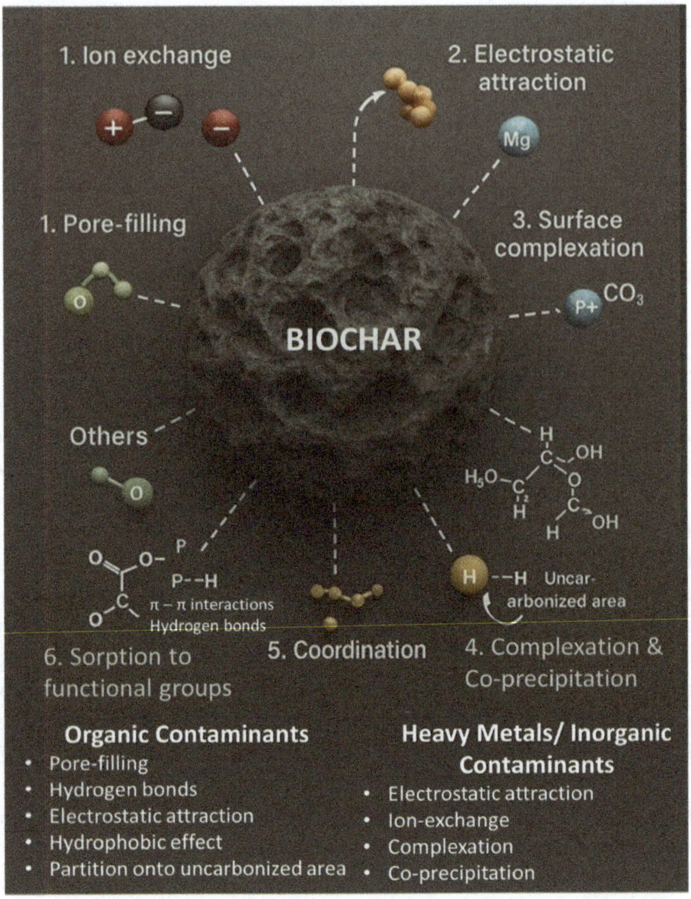

Fig. 3 Illustration of the adsorption mechanism of various pollutants on biochar

2.3.3 Environmental Factors Affecting Interactions Between Pollutants

- Several environmental factors dramatically influence the impact of pollutants when interacting with biochar, and sorption becomes generally less predictable under real conditions. The environmental variables of major importance are:
- · pH: Influences the ionization of biochar surface functional groups and pollutants, which may bring about modifying sorption mechanisms such as electrostatic attraction/repulsion (Qiu et al. 2022).
- Ionic Strength: Excessive levels of background salts may protect electrostatic interactions or cause desorption by capitalistic ion exchange (Kah et al. 2017).
- Temperature: Affects diffusion rates, adsorption equilibrium, and reaction kinetics and, as a result, the total capacity of adsorption (Kah et al. 2017).

- Natural Organic Matter (NOM): It can potentially block active sorption sites on biochar, pre-adsorb target pollutants, or alter the surface charge and hydrophobicity of the biochar (Lin et al. 2017).

Although this is very relevant data, these aspects have rarely been experimentally analyzed, thus undermining the understanding of the strength and utility of biochar in variable field environments. To address these gaps in knowledge, prime importance should be given to the follow-up research with more emphasis on systematic field-based studies that test the performance of biochar under diverse realistic environmental conditions—variations in pH, ionic strengths, temperatures, and natural organic presence. Investigations of this type will help identify biochar stability, flexibility, and efficacy over varied ecosystems and ultimately lead to the development of more secure and predictive models for its use in the environment.

2.4 Long-Term Stability and Environmental Fate of Biochar

While extensive laboratory-based data strongly substantiates the short-term efficacy of biochar, the long-term environmental fate and stability of biochar are critical areas needing additional effort in today's science. Poorly defined concerns regarding biochar in real-world applications are persistence, transformation, mobility, and final degradation processes in a wide range of environmental media (Yang et al. 2024).

2.4.1 Uncertainty in Biochar Aging Mechanisms

Aging procedures are considered among the most relevant aspects to be studied in relation to biochar long-term performance (Wang et al. 2020). When integrated into soils or water networks, the biochar undergoes physical, chemical, and biological transformations that result in changes to its properties over time. This aging can consist of, but not be limited to, surface oxidation, microbe-staged degradation or mineralization, and reaction with natural organic material. Conversely, aging can enhance the properties of certain biochar traits, including the number of surface functional groups (ex: hydroxyl, carboxyl), which are known to increase its cation exchange capacity and adsorptivity (Fig. 4). As an example, oxidative weathering and microbial activity may insert groups containing oxygen and increase the binding stability of contaminants (Xiang et al. 2021). Yet, aging depletion of porosity, pH buffering capacity, and structural stability can also be occurring if biochar used in an environmental application should keep its performance over time. Although the aging of biochar can be achieved by laboratory procedures such as thermal oxidation, chemical treatment, or microbial incubation, these conventional methods are usually less efficient at simulating aging in the dynamic and interactive environments under which biochar ages (Wang et al. 2020). Thus, the laboratory simulations provide inconclusive evidence of the biochar behavior over time in the real-life conditions.

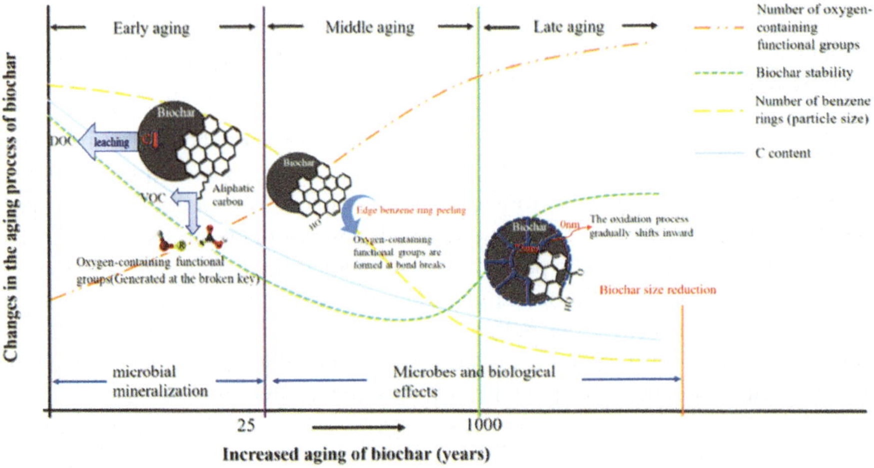

Fig. 4 Illustration of structural changes during aging and oxidation of biochar (Xiang et al. 2021). Reproduced with permission from Elsevier, 2021 with licence no. 6017920209761

Further research was required to achieve higher predictive performance as a means of examining natural aging conditions in simulated forms on long-term field tests and as a means of examining standardized procedures to measure the effect of aging on the long-term environmental behavior of biochar. Additionally, this investigation will explore more realistic expectations of biochar's durability, function, and environmental performance.

2.4.2 Incomplete Concept of Environmental Mobility and Transport

Another huge gap in the current investigations refers to the macro- and micro-scale mobilization and transport of biochar particles across environmental matrices. Although biochar, in general, is considered relatively immobile, under certain conditions, it leaches off small biochar particles and dissolved organic substances (possibly harmful or toxic) into groundwater or runoff with higher concentrations in the surrounding ecosystems (Smith et al. 2016). This provides excellent potential to mimic the risk of secondary pollution, especially where biochar is loaded with residual toxic pollutants like polycyclic aromatic hydrocarbons, heavy metals, or persistent organic pollutants taken up during and after production. Transport models of biochar at present are narrow in focus and mainly lab-scale oriented, and therefore tend to ignore the role of shifting soil texture, hydraulic conductivity, vegetation cover, and climatic factors in the transport of biochar (Aller 2016). The lack of long-term monitoring has thereby largely disabled our ability to predict the environmental behavior of biochar under different uses and environments (Mandal et al. 2021).

2.4.3 Uncertainty Regarding Persistence of Sorbed Pollutants

Biochar has been applied to immobilize contaminants by adsorption or complexation (Li et al. 2021); however, the long-term stability of these interactions is unclear. Environmental fluctuations, including pH change, redox conditions, and microbial communities, could result in the desorption or remobilization of sequestered contaminants. In such a manner, bioavailability applies to heavy metals, including arsenic or cadmium, with appropriate redox shift whereas organic contaminants can be transformed into more detrimental forms after they have leached out of the biochar matrix (Olaniran et al. 2013). There is a complete paucity of long-term desorption and field leaching tests, and this introduces a significant amount of uncertainty. Field research and lifecycle assessments (LCAs) of biochar have wide gaps (Mohammadi et al. 2020). A significant portion of the recent evidence on environmental fate of biochar has been obtained from well-controlled interventional lab experiments that do not simulate diversity and complexity of the real environment (Mahmood et al. 2024). Long-term field studies are challenging and data on the persistence of biochar, degradation products, and ecological interactions are severely lacking. Similarly, the LCAs of biochar rarely consider any end-of-life situations or unanticipated environmental effects beyond carbon sequestration.

2.5 *Reusability and Regeneration Challenges*

Biochar has proved to be an affordable and practically sustainable approach to adsorption of environmental pollutants in a wide range of applications, including, but not limited to, heavy metals, organic pollutants, nutrients, and new contaminants (Huang et al. 2019). Its short-term performance is very often striking, but the long-term sustainability and cost-effectiveness of biochar-based systems ultimately rely on how the material can be regenerated and reused without major loss of functionality. Nevertheless, existing research on the reusability and regeneration of biochar is restricted, conflicting, and often inconclusive (Mahmood et al. 2024). This is a main bottleneck for applied biochar utilization in large-scale environmental cleanup.

2.5.1 Inadequacy of Proper and Standardized Activation Regeneration Methods

Among the challenges is the inadequacy of standardized and efficient biochar regeneration methods after the saturation of contaminants. Studies on biochar regeneration through chemical washing, thermal treatment, and solvent extraction have been conducted to some extent, but a method that is both efficient and eco-friendly has yet to be developed (Alsawy et al. 2022; Dutta et al. 2019). Chemical regeneration using acids, bases, or chelating compounds is reported to recover some adsorption capacity, although at the expense of altering its surface chemistry, destroying the porous network, or, worse, leaching toxic residues into the environment (Alsawy et al. 2022). Thermal regeneration can work for certain applications, but it causes partial combustion or degradation of the carbon matrix, leading to a reduction in the strength and performance of the biochar (Kim and Ko 2020). The large energy input and resulting greenhouse gas emissions make thermal processes highly unsustainable. Performance decreases slowly over time due to multiple regeneration cycles, making the biochar less desirable for long-term use compared to other regenerable adsorbents like activated carbon.

2.5.2 Reducing Adsorption Efficiency with Reuse

A principal limitation in biochar reuse is the dramatic loss of adsorption capacity with every subsequent cycle. The loss may be due to the incomplete desorption of adsorbed contaminants, clogging of active sites, change of the surface charge or chemistry, and physical attrition of the material itself. For example, the metal-loaded biochar is known to usually display an irreversible binding for certain cations, Pb^{2+} or Cu^{2+}, thereby making most adsorption sites unavailable for further cycles (Azadian and Gilani 2023). For instance, hydrophobic fouling and clogging by organic contaminants may lead to desorption becoming impossible. Simple washing several times may be enough to desorb loosely bound contaminants but cannot restore or regenerate the original reactivity of the surface from within the deeper pore depths. The mechanism of these degradation processes will become significant when preparing more effective regeneration processes and forecasting the durability of biochar-based adsorbents in practical uses (Biswal and Balasubramanian 2023). The main focus of the reported research was how (1) the reusable biochar would function or re-function under ideal laboratory conditions with a single solute, and quite differently in environments found in target end-users including industrial wastewater treatment, municipal effluent management, or stormwater runoff (Leng et al. 2019a, b). Here, biochar is faced with complex and dynamic mixtures of pollutants at varying timescales and severe conditions, which diminish the renewal capacity and reuse efficiency of biochar (Venkatachalam et al. 2023). They are more likely to lead to faster biochar saturation and adsorption competition as well as biofouling. Very little research has examined the impact of matrix complexity on biochar regeneration performance (Jia et al. 2023; Alsawy et al. 2022). Long-term cycling experiment times that continue over more than 3 or 4 regeneration cycles are also very infrequently performed, and therefore, a very large gap in our knowledge about the operational stability of biochars exists. Without this knowledge, statements on biochar sustainability must be viewed as elements of speculation.

2.5.3 Limited Studies on Reusability in Complex Matrices

As discussed earlier, majority of research on biochar reusability has been conducted under simplified laboratory conditions involving single-solute systems with controlled environmental variables. However, in scenarios that involve the treatment of industrial wastewater, municipal effluents, or stormwater runoff, biochar confronts a complex mixture of contaminants and fluctuating environmental conditions (Leng et al. 2019a). These factors significantly impact the regeneration capability and reusability of biochar, often resulting in faster saturation, competitive adsorption, and biofouling. Few studies have examined the effect of matrix complexity on biochar regeneration performance (Sun et al. 2022; Jia et al. 2023) this issue is further complicated by the rarity of long appointment cycling experiments that extend beyond three or four regeneration rounds. Consequently, this creates a substantial gap in our understanding of biochar's operational durability. Until such information becomes available, the sustainability claims will remain largely unsubstantiated and speculative.

2.5.4 Economic and Environmental Trade-Offs

Even if regeneration is technically possible, the expense of implementing it and the resulting environmental cost of undertaking such a process may surpass the overall benefit that nature receives through the application of this technology. For instance, the inefficient use of extremely demanding chemicals, high energy requirements, or abundant washing water amount can offset household products of biochar application (Panwar et al. 2019). Beyond this activity, there remains a need for taking care of spent regeneration solution, which contains desorbed pollutants-thereby introducing additional complexity and cost. Thus, most of the environmental and economic aspects of biochar-retaining systems in long-term use are still poorly understood. Optimizing regeneration processes toward low-cost, low-waste, and scalable opportunities will then be necessary to unlock the fullest potential of biochar (Kumar and Bhattacharya 2021) (Fig. 5).

3 Regulatory and Environmental Concerns

3.1 Existing Regulatory Frameworks and Guidelines for Biochar Use

Despite growing interest in biochar as a clean technology soil amendment and carbon sequestration tool, the regulation of its production, quality, and use is varied and still in development (Panwar et al. 2019). There is country-specific legislation or no regulatory provision for the application in some areas. An attempt to develop standards is already being pursued by the European Union (EU) and the United States (US), but the process is ongoing (Thomsen 2024).

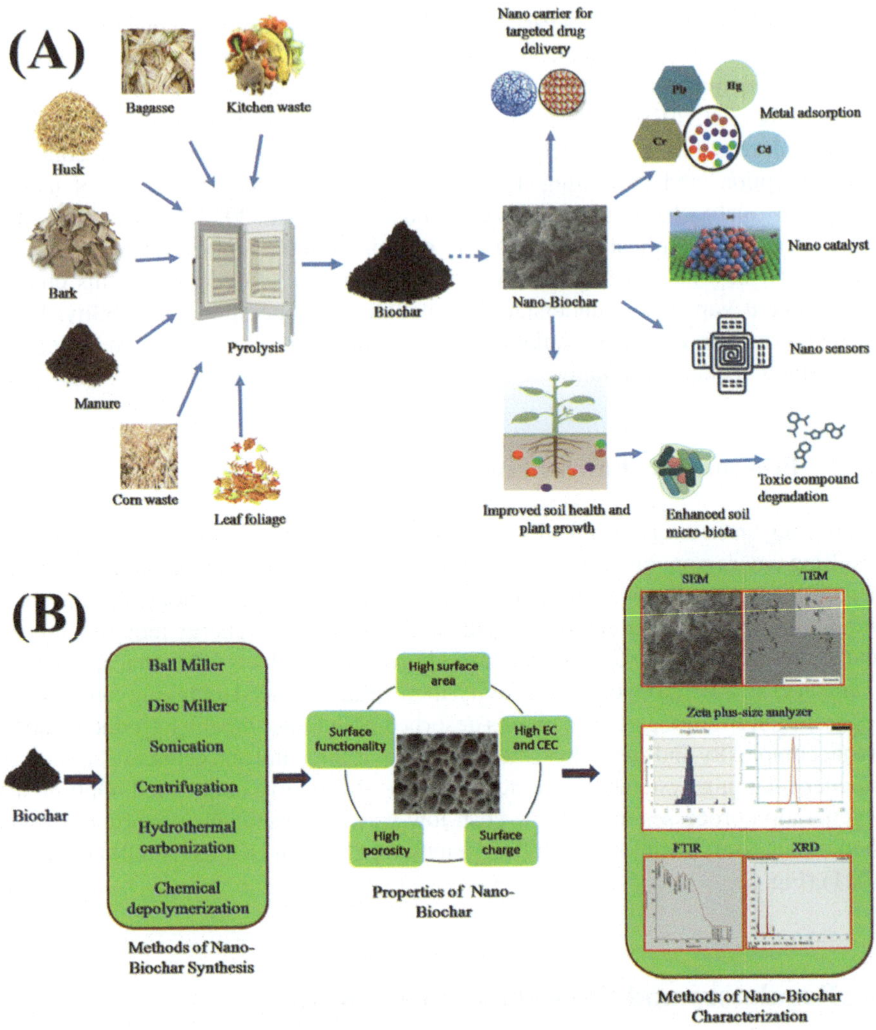

Fig. 5 **A** Various feedstocks used for biochar synthesis and its different applications; **B** Schematic diagram on various synthesis methods of biochar, its properties and characterization techniques. Adopted from Bhandari et al. 2023

Biochar is a fertilizer according to the Fertilising Products Regulation (FPR) 2019/1009 (Štrubelj 2022). The regulation imposes certain conditions on the utilization of biochar as an agricultural product, e.g., the upper limits for impurities like heavy metals and certain polycyclic aromatic hydrocarbons (PAHs). The standard is not global, and its use is subject to the member state. These voluntary certification programs are provided by the European Biochar Certificate (EBC) and the International Biochar Initiative (IBI), though voluntary adoption comes into play (Thomsen 2024).

In the United States, the use of biochar as a soil amendment self-declaration with pesticidal purposes is made possible under the Federal Insecticide, Fungicide, and Rodenticide Act (FIFRA) by the Environmental Protection Agency (EPA) (Rodriguez Franco et al. 2024). The remaining uses are typically in the purview of the state, resulting in patchy responses. As there is no universal global standard, this effectively prevents the potential for large-scale commercialization and raises issues regarding inconsistent product safety.

3.2 Environmental Risks Associated with Biochar Production and Use

Even if widely promoted as an environmentally friendly product, the use and production of biochar inescapably involve environmental hazards. Greenhouse gases such as carbon monoxide (CO), methane (CH_4), and nitrous oxide (N_2O) would be emitted if the process of pyrolysis is not completed, thus negating the carbon sequestration advantage (Li 2024). Incomplete burning would also create particulates, hence air pollution.

In addition, there is still insufficient knowledge of the effect of biochar on the environment, depending on its uses in soils or aquatic systems (Brtnicky et al. 2021). Over a longer term, depending on its production conditions and its feedstock, biochar may be able to alter soil microbial processes and nutrient cycling or leach some toxic constituents (Xiang et al. 2021). The interaction of biochar with the existing soil pollutants and/or agrochemicals will yield some unknown ramifications that can impact yield or environmental integrity. The technology of biochar can be classified under different policy areas in the EU: environment, waste policy, agricultural policy, climate change policy, development aid, research policy, industry, and energy (Van Laer et al. 2015). The greatest potential obstacle in this context is to combine those central benefits with the effects into viable systems, determining the most optimal and the least expensive solutions, and maximizing the advantages and not exceeding the limits of both the intra- and international law and rules (Kumar and Bhattacharya 2021).

3.3 Potential Release of Toxic Byproducts

Biochar production is also an issue of concern in terms of the possible formation and release of toxic byproducts, including PAHs, dioxins, furans, and heavy metals (Han et al. 2022). These contaminants may be present in the feedstock or may be formed during pyrolysis, especially when performed under suboptimal conditions, such as lacking good control of oxygen or excessive temperature. The presence of PAHs in biochar is undesirable since they are established carcinogens and harmful to human health and the environment (Wang et al. 2018). Similarly, feedstock biochars contaminated with sewage sludge, industrial waste, etc., could contain elevated levels of heavy metals, such as lead, arsenic, or cadmium (Lu et al. 2016). Without screening or treatment, their application can lead to secondary contamination of soil, groundwater, or crops.

Notwithstanding the knowledge of such harmful substances, the majority of research has been directed solely to the positive characteristics of biochar and has hardly touched on or inadequately addressed the negative characteristics of byproducts. This resultant lack of credibility and non-acceptance from the public constrains the acceptance through regulatory channels.

3.4 Requirement for Standardized Tests and Risk Assessment Procedures

The key restriction to the safe use of biochar is the lack of standardized science-based testing and risk assessment protocols.

The protocols to determine biochar characteristics such as surface area, pH, and simple analysis are presently in flux, with researchers using different methods and treatments on samples, rendering comparisons virtually unachievable (Igalavithana et al. 2017). Even less common are comprehensive environmental assessments including ecotoxicity tests, leach tests, and field observation in long-term real soils. Absence of standardized test standards is also c as obstacles in the safe utilisation and on the contrary impact obstacle to its incorporation within them conventional environmental management systems (Pierson et al. 2024). Any attempt at the design of such a class of protocols must provide for the great diversity among biochar types, feedstocks, and end-uses. There will be no one-size-fits-all; this brings the need for flexibility and strictness in control systems that balance efficiency and safety.

4 Scaling up and Industrial Implementation

Using biochar as the multipurpose material would presuppose its usage in pollutant adsorption, fertilization of soil, storage of carbon, and regulation of the climate. Although the effectiveness of biochar has been established in the laboratory and pilot tests, large-scale industrial application of biochar technologies is not yet available. Numerous scientific, economic, logistical, and regulatory obstacles face biochar technologies in making the transition from small-scale research to full-scale production and incorporation into existing infrastructure (Prochnow et al. 2024). The impediments to scaling up biochar application and the commercialization prospects, and policy support highlight the key issues.

4.1 Logistical and Economic Issues of Large-Scale Biochar Production

While there are countless demonstrated advantages of biochar, ranging from carbon sequestration and soil health improvement to waste reduction, expanding production to capacities adequate to meet these advantages is hindered by substantial economic and logistical challenges. The most important challenge lies in capital and operational expenses entailed with having industrial-sized pyrolysis units. While sophisticated pyrolysis reactors are very costly and therefore not easily accessible to small-scale farmers and businesses, they are shown to be able to yield quality biochar (Heinrich et al. 2023). Another reason why the supply chain logistics are complicated is that feedstock is highly variable in terms of availability as well as quality, contributing to the cost of transport and storage (Miao et al. 2012).

Another issue is energy efficiency. Slow pyrolysis produces more biochar than fast pyrolysis, which is more oriented toward bio-oils production, although it is usually inefficient (Vuppaladadiyam et al. 2023). There lies the trade-off between biochar yield and energy input, which influences the economic viability of large operations. In addition to that, because of the absence of any biochar market, the private sector is discouraged from investing, as there is no guarantee of profitability without government subsidization and/or carbon credit incentives (Salma et al. 2024).

4.2 Potential for Co-installing Biochar with Existing Waste Treatment Systems

Successful laboratory demonstration is primarily the requirement for those putting forward a feasibility study on integration with existing treatment systems. Some other factors may not be so straightforward except the price or the proof of efficiency in a conventional wastewater treatment. The compatibility with installed infrastructure and user friendliness, including a consideration of legal aspects, have the highest priorities. As an example, biochar should be included inthe design of either filtration beds or adsorption columns or constructed wetlands for wastewater treatment at stages that do not intefere with flow rates, maintenance activitiesand microbial population (Bui et al. 2024) Unlike activated carbon, which can be regenerated easily with well designed common protocols, loss of adsorption capacity is often irreparable for biochar, or in more complex and costly reactivation methods. When biochar is soppy with poisonous contaminants, it raises good issues regarding its disposal or reuse as well, leading to questions regarding environmental impact over its lifecycle (Shoudho et al. 2024). However, while biochar production could be linked to existing biomass and organic waste treatment systems around the management of crop residues, the processing of forestry waste, and municipal solid waste (MSW) treatment systems, some hindrances to effective linking in due time remain (Zhang et al. 2024).

4.2.1 Feedstock Preprocessing

Contamination of the waste streams, such as plastics in MSW or heavy metals in sewage sludge, makes preprocessing stringent to avoid producing toxic byproducts in biochar (Razzak 2024). Current technologies for sorting and cleaning are prohibitively expensive and not widely accessible.

4.2.2 Compatibility with Existing Infrastructure

Most of the facilities for waste treatment are not intended for pyrolysis activity and would require changes or new construction, which add to capital expenditure and would make regulatory approval more difficult to navigate (Chen et al. 2014).

4.2.3 Regulatory and Public Acceptance

Concerns over emissions from pyrolysis plants and potential soil contamination from biochar application can delay permits and community acceptance, particularly in urban areas (Schwartz et al. 2020).

4.3 Future Perspectives on Commercialization and Policy Interventions

Overcoming these challenges would require the integration of continued scientific novelties with industrial collaboration and policy intervention. From a technological perspective, a crucial aspect would be standardized, energy-efficient pyrolysis systems specifically tailored for different sources and end uses. Modular and mobile biochar units would also resolve inefficiencies in transport and facilitate models for community-based production (Joseph et al. 2015). Policy intervention will also play a transformative role (Amin et al. 2016). Incentives might involve carbon credits, subsidies for the development of renewable materials, and requirements to manage sustainably the waste produced in new applications. Such recommendations, providing for well-defined quality standards under certification schemes issued for different purposes, use-by-catchment, conversion, and utilization-focused, are an essential means to gain trust of the stakeholders (Rodriguez Franco et al. 2024).

Moreover, this can be done by pilot demonstration projects as well as through public–private partnerships. Such collaborations may also provide the opportunity for lifecycle assessments and cost–benefit analyses essential for justifying the introduction of biochar into industrial and municipal systems.

5 Solutions and Future Research to Overcome the Challenges

5.1 Knowledge Gaps in Biochar

5.1.1 Partial Understanding of Biochar–Microbe Interaction

Microbial interaction with biochar is likely the least-studied area of research. Biochars shape microbial function, diversity, and activity in the soil and, by extension, plant growth, nutrient cycling, and contaminant biodegradation. Mechanisms involved in such interaction are only vaguely well understood. Surface chemistry, porosity, and pH of biochars can create a variety of microhabitats with possible selection of specific microbial consortia (Bolan et al. 2023).

The least well-researched field of investigation perhaps is microbial interaction with biochar. Biochars have a form of control on microbial activity, diversity, and functionality in the soil and consequently plant growth, nutrient cycling, and biodegradation of contaminants. Mechanisms which pertain to such interaction are vaguely understood. Microhabitats with potential selection of certain microbial consortia can be generated by surface chemistry, porosity, and pH of biochars (Bolan et al. 2023).

Biochar too would create habitats, act as an electron shuttle, and be a source of labile organic matter to be used by microbes in processes like denitrification, methanogenesis, and biodegradation of organic contaminants (Dorner et al. 2022). It is this insight into the intricate interaction between microbes and biochar that is the key to the optimization of the production of biochar to fit certain environmental applications (Chauhan et al. 2024). Further studies ought to welcome the application of the metagenomic, transcriptomic, and ecological approaches in clarifying these multifaceted dynamics.

5.1.2 Lack of Effective Large-Scale Field Application Information

Despite some evidence from lab and pilot-scale trials of the mechanism of pollutant adsorption by biochar (and certain soil amendments such as biochar), very few large-scale field studies have been or are likely to be conducted. Field situations are very heterogeneous and dynamic: climate, soil, land usage management, and contaminant inputs are different; this difference makes it more difficult to replicate and scale up results from the laboratory. There is an urgent need for long-term, multidisciplinary field experiments where biochar is subjected to real conditions. These should not only quantify the chemisorptive removal of pollutants but also different impacts on ecology, economy, and interpersonal acceptability. Constructing right life cycle assessments and cost–benefit analyses should present a balanced perspective of the viability and limitations of biochar (You and Wang 2019).

5.2 *Development of Advanced Characterization Techniques*

Characterization methods of biochar used today suffer from shortcomings with respect to measurement of chemical and structural heterogeneity of biochar. Most studies employ basic parameters such as surface area (BET), elemental analysis, and pH, which indeed are not indicative of the functional efficacy of biochar in a complex system straightaway (Amin et al. 2016). Some of the more complex methods, including X-ray photoelectron spectroscopy (XPS), nuclear magnetic resonance (NMR), X-ray absorption spectroscopy (XAS) at synchrotrons, and secondary ion mass spectrometry (TOF–SIMS) may provide a more detailed view of the chemical surface, surrounding of bonding, and compositional distribution of functional groups (Bolan et al. 2022). New accounting algorithms or computational models may be used to build a relationship between the biochar properties with adsorption behavior before predicting optimum conditions for specific process fields (Jaffari et al. 2023). Adoption of such standardized processes by the research community would offer a short-term measurement comparison data, while accelerating the application-based biochar production.

5.3 *Proper Selection of Feedstock*

To facilitate and improve selecting feedstocks for specific biochar applications, it is suggested to construct practical tools such as a feedstock selection matrix, decision trees, or color-coded compatibility charts that link biomass properties (e.g., lignin content, nutrient content, and ash content) with anticipated effects, e.g., on soil fertility, carbon sequestration, and contaminant adsorption (Vidakis et al. 2025; Haarlemmer et al. 2024). These tools can help users blend alternative feedstocks (for example, high-lignin wood and nutrient-rich manure) to create synergistic results through co-pyrolysis. Additionally, standard checklists for feedstock characterization and, in the long run, software or mobile applications can aid in predicting biochar performance (Supraja et al. 2023), reducing trial-and-error, and maximizing consistency between research and field applications, ultimately promoting more efficient and goal-oriented biochar production in line with environmental and agronomic requirements.

5.4 Exploring New Functionalization and Modification Strategies

Physiochemical property manipulation of biochar is crucial in optimizing its performance and selectivity. Different forms of functionalization processes, such as acid/base treatment, metal impregnation, polymer coating, and doping with nanoparticles, have been applied to improve adsorption capacity, reusability, or selectivity toward target pollutants in biochar (Zeghioud et al. 2022).

Future research should design multifunctional biochars that can capture different types of pollutants, such as heavy metals, organics, and emerging contaminants. Green, low-cost modification processes should be promoted to enable environmental sustainability and economic feasibility. New technologies are emerging as nanotechnology is merged with bioengineering in biochar; i.e., biochar-engineered nanocomposites or biosensors based on biochar can be used as new-generation environmental sensing and clean-up materials. Functionalization can further target improved biochar interactions with microbial consortia for synergistic biodegradation of pollutants (Fig. 6).

6 Conclusion

The overall assessment of biochar limitations and gaps in research that the chapter provides therefore points to the same unexamined promise and significant obstacles on the road to its broader use. Despite the fact that biochar can have become a possible solution to carbon sequestration, soil improvement, and waste recovery, scientific, technical, economical, and regulatory barriers to implementing the research in practice are not yet removed. These fortes are severely multidisciplinary and develop in the fields of materials science, environmental engineering, policy frameworks, and market development. Past studies have not yet fully solved many of the more fundamental questions regarding long-term biochar stability in a variety of soil contexts, its effects on different farm systems, and many other potential environmental risks present.

There continue to be no standard protocols for characterizing biochar, application rates, nor contaminant analysis that allow consistent results of value. More investigating into how pyrolysis conditions, feedstock properties, and biochar functionality are interrelated will help maximize production for specific uses. Biochar has encountered numerous commercial barriers—looming fuel costs for pyrolysis reactors, feedstock availability, and farmers not getting a return on their investment

Fig. 6 Current and future research prospects of biochar (Amalina et al. 2023). Reproduced with permission from Elsevier, 2023. With licence no. 6017920404697

in biochar, to name just a few. In the majority of cases, mass deployment would not be financially viable to most of the parties without a route of carbon credits or government subsidies. Additionally, it will necessitate subsequent advances in terms of technology to boost efficiency and reduce the price of integrating the production of biochar in the existing waste management and energy systems. This renders the commercialization and export/import of biochar difficult on account of the absence of universally harmonized standards for the quality, safety, and application of biochar. The European Union and a few other regions have begun taking steps through the Fertilising Products Regulation. But the majority of the world lacks such legislation. Policymakers will need to collaborate with industry and scientists to establish a new set of evidence-based regulations that provide environmental protection but facilitate the uptake of biochar.

Its environmental impact will be based on the origin of the feedstock, production methods, and application context. For example, while biochar holds immense potential for climate change mitigation, unsustainable biomass removal may lead to further deforestation or soil nutrients being lost instead of benefits gained elsewhere, rendering the benefits meaningless. Experiencing highly poor public perception and farmers' hostility due to misunderstanding of the efficacy and safety of the material is also among the other limitations.

Future research will have to cross between laboratory and short-term field trial experiments to long-term, large-scale experiments under various types of climate and soil conditions. These should examine the aging of biochar under carbon sequestration stability and with soil microbiomes to generate hard data for evidence-based advice. Next-generation pyrolysis technologies, modular energy-efficient reactor designs, and AI-driven process improvements can all enable low-cost and scalable manufacturing. Biochar co-processing with other waste-to-energy-type systems (e.g., biogas and biofuel production) is likely to render such systems more commercially feasible with added revenue streams. International organizations and governments should come together to offer unified biochar standards that will help in safeguarding the quality of products. Adoption incentives may also be offered in the form of financial instruments in terms of carbon pricing and tax incentive or subsidies. Connecting it to carbon offset schemes, such as using Verra or Puroearth-can, can also trigger an economic incentive scenario to manufacturers, and help meet the global climate objectives. Extension programs, certification schemes, and demonstration plots are just some examples which will help create confidence and communicate the agronomic benefits of biochar. Public–private partnerships could, in turn, help information transfer and investment incentives for biochar infrastructure. Biochar has reached a stage where advances in science, policy, and market development need to meet to shape its future.

References

Ahmed MJ, Hameed BH (2020) Insight into the co-pyrolysis of different blended feedstocks to biochar for the adsorption of organic and inorganic pollutants: a review. J Clean Prod 265:121762

Ahmed SF, Mehejabin F, Chowdhury AA, Almomani F, Khan NA, Badruddin IA, Kamangar S (2024) Biochar produced from waste-based feedstocks: mechanisms, affecting factors, economy, utilization, challenges, and prospects. GCB Bioenergy 16(8):e13175

Aller MF (2016) Biochar properties: transport, fate, and impact. Crit Rev Environ Sci Technol 46(14–15):1183–1296

Al-Rumaihi A, Shahbaz M, Mckay G, Mackey H, Al-Ansari T (2022) A review of pyrolysis technologies and feedstock: a blending approach for plastic and biomass towards optimum biochar yield. Renew Sustain Energy Rev 167:112715

Alsawy T, Rashad E, El-Qelish M, Mohammed RH (2022) A comprehensive review on the chemical regeneration of biochar adsorbent for sustainable wastewater treatment. NPJ Clean Water 5(1):29

Amalina F, Krishnan S, Zularisam AW, Nasrullah M (2023) Recent advancement and applications of biochar technology as a multifunctional component towards sustainable environment. Environ Dev 46:100819

Ambaye TG, Vaccari M, van Hullebusch ED, Amrane A, Rtimi SJIJOES (2021) Mechanisms and adsorption capacities of biochar for the removal of organic and inorganic pollutants from industrial wastewater. Int J Environ Sci Technol 18(10):3273–3294

Amin FR, Huang Y, He Y, Zhang R, Liu G, Chen C (2016) Biochar applications and modern techniques for characterization. Clean Technol Environ Policy 18:1457–1473

Amrutha JG, Girish CR, Prabhu B, Mayer K (2023) Multi-component adsorption isotherms: review and modeling studies. Environ Process 10(2):38

Azadian M, Gilani HG (2023) Adsorption of Cu^{2+}, Cd^{2+}, and Zn^{2+} by engineered biochar: preparation, characterization, and adsorption properties. Environ Prog Sustain Energy 42(4):e14088

Barquilha CE, Braga MC (2021) Adsorption of organic and inorganic pollutants onto biochars: challenges, operating conditions, and mechanisms. Bioresour Technol Rep 15:100728

Bardhan S, Roy S, Chanda DK, Mondal D, Das S, Das S (2021) Flexible and reusable carbon dot decorated natural microcline membrane: a futuristic probe for multiple heavy metal induced carcinogen detection. Microchim Acta 188(4):134

Beusch C (2021) Biochar as a soil ameliorant: how biochar properties benefit soil fertility—A review. J Geosci Environ Prot 9(10):28–46

Bhandari G, Gangola S, Dhasmana A, Rajput V, Gupta S, Malik S, Slama P (2023) Nano-biochar: recent progress, challenges, and opportunities for sustainable environmental remediation. Front Microbiol 14:1214870

Biswal BK, Balasubramanian R (2023) Use of biochar as a low-cost adsorbent for removal of heavy metals from water and wastewater: a review. J Environ Chem Eng 11(5):110986

Bolan N, Hoang SA, Beiyuan J, Gupta S, Hou D, Karakoti A, Van Zwieten L (2022) Multifunctional applications of biochar beyond carbon storage. Int Mater Rev 67(2):150–200

Bolan S, Hou D, Wang L, Hale L, Egamberdieva D, Tammeorg P, Bolan N (2023) The potential of biochar as a microbial carrier for agricultural and environmental applications. Sci Total Environ 886:163968

Brar B, Saharan BS, Seth CS, Kamboj A, Bala K, Rajput VD, Duhan JS (2024) Nanobiochar: soil and plant interactions and their implications for sustainable agriculture. Biocatal Agric Biotechnol, p 103077

Brtnicky M, Datta R, Holatko J, Bielska L, Gusiatin ZM, Kucerik J, Pecina V (2021) A critical review of the possible adverse effects of biochar in the soil environment. Sci Total Environ 796:148756

Bui VKH, Nguyen TP, Tran TP, Nguyen TN, Duong TN, Nguyen VT, Nguyen XC (2024) Biochar-based fixed filter columns for water treatment: a comprehensive review. Sci Total Environ, p 176199

Campion L, Bekchanova M, Malina R, Kuppens T (2023) The costs and benefits of biochar production and use: a systematic review. J Clean Prod 408:137138

Castiglioni M, Rivoira L, Ingrando I, Del Bubba M, Bruzzoniti MC (2021) Characterization techniques as supporting tools for the interpretation of biochar adsorption efficiency in water treatment: a critical review. Molecules 26(16):5063

Chauhan PK, Upadhyay SK, Rajput VD et al (2024) Fostering plant growth performance under drought stress using rhizospheric microbes, their gene editing, and biochar. Environ Geochem Health 46:41. https://doi.org/10.1007/s10653-023-01823-1

Chauhan PK, Upadhyay SK, Chauhan A, Bhutiani R, Sikarwar RLS, Tiwari MK (2025) A review on simultaneous use of microbes and biochar for removal of toxic heavy metals: recent updates and future outlooks. Physiol Mol Biol Plants 31(3):1–15. https://doi.org/10.1007/s12298-025-01649-4

Chen D, Yin L, Wang H, He P (2014) Pyrolysis technologies for municipal solid waste: a review. Waste Manage 34(12):2466–2486

Cheng N, Wang B, Wu P, Lee X, Xing Y, Chen M, Gao B (2021) Adsorption of emerging contaminants from water and wastewater by modified biochar: a review. Environ Pollut 273:116448

DeLuca TH, Gundale M, MacKenzie MD, Gao S, Jones D (2024) Biochar effects on soil nutrient transformations. In: Biochar for environmental management. Routledge-Taylor and Francis, p 41

Dickman RA, Aga DS (2022) A review of recent studies on toxicity, sequestration, and degradation of per-and polyfluoroalkyl substances (PFAS). J Hazard Mater 436:129120

Dorner M, Lokesh S, Yang Y, Behrens S (2022) Biochar-mediated abiotic and biotic degradation of halogenated organic contaminants—A review. Sci Total Environ 852:158381

Dutta T, Kim T, Vellingiri K, Tsang DC, Shon JR, Kim KH, Kumar S (2019) Recycling and regeneration of carbonaceous and porous materials through thermal or solvent treatment. Chem Eng J 364:514–529

Esso SBE, Xiong Z, Chaiwat W, Kamara MF, Longfei X, Xu J, Xiang J (2022) Review on synergistic effects during co-pyrolysis of biomass and plastic waste: significance of operating conditions and interaction mechanism. Biomass Bioenerg 159:106415

Fairhead J, Leach M, Amanor K (2012) Anthropogenic dark earths and Africa: a political agronomy of research disjunctures. In: Contested agronomy. Routledge, pp 64–85

Fridahl M, Haikola S, Rogers PM, Hansson A (2020) Biochar deployment drivers and barriers in least developed countries. In: Handbook of climate change management: research, leadership, transformation, pp 1–30

Gęca M, Khalil AM, Tang M, Bhakta AK, Snoussi Y, Nowicki P, Chehimi MM (2023) Surface treatment of biochar—methods, surface analysis and potential applications: a comprehensive review. Surfaces 6(2):179–213

Genuino DAD, de Luna MDG, Capareda SC (2018) Improving the surface properties of municipal solid waste-derived pyrolysis biochar by chemical and thermal activation: optimization of process parameters and environmental application. Waste Manage 72:255–264

Haarlemmer G, Matricon L, Roubaud A (2024) Comprehensive review of hydrothermal liquefaction data for use in machine-learning models. Biofuels, Bioprod Biorefin 18(5):1782–1798

Han H, Buss W, Zheng Y, Song P, Rafiq MK, Liu P, Li X (2022) Contaminants in biochar and suggested mitigation measures—A review. Chem Eng J 429:132287

Heinrich T, Park H, Orozco R, Ding Z, Álvarez-López V, Mosquera-Losada MR, Hoffmann T (2023) Biochar production from late-harvest grass–Challenges and potential for farm-scale implementation. Sustain Prod Consum 37:256–267

Huang Q, Song S, Chen Z, Hu B, Chen J, Wang X (2019) Biochar-based materials and their applications in removal of organic contaminants from wastewater: state-of-the-art review. Biochar 1:45–73

Igalavithana AD, Mandal S, Niazi NK, Vithanage M, Parikh SJ, Mukome FN, Ok YS (2017) Advances and future directions of biochar characterization methods and applications. Crit Rev Environ Sci Technol 47(23):2275–2330

Ippolito JA, Cui L, Kammann C, Wrage-Mönnig N, Estavillo JM, Fuertes-Mendizabal T, Borchard N (2020) Feedstock choice, pyrolysis temperature and type influence biochar characteristics: a comprehensive meta-data analysis review. Biochar 2:421–438

Jaffari ZH, Jeong H, Shin J, Kwak J, Son C, Lee YG, Cho KH (2023) Machine-learning-based prediction and optimization of emerging contaminants' adsorption capacity on biochar materials. Chem Eng J 466:143073

Jia L, Cheng P, Yu Y, Chen SH, Wang CX, He L, Jin Y (2023) Regeneration mechanism of a novel high-performance biochar mercury adsorbent directionally modified by multimetal multilayer loading. J Environ Manage 326:116790

Joseph S, Anh ML, Clare A, Shackley S (2015) Socio-economic feasibility, implementation and evaluation of small-scale biochar projects. In: Biochar for environmental management. Routledge, pp 853–879

Kah M, Sigmund G, Xiao F, Hofmann T (2017) Sorption of ionizable and ionic organic compounds to biochar, activated carbon and other carbonaceous materials. Water Res 124:673–692

Kenney KL, Smith WA, Gresham GL, Westover TL (2013) Understanding biomass feedstock variability. Biofuels 4(1):111–127

Kim DG, Ko SO (2020) Effects of thermal modification of a biochar on persulfate activation and mechanisms of catalytic degradation of a pharmaceutical. Chem Eng J 399:125377

Kochanek J, Soo RM, Martinez C, Dakuidreketi A, Mudge AM (2022) Biochar for intensification of plant-related industries to meet productivity, sustainability and economic goals: a review. Resour Conserv Recycl 179:106109

Kumar A, Bhattacharya T (2021) Biochar: a sustainable solution. Environ Dev Sustain 23:6642–6680

Kumar R, Sharma P, Sharma PK, Rose PK, Singh RK, Kumar N, Pandey A (2023a) Rice husk biochar-A novel engineered bio-based material for transforming groundwater-mediated fluoride cycling in natural environments. J Environ Manage 343:118222

Kumar R, Verma A, Rakib MRJ, Gupta PK, Sharma P, Garg A, Aminabhavi TM (2023b) Adsorptive behavior of micro (nano) plastics through biochar: co-existence, consequences, and challenges in contaminated ecosystems. Sci Total Environ 856:159097

Kuppusamy S, Palanisami T, Megharaj M, Venkateswarlu K, Naidu R (2016) Ex-situ remediation technologies for environmental pollutants: a critical perspective. Rev Environ Contam Toxicol 236:117–192

Kwon G, Cho DW, Jang H, Lam SS, Song H (2022) Synergistic effects of blending seafood wastes as Co-pyrolysis feedstock on syngas production and biochar properties. Chem Eng J 429:132487

Lee HS, Shin HS (2021) Competitive adsorption of heavy metals onto modified biochars: comparison of biochar properties and modification methods. J Environ Manage 299:113651

Lee M, Lin YL, Chiueh PT, Den W (2020) Environmental and energy assessment of biomass residues to biochar as fuel: a brief review with recommendations for future bioenergy systems. J Clean Prod 251:119714

Leng L, Huang H, Li H, Li J, Zhou W (2019a) Biochar stability assessment methods: a review. Sci Total Environ 647:210–222

Leng L, Xu X, Wei L, Fan L, Huang H, Li J, Zhou W (2019b) Biochar stability assessment by incubation and modelling: methods, drawbacks and recommendations. Sci Total Environ 664:11–23

Leng L, Xiong Q, Yang L, Li H, Zhou Y, Zhang W, Huang H (2021) An overview on engineering the surface area and porosity of biochar. Sci Total Environ 763:144204

Li S (2024) Reviewing air pollutants generated during the pyrolysis of solid waste for biofuel and biochar production: toward cleaner production practices. Sustainability 16(3):1169

Li S, Harris S, Anandhi A, Chen G (2019) Predicting biochar properties and functions based on feedstock and pyrolysis temperature: a review and data syntheses. J Clean Prod 215:890–902

Li Y, Yu H, Liu L, Yu H (2021) Application of co-pyrolysis biochar for the adsorption and immobilization of heavy metals in contaminated environmental substrates. J Hazard Mater 420:126655

Lin L, Jiang W, Xu P (2017) Comparative study on pharmaceuticals adsorption in reclaimed water desalination concentrate using biochar: Impact of salts and organic matter. Sci Total Environ 601:857–864

Liu Z, Niu W, Chu H, Niu Z (2018) Process optimization for straws pyrolysis and analysis of biochar physiochemical properties. Trans Chin Soc Agric Eng 34(5):196–203

Liu C, Bolan N, Rajapaksha AU, Wang H, Balasubramanian P, Zhang P, Li F (2024) Critical review of biochar for the removal of emerging inorganic pollutants from wastewater. Chin Chem Lett, p 109960

Lu T, Yuan H, Wang Y, Huang H, Chen Y (2016) Characteristic of heavy metals in biochar derived from sewage sludge. J Mater Cycles Waste Manage 18:725–733

Mahmood R, Arooj A, Sarfraz R (2024) Potential limitations of biochar. In: Biochar-solid carbon for sustainable agriculture. Bentham Science Publishers, pp 126–138

Mandal S, Pu S, Adhikari S, Ma H, Kim DH, Bai Y, Hou D (2021) Progress and future prospects in biochar composites: application and reflection in the soil environment. Crit Rev Environ Sci Technol 51(3):219–271

Manyà JJ (2012) Pyrolysis for biochar purposes: a review to establish current knowledge gaps and research needs. Environ Sci Technol 46(15):7939–7954

Mašek O, Ronsse F, Dickinson D (2016) Biochar production and feedstock. Biochar Eur Soils Agric, pp 17–40

Miao Z, Shastri Y, Grift TE, Hansen AC, Ting KC (2012) Lignocellulosic biomass feedstock transportation alternatives, logistics, equipment configurations, and modeling. Biofuels, Bioprod Biorefin 6(3):351–362
Mishra RK, Kumar DJP, Narula A, Chistie SM, Naik SU (2023) Production and beneficial impact of biochar for environmental application: a review on types of feedstocks, chemical compositions, operating parameters, techno-economic study, and life cycle assessment. Fuel 343:127968
Mohammadi A, Cowie AL, Mai TLA, Brandao M, de la Rosa RA, Kristiansen P, Joseph S (2017) Climate-change and health effects of using rice husk for biochar-compost: comparing three pyrolysis systems. J Clean Prod 162:260–272
Mohammadi A, Khoshnevisan B, Venkatesh G, Eskandari S (2020) A critical review on advancement and challenges of biochar application in paddy fields: environmental and life cycle cost analysis. Processes 8(10):1275
Mondal D, Bardhan S, Das N, Roy J, Ghosh S, Maity A, Roy S, Basu R, Das S (2022) Natural clay-based reusable piezo-responsive membrane for water droplet mediated energy harvesting, degradation of organic dye and pathogenic bacteria. Nano Energy 104:107893
Murtaza G, Ahmed Z, Eldin SM, Ali I, Usman M, Iqbal R, Tariq A (2023) Biochar as a green sorbent for remediation of polluted soils and associated toxicity risks: a critical review. Separations 10(3):197
Olaniran AO, Balgobind A, Pillay B (2013) Bioavailability of heavy metals in soil: impact on microbial biodegradation of organic compounds and possible improvement strategies. Int J Mol Sci 14(5):10197–10228
Panwar NL, Pawar A, Salvi BL (2019) Comprehensive review on production and utilization of biochar. SN Applied Sciences 1:1–19
Pariyar P, Kumari K, Jain MK, Jadhao PS (2020) Evaluation of change in biochar properties derived from different feedstock and pyrolysis temperature for environmental and agricultural application. Sci Total Environ 713:136433
Pathak HK, Chauhan PK, Seth CS, Dubey G, Upadhyay SK (2024a) Mechanistic and future prospects in rhizospheric engineering for agricultural contaminants removal, soil health restoration, and management of climate change stress. Sci Total Environ 927:172116. https://doi.org/10.1016/j.scitotenv.2024.172116
Pathak HK, Seth CS, Chauhan PK, Dubey G, Singh G, Jain D, Upadhyay SK, Dwivedi P, Khoo KS (2024b) Recent advancement of nano-biochar for the remediation of heavy metals and emerging contaminants: mechanism, adsorption kinetic model, plant growth and development. Environ Res 255:119136. https://doi.org/10.1016/j.envres.2024.119136
Peng B, Chen L, Que C, Yang K, Deng F, Deng X, Wu M (2016) Adsorption of antibiotics on graphene and biochar in aqueous solutions induced by π-π interactions. Sci Rep 6(1):31920
Pierson D, Anderson N, Brewen J, Clark N, Hardy MC, McCollum D, Tirocke J (2024) Beyond the basics: a perspective on barriers and opportunities for scaling up biochar production from forest slash. Biochar 6(1):1
Prochnow FD, Cavali M, Dresch AP, Belli IM, Libardi N, de Castilhos AB (2024) Biochar: from laboratory to industry scale—An overview of scientific and industrial advances, opportunities in the brazilian context, and contributions to sustainable development. Processes 12(5):1006
Qiu B, Shao Q, Shi J, Yang C, Chu H (2022) Application of biochar for the adsorption of organic pollutants from wastewater: Modification strategies, mechanisms and challenges. Sep Purif Technol 300:121925
Rao D, Rajput P, Choudhary R, Yadav S, Yadav SK, Rajput VD, Matić S (2024) Multifaceted characteristics of biochar and its implementation in environmental management in a sustainable way. Environ Qual Manage 34(1):e22305
Roy S, Bardhan S, Mondal D, Saha I, Roy J, Das S, Chanda DK, Karmakar P, Das S (2021) Polymeric carbon dot/boehmite nanocomposite made portable sensing device (Kavach) for non-invasive and selective detection of Cr (VI) in wastewater and living cells. Senss Actuators B Chem 348:130662

Razzak SA (2024) Municipal solid and plastic waste derived high-performance biochar production: a comprehensive review. J Anal Appl Pyrolysis, p 106622

Rodriguez Franco C, Page-Dumroese DS, Pierson D, Miller M, Miles T (2024) Policy and regulations for mobile biochar production in the United States of America. Forests 15(1):192

Salma A, Fryda L, Djelal H (2024) Biochar: a key player in carbon credits and climate mitigation. Resources 13(2):31

Saxena V (2025) Water quality, air pollution, and climate change: investigating the environmental impacts of industrialization and urbanization. Water Air Soil Pollut 236(2):1–40

Schwartz NR, Paulsen AD, Blaise MJ, Wagner AL, Yelvington PE (2020) Analysis of emissions from combusting pyrolysis products. Fuel 274:117863

Shoudho KN, Khan TH, Ara UR, Khan MR, Shawon ZBZ, Hoque ME (2024) Biochar in global carbon cycle: towards sustainable development goals. Curr Res Green Sustain Chem, p 100409

Smith CR, Hatcher PG, Kumar S, Lee JW (2016) Investigation into the sources of biochar water-soluble organic compounds and their potential toxicity on aquatic microorganisms. ACS Sustain Chem Eng 4(5):2550–2558

Štrubelj L (2022) Waste, fertilising product, or something else? EU regulation of biochar. J Environ Law 34(3):529–540

Sun H, Feng D, Zhang Y, Sun S, Zhao Y, Zhang F (2022) Regeneration of deactivated biochar for catalytic tar reforming by partial oxidation: effect of oxygen concentration and regeneration time. Fuel 330:125572

Supraja KV, Kachroo H, Viswanathan G, Verma VK, Behera B, Doddapaneni TRKC, Jain R (2023) Biochar production and its environmental applications: recent developments and machine learning insights. Biores Technol 387:129634

Tan XF, Zhu SS, Wang RP, Chen YD, Show PL, Zhang FF, Ho SH (2021) Role of biochar surface characteristics in the adsorption of aromatic compounds: pore structure and functional groups. Chin Chem Lett 32(10):2939–2946

Teja NB, Bhaskar C, Nagamalleswari D, Halder S, Kannan M (2024) Harnessing biochar, biofuels, and solar cells: transformative solutions for environmental impact. In: Carbon-based materials and environmental remediation: graphene, biochar, and more. IGI Global, pp 339–368

Thomsen T (2024) Short survey: political regulation of biochar production and use in agricultural soils

Tisserant A, Cherubini F (2019) Potentials, limitations, co-benefits, and trade-offs of biochar applications to soils for climate change mitigation. Land 8(12):179

Tomczyk A, Sokołowska Z, Boguta P (2020) Biochar physicochemical properties: pyrolysis temperature and feedstock kind effects. Rev Environ Sci Bio/technol 19(1):191–215

Tong Y, McNamara PJ, Mayer BK (2019) Adsorption of organic micropollutants onto biochar: a review of relevant kinetics, mechanisms and equilibrium. Environ Sci: Water Res Technol 5(5):821–838

Upadhyay V, Choudhary KK, Agrawal SB (2024) Use of biochar as a sustainable agronomic tool, its limitations and impact on environment: a review. Discover Agriculture 2(1):20

Van Laer T, De Smedt P, Ronsse F, Ruysschaert G, Boeckx P, Verstraete W, Lavrysen LJ (2015) Legal constraints and opportunities for biochar: a case analysis of EU law. Gcb Bioenergy 7(1):14–24

Venkatachalam CD, Sekar S, Sengottian M, Ravichandran SR, Bhuvaneshwaran P (2023) A critical review of the production, activation, and morphological characteristic study on functionalized biochar. J Energy Storage 67:107525

Vidakis N, Petousis M, Sagris D, David C, Mountakis N, Spiridaki M, Stratakis E (2025) Enhancing biocomposite critical quality indicators (CQIs): the impact of biochar content in additive manufacturing. Biochar 7(1):1–18

Vieira Y, Lima EC, Foletto EL, Dotto GL (2021) Microplastics physicochemical properties, specific adsorption modeling and their interaction with pharmaceuticals and other emerging contaminants. Sci Total Environ 753:141981

Vijayaraghavan K (2019) Recent advancements in biochar preparation, feedstocks, modification, characterization and future applications. Environ Technol Rev 8(1):47–64

Vuppaladadiyam AK, Vuppaladadiyam SSV, Sahoo A, Murugavelh S, Anthony E, Bhaskar T, Leu SY (2023) Bio-oil and biochar from the pyrolytic conversion of biomass: a current and future perspective on the trade-off between economic, environmental, and technical indicators. Sci Total Environ 857:159155

Wang J, Xia K, Waigi MG, Gao Y, Odinga ES, Ling W, Liu J (2018) Application of biochar to soils may result in plant contamination and human cancer risk due to exposure of polycyclic aromatic hydrocarbons. Environ Int 121:169–177

Wang L, O'Connor D, Rinklebe J, Ok YS, Tsang DC, Shen Z, Hou D (2020) Biochar aging: mechanisms, physicochemical changes, assessment, and implications for field applications. Environ Sci Technol 54(23):14797–14814

Xiang L, Liu S, Ye S, Yang H, Song B, Qin F, Tan X (2021) Potential hazards of biochar: the negative environmental impacts of biochar applications. J Hazard Mater 420:126611

Xie Y, Zhou G, Huang X, Cao X, Ye A, Deng Y, Zhang R (2022) Study on the physicochemical properties changes of field aging biochar and its effects on the immobilization mechanism for Cd^{2+} and Pb^{2+}. Ecotoxicol Environ Saf 230:113107

Xu P, Gao Y, Cui Z, Wu B, Yan B, Wang Y, Xue W (2023) Research progress on effects of biochar on soil environment and crop nutrient absorption and utilization. Sustainability 15(6):4861

Yang X, Hou R, Fu Q, Li T, Li M, Cui S, Liu M (2024) A critical review of biochar as an environmental functional material in soil ecosystems for migration and transformation mechanisms and ecological risk assessment. J Environ Manage 360:121196

You S, Wang X (2019) On the carbon abatement potential and economic viability of biochar production systems: cost-benefit and life cycle assessment. In: Biochar from biomass and waste. Elsevier, pp 385–408

Yuan H, Lu T, Wang Y, Chen Y, Lei T (2016) Sewage sludge biochar: nutrient composition and its effect on the leaching of soil nutrients. Geoderma 267:17–23

Zeghioud H, Fryda L, Djelal H, Assadi A, Kane A (2022) A comprehensive review of biochar in removal of organic pollutants from wastewater: characterization, toxicity, activation/functionalization and influencing treatment factors. J Water Process Eng 47:102801

Zhang M, Song G, Gelardi DL, Huang L, Khan E, Mašek O, Ok YS (2020) Evaluating biochar and its modifications for the removal of ammonium, nitrate, and phosphate in water. Water Res 186:116303

Zhang Z, Chen Z, Zhang J, Liu Y, Chen L, Yang M, Yap PS (2024) Municipal solid waste management challenges in developing regions: a comprehensive review and future perspectives for Asia and Africa. Sci Total Environ, p 172794

Zhao J, Shen XJ, Domene X, Alcañiz JM, Liao X, Palet C (2019) Comparison of biochars derived from different types of feedstock and their potential for heavy metal removal in multiple-metal solutions. Sci Rep 9(1):9869

Zhao R, Wang B, Theng BK, Wu P, Liu F, Lee X, Sun J (2021) Fabrication and environmental applications of metal-containing solid waste/biochar composites: a review. Sci Total Environ 799:149295

The manufacturer's authorised representative in the EU is Springer Nature Customer Service Centre GmbH, Europaplatz 3, 69115 Heidelberg, Germany. If you have any concerns regarding our products, please contact ProductSafety@springernature.com

Printed and bound by CPI Group (UK) Ltd, Croydon, CR0 4YY
05/10/2026
02228016-0011